High Entropy Materials

High Entropy Materials covers the fundamental concepts of these materials and their emerging applications. To fulfill growing energy demand, scientists are looking for novel materials which can be used for the fabrication of high-performance energy devices. Many materials such as graphene, carbon nanotubes, and metal oxides are used in energy production and storage. A new class of metal oxides, multicomponent metal oxides, known as high entropy materials, have attracted considerable attention not only for their energy applications but also for other emerging applications such as use in sensors, catalysts, and CO_2 absorption.

Key Features:

- Reviews state-of-the-art developments;
- Provides new directions to scientists, researchers, and students to better understand the principles, technologies, and applications of high entropy materials;
- Discusses ongoing challenges and visions for the future.

Dr. Anuj Kumar is an Assistant Professor at GLA University, Mathura, India. His research focus is on molecular as well as M-N-C electrocatalysts for H_2, O_2, and CO_2 involving electrocatalysis, nanomaterials, nanocomposites, fuel cells, water electrolyzers, nanosensors, bio-inorganic chemistry, and macrocyclic chemistry. He is serving as a section editor, a guest editor, and an editorial board member for various journals.

Dr. Ram K. Gupta is a Professor at Pittsburg State University. Before joining Pittsburg State University, he worked as an Assistant Research Professor at Missouri State University, Springfield, MO, and then as a senior research scientist at North Carolina A&T State University, Greensboro, NC. He is serving as an associate editor, a guest editor, and an editorial board member for various journals.

High Entropy Materials
Fundamentals to Emerging Applications

Edited by

Anuj Kumar and Ram K. Gupta

CRC Press
Taylor & Francis Group
Boca Raton London New York

CRC Press is an imprint of the
Taylor & Francis Group, an **informa** business

Designed cover image: Shutterstock_285331037

First edition published 2024
by CRC Press
2385 NW Executive Center Drive, Suite 320, Boca Raton FL 33431

and by CRC Press
4 Park Square, Milton Park, Abingdon, Oxon, OX14 4RN

CRC Press is an imprint of Taylor & Francis Group, LLC

ISBN: 9781032489100 (hbk)
ISBN: 9781032489124 (pbk)
ISBN: 9781003391388 (ebk)

DOI: 10.1201/9781003391388

Typeset in Times
by codeMantra

Contents

Contributors

Md. Imteyaz Ahmad
Indian Institute of Technology (BHU)
Varanasi, India

Molla Asmare Alemu
Bahir Dar University
Bahir Dar, Ethiopia

Anitha S
PSD Institute of Technology and Applied Research
Neelambur, India

Anuj Kumar
GLA University
Mathura, India

Vanitha M Archana
PSG College of Arts and Science
Coimbatore, India

Athira S
Sree Narayana College
Kollam, India

Delele Worku Ayele
Bahir Dar University
Bahir Dar, Ethiopia

Kailash Barman
Central Institute of Technology Kokrajhar
Kokrajhar, India

Pranjal K. Baruah
Gauhati University
Guwahati, India

Guolin Cao
Nanjing University of Science and Technology
Nanjing, China

Xuzi Cong
Shandong University of Technology
Zibo, China

Sheetal Kumar Dewangan
Ajou University
Suwon, South Korea

Rekha Gaba
DAV University
Jalandhar, India

Ashwani Gautam
Indian Institute of Technology (BHU)
Varanasi, India

Omer Guler
Munzur University
Tunceli, Turkey

Seval H. Guler
Munzur University
Tunceli, Turkey

Amit K Gupta
Central University of South Bihar
Gaya, India

Ram K. Gupta
Pittsburg State University
Kansas, United States

Yixuan Hu
Shanghai Jiao Tong University
Shanghai, China

L. J. Jing
Harbin Institute of Technology
Shenzhen, China

Bhagyashri. B. Kamble
Shivaji University
Kolhapur, India

Arun Karmakar
Academy of Scientific and Innovative Research and Electrochemical Research Institute
Karaikudi, India

Ramesh Kataria
Panjab University
Chandigarh, India

Vinod Kumar
Indian Institute of Technology (IIT)
Indore, India

Priyanka Kumari
Central University of South Bihar
Gaya, India

Subrata Kundu
Academy of Scientific and Innovative Research and Electrochemical Research Institute
Karaikudi, India

Shuo Li
Shandong Laboratory of Yantai Advanced Materials and Green Manufacturing
Yantai, China

Wenjie Li
The Pennsylvania State University
University Park, United States

Zhongfang Li
Shandong University of Technology
Zibo, China

Yichao Lin
Chinese Academy of Sciences
Zhejiang, China

Wei Liu
Chinese Academy of Sciences
Changchun, China

Yuanwei Ma
Shandong University of Technology
Zibo, China

Cheenepalli Nagarjuna
Ajou University
Suwon, South Korea

Priyadharshini P
PSG College of Arts and Science
Coimbatore, India

Bed Poudel
The Pennsylvania State University
University Park, United States

Lavanya Raman
The Pennsylvania State University
University Park, United States

Abhijit Ray
Pandit Deendayal Energy University
Gandhinagar, India

Kolan Madhav Reddy
Shanghai Jiao Tong University
Shanghai, China

Ji-Chang Ren
Nanjing University of Science and Technology
Nanjing, China

Rijith S
Sree Narayana College
Kollam, India

Sarika S
Sree Narayana College
Kollam, India

Partha P. Sarma
Gauhati University
Guwahati, India

Rohit R. Shahi
Central University of South Bihar
Gaya, India

Chenyang Shao
Nanjing University of Posts & Telecommunications (NUPT)
Nanjing, China

Telem Simsek
Hacettepe University
Ankara, Turkey

Tuncay Simsek
Kirikkale University
Kirikkale, Turkey

Aryan Singh
Central University of South Bihar
Gaya, India

Babak Sokouti
Biotechnology Research Center and University of Medical Sciences
Tabriz, Iran

Soumya Sridar
University of Pittsburgh
Pittsburgh, United States

Sumi V S
Government College
Attingal, India

Biniyam Zemene Taye
School of Energy Science and Engineering, University of Science and Technology of China,
Hefei, 230026, P.R. China

Minbale Admas Teshager
Bahir Dar University
Bahir Dar, Ethiopia

Haifeng Wang
Northwestern Polytechnical University
Xi'an, China

Likai Wang
Shandong University of Technology
Zibo, China

Xiaodong Wang
Shanghai Jiao Tong University
Shanghai, China

Ababay Ketema Worku
Bahir Dar University
Bahir Dar, Ethiopia

Huimin Xiang
Zhengzhou University
Zhengzhou, China
and
Aerospace Institute of Materials and Processing Technology
Beijing, China

Mingda Xie
Northwestern Polytechnical University
Xi'an, China

Yan Xing
Nanjing University of Posts & Telecommunications (NUPT)
Nanjing, China

Xueqing Xu
School of Energy Science and Engineering, University of Science and Technology of China,
Hefei, 230026, P.R. China

T. Yang
City University of Hong Kong
Hong Kong, China

J.Y. Zhang
City University of Hong Kong
Hong Kong, China

Y. L. Zhao
Harbin Institute of Technology
Shenzhen, China

Yayun Zhao
Chinese Academy of Sciences
Zhejiang, China

Yumeng Zhang
Shanghai Jiao Tong University
Shanghai, China

Qing Zhou
Northwestern Polytechnical University
Xi'an, China

Yanchun Zhou
Zhengzhou University
Zhengzhou, China

Hangtian Zhu
Chinese Academy of Sciences
Beijing, China

1 Introduction to High Entropy Materials

Anuj Kumar and Ram K. Gupta

1.1 INTRODUCTION

In popular usage, high entropy alloys, also known as multi-principle element alloys, are commonly referred to as HEAs. In 2004, high entropy alloys were published for the first time, and ever since, a significant amount of research has been done to investigate these materials, both for fundamental knowledge and for a variety of applications. This type of material is a multicomponent solid solution material that comprises five or more elements located in nearly equiatomic proportion [1]. These elements, in turn, contribute to the structural stabilization process by optimizing the degree to which the configurational entropy is maximized [2–4]. They offer the possibility of a limitless combination of phase stability in addition to mechanical performance [5], which is due to the non-traditional nature of their structure. This, in particular, helps to overcome the strength-ductility trade-off [6]. Although it is not known whether or not entropy [7] plays the most essential function in the stabilization of multicomponent alloys, HEAs are still being utilized as a result of the frequency with which they are discussed in the relevant literature [8]. The high entropic effect, the lattice distortion effect, the sluggish diffusion effect, and the cocktail effect are the four so-called "core effects" that researchers generally believe contribute to the outstanding performance of HEAs [7]. Both the notion of a signature and the high entropy effect are defining characteristics for HEAs since they ensure that it is possible to combine five or more elements to generate a solid solution as opposed to an intermetallic complex. This is because of the signature notion and the high entropy effect. The antisite disordering of atomic types in a systematic crystal appears to be the most essential structural component of HEAs, which can be extended to compounds and diverse applications. HEAs can be used in a wide variety of contexts. Rock salt $(NiCoCuMnMg)O_x$ solid solution was reportedly manufactured by Rost et al. [9]. Since high configurational entropy also contributes to the oxides' stability, they were given the moniker "high entropy oxides," or HEOs for short [10].

Till now, the hugely developed list includes many functional HEOs, including high entropy metallic gases (HEMGs) NPs [11], high entropy borides (HEBs) [12], high entropy sulfides (HESs) [13], high entropy carbides (HECs) [14], and high entropy nitrides (HENs) [15]. HENs, along with the above materials, are known as solid solution materials consisting of quasi-equimolar multicomponents. Recently, reported works highlight the utilization of HEMs in energy-relevant applications such as catalysts for NH_3 oxidation and decomposition [16]; electrocatalysts for methanol oxidation [17]; carbon monoxide [15]; evolution (oxygen [18], hydrogen [13]); reduction (carbon dioxide [19], oxygen [5]); as electrodes for batteries [20]; as hydrogen storage [13]; as supercapacitors [21], etc. The possibility to adopt structural tailor ability in HEMs makes them highly useful in energy-related applications. For instance, the enormous atomic sites on the surface of HEMs offer an unlimited chance to tailor the surface electronic structure along with adsorption energy and catalytic activity. The nanoporous alloys, or NPs, are prepared for energy-related applications to extend their surface area to the maximum possible extent to promote chemical kinetics. Following wet chemistry, nanosized HEMs can be produced along with thermodynamic stability. Single-phase solid solutions of elements of different sizes may result in lattice distortion, which in turn enhances the activation barrier of diffusion (the sluggish diffusion effect) and better kinetic stability [22].

DOI: 10.1201/9781003391388-1

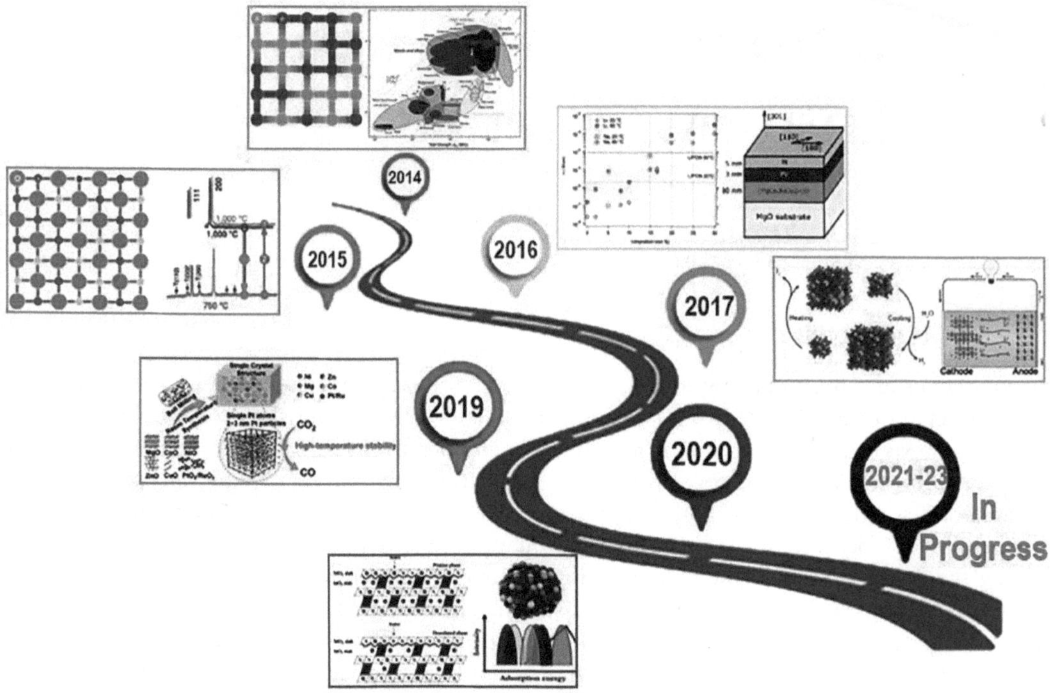

FIGURE 1.1 Roadmap for the development of HEMs toward energy conversion and storages [9,11,23–25].

Studies on HEMs for catalysis or energy storage began less than a decade ago, requiring an extensive investigation to understand their structure-performance connection. Here, we focus on the compilation of research advances toward energy-related uses of HEMs. In addition, synthesis strategies, including the structure and theory of HEMs, are also introduced. A roadmap toward the application of nanosized HEMs with respect to their structure is shown in Figure 1.1. Before conclusion, challenges and future prospects, including future research aspects, are also discussed.

1.2 ARCHITECTURAL ASPECTS OF HIGH ENTROPY MATERIALS

Cubical structures (Figure 1.2a and b), hexagonal structures (Figure 1.2c), and other shapes are possible for HEAs that feature a bcc or fcc lattice [26]. Until then, HEAs that have fcc lattices are being explored in great detail [26,27]. Nano HEAs such as CoMoFeNiCu and CoFeLaNiPt, both of which have fcc lattices and a lattice parameter of 23.6 Å, are being invested in for energy-related applications. Heavy metals containing HEAs (Os, Re, Ru, and Zn) are able to have hcp structures [8]. According to Yusenko et al.'s [27] report, they were able to synthesize $Ir_{0.19}Os_{0.22}Re_{0.12}Rh_{0.20}Ru_{0.19}$ HEA with the space group P63/mmc and the lattice parameter values of $a = 2.728$ Å and $c = 4.338$ Å.

At the moment, there is a lack of clarity on the fundamental connectivity that exists between the composition of HEAs and their phases; however, this comprehension is crucial for the continuation of growth. Batchelor et al. [10] made use of theory in order to realize the activity of HEAs and optimize employing IrPdPtRhRu HEAs for ORR. Following a DFT calculation of OH adsorption energies on catalytic sites of the surface of this material, a prediction was given regarding the possibility of reaching near continuous adsorption energy through the customization of the surface structure. Other than the surface structure, numerous local electronic structures affect the properties of HEMs. These structures affect things such as ionic and electronic conductivity, thermal conductivity, charge distribution, magnetic properties, and other properties. These features, in turn,

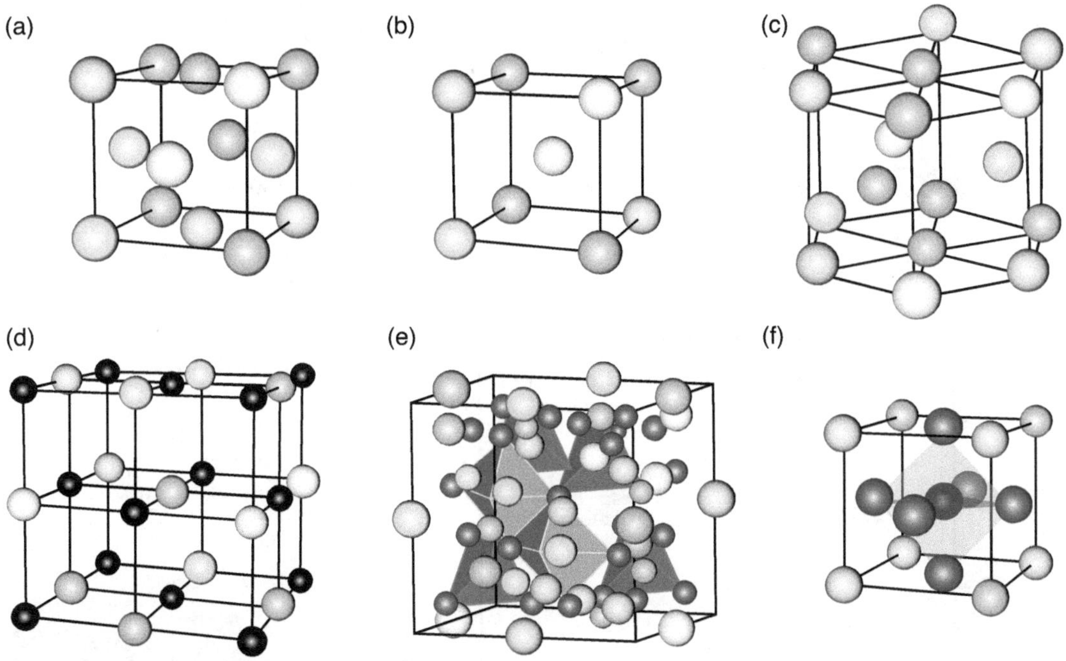

FIGURE 1.2 Schematic representation for the various lattice structures, including (a) fcc, (b) bcc, (c) hexagonal, (d) rock salt, (e) spinel, and (f) perovskite. Adapted with permission [25]. Copyrights 2021, Elsevier.

may affect electrocatalytic capability via mass transfer and charge transfer. It is believed that a high level of atomic disorder is responsible for the stabilization of HEMs. The likelihood of the surface structure of HEMs being preserved while they are in functioning condition is uncertain.

1.3 STRATEGIES TO PREPARE HIGH ENTROPY MATERIALS

The synthesis of HEAs often takes place in nanostructures as nonporous NPs in order to acquire the largest surface area for use in applications relating to energy. To acquire increased surface area and homogeneity in pore structure while preparing nanoporous HEAs, the most frequent method employed is the de-alloying of bulk alloy. This is done so that the electrocatalytic performance can be improved [28]. Wet chemistry is the approach most commonly used to synthesize NPs. This process involves the reduction of a metallic precursor in solution, and it is possible that this method is not the best one to use in order to create homogeneous multicomponent nano HEAs. Liu et al. [28] established a simple ultrasonication-assisted method for the production of HEA nanoparticles. The temperature of ultrasonication cavities used to be higher than 5000°C, which prompted the beginning of the process of metal ion reduction to establish an entropy-maximizing state in order to form desired HEAs. In a study that was quite similar to this one, Yao et al. [29] proved how to manufacture multicomponent HEA NPs using the CTS approach. The metal salt mixture was subjected to thermal stunning at 2000 K for 55 ms (Figure 1.3a), which resulted in the alloying of eight different elements into single-phase solid solution nanoparticles (PtPdCoNiFeCuAuSn, etc.) on carbon support. Although this technique makes it possible to manufacture a vast range of multicomponent NPs with the appropriate composition and size via monitoring the CTS parameter to prepare low-cost catalysts, the productivity of HEA NPs is extremely low.

It was stated by Gao et al. that they have invented a process called rapid moving bed pyrolysis (FMBP) to manufacture HEAs with the help of a device. For example, solvothermal, co-sputtering,

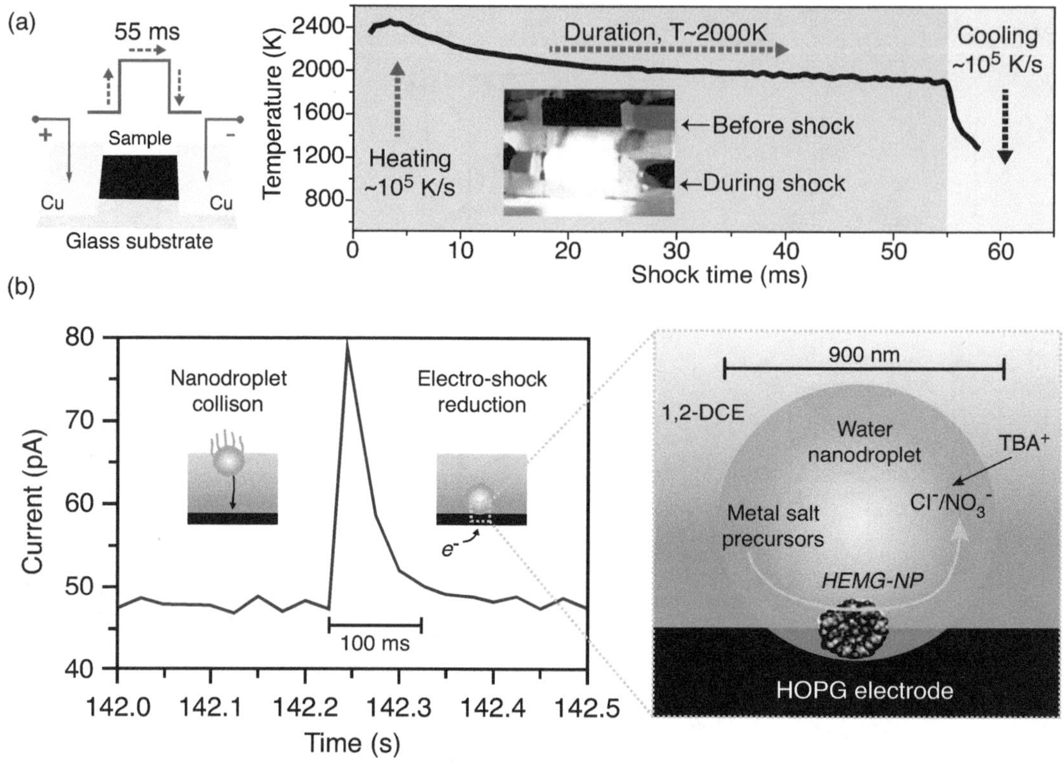

FIGURE 1.3 (a) Carbothermal preparation illustration for HEA-NPs@ carbon templet. Taken with permission [33]. Copyright (a) 2018, American Association for the Advancement of Science. (b) Representation for the nanodroplet-mediated electrodeposition strategy to prepare HEMs. Adapted with permission [11]. Copyright 2019, Springer Nature.

and deposition are all being developed as methods for the fabrication of HEA NPs [29]. It is possible to use electrodeposition to produce multicomponent HEMGs such as PtCoFeLaNi by adding various metal salt precursors to water nanodroplets emulsified in dichloroethane and then subjecting the mixture to electroshock lasting for 100 ms (Figure 1.3b) [11]. According to Bondessgaard et al.'s study [30], they established a universal solvothermal approach that may be used to manufacture PtPdIrRhRu HEA electrocatalyst. However, it is always presumed that the reduction of all components at the same time is the key aspect in place of cooling rate, as revealed by Gao et al. [31] and Weu et al. [32] demonstrating that a slow cooling rate is also capable of initiating the formation of HEAs.

1.4 HIGH ENTROPY MATERIALS FOR ENERGY CONVERSION REACTIONS

1.4.1 Hydrogen Evolution Reaction

As a result of the local disordered chemistry, hydrogen evolution reactions (HEAs) exhibit greater resistance against corrosion when compared to conventional alloys. Therefore, it was expected that HEAs would be highly active and stable electrocatalysts toward HER in both acidic and basic conditions. For instance, it has been observed that the $Ni_{20}Fe_{20}Mo_{10}Co_{35}Cr_{15}$ catalyst has an overpotential of 107 mV in an acidic electrolyte and 172 mV in a basic condition. This catalyst demonstrates superior activity when compared to dual-phase electrocatalysts and activity that is comparable to that of Pt sheets. It is believed that partially filled d-orbitals are fitted for gaining as well as losing an electron and are responsible for the augmentation of the HER activity of HEAs by boosting the adsorption of hydrogen [13]. This system (Figure 1.4a and b) benefited from the synergistic activity of the Pt site as well as the transition metal's high HER overpotential of 555 G 2 mV and high oxygen evolution reaction (OER) overpotential of 377 G 2 mV.

According to Wu et al.'s research [32], they also built a RuRhPdIrPt HER catalyst, which showed a broad valence band spectrum with fewer features. Another perspective is that feature-less XPS spectra indicate atomic configuration randomly with diverse electronic structures as well as a

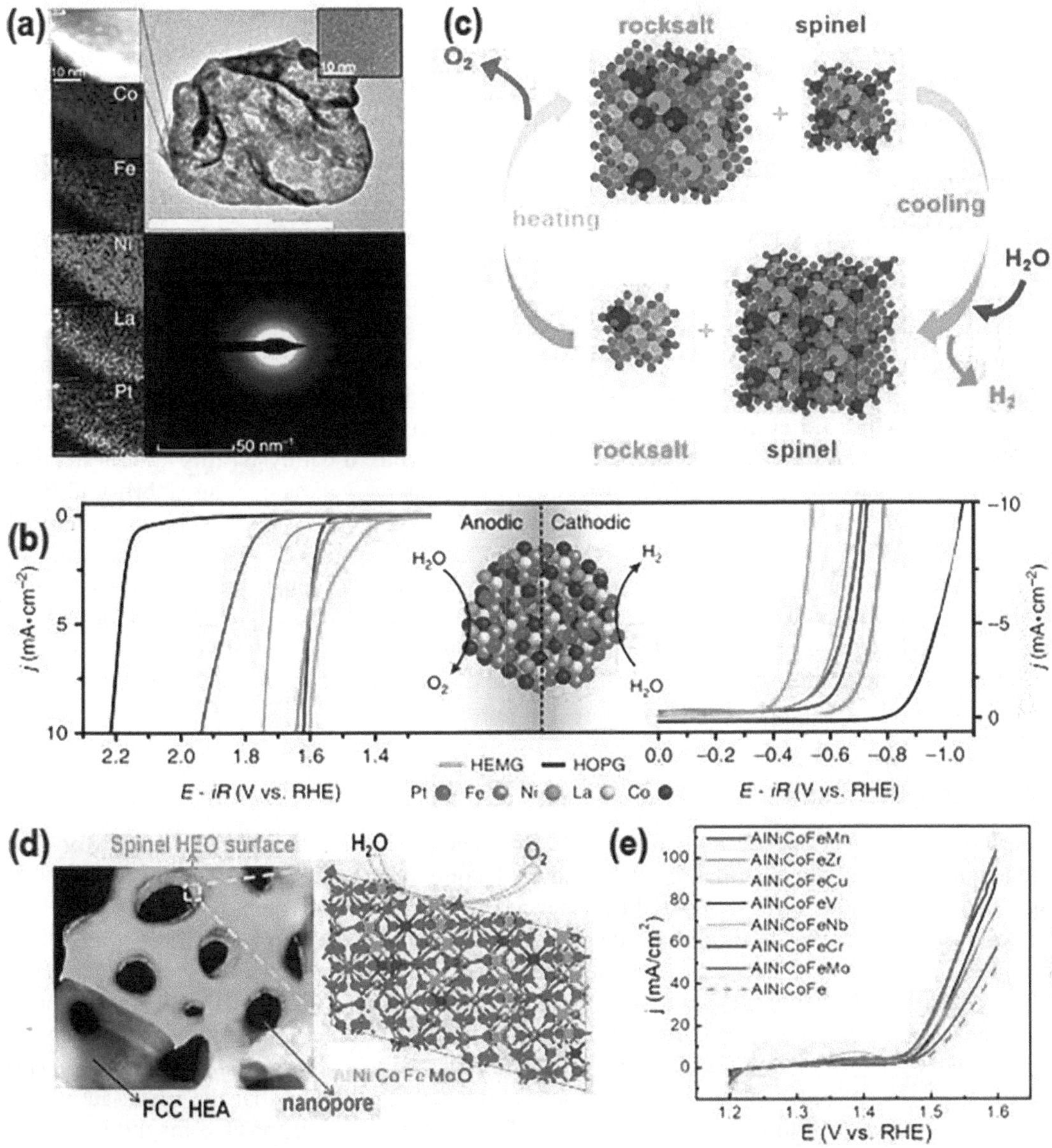

FIGURE 1.4 (a) TEM along with elemental mapping pictures, and (b) HER and OER evaluation for the prepared CoFeLaNiPt electrocatalyst. Adapted with permission [11]. Copyright 2019, Springer Nature. (c) Representation of water splitting on prepared $(FeMnCoNi)_xO$ HEO. Adapted with permission [34]. Copyright 2018, The Royal Society of Chemistry. (d) HAADF picture, (e) OER LSV, for the prepared AlNiCoFeMo HEA. Adapted with permission [18]. Copyright 2019, American Chemical Society.

unique local density state in HEAs. Zhai et al. [34] showed that a HEO combination with the structure of rock salt and a spinal structure called (FeMgCoNi)O_x($x_{1.2}$) is able to thermochemically divide water into H_2 and O_2 into double stages. More than that, Zhai et al. [34] reported that this HEO mixture was able to accomplish so. During the reaction of the HEO mixture, it was found that the phase switched back and forth between rock salt and spinal phases reversibly (Figure 1.4c). The Fe ion is exhibited in only Fe^{3+}, presuming to have a larger redox capacity compared to spinal or rock salt two-phase systems where Fe^{2+} and Fe^{3+} are found. Although only the Fe component was shown in HEO by XAS analysis, the Fe ion is shown only in Fe^{3+}. According to Zhao et al. [35], they were able to construct a CrMnFeCoNi XPHEMP nanosheet with improved catalytic activity, which may have been caused by the components present there working together in a synergistic manner.

1.4.2 Oxygen Evolution Reaction

As OER electrocatalysts, nanoporous AlNiCoFeX (X = Mo, Nb, Cr) HEAs, which were made from NiFe or NiCoFe nano-alloys, have low impedance resistance and great durability (Figure 1.4d and e) [18]. Chemical etching was the method of synthesis of this catalyst. Chemical etching is a top-down approach that is widely regarded as being extremely reproducible and adaptable for use in other HEA catalyst synthesis processes. To improve the overall performance of the OER, nanoporous AlNiCoFeX HEAs were produced by adding various elements such as Cr/Nb, V, Mn, and Cu. Whereas V, Zr, and Cu showed only a moderate degree of enhancement [18], Cr, Nb, and Mo all contributed by displaying substantial enhancement. Through the process of de-alloying, Jin et al. [36] were able to produce a nonporous AlNiColrMo MEA with nanopores that were aligned at 2 nm. In comparison to Ir-based catalysts, this system demonstrated exceptionally high levels of electrocatalytic activity as well as cycling stability [37]. In alkaline environments, Wang et al. [38] suggested that a high entropy perovskite fluoride of K(MgMnFeCoNi)Fe_3 would be a good OER catalyst. OER performance can be increased further by substituting K for Na. For example, $K_{0.8}Na_{0.2}$ (MgMnFeCoNi)Fe_3 displays a reduced overpotential of 314 mV at 10 mA/cm^2 current density, which is significantly smaller than IrO_2 or $Ba_{0.5}Sr_{0.5}Co_{0.8}Fe_{0.2}O_3$. Electrocatalytic performance could be significantly improved by a number of factors, including highly scattered active sites, low charge transfer resistance, and rapid mass transfer due to high entropy.

1.4.3 CO_2 Reduction Reaction

Pedersen et al. [39] proposed a method to create a selective and active electrocatalyst for CO_2 Reduction Reaction (CO_2RR) that incorporated DFT computation and supervised machine learning (Figure 1.5a). This work computed the energies of CO and H adsorption on the surfaces of CoCuGaZnNi and AgAuCuPdPt HEAs and hypothesized that HEAs with low H adsorption might inhibit the synthesis of molecular hydrogen, whereas HEAs with good CO adsorption could kick off the reduction of CO. At low applied potential (0.3 V vs. RHE) and high CO2RR, Nellaiappan et al. [19] showed that AuAgPtPdCu exhibited near 100% FE toward gaseous products. Adsorption trends for *OCH3 and *O on Cu (111) and HEA are reversed when compared with pure Cu(111) (Figure 1.5b), which may explain the increased activity. It is believed that the redox-active Cu metal, in conjunction with the synergistic effect of other metals, is responsible for the increased activity. High electrocatalytic activity in the conversion of CO_2 to CO (Figure 1.5c) was demonstrated by Chen et al. using a synthesized (NiMgCuZnCo)O_x catalyst decorated with a single atom or nanocluster of Pt/Ru. Although HEOs are anticipated to be thermodynamically stable at low temperatures, the Gibbs free energy and catering space distribution could be enhanced by increasing the entropy.

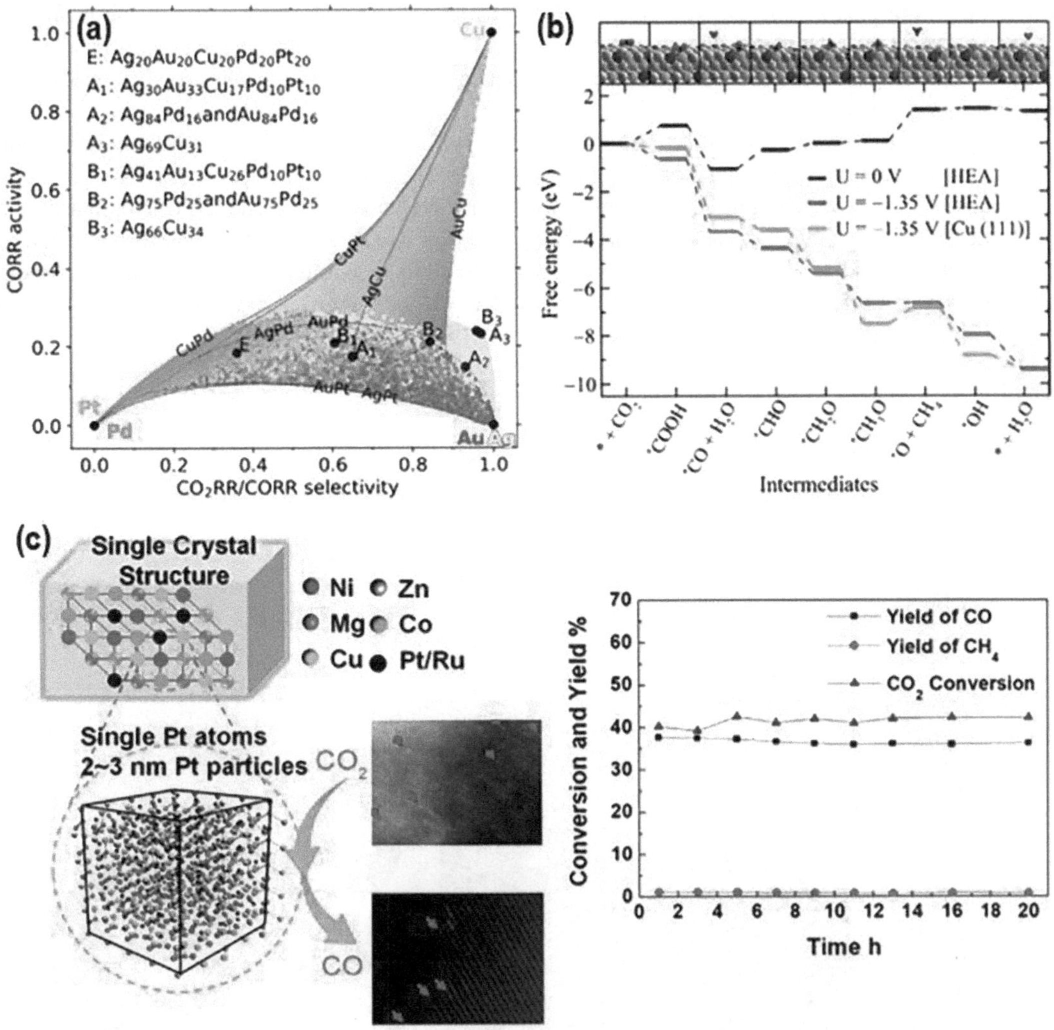

FIGURE 1.5 (a) CO_2RR performance for AgAuCuPdPt NP catalyst. Adapted with permission [39]. Copyright 2020, American Chemical Society. (b) CO_2RR Gibbs free energy diagram for AuAgPtPdCu. Adapted with permission [19]. Copyright 2020, American Chemical Society. (c) CO_2 hydrogenation curve for the Pt-(NiMgCuZnCo)O_x. Adapted with permission [40]. Copyright 2019, American Chemical Society.

1.5 HIGH ENTROPY MATERIALS FOR ENERGY STORAGE

HEOs are currently undergoing testing to see whether they are suitable for use as electrode material for Li/Na-ion batteries, solid electrolytes, and electrode material for Li-batteries, all of which require high capacity and either high electronic conductivity or ionic conductivity, respectively. HEOs that have the ability to tune their cations satisfy such a need.

1.5.1 For Li-ion Batteries

As can be seen in Figure 1.6a, HEOs are versatile enough to be used in either the anode or the cathode of a lithium-ion battery (LIB). To design high-performance electrode materials (HEOs), a technique that combines active and inactive cations is now being utilized as part of the evaluation process for high-performing electrode substances. Rock salt was reportedly used in the design of a high entropy LIB cathode that was carried out by Lun et al. [41] using DFT calculations. Thirty percent of the available extra is being utilized in order to achieve good Li transport without severely compromising capacity.

Energy density and rate capability were found to rise with the number of metals present in a group of rock salt containing 2,4,6 transition metal (TM6: $Li_{1.3}Mn^{2+}{}_{0.1}Co^{2+}{}_{0.1}Mn^{3+}{}_{0.1}Cr^{3+}{}_{0.1}Ti_{0.1}Nb_{0.2}O_{1.7}F_{0.3}$). This was discovered by comparing the compositions of the various salts. The TM6 demonstrated a great rate of 307 mAh/g at a rate of 20 mA/g and 170 mAh/g at a rate of 2000 mA/g, respectively, as shown in Figure 1.6b. Following a mechano-chemistry technique, Wang et al. claimed that they were able to manufacture Li-containing HEOF of $Lix(Co_{0.2}Cu_{0.2}Mg_{0.2}Ni_{0.2}Zn_{0.2})OF_x$. The resulting material

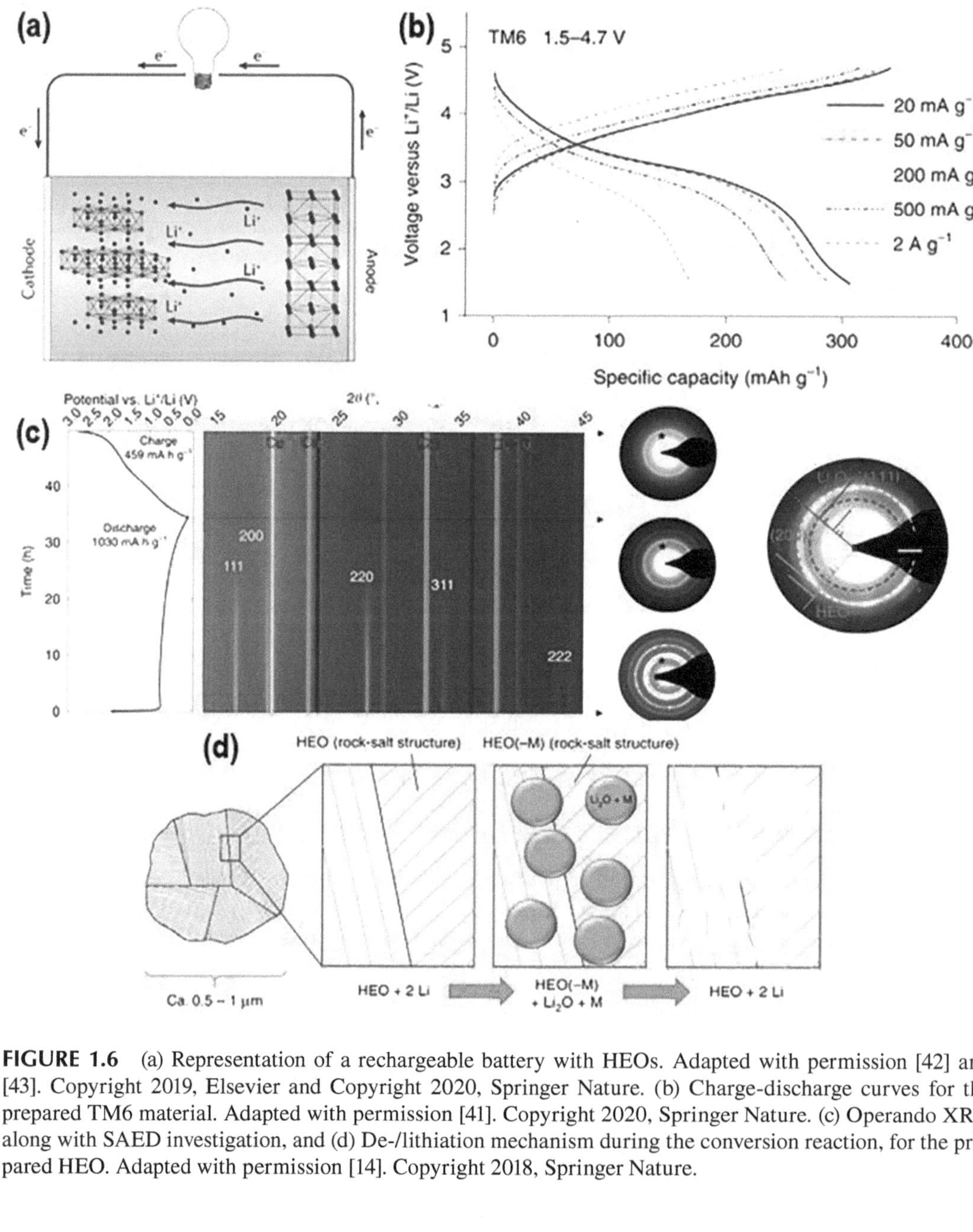

FIGURE 1.6 (a) Representation of a rechargeable battery with HEOs. Adapted with permission [42] and [43]. Copyright 2019, Elsevier and Copyright 2020, Springer Nature. (b) Charge-discharge curves for the prepared TM6 material. Adapted with permission [41]. Copyright 2020, Springer Nature. (c) Operando XRD along with SAED investigation, and (d) De-/lithiation mechanism during the conversion reaction, for the prepared HEO. Adapted with permission [14]. Copyright 2018, Springer Nature.

demonstrated a 3.4 V potential versus Li^+/Li and was suitable for use as a cathode active material as a result of entropy stabilization. Following the discharge of $(Co_{0.2}Cu_{0.2}Mg_{0.2}Ni_{0.2}Zn_{0.2})O_x$ HEO, Qiu et al. [44] reported by proposing the production of a small quantity of MgO. This regulates the volume change of the HEO anode while also preserving the active nanosized crystals separated. Wang et al. [43] reported that they produced a complete cell with an anode consisting of $(Co_{0.2}Cu_{0.2}Mg_{0.2}Ni_{0.2}Zn_{0.2})O_x$ and a cathode consisting of $LiNi_{1/3}Co_{1/3}Mn_{1/3}O_2$, demonstrating good results with an initial capacity of 446 mAh/g and a capacity of 256 mAh/g after 100 cycles (Figure 1.6c and d).

1.5.2 Na-Ion Batteries (SIBs)

HEOs are another type of electrode that can be used in place of LIBs. According to Zhao et al.'s report [24], they prepared a Q3 type through a reaction in solid form, NaNi contains 0.12% Cu, 0.12% Mg, 0.12% Fe, 0.15% Co, 0.15% Mn, 0.1% Ti, 0.1% Sn, 0.1% Sb, and 0.04% O_2 (Figure 1.7a). One of the most widely held beliefs is that various kinds of cations play a variety of roles. Mg^{2+} and Ti^{4+} may work to stabilize the host structure during cycling; Mn^{2+} may assist in structure creation; and Sn^{4+} and Sb^{5+} may improve the average voltage [45]. For instance, Ni^{2+}, Cu^{2+}, Fe^{3+}, and Co^{3+} can offer capacity; Mg^{2+} and Ti^{4+} may act to stabilize host structure during cycling; Ni^{2+}, Cu^{2+}, and Fe^{3+} and Co^{3+} can offer capacity. The high reversible O3-P3 phase transition is responsible for the improvement, which can be seen in Figure 1.7b and c. HEO demonstrated long cycling stability as high as 83% after 500 cycles, in addition to excellent rate capability. It has been hypothesized

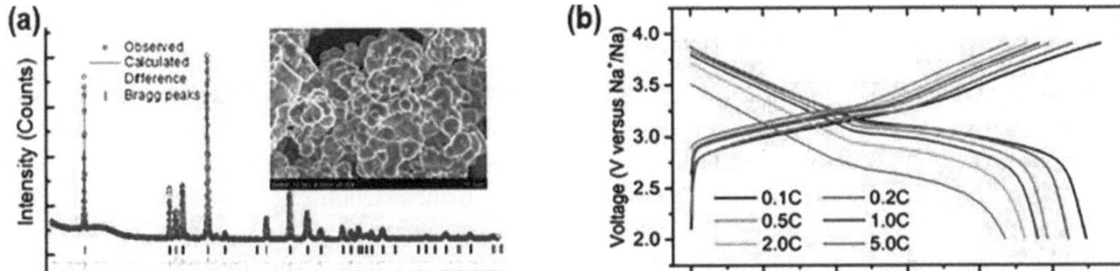

FIGURE 1.7 (a) XRD pattern along with O_3-type crystal structure, (b) Charge-discharge curves, (c) Capacity retention and coulombic efficiency, and (d) Schematic of possible Na^+ (de)intercalation process, for the prepared NaNi HEO. Adapted with permission [24]. Copyright 2020, Wiley-VCH.

that numerous transition metals have the ability to accommodate band charges at the moment of (de) intercalation, which results in a varied reaction of TMO_2 with NaO_2 slab (Figure 1.7d). This hypothesis has been put forward. Changes can be made to both the size of the transition metal and the oxidation state it is in. This metal is responsible for charge compensation. A high entropic composition and careful management of the local alteration work together to encourage a generally homogenous distribution of redox components. On the other hand, it may help sustain the O3-type structure, which would enable longer cycling stability in addition to superior rate performance [24].

1.5.3 Supercapacitor

Through the process of dissolving key alloys in H_2SO_4, Kong et al. [21] reported the formation of a number of nonporous HEs. The resultant AlCoCrFeNi can be employed as a binder-free electrode for supercapacitors. It exhibited a volumetric capacitance of 700 F/cm^3 and an exceptional cyclic capacity of more than 3000 cycles. The impressive performance may be due to the material's enlarged surface area. In addition to that, numerous HEMs that have a large surface area have been reported to be used in supercapacitors. For instance, Jin et al. demonstrated (VCrNbMoZr)N_x HEN by using a mechanochemically supported approach, and they reported a specific capacitance of 78 F/g when scanned at 100 mV/s in a solution of 1 M KOH [37]. Xu et al. [46] also reported the preparation of FeNiCoMnMg and FeNiCoMnCu NPs. The second one had a larger capacitance as well as a specific current density.

1.6 CONCLUSIONS AND REMARKS

In the past few years, HEMs have evolved for the purpose of energy-related utilization, particularly in the fields of catalysis and energy storage. To improve performance, HEMs have provided an amazing platform for adjusting the antisite disordering of atoms within crystalline structures. On the other hand, the association between performance and surface structure on an atomic level is not yet well understood. The development of HEMs intends to achieve structural property correlation and precisely monitor their structures along with the local electronic structures of their environments. The following are some of the difficulties and future directions that, at the same time, present a chance to investigate HEMs:

i. Techniques for designing HEMs: Theoretical investigations, molecular dynamics, and other methods may offer appropriate guidance for the design of HEMs. On the other hand, atomic-type disordering may offer a significant challenge when attempting to create an atomic model and identify constituents for applications relating to energy. To accomplish this goal, we will need a significant amount of labor, and a strategy that combines machine learning and big data could be helpful in managing the enormous data collection.
ii. Precise control of the HEM's structural components: To conduct experiments successfully, you need to have regulated synthesis procedures, systems that are simple to operate, and manufacturing on an industrial scale. The protocol for acquiring the necessary size, shape, and composition in order to make HEMs that are analogs to well-defined metals as well as semiconductor metals has to have its scope expanded.
iii. A very advanced form of characterization: For the purpose of structural characterization, sophisticated characterization methods, including X-ray-based techniques, scanning electron microscopy, and transmission electron microscopy (SEM and TEM), are used. These instruments provide accurate structural and chemical information regarding HEMs. The job of establishing a three-dimensional method to analyze the atomic structure of HEMs is a tough one. A combination of APT and 3D electron tomography might be helpful in this regard.

iv. The HEMs' capacity to withstand heat: High entropy is more common in the scientific literature, but configurational entropy, which stabilizes a single-phase solid solution, is more widely acknowledged these days. The actual contributions of entropy and enthalpy to Gibbs free energy as a whole are notoriously difficult to ascertain. Surface strain is yet another factor that needs to be taken into consideration.

v. The stability of the working environment's surface: The necessary surface structure as well as stability at the time of the reaction are requirements for catalysis; however, the surface structure of HEMs is likely to be more intricate compared to the catalyst that is currently being employed.

vi. Investigation into novel categories of HEMs: The investigation of HEMs and the utilities associated with energy is just getting started at this point. If more than five different components can be arranged in one lattice, then definite combinations are possible. The way that antisite disordering interacts with several types of physical parameters, including strain, 2D geometry, and magnetism, has to be researched. It is based on the belief that HEMs should become the future's significant platform in addition to other functional devices.

REFERENCES

[1] B. Cantor, I. Chang, P. Knight, A. Vincent, Microstructural development in equiatomic multicomponent alloys, *Materials Science and Engineering: A*, 375 (2004) 213–218.

[2] Q. Ding, Y. Zhang, X. Chen, X. Fu, D. Chen, S. Chen, L. Gu, F. Wei, H. Bei, Y. Gao, Tuning element distribution, structure and properties by composition in high-entropy alloys, *Nature*, 574 (2019) 223–227.

[3] B. Gludovatz, A. Hohenwarter, D. Catoor, E.H. Chang, E.P. George, R.O. Ritchie, A fracture-resistant high-entropy alloy for cryogenic applications, *Science*, 345 (2014) 1153–1158.

[4] Z. Lei, X. Liu, Y. Wu, H. Wang, S. Jiang, S. Wang, X. Hui, Y. Wu, B. Gault, P. Kontis, Enhanced strength and ductility in a high-entropy alloy via ordered oxygen complexes, *Nature*, 563 (2018) 546–550.

[5] T. Löffler, H. Meyer, A. Savan, P. Wilde, A. Garzón Manjón, Y.T. Chen, E. Ventosa, C. Scheu, A. Ludwig, W. Schuhmann, Discovery of a multinary noble metal-free oxygen reduction catalyst, *Advanced Energy Materials*, 8 (2018) 1802269.

[6] E.P. George, D. Raabe, R.O. Ritchie, High-entropy alloys, *Nature Reviews Materials*, 4 (2019) 515–534.

[7] D.B. Miracle, O.N. Senkov, A critical review of high entropy alloys and related concepts, *Acta Materialia*, 122 (2017) 448–511.

[8] J.W. Yeh, S.K. Chen, S.J. Lin, J.Y. Gan, T.S. Chin, T.T. Shun, C.H. Tsau, S.Y. Chang, Nanostructured high-entropy alloys with multiple principal elements: Novel alloy design concepts and outcomes, *Advanced Engineering Materials*, 6 (2004) 299–303.

[9] C.M. Rost, E. Sachet, T. Borman, A. Moballegh, E.C. Dickey, D. Hou, J.L. Jones, S. Curtarolo, J.-P. Maria, Entropy-stabilized oxides, *Nature Communications*, 6 (2015) 8485.

[10] T.A. Batchelor, J.K. Pedersen, S.H. Winther, I.E. Castelli, K.W. Jacobsen, J. Rossmeisl, High-entropy alloys as a discovery platform for electrocatalysis, *Joule*, 3 (2019) 834–845.

[11] M.W. Glasscott, A.D. Pendergast, S. Goines, A.R. Bishop, A.T. Hoang, C. Renault, J.E. Dick, Electrosynthesis of high-entropy metallic glass nanoparticles for designer, multi-functional electrocatalysis, *Nature Communications*, 10 (2019) 2650.

[12] J. Gild, Y. Zhang, T. Harrington, S. Jiang, T. Hu, M.C. Quinn, W.M. Mellor, N. Zhou, K. Vecchio, J. Luo, High-entropy metal diborides: A new class of high-entropy materials and a new type of ultrahigh temperature ceramics, *Scientific Reports*, 6 (2016) 1–10.

[13] G. Zhang, K. Ming, J. Kang, Q. Huang, Z. Zhang, X. Zheng, X. Bi, High entropy alloy as a highly active and stable electrocatalyst for hydrogen evolution reaction, *Electrochimica Acta*, 279 (2018) 19–23.

[14] A. Sarkar, L. Velasco, D. Wang, Q. Wang, G. Talasila, L. de Biasi, C. Kübel, T. Brezesinski, S.S. Bhattacharya, H. Hahn, High entropy oxides for reversible energy storage, *Nature Communications*, 9 (2018) 3400.

[15] S. Dai, Across the board: Sheng dai on catalyst design by entropic factors, *ChemSusChem*, 13 (2020) 1915–1917.

[16] P. Xie, Y. Yao, Z. Huang, Z. Liu, J. Zhang, T. Li, G. Wang, R. Shahbazian-Yassar, L. Hu, C. Wang, Highly efficient decomposition of ammonia using high-entropy alloy catalysts, *Nature Communications*, 10 (2019) 4011.

[17] A.-L. Wang, H.-C. Wan, H. Xu, Y.-X. Tong, G.-R. Li, Quinary PdNiCoCuFe alloy nanotube arrays as efficient electrocatalysts for methanol oxidation, *Electrochimica Acta*, 127 (2014) 448–453.

[18] H.-J. Qiu, G. Fang, J. Gao, Y. Wen, J. Lv, H. Li, G. Xie, X. Liu, S. Sun, Noble metal-free nanoporous high-entropy alloys as highly efficient electrocatalysts for oxygen evolution reaction, *ACS Materials Letters*, 1 (2019) 526–533.

[19] S. Nellaiappan, N.K. Katiyar, R. Kumar, A. Parui, K.D. Malviya, K. Pradeep, A.K. Singh, S. Sharma, C.S. Tiwary, K. Biswas, High-entropy alloys as catalysts for the CO2 and CO reduction reactions: Experimental realization, *ACS Catalysis*, 10 (2020) 3658–3663.

[20] N. Dragoe, D. Bérardan, Order emerging from disorder, *Science*, 366 (2019) 573–574.

[21] K. Kong, J. Hyun, Y. Kim, W. Kim, D. Kim, Nanoporous structure synthesized by selective phase dissolution of AlCoCrFeNi high entropy alloy and its electrochemical properties as supercapacitor electrode, *Journal of Power Sources*, 437 (2019) 226927.

[22] N.J. Usharani, A. Bhandarkar, S. Subramanian, S.S. Bhattacharya, Antiferromagnetism in a nanocrystalline high entropy oxide (Co, Cu, Mg, Ni, Zn) O: Magnetic constituents and surface anisotropy leading to lattice distortion, *Acta Materialia*, 200 (2020) 526–536.

[23] D. Bérardan, S. Franger, A. Meena, N. Dragoe, Room temperature lithium superionic conductivity in high entropy oxides, *Journal of Materials Chemistry A*, 4 (2016) 9536–9541.

[24] C. Zhao, F. Ding, Y. Lu, L. Chen, Y.S. Hu, High-entropy layered oxide cathodes for sodium-ion batteries, *Angewandte Chemie International Edition*, 59 (2020) 264–269.

[25] M. Fu, X. Ma, K. Zhao, X. Li, D. Su, High-entropy materials for energy-related applications, *Iscience*, 24 (2021) 102177.

[26] Y. Ye, Q. Wang, J. Lu, C. Liu, Y. Yang, High-entropy alloy: Challenges and prospects, *Materials Today*, 19 (2016) 349–362.

[27] K.V. Yusenko, S. Riva, P.A. Carvalho, M.V. Yusenko, S. Arnaboldi, A.S. Sukhikh, M. Hanfland, S.A. Gromilov, First hexagonal close packed high-entropy alloy with outstanding stability under extreme conditions and electrocatalytic activity for methanol oxidation, *Scripta Materialia*, 138 (2017) 22–27.

[28] D. Liu, X. Peng, J. Liu, L. Chen, Y. Yang, L. An, Ultrafast synthesis of entropy-stabilized oxide at room temperature, *Journal of the European Ceramic Society*, 40 (2020) 2504–2508.

[29] Y. Yao, Z. Liu, P. Xie, Z. Huang, T. Li, D. Morris, Z. Finfrock, J. Zhou, M. Jiao, J. Gao, Computationally aided, entropy-driven synthesis of highly efficient and durable multi-elemental alloy catalysts, *Science Advances*, 6 (2020) eaaz0510.

[30] M. Bondesgaard, N.L.N. Broge, A. Mamakhel, M. Bremholm, B.B. Iversen, General solvothermal synthesis method for complete solubility range bimetallic and high-entropy alloy nanocatalysts, *Advanced Functional Materials*, 29 (2019) 1905933.

[31] S. Gao, S. Hao, Z. Huang, Y. Yuan, S. Han, L. Lei, X. Zhang, R. Shahbazian-Yassar, J. Lu, Synthesis of high-entropy alloy nanoparticles on supports by the fast moving bed pyrolysis, *Nature Communications*, 11 (2020) 2016.

[32] D. Wu, K. Kusada, T. Yamamoto, T. Toriyama, S. Matsumura, I. Gueye, O. Seo, J. Kim, S. Hiroi, O. Sakata, On the electronic structure and hydrogen evolution reaction activity of platinum group metal-based high-entropy-alloy nanoparticles, *Chemical Science*, 11 (2020) 12731–12736.

[33] Y. Yao, Z. Huang, P. Xie, S.D. Lacey, R.J. Jacob, H. Xie, F. Chen, A. Nie, T. Pu, M. Rehwoldt, Carbothermal shock synthesis of high-entropy-alloy nanoparticles, *Science*, 359 (2018) 1489–1494.

[34] S. Zhai, J. Rojas, N. Ahlborg, K. Lim, M.F. Toney, H. Jin, W.C. Chueh, A. Majumdar, The use of poly-cation oxides to lower the temperature of two-step thermochemical water splitting, *Energy & Environmental Science*, 11 (2018) 2172–2178.

[35] X. Zhao, Z. Xue, W. Chen, Y. Wang, T. Mu, Eutectic synthesis of high-entropy metal phosphides for electrocatalytic water splitting, *ChemSusChem*, 13 (2020) 2038–2042.

[36] Z. Jin, J. Lv, H. Jia, W. Liu, H. Li, Z. Chen, X. Lin, G. Xie, X. Liu, S. Sun, Nanoporous Al-Ni-Co-Ir-Mo high-entropy alloy for record-high water splitting activity in acidic environments, *Small*, 15 (2019) 1904180.

[37] T. Jin, X. Sang, R.R. Unocic, R.T. Kinch, X. Liu, J. Hu, H. Liu, S. Dai, Mechanochemical-assisted synthesis of high-entropy metal nitride via a soft urea strategy, *Advanced Materials*, 30 (2018) 1707512.

[38] T. Wang, H. Chen, Z. Yang, J. Liang, S. Dai, High-entropy perovskite fluorides: A new platform for oxygen evolution catalysis, *Journal of the American Chemical Society*, 142 (2020) 4550–4554.

[39] J.K. Pedersen, T.A. Batchelor, A. Bagger, J. Rossmeisl, High-entropy alloys as catalysts for the CO2 and CO reduction reactions, *Acs Catalysis*, 10 (2020) 2169–2176.

[40] H. Chen, W. Lin, Z. Zhang, K. Jie, D.R. Mullins, X. Sang, S.-Z. Yang, C.J. Jafta, C.A. Bridges, X. Hu, Mechanochemical synthesis of high entropy oxide materials under ambient conditions: Dispersion of catalysts via entropy maximization, *ACS Materials Letters*, 1 (2019) 83–88.

[41] Z. Lun, B. Ouyang, D.-H. Kwon, Y. Ha, E.E. Foley, T.-Y. Huang, Z. Cai, H. Kim, M. Balasubramanian, Y. Sun, Cation-disordered rocksalt-type high-entropy cathodes for Li-ion batteries, *Nature Materials*, 20 (2021) 214–221.

[42] C. Oses, C. Toher, S. Curtarolo, High-entropy ceramics, *Nature Reviews Materials*, 5 (2020) 295–309.

[43] Q. Wang, A. Sarkar, Z. Li, Y. Lu, L. Velasco, S.S. Bhattacharya, T. Brezesinski, H. Hahn, B. Breitung, High entropy oxides as anode material for Li-ion battery applications: A practical approach, *Electrochemistry Communications*, 100 (2019) 121–125.

[44] N. Qiu, H. Chen, Z. Yang, S. Sun, Y. Wang, Y. Cui, A high entropy oxide (Mg0. 2Co0. 2Ni0. 2Cu0. 2Zn0. 2O) with superior lithium storage performance, *Journal of Alloys and Compounds*, 777 (2019) 767–774.

[45] M. Sathiya, Q. Jacquet, M.L. Doublet, O.M. Karakulina, J. Hadermann, J.M. Tarascon, A chemical approach to raise cell voltage and suppress phase transition in O3 sodium layered oxide electrodes, *Advanced Energy Materials*, 8 (2018) 1702599.

[46] X. Xu, Y. Du, C. Wang, Y. Guo, J. Zou, K. Zhou, Z. Zeng, Y. Liu, L. Li, High-entropy alloy nanoparticles on aligned electronspun carbon nanofibers for supercapacitors, *Journal of Alloys and Compounds*, 822 (2020) 153642.

2 A Brief History of High Entropy Materials

Partha P. Sarma, Kailash Barman, and Pranjal K. Baruah

2.1 INTRODUCTION

Metallic substances have occupied a special position in the evolution of human civilizations as basic materials for human existence and development worldwide. They serve as the key materials for human evolution and survival and can aid in the advancement of society. Any leading breakthrough in materials science can bring a radical change to the socioeconomic scenario of human civilization [1]. Although the widespread use of advanced materials such as ceramics, polymers, plastics, metals, alloys, semiconductors, composites, superconductors, and nanomaterials represents the current era of materials, however, the Stone Age began well before 10,000 B.C. marks the beginning of materials history [2,3]. To meet the various needs of people with changing times and in parallel with the development of science and technology, the world of metallic materials has undergone drastic changes from simple to complicated composite materials. This results in the evolution of several types of special materials with enhanced efficiency [4,5]. From prehistoric times, starting with stone and then copper and then, during the 19th century, iron and steel, engineers had perhaps access to a few hundred of different materials. But with the tremendous scientific revolution, nowadays this number has increased by more than a thousand times, and most of them have been discovered in the last century [6]. The rate of advancement of metals has accelerated more than ever in the last century. Through persistent work and tireless effort, researchers have already made the application of metallic materials wide and fascinating [1]. The growing interest in sustainability and energy efficiency demands the continuous development of advanced materials with enhanced performance [7]. An age-old technique for enhancing materials' performance is alloying, which channels properties through mixing. Normally, a mixture should produce characteristics that are an average of those of its constituent parts, but occasionally, improved or completely new characteristics can emerge [8].

Traditionally, alloy fabrication was restricted to incorporating a small proportion of another component into a primary constituent, which is considered the base element to develop a novel composite material with enhanced properties [9]. Numerous alloy systems based on a single basic metal have already been synthesized and commercialized. Most conventional alloy materials are typically based on Fe, Al, Mg, Ti, Ni, and Cu as principal materials. Stainless steel, satellite alloys, aluminum alloys, silicon alloys, superalloys, high-speed steels, alnico alloys, titanium alloys, permalloys, metallic glasses, intermetallic compounds, Ni–Al alloys, Ti–Al alloys, etc. are some eminent industrial alloys that are playing a major role in the amelioration of human society worldwide [3]. However, other important classes of alloys such as Mg alloys, Al alloys, and Ti alloys have emerged in the last century [4]. After substantial achievements in traditional alloy materials, nowadays the developments in this field reaching their limits. The traditional strategy for alloy designing allows for the development of a limited number of alloys, the majority of which have already been explored and identified [9]. Enormously growing demands and comforts of the people significantly demand newer materials possessing specific applications at a reasonable price with significantly enhanced life, which can be produced and used easily. To satisfy the rising demand, it is essential to explore unfamiliar alloys. With the development of high entropy alloys (HEAs), which can also be regarded as HEMs, a novel and cutting-edge concept has emerged to address such needs [5].

 DOI: 10.1201/9781003391388-2

2.2 HEMs – A BRIEF OVERVIEW

In 2004, a novel method of alloying was proposed independently by Yeh et al. and Cantor et al. introducing solid solution single-phased alloys consisting of five or more fundamental elements with almost equimolar compositions [9–11]. This new strategy for alloy design involves simultaneous alloying of multiple principal elements to achieve high entropy of mixing leading to the development of numerous innovative concentrated multicomponent alloy materials termed as HEMs [12,13]. The presence of multiple elements imparts unique properties to the HEAs. Superior hardness, high strength, and great water resistance are just a few of the traditional qualities that make HEAs so appealing in a broad field of applications such as thermoelectric applications, catalysis, superconducting materials, magnetocaloric, energy storage, etc. A wide variety of HEMs including oxides, alloys, nitrides, carbides, borides, sulfides, phosphides, and oxyfluorides have been reported with wide and fascinating applications [4]. High entropy imparts a variety of superior properties to these HEMs as compared to their constituent elements [8,14,15]. These unique structures provide exceptional phase stability and enhance mechanical properties like strength, ductility, etc. [16]. The field of high-entropy alloy materials is diverse and engrossing. The addition of different species leads to the emergence of unexpected properties, which can be altered by varying the concentrations. Using effective high-throughput techniques and enhanced artificial intelligence techniques, the enormous prospects for improvising innovative multicomponent alloy systems can be addressed [17,18].

2.3 DEFINITION

The introduction of the new approach of alloying involving solid solution single-phase alloy comprising five or more elements, known as HEAs, has broadened up the world of metallic materials. The concept behind the development of HEAs is mainly based on configurational entropy. The mixing entropy of the alloy system increases with the addition of other basic elements in an equimolar ratio [9]. The increase in configurational entropy can promote the formation of single-phase solid solutions with enhanced stability over alloy systems possessing intermetallic phases more prominently at high temperatures [19,20]. However, no definition has been provided for multicomponent alloys, but there exist several definitions regarding HEAs, which are also addressed as HEMs. Some definitions of HEA create some confusion, fueling debates about whether certain alloys can be called HEAs [19]. There are mainly two definitions to understand HEAs, one is based on composition and the other is based on entropy, both were widely accepted and are discussed below.

2.3.1 Composition-Based Definition

The earliest definition of HEAs based on compositional specifications was introduced in 2004 [10,11]. High entropy materials or alloys may be specified as solid-solution multicomponent alloys comprising at a minimum five principal elements each possessing an atomic composition of 5%–35% [3,21–23]. This definition indicates that if there are any minor elements present, then their atomic percentage must be less than 5% [24]. Based on this definition, it can be addressed that a high entropy alloy can consist of elements in equimolar ratio, a variety of nonequimolar components, or a variety of minor elements [21]. This proposed definition only specifies elemental concentrations and assigns no limits on the magnitude of configurational entropy. Furthermore, the existence of a single-phase solid solution is not required by this definition [19].

2.3.2 Entropy-Based Definition

HEAs or HEMs can be defined as the class of alloys or materials that possess a configurational entropy or in other words mixing entropy value greater than 1.5R in solid solution phase regardless of whether they are single or multiple phases at room temperature.

Although there are four contributing factors to the total entropy of mixing which includes configurational entropy, magnetic dipole entropy, vibrational entropy, and electronic randomness entropy, the configurational entropy acts as the dominating factor. Thus, the configurational entropy can often be regarded as the mixing entropy for simplicity [3,4].

In thermodynamics, entropy is regarded as a state function that gives characteristics and information about the randomness of the system. From the statistical thermodynamics, according to Boltzmann's distribution, the configurational entropy of mixing for a particular system can be represented by the following equation [3,24]

$$S = k_B \ln W \tag{2.1}$$

where k_B represents Boltzmann's constant, W represents the thermodynamic probability which is the number of ways that the system's particles can combine or distribute the available energy.

In the case of a random solid solution comprising n components, the ideal configurational entropy per mole can be calculated by the following equation [3,24]:

$$\Delta S(\text{conf}) = -R\sum_{i=1}^{n}(x_i \ln x_i) \tag{2.2}$$

where R denotes the ideal gas constant and x_i denotes the mole fraction of the ith component. However, for an alloy system comprising n components with an equimolar ratio in its solid solution, the equation for configurational entropy per mole becomes [9,24,25]

$$\Delta S_{\text{conf}} = -R\left(\frac{1}{n}\ln\frac{1}{n} + \frac{1}{n}\ln\frac{1}{n} + \frac{1}{n}\ln\frac{1}{n} + \cdots + \frac{1}{n}\ln\frac{1}{n}\right) = -\text{R}\ ln\frac{1}{n} = \text{R}\ \ln n \tag{2.3}$$

The configurational entropy depends on the number of elements incorporated [24]. As the number of elements increases, the configurational entropy of an alloy system increases. The increase in configurational entropy is because of the different arrangements of atoms of various elements present in the alloy system [21]. The entropy-based definition assumes that an alloy can be depicted as a state of a liquid solution or a high-temperature solid solution within which thermal energy is adequately high, which leads different elements to occupy random positions in the structure [19].

2.3.3 Classification of Alloys Based on Configurational Entropy (ΔS_{CONF})

From the above-mentioned entropy-based definition, based on the value of the configurational entropy or more simply mixing entropy, the materials world can be divided into four different classes [9,20,22,26]. A diagrammatical representation of the different classes of alloy materials is depicted in Figure 2.1.

i. *Ultrapure materials*: These are the materials for which the ΔS_{conf} value approaches zero. These are also known as zero entropy alloys or high-purity materials.
ii. *Low entropy alloys*: These are the alloy materials for which $\Delta S_{\text{conf}} \leq 1\text{R}$. Such an alloy system possesses one or two basic elements. Traditional alloys fall under this category.
iii. *Medium entropy alloys*: These are the alloy materials for which the ΔS_{conf} value lies between 1R and 1.5R. Such an alloy system possesses two to four basic elements.
iv. *HEAs*: For HEAs the $\Delta S_{\text{conf}} \geq 1.5\text{R}$. These alloy systems contain at least five basic elements.

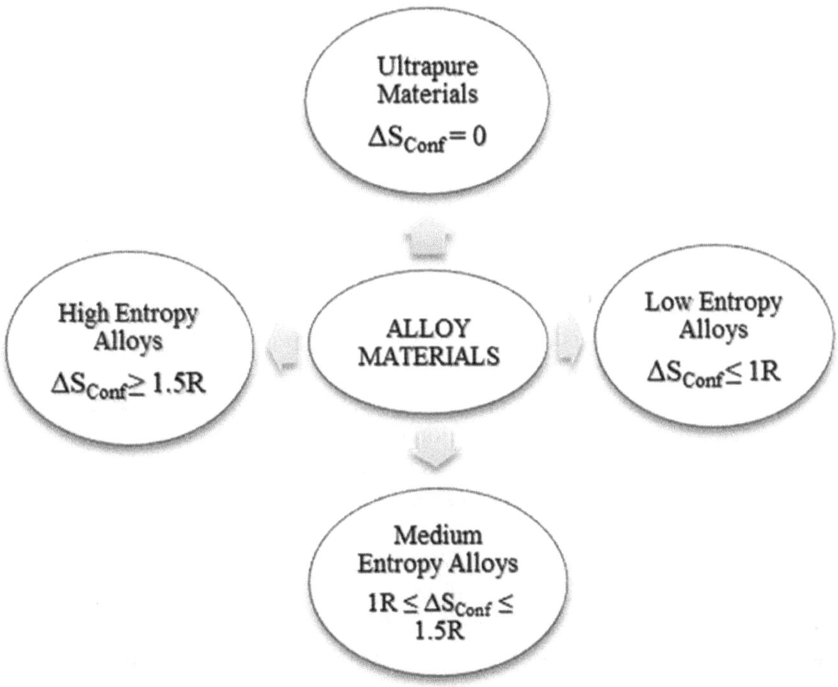

FIGURE 2.1 Classification of alloys based on configurational entropy.

2.4 FOUR CORE EFFECTS

From the thermodynamic point of view, it has been assumed that the equilibrium state of HEA will result in several intermetallic compounds with complex structures. With the progress in research about HEAs, it was found that most HEA systems do not yield intermetallic phases as well as any complicated structural phases over their solidification [21]. On the contrary, such HEA systems frequently end up with a single solid solution phase possessing a simple structure. Furthermore, in contrast to traditional alloy materials, HEMs possess distinct kinetics along with microstructural and functional properties, which are influenced by a variety of factors [27]. Among these high entropy effects, lattice distortion effect, sluggish diffusion effect, and cocktail effects are the four "core effects" that are most fundamental as well as responsible for excellent performances and are used to describe the HEMs [20,27].

2.4.1 The High Entropy Effect

The high entropy effect, which is the leading idea behind HEMs, postulates that higher configurational entropy in alloys consisting of five or more elements that are nearly equimolar can promote single solid solution phases instead of forming intermetallic compounds [22]. High entropy, particularly at high temperatures, can minimize the Gibbs free energy promoting the development of a solid solution phase [27]. High entropy can also decrease the electronegativity difference and prevent phase separation. When the mixing entropy is high, the number of developed phases is much lower than the optimum amount estimated by the Gibbs phase law, which enhances component compatibility that makes it easier to form stable and simple phases [21]. Thus, the elevated configuration entropy is the prime contributing factor to the exceptional stability of the HEMs [27,28]. However, due to strong bonding among some of the metallic elements, the formation of intermetallic phases may occur in the alloy system, and such phases may frequently include many additional elements which results significantly reduced degree of orderliness [29,30].

2.4.2 The Lattice Distortion Effect

Due to the presence of multiple elements in HEAs, each atom in the solid solution phase is surrounded by several different types of atoms, which impart lattice strain and stress because of the large variation in atomic size known as the lattice distortion effect. This lattice distortion effect can influence the electrical, thermal, chemical, and optical properties of the materials [29]. A variety of elements with varying atomic sizes contribute to the lattice structure of HEMs. Such dimensional variations inevitably caused lattice distortion. Whereas smaller atoms have voids surrounding them, larger atoms push aside the neighboring atoms [21]. In addition to the atomic size variation, it is anticipated that dissimilarities in the crystal structures across the constituent elements, as well as variations in bonding energies, also act as contributing factors to the higher lattice distortion [20]. In HEAs, lattice distortion effects result in a significant increase in hardness with enhanced thermal and electrical resistance. The distorted lattice of atoms boosted X-ray diffuse scattering phenomena, which lead to a lower intensity XRD peak [27,29]. However, higher electron and phonon scattering due to the lattice distortion significantly decreases the electrical as well as thermal conductivity of the HEMs. In the case of adequately high differences in atomic sizes, it might be assumed that distorted lattice is likely to crumble into an amorphous state because of very high lattice distortion energy for maintaining a crystalline structure [27]. Furthermore, the lattice distortion of HEAs may impede the atomic movement as well as substitutions resulting in another phenomenon known as sluggish diffusion [9].

2.4.3 The Sluggish Diffusion Effect

The sluggish diffusion effect can be regarded as the consequence of the lattice distortion effect. Each lattice position of HEAs has different neighboring atoms, and each of them possesses different diffusion rates [4]. It has been found that in HEAs each element possesses a much lower self-diffusion coefficient as compared to other conventional alloys, which indicates that the atoms present in HEAs have a smaller rate of diffusion as compared to the other alloys. This results from interactions between various atoms and distorted lattice structures, both of which have a significant impact on the diffusion rates of the different atoms. In HEAs, the sluggish diffusion effect has a great impact on new phase formation. During phase transformations which are strictly dependent on atomic diffusion, to achieve an equilibrium partitioning between the phases, elements must be cooperatively diffused together. This along with the lattice distortion effect that restricts the atomic movement constrained the rate of diffusion in HEAs [10,31]. It has been proposed that the significant variation in lattice potential energy (LPE) among crystal lattice sites results in the slow diffusion rate as well as the higher activation energy of HEAs. Numerous low LPE sites could act as traps for impeding atom diffusion, contributing to the sluggish diffusion effect [4]. The slow diffusion effect could offer numerous benefits including supersaturated states and fine precipitation, increased recrystallization temperature, minimized grain coarsening, and slower grain growth [32]. These advantages contribute to microstructure and property control, as well as high performances including improved strength and toughness, thermal stability, and corrosion resistance [4,33,34].

2.4.4 The Cocktail Effect

Each of the components of HEAs has distinct properties, and the interaction between them causes the HEAs to acquire a composite effect, known as the "cocktail" effect, which was initially proposed by the Indian scholar Ranganatha [21]. The atomic-scale interaction and impact of the alloy components will eventually be displayed in the extensive macroscopic attributes of the alloy and may even introduce additional effects [21]. The cocktail effect is utilized in HEAs to highlight the advancement of properties by multiple constituent elements. As the HEAs can possess single or multiple phases based on composition and fabrication, hence the overall properties are derived from the contribution of the entire constituent phases. This correlates with the phase boundaries, distribution, phase size and shape, and the characteristics of each phase [20]. Furthermore, each phase in a HEAs is a solid solution multicomponent system and can be considered as an atomic scale composite where the overall properties are not only derived from the fundamental properties of the constituent elements but also considering the mutual interactions between each element and from the major lattice distortions [27]. The synergistic interaction between the different elemental components of HEA will result in a number of unusual properties such as thermal, electrical, magnetic, and mechanical properties and make them useful to serve as an effective material in energy conversion and storage systems [9,29].

2.5 BRIEF HISTORY OF HEMs

Both Cantor and Yeh along with their co-workers independently developed the concept of fabricating equiatomic or near equiatomic alloys with multiple components, causing a significant deviation from conventional alloy designing methods. Yeh et al. promoted this class of innovative alloys as HEAs by highlighting the well-known thermodynamic phenomena that, as the number of elements increases the optimum configurational entropy in any system also increases and in the case of the binary alloys, configurational entropy becomes maximum when the constituent elements are present in equiatomic proportions [10].

Along with studying the fundamental phases and the microstructure of various equiatomic multicomponent alloys with a range of 5–20 components, another approach of equiatomic substitution of constituents in a binary alloy was introduced by Cantor et al. in which various identical elements were substituted in the same proportions in the binary system. For instance, replacing 75% of the Zr in the $Zr_{50}Cu_{50}$ alloy with Ti, Nb, and Hf so that all four elements have equal atomic proportions [11,35].

At the earliest stage of development, typical HEMs were interpreted as a multicomponent alloy system comprising a minimum of five elements with equimolar proportion. However, with the scientific revolution in this field, new classes of HEMs possessing nonequal atomic proportions with enhanced properties as compared to the HEMs with equimolar mixture have been developed. This led to the wide extension of the concept of HEMs introducing more flexibility in the compositional section. A wide range of metals and non-metals including Al, B, Mg, Si, etc. from group IIA, IIIA, and IVA of main group elements, all subgroup elements of the fourth period such as Cr, Ti, Mn, Fe, Ni, etc., and several subgroup elements of fifth period such as Mo, Zr, Nb, etc., are involved in the formation of HEMs [1].

2.5.1 Contribution from B. Cantor

B. Cantor began exploring the field of equiatomic multicomponent alloy materials along with Alain Vincent, an undergraduate student in 1981. They have developed a huge number of multicomponent alloy systems by considering a wide range of elements in equiatomic proportions. However, only one alloy system with the composition $Co_{20}Cr_{20}Fe_{20}Mn_{20}Ni_{20}$ was found to form a single-phase solid solution with an FCC lattice structure. One of their experimental alloys consisted of 20 different components each possessing an atomic percentage of 5%. However, this work remained unpublished. After this study in 1998, Peter Knight, who was also an undergraduate student, repeated a similar study on several different types of alloys and obtained similar kinds of results [3]. Finally, in around 2000, Isaac Chang performed similar work at Oxford and published the findings with several significant conclusions. Two different important alloy systems one comprising 20 components each possessing an atomic percentage of 5 and another having 16 components each possessing an atomic percentage of 6.25 were reported with multiphase, crystalline, and brittle nature. However, the number of total phases is always considerably lower than the optimum equilibrium number predicted by the Gibbs phase rule and much lower than the highest number permitted during non-equilibrium solidification conditions as well [11].

Cantor et al. reported in their study in 2002 about the development of several unique multicomponent amorphous alloy systems through equiatomic alloy substitution with a general formula of $A_xB_yC_z$ where A and B represent the early transition metals such as Ti, Zr, Hf, etc., and the later transition metals such as Cu, Ni, etc., respectively. C denotes a nontransition simple metal such as Al, Be, etc. with $x=50$–70%, $y=15$–30%, and $z=10$–30% of alloy content. They have fabricated three different series of alloys including base $Ti_{33}Zr_{33}Hf_{33}$ $Ni_{50}Cu_{50}Al_{10}$ alloys, Ag containing $Ti_{33}Zr_{33}Hf_{33}Ni_{33}Cu_{33}Ag_{33}Al_{10}$ and Nb containing $Ti_{25}Zr_{25}Hf_{25}Nb_{25}Ni_{50}Cu_{50}Al_{10}$ alloys. For the early transition metal composition Zr, Ti, Hf, and Nb were regarded as chemically similar for the equiatomic alloy substitution. However, Cu, Ag, and Ni were regarded as chemically similar in the case of later transition metal composition for equiatomic alloy substitution. However, no information about the entropy of the systems was provided in their study [35].

2.5.2 Contribution from J. W. Yeh

Yeh started exploring multi-principle element alloy by the year 1995 based on the assumption that high entropy of mixing can improve the mixing among the components and minimize the total number of phases [3]. At the start of his research along with his co-workers, he synthesized approximately 40 different alloys with equiatomic composition using five to nine commonly used constituent elements with or without the addition of a minor element through arc melting and different properties such as hardness, corrosion resistance, and microstructural properties were investigated. Increased hardening due to significant lattice distortion and an increase in bond strength were proposed during the discussion of these trends. In general, all these alloys demonstrated excellent corrosion resistance. Corrosion resistance was assumed to be aided by the incorporation of passive components and the advantage of lower free energy caused by high entropy. Thus, this study provided helpful suggestions regarding the slow diffusion effect, lattice distortion effect, and high entropy effect [3].

Yeh et al. in their study in 2004, reported on the synthesis of four different alloy systems containing ≥5 principal metal elements through sputtering, casting, and splat quenching methods. From the structural investigations of these alloys, they have found the formation of a solid solution possessing with crystal structure with a BCC or FCC lattice. The four alloy systems that they developed were CuCoNiCrFe, $CuCoNiCrAl_{0.5}Fe$, $Cu_{0.5}CoNiCrAl$, and CuCoNiCrAlFeTiV [36]. Yeh et al., in another study, developed a new concept for alloy designing and prepared a multicomponent alloy system of $AlSiTiCrFeNiMo_{0.5}$ and $AlSiTiCrFeCoNiMo_{0.5}$ through arc melting which was termed as multi-principle element alloys [37]. Again, in another study, Yeh et al. reported that to a previously synthesized FCC $CuCoNiCrAl_{0.5}Fe$ high entropy alloy, the addition of boron, which results in $CuCoNiCrAl_{0.5}FeB_x$ alloy material makes enhancement in its properties such as wear resistance, increase hardness, and improved high-temperature strength [38]. However, in another study, Yeh et al. focused on the development of HEAs using multiple principal metal elements in an equimolar ratio. They have developed two different nanostructured alloy systems $CuCoNiCrAl_xFe$ and CuCoNiCrAlFeMoTiVZr one consisting of six different metal elements and another one comprising ten principal elements in equimolar ratio. Both possess a simple solid solution phase. These nanostructured alloy materials exhibited high wear resistance, high oxidation resistance, higher thermal stability, and increased strength [10].

These were the first reported studies to provide experimental findings along with associated theories that illuminate the idea of HEAs and can be regarded as a milestone in the history of the development of HEMs. Several promising applications in diverse fields have been shown in these reports. Yeh and his co-worker had published 111 research articles on HEAs by the end of the year 2014 [3]. Employing these HEA concepts, advances in the world of materials science have been expanded to HEMs nowadays.

2.5.3 Other Historical Works

Although well published in 2004, a variety of multicomponent alloys have been investigated in different studies before that. For example, in 1996, Xing et al. reported in their study about the synthesis of a Zr-based multicomponent amorphous alloy system $Zr_{62-x}Ti_xCu_{20}Ni_{18}Al_{10}$ ($3 \leq x \leq 5$). Such an amorphous alloy system possessing good ductility and higher strength was developed through the deposition of icosahedral quasicrystals which led to the formation of nano-scaled quasicrystals incorporated in an amorphous matrix in the initial step [26,35].

It should be noted that Franz Karl Achard, a German scientist, and metallurgist, explored multicomponent equimolar alloys containing five to seven different elements at the end of the eighteenth century. He was likely the first person who studied multi-principal alloys consisting of five to seven elements [39]. He has carried out his study in more than 900 alloy compositions within 11 different metals. He developed representative alloys of different compositions of components to a maximum of seven components. In addition to different binary, ternary, and quaternary alloy systems, he also developed several quinary, and septenary alloy systems only in similar weight proportions and investigated numerous physical and mechanical properties including strength, density, hardness, abrasion resistance, ductility, etc. He emphasized that the properties of alloys differ significantly when compared to pure metals as well as are unpredictable. However, he has only reported the experimental results without further discussions [39].

2.6 DIFFERENT CLASSES AND WIDE APPLICATIONS OF HEMs – HISTORICAL SKETCH

Multi-principal element alloys, commonly referred to as HEAs, were introduced in academic literature in 2004 and the knowledge, applications, and research were subsequently expanded since then. These nontraditional architectures offer new prospects to achieve unusual combinations of mechanical properties and phase stability, particularly by enhancing the strength and ductility of the materials. The concept of "High Entropy" was initially proposed only for metallic alloys. As the research progressed, numerous new classes of materials were started to be discovered and added to the world of HEAs at different times. Such a new class of HEMs includes high entropy oxides, ceramics, carbides, sulfides, nitrides, nanoparticles, and borides [40–45].

At the early stage of development, the applications of the HEMs were very limited. HEMs were mostly employed as structural materials before 2015. Rost et al., in 2015, derived rock-salt structured solid solution oxide, which is termed as high-entropy oxide (HEO) or entropy-stabilized oxide since they are also predominantly stabilized by large configurational entropy. They have extended the concept of entropy to the five-component oxides that include MgO, CuO, NiO, CoO, and ZnO. They have introduced a new class of HEAs consisting of mixed oxides that not only possess high entropy but indeed also completely entropy stabilized. Starting with this nowadays different HEOs were fabricated with numerous applications [40].

A new kind of ultra-high temperature ceramic as well as a new class of HEMs termed as high-entropy metal diborides or simply high entropy borides (HEBs) was introduced by Glid et al. in 2016. They have fabricated seven different HEBs using five components in equimolar ratio possessing a unique hexagonal and layered crystal structure. These HEBs possess higher hardness as well as high oxidation resistance as compared to the borides of individual metals [41]. Jin et al. in 2018 made a new addition to the field of HEMs by proposing a new approach to developing high-entropy metal nitride through mechanochemical synthesis. They had developed High Entropy Metal Nitride (HENs) consisting of five different highly dispersed principal metal components that include Cr, V, Mo, Nb, and Zr exhibiting cubic crystalline metal nitride structures. These emerging HENs are regarded as promising for supercapacitor applications [44].

High entropy sulfides are also a recently developed important class of HEMs which was first introduced by Zhang et al. in 2018. Two different types of single-phase thermoelectric materials including a semiconducting $Cu_3SnMgInZnS_7$ and a metallic $Cu_5SnMgGeZnS_9$ have been designed which were termed high entropy sulfides [43]. Sarker et al. in 2018 introduced another new class of HEMs known as high entropy carbides with the development of six different metal carbides including $MoNbTaVWC_5$, $HfNbTaTiVC_5$, $HfNbTaTiZrC_5$, $HfNbTaTiWC_5$, $HfTaTiWZrC_5$, and $NbTaTiVWC_5$ which are regarded as promising prospects for the high-hardness applications [42].

Glasscott et al. in 2019 developed a new strategy for synthesizing high-entropy metallic glasses nanoparticles (HEMG-NPs) with a maximum of eight components in an equimolar ratio. They have highlighted the potential applications of these HEMG-NPs in the field of energy conversion and water-splitting reactions. However, they have demonstrated that CoFeLaNiPt HEMG-NP system that was synthesized by them possesses multifunctional properties and acts as an efficient electrocatalyst in the oxygen evolution reaction (OER) as well as in the hydrogen evolution reaction (HER) [45].

These recent developments have highlighted the wide application of HEMs in numerous fields including energy-related applications, catalysis, electrodes for batteries, supercapacitors, water splitting, energy storage and conversion, and hydrogen storage nowadays [9,16,22]. The structural flexibility of HEMs makes them attractive to energy-related applications. For instance, tuning the surface electronic frameworks, adsorption energies, and ultimately the catalytic performance of HEMs is greatly facilitated by the increased number of atomic sites on their surfaces [16,22].

HEM catalysts with enhanced catalytic activities can be achieved by rationally altering the geometry and composition of HEMs. Due to the presence of multiple elements, many atom arrangements are possible across the surface of the HEMs which introduce various adsorption modes for the reactants and the intermediates [22]. Through the combination of reactive metals, the fundamental synergistic effect of HEMs also allows for enhancement in the catalytic activity [1,19]. Presently, HEMs are being widely used as highly efficient heterogeneous catalysts in versatile electrocatalytic and thermal-driven reactions with surprising performance. A brief overview of different applications of HEMs is represented below (Figure 2.2).

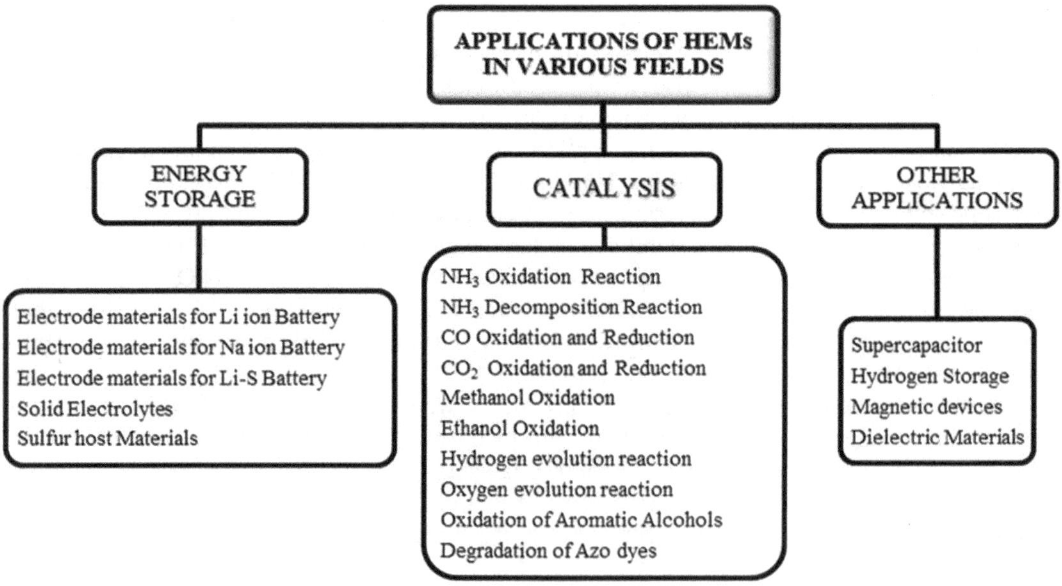

FIGURE 2.2 Various ongoing applications of HEMs.

2.7 ADVANTAGES OF HEMs OVER TRADITIONAL ALLOYS

HEMs are predicted to have the ability to overcome numerous bottlenecks associated with conventional materials with several promising revolutionary applications. The evolution of HEMs with excellent mechanical, chemical, and physical properties, high thermal and chemical stability, oxidation, and corrosion resistance has led to significant advancements in the field of catalysis, energy storage, and conversion technologies [16,46]. HEAs innovated a novel category of materials that vary significantly from traditional alloys in consisting of multiple primary elements incorporated in the crystal structure. The cocktail effect associated with HEAs is responsible for the key benefits of HEAs over conventional alloys. Taking advantage of the individual constituent elements, the HEAs system can be considered as an "atomic composite," frequently displaying unusual properties. In contrast with the traditional alloys, the formation of HEAs leads to substantial lattice distortion because of the incorporation of metal elements with varying atomic sizes. This is favorable for producing more suitable active sites, leading to the development of promising catalytic properties [24]. Recently, emerging uses of HEMs as functional materials have attracted interest from both experimental and theoretical points of view, particularly as heterogeneous catalysts in a wide variety of reactions. The unique properties of HEAs over traditional alloys, as well as the versatile compositions that may introduce new phenomena and functionality, make these innovative HEMs attractive not only in the field of research but also in a wide range of industrial applications. Despite limited research on HEM catalysis, HEM-based catalysts have demonstrated excellent catalytic performance in a variety of thermal and electrocatalytic reactions as well as in energy storage.

2.8 CHALLENGES ASSOCIATED WITH HEMs

Despite all the potential benefits of HEMs over conventional alloys, in the case of many HEOs along with other HEMs, there are many challenges concerning their practical applications in different fields. Although studies indicate that compositional modification can be used to tailor the characteristics of HEMs, however, detection of the functional units and realizing the function of individual components in various applications is very challenging. The design, manufacture, modification, and subsequent advancement of metallic catalysts are directly influenced by the accurate detection of active centers [22]. The research and developments that have been done till now are still in their infancy, and it seems, in most cases, tailoring is done in a random manner. Hence, the rational designing of HEMs is still very challenging. An advanced and deeper understanding of the high entropy concept is also another challenge associated with HEMs. Despite the recent extensive exploration of bulk and nanostructured HEMs, application-specific novel synthesis techniques are still critically needed. The most significant task in the development of HEMs is to accomplish this structure-property relationship to control their framework and specific electronic structure [24]. Regardless of the challenges, the progress made at present is very inspiring and introduced numerous possibilities regarding future applications of this innovative class of materials in renewable energy.

2.9 CONCLUSION

These freshly developed materials with enhanced functional properties can make a significant contribution to accelerating the development of human civilization worldwide. Recent progresses in research have shown that the relatively young HEMs have enormous potential in energy-related applications. By tailoring the constituent elements of high-entropy systems, promising functional attributes such as gas absorption ability, electrocatalytic activity, catalytic activity, electrochemical charge storage, etc. can be achieved [24]. HEMs in the form of nanoparticles acquire large surface area which stimulates the reaction kinetics. The principle of high entropy has now extended to a wide variety of inorganic as well as organic materials, and it can be effectively employed to design and development of a broad range of novel materials with the desired characteristics [9]. Due to the entropy-driven impacts along with structural and chemical diversities, HEMs hold great potential in sustainable energy storage as well as conversions [24]. Correlations in HEMs between their composition, methods, characteristics, and microstructure offer several research areas for both basic and applied sciences as well as industry. The field of HEMs is growing and offers enormous prospects for research and inventions.

REFERENCES

[1] Y. Zhang, *History of high-entropy materials, high-entropy materials* (2019) 1–33 Springer, Singapore.

[2] L.E. Murr, *A brief history of metals, handbook of materials structures, properties, processing and performance* (2015) 3–9 Springer International Publishing, Switzerland.

[3] J.W. Yeh, *Overview of high-entropy alloys, high-entropy alloys* (2016) 1–19 Springer International Publishing Switzerland.

[4] X. Wang, W. Guo, Y. Fu, High-entropy alloys: Emerging materials for advanced functional applications, *J. Mater. Chem. A* 9 (2021) 663–701.

[5] C.K. Reddy, M.G. Krishna, P. Srikant, Brief evolution story and some basic limitations of high entropy Alloys (HEAs) - A Review, *Mater. Today Proc.* 18 (2019) 436–439.

[6] D. Banerjee, J.C. Williams, Materials and engineering: An evolving landscape, *MRS Bull.* 40 (2015) 999–1008.

[7] Z.P. Lu, H. Wang, M.W. Chen, I. Baker, J.W. Yeh, C.T. Liu, T.G. Nieh, An assessment on the future development of high-entropy alloys: Summary from a recent workshop, *Intermetallics* 66 (2015) 67–76.

[8] C. Oses, C. Toher, S. Curtarolo, High-entropy ceramics, *Nat. Rev. Mater.* 5 (2020) 295–309.

[9] A. Amiri, R. Shahbazian-Yassar, Recent progress of high-entropy materials for energy storage and conversion, *J. Mater. Chem. A* 9 (2021) 782–823.

[10] J.W. Yeh, S.K. Chen, S.J. Lin, J.Y. Gan, T.S. Chin, T.T. Shun, C.H. Tsau, S.Y. Chang, Nanostructured high-entropy alloys with multiple principal elements: Novel alloy design concepts and outcomes, *Adv. Eng. Mater.* 6 (2004) 299–303.

[11] B. Cantor, I.T.H. Chang, P. Knight, A.J.B. Vincent, Microstructural development in equiatomic multicomponent alloys, *Mater. Sci. Eng. A* 375-377 (2004) 213–218.

[12] D. Miracle, J. Miller, O. Senkov, C. Woodward, M. Uchic, J. Tiley, Exploration and development of high entropy alloys for structural applications, *Entropy* 16 (2014) 494–525.

[13] M.H. Tsai, J.W. Yeh, High-entropy alloys: A critical review, *Mater. Res. Lett.* 2 (2014) 107–123.

[14] M. Widom, Modeling the structure and thermodynamics of high-entropy alloys, *J. Mater. Res.* 33 (2018) 2881–2898.

[15] E.P. George, D. Raabe, R.O. Ritchie, High-entropy alloys, *Nat. Rev. Mater.* 4 (2019) 515–534.

[16] M. Fu, X. Ma, K. Zhao, X. Li, D. Su, High-entropy materials for energy-related applications, *iScience* 24 (2021) 102177.

[17] I. Takeuchi, O.O. Famodu, J.C. Read, M.A. Aronova, K.S. Chang, C. Craciunescu, S.E. Lofland, M. Wuttig, F.C. Wellstood, L. Knauss, A. Orozco, Identification of novel compositions of ferromagnetic shape-memory alloys using composition spreads, *Nat. Mater.* 2 (2003) 180–184.

[18] S. Curtarolo, G.L.W. Hart, M.B. Nardelli, N. Mingo, S. Sanvito, O. Levy, The high-throughput highway to computational materials design, *Nat. Mater.* 12 (2013) 191–201.

[19] D.B. Miracle, O.N. Senkov, A critical review of high entropy alloys and related concepts, *Acta Mater.* 122 (2017) 448–511.

[20] B.S. Murty, J.W. Yeh, S. Ranganathan, P.P. Bhattacharjee, *High-entropy alloys: basic concepts, High-Entropy Alloys* 2019 13–30 Elsevier, New York.

[21] Y. Zhang, Materials design of high-entropy materials, *High-Entropy Materials* 2019 35–63.

[22] Y. Xin, S. Li, Y. Qian, W. Zhu, H. Yuan, P. Jiang, R. Guo, L. Wang, High-entropy alloys as a platform for catalysis: Progress, *Challenges, and Opportunities, ACS Catal.* 10 (2020) 11280–11306.

[23] W. Zhang, P.K. Liaw, Y. Zhang, Science and technology in high-entropy alloys, *Sci. China Mater.* 61 (2018) 2–22.

[24] Y. Ma, Q. Wang, S. Schweidler, M. Botros, T. Fu, H. Hahn, T. Brezesinski, B. Breitung, High-entropy energy materials: Challenges and new opportunities, *Energy Environ. Sci.* 14 (2021) 2883–2905.

[25] M.C. Gao, J. Yeh, P.K. Liaw, Y. Zhang, *High-Entropy Alloys* (2016) Springer International Publishing, Cham.

[26] L.Q. Xing, J. Eckert, W. Loser, L. Schultz, High-strength materials produced by precipitation of icosahedral quasicrystals in bulk Zr-Ti-Cu-Ni-Al amorphous alloys, *Appl. Phys. Lett.* 74 (1999) 664–666.

[27] J.W. Yeh, Alloy design strategies and future trends in high-entropy alloys, *JOM* 65 (2013) 1759–1771.

[28] Y. Zhang, T.T. Zuo, Z. Tang, M.C. Gao, K.A. Dahmen, P.K. Liaw, Z.P. Lu, Microstructures and properties of high-entropy alloys, *Prog. Mater. Sci.* 61 (2014) 1–93.

[29] J.W. Yeh, Recent progress in high-entropy alloys, *Annales de Chimie Science des Materiaux* 31 (2006) 633–648.

[30] B.E. MacDonald, Z. Fu, B. Zheng, W. Chen, Y. Lin, F. Chen, L. Zhang, J. Ivanisenko, Y. Zhou, H. Hahn, E.J. Laverina. Recent progress in high entropy alloy research, *JOM* 69 (2017) 2024–2031.

[31] C.J. Tong, Y.L. Chen, J.W. Yeh, S.J. Lin, S.K. Chen, T.T. Shun, C.H. Tsau, S.Y. Chang, Microstructure characterization of AlxCoCrCuFeNi High-entropy alloy system with Multiprincipal Elements, *Metall. Mater. Trans. A: Phys. Metall. Mater. Sci.* 36 (2005) 881–893.

[32] W.H. Liu, Y. Wu, J.Y. He, T.G. Nieh, Z.P. Lu, Grain growth and the Hall-Petch relationship in a high-entropy FeCrNiCoMn alloy, *Scr. Mater.* 68 (2013) 526–529.

[33] O.N. Senkov, G.B. Wilks, J.M. Scott, D.B. Miracle, Mechanical properties of Nb25Mo25Ta25W25 and V20Nb20Mo20Ta20W20 refractory high entropy alloys, *Intermetallics* 19 (2011) 698–706.

[34] K.Y. Tsai, M.H. Tsai, J.W. Yeh, Sluggish diffusion in Co-Cr-Fe-Mn-Ni high-entropy alloys, *Acta Mater.* 61 (2013) 4887–4897.

[35] B. Cantor, K.B. Kim, P.J. Warren, Novel Multicomponent Amorphous Alloys, *JMNM* 13 (2002) 27–32.

[36] J.W. Yeh, S.K. Chen, J.Y. Gan, S.J. Lin, T.S. Chin, T.T. Shun, C.H. Tsau, S.Y. Chang, Formation of Simple Crystal Structures in Cu-Co-Ni-Cr-Al-Fe-Ti-V Alloys with Multiprincipal Metallic Elements, Metall. *Mater. Trans. A: Phys. Metall. Mater. Sci.* 35 (2004) 2533–2536.

[37] P.K. Huang, J.W. Yeh, T.T. Shun, S.K. Chen, Multi-principal-element alloys with improved oxidation and wear resistance for thermal spray coating, *Adv. Eng. Mater.* 6 (2004) 74–78.

[38] C.Y. Hsu, J.W. Yeh, S.K. Chen, T.T. Shun, Wear resistance and high-temperature compression strength of FCC CuCoNiCrAl0.5Fe alloy with Boron addition, Metall. *Mater. Trans. A: Phys. Metall. Mater. Sci* 35 (2004) 1465–1469.

[39] B.S. Murty, J.W. Yeh, S. Ranganathan, P.P. Bhattacharjee, A brief history of alloys and the birth of high-entropy alloys, *High-Entropy Alloys* (2019) 1–12 Elsevier.

[40] C.M. Rost, E. Sachet, T. Borman, A. Moballegh, E.C. Dickey, D. Hou, J.L. Jones, S. Curtarolo, J.P. Maria, Entropy-stabilized oxides, *Nat. Commun* 6 (2015) 8485.

[41] J. Gild, Y. Zhang, T. Harrington, S. Jiang, T. Hu, M.C. Quinn, W.M. Mellor, N. Zhou, K. Vecchio, J. Luo, High-entropy metal diborides: A new class of high-entropy materials and a new type of ultrahigh temperature ceramics, *Sci. Rep.* 6 (2016) 37946.

[42] P. Sarker, T. Harrington, C. Toher, C. Oses, M. Samiee, J.P. Maria, D.W. Brenner, K.S. Vecchio, S. Curtarolo, High-entropy high-hardness metal carbides discovered by entropy descriptors, *Nat. Commun.* 9 (2018) 4980.

[43] R.Z. Zhang, F. Gucci, H. Zhu, K. Chen, M.J. Reece, Data-driven design of ecofriendly thermoelectric high-entropy sulfides, *Inorg. Chem.* 57 (2018) 13027–13033.

[44] T. Jin, X. Sang, R.R. Unocic, R.T. Kinch, X. Liu, J. Hu, H. Liu, S. Dai, Mechanochemical-assisted synthesis of high-entropy metal nitride via a soft urea strategy, *Adv. Mater.* 30 (2018) 1707512.

[45] M.W. Glasscott, A.D. Pendergast, S. Goines, A.R. Bishop, A.T. Hoang, C. Renault, J.E. Dick, Electrosynthesis of high-entropy metallic glass nanoparticles for designer, multi-functional electrocatalysis, *Nat. Commun.* 10 (2019) 2650.

[46] J.W. Yeh, S.J. Lin, Breakthrough applications of high-entropy materials, *J. Mater. Res.* 33 (2018) 3129–3137.

3 Synthesis Strategies for the Preparation of High-Entropy Materials

Yanchun Zhou and Huimin Xiang

3.1 INTRODUCTION

High-entropy materials (HEMs) are single-phase solid solutions wherein the contribution of configuration entropy to the Gibbs' free energy dominates at higher temperatures, which may even be in excess of the positive enthalpy of formation. Thus, the basic principle for the synthesis of HEMs is the formation of homogeneous solid solutions at high temperatures followed by rapid cooling to maintain the solid solution phases at room temperature [1]. Since synthesis is the first step to gaining insight into the microstructure and properties, this chapter will focus on the synthesis strategies of HEMs. To benefit the readers, the advantages and typical applications of these strategies will also be discussed. It is worth mentioning that there are several review papers and book chapters that have already been published concerning the synthesis of HEMs [2–6]. To make a difference from these early review papers and book chapters, this chapter emphasizes the strategies for the synthesis of different types and forms of HEMs: powders, thin films, and thick coatings, dense and porous bulks as well as single crystals.

3.2 STRATEGIES FOR THE SYNTHESIS OF HEM POWDERS

The strategies for the synthesis of HEM powders include mechanical alloying (MA) [7], atomization [8], microwave heating [9], solvothermal synthesis [10], fast-moving bed pyrolysis [11], sputter deposition [12], carbothermal shock [13], etc.

MA is normally a dry, high-energy ball milling technique [14], which is described by the repeated welding and fracturing of powder particles entrapped between milling media, the extent of which depends on the mechanical attributes of powder constituents. Using the MA process, HE AlFeTiCrZnCu [15] was synthesized. Then the method was extended to CuNiCoZnAlTi [16], CuNiAlCoCrFeTiMo [17], FeSiBAlNi [18], $Al_2NbTi_3V_2Zr$ [19], and AlCoCrFeNi [20] alloys. The advantages of MA include increased solid solubility, nanocrystalline structure, and greater homogeneity [21,22]. MA was used for the synthesis of HEC powders like (HfTaTiNbZr)C and (HfTaTiNbMo)C [23] and high-entropy metal boron carbonitride powders [24]. Mechanochemical synthesis (MS) is a variation of MA, which shows the advantages of room temperature synthesis, quick solid-state reaction, and with little-to-no solvent use. MS was used in the synthesis of HE (VCrNbMoZr)N [25] and catalytically active Pt/Ru-(NiMgCuZnCo)O powders [26].

Atomization involves the formation of powders from molten metal using a spray of either a gas or liquid, which has been established as a promising technology for the fabrication of spherical metal powders. Two typical methods, i.e., gas atomization and water atomization, have been used to prepare HEA powders. Inert-gas atomization was used to produce HE powders of AlCoCrFeNi [8] CrMnFeCoNi [27], $AlCoCrFeNi_{2.1}$ [28], AlCoCrFeNiCu [29], and $Al_{0.5}CoCrFeNi$ [30]. One of the most important advantages of the gas atomization process is the high cooling rate, which favors the formation of single-phase solid solutions.

 DOI: 10.1201/9781003391388-3

Hydrogenation-dehydrogenation is a commonly used process to make ductile metallic powders. VNbMoTaW powders [31] were prepared by first exposing the ingot to a hydrogen atmosphere and then the hydrogen-annealed ingot was crushed into particles, and finally, the powders were annealed in a high-vacuum atmosphere to eliminate hydrogen. The advantage of this method is that it is scalable and the prepared HEA powders can be stored before dehydrogenation.

High-pressure torsion (HPT) refers to a process in which a thin disk is subjected to severe torsional strain under a high hydrostatic pressure. This method can also be employed to synthesize HEMs powders. For example, TiZrHfNbTa HEA powders were prepared by HPT first, and then the alloyed powders were transformed into oxide and oxynitride nanoparticles by oxidizing and nitriding treatment [32,33] (Table 3.1).

Solid-state reaction is also a commonly used method to obtain HEM powders, which shows the advantages of simplicity, and large-scale production, and has been used for the synthesis of HE oxides, silicates, carbides, borides, silicide carbides, and sulfides powders. Rost et al. [34] prepared the first entropy-stabilized rock salt structured (MgNiCoCuZn)O using the solid-state reaction method. Then this method was extended to $(CoCrFeMnNi)_3O_4$ [35], $(Na_{0.2}Bi_{0.2}Ba_{0.2}Sr_{0.2}Ca_{0.2})TiO_3$ [36], $(Y_{0.2}Yb_{0.2}Lu_{0.2}Eu_{0.2}Er_{0.2})_3Al_5O_{12}$ [37] and γ-type $(Gd_{1/6}Tb_{1/6}Dy_{1/6}Tm_{1/6}Yb_{1/6}Lu_{1/6})_2Si_2O_7$ [38]. High-entropy borides and carbides [39–41] can also be synthesized using the solid-state reaction method. Zhang et al. [42] synthesized five high-entropy rare earth hexaborides powders and five HE

TABLE 3.1
Strategies for the Synthesis of HEM Powders

Strategy	Advantages	Suitable Materials
Mechanical alloying (MA)	Room-temperature processing, nanocrystalline structure, increased solid solubility	HEAs HECs
Mechanochemical synthesis (MS)	Room-temperature synthesis, quick reaction	HECs
Atomization	High cooling rate, spherical morphology, and high purity	HEAs
Hydrogenation-dehydronation	Easy crushing and milling, large-scale production, easy storage	HEAs
High-pressure torsion	Room temperature processing, small grain size	HEAs HECs
Solid-state reaction	Simplicity, large-scale production	HECs
Microwave heating	Rapid heating and cooling, uniform elemental distribution	HEAs HECs
Flash sintering	Rapid heating and cooling, uniform elemental distribution	HEAs HECs
Wet chemical synthesis (aqueous solutions)	Low synthesis temperature, atomic level mixing, homogeneous distribution of elements	HEAs HECs
Solvothermal synthesis (nonaqueous solutions)	Low synthesis temperature, ultrafine particle size	HEAs HECs
Nebulized spray pyrolysis	Fast heating and cooling rate, low-cost, highly efficient, ultrafine grain size	HECs
Flame spray pyrolysis	Fast heating and cooling rate, low-cost, highly efficient, ultrafine grain size	HECs
Fast-moving bed pyrolysis	Fast heating and cooling rate, ultrafine particle size	HEAs HECs
Carbothermal shock synthesis	Fast heating and cooling rate, ultrafine particle size	HEAs HECs
Laser ablation synthesis	Rapid synthesis, low environmental temperatures, nanocrystalline sizes, increased solid solubility	HEAs HECs

REB_6/HE $REBO_3$ composite powders with good electromagnetic wave absorption properties. They also synthesized HE rare earth hexaborides/tetraborides composite powders [43] using a similar strategy. For HE carbides and carbonitrides, the synthesis temperature is strongly dependent on the carbon (nitrogen) vacancy concentration [44,45] due to the carbon (nitrogen) vacancy-induced acceleration of mass transfer [46]. Complex structured materials like high-entropy rare earth silicide carbides were also synthesized using the solid-state reaction method [47].

Microwave heating is a process whereby microwaves produced by magnetrons are directed toward reactants or heating medium, which absorb the electromagnetic energy volumetrically to achieve self-heating uniformly and rapidly. Using this process, Qiao et al. [9] synthesized PtPdFeCoNi nanoparticles with an average particle size of *ca* 12 nm and uniform elemental distribution. Like microwave heating, flash sintering is also an ultrafast synthesis method and was used to fabricate HEC powders and dense bulks [48–50]. To achieve homogeneous distribution of the constituting elements at atomic level, wet chemical synthesis was developed. NiFeCrCuCo [51] and CoCrCuNiAl [52] nanoparticles were synthesized by the wet chemical process. This synthesis strategy was also widely used to prepare HEC powders [53–59]. Solvothermal synthesis is a chemical reaction that occurs in nonaqueous solutions to produce inorganic nanoparticles and was used for the synthesis of HEM powders [10,60].

Wet chemistry is a perfect method to achieve chemical homogeneity at an atomic level. However, post high temperature treatment is always needed to form a homogeneous solid solution phase. Thus, a combination of wet chemistry with fast pyrolysis is unique for synthesizing HEMs. Spray pyrolysis techniques including nebulized spray pyrolysis (NSP) and flame spray pyrolysis (FSP) in combination with the wet chemistry method are widely used to synthesize HEM powders with uniform size and spherical morphology. The advantages of spray pyrolysis techniques include low cost, high efficiency, fast heating and cooling, and the obtained powders exhibit small particle size (<1 μm), narrow size distribution, high purity, and large surface area. HE perovskite-type oxides [61], rare earth oxides [62], and Li-containing spinel-type oxides were prepared using NSP.

Synthesis of HEA nanoparticles (NPs) with uniform dispersion on granular supports such as carbon materials, γ-Al_2O_3, and zeolite, is vital to their catalytic applications. To achieve such a goal, a fast-moving bed pyrolysis strategy (FMBP) following wet impregnation for the preparation of highly dispersed HEA-NPs via the pyrolysis of the mixed metal chlorides precursors loading on various granular supports was reported [11]. This strategy ensures the mixed metal precursors pyrolyzed rapidly and simultaneously at high temperatures, resulting in nanometer-size particles.

Another strategy for the synthesis of HEAs nanoparticles is carbothermal shock synthesis (CTS), which relies on rapid Joule heating and quenching [63]. CTS has shown great promise in synthesizing HE alloys and oxide nanoparticles, which were known to be immiscible at the bulk scale by conventional methods [13,63]. The advantages of this method include fast heating and cooling, and tunable composition, particle size, and phases [63]. Laser ablation synthesis is an effective means to synthesize nanocrystals, which is [64] a chemically "simple and clean" synthesis strategy that operates at ambient conditions. Using laser scanning ablation, Wang et al. [65] synthesized HE alloy and ceramic nanoparticles. The extremely rapid process ensures that dissimilar metallic elements combine to form a solid solution regardless of their thermodynamic solubility.

3.3 STRATEGIES FOR THE SYNTHESIS OF HEM FILMS AND COATINGS

Strategies for the synthesis of HEM films and coatings include gas-state methods such as magnetron sputtering, beam epitaxy, vapor deposition, pulsed laser deposition (PLD), plasma spraying, thermal spraying, and liquid-state methods such as electrochemical deposition, laser cladding (LC), slurry paste coating, etc. (Table 3.2).

Sputtering depositions have been widely used for synthesizing HEM films. $Cu_{0.5}CoNiCrAl$ [66], AlMoNbSiTaTiVZr [67], TiTaHfNbZr [68], $AlCoCrCu_{0.5}FeNi$ [69], and $CoCrFeNiAl_{0.3}$ [70] films were deposited by RF or direct current (DC) magnetron sputtering. The advantage of the sputtering

TABLE 3.2
Strategies for the Synthesis of HEM Films and Coatings

Strategy	Advantages	Suitable Materials
Magnetron sputtering	High deposition rate, sizable sputtering area	HEAs HECs
Reactive magnetron sputtering	Easy formation of oxides, carbides and nitrides	HECs
Pulsed laser deposition	High deposition rate, low substrate temperature	HEAs HECs
Electrochemical deposition	Low deposition temperature, low cost, easy control of composition and microstructure	HEAs
Slurry paste process	Low cost and high efficiency	HECs
Laser cladding	Extremely high cooling rate, suitable for any shape and structure	HEAs HECs
Atmospheric plasma spraying	Ultrahigh flame temperature and ultrahigh cooling rate, high-volume production, low cost	HEAs HECs
High-velocity oxy-fuel (HVOF) spraying	Higher particle impact rate, less in-flight exposure time, low oxygen and porosity content, high bond strength	HEAs HECs
Cold spray	Low or no heat input, high efficiency, high residual stress, devoid of oxidation and phase transformation	HEAs HECs

methods is that the microstructure and phase composition of the HEA films are tunable by controlling the working pressure, substrate bias, target-substrate distance, substrate temperature, and sputtering power [71–74]. High-entropy $(ZrTiHfVTa)B_2$ film [75] was also deposited by non-reactive magnetron sputtering (RMS).

RMS is a well-known process used to synthesize nitrides, carbides, and oxides films using a metallic target and a reactive gas mixture. In this strategy, HEAs can be used as targets, and HE nitrides, carbides, and oxides ceramic films can be formed through RMS in $Ar–N_2$, $Ar–CH_4$ or $Ar–O_2$ atmosphere. HE nitride films were deposited in nitrogen atmosphere using FeCoNiCrCuAlMn, $FeCoNiCrCuAl_{0.5}$, and AlCrMoSiTi alloy targets [76,77]. PLD utilizes intense laser pulses to vaporize the target material, which is evaporated to create a plasma plume and deposits as a thin film on a substrate. This process was used to deposit AlCrFeCoNiCu [78] and $CoCrFeNiAl_{0.3}$ [79] alloy thin films and $(Mg_{0.25(1-x)}Co_xNi_{0.25(1-x)}Cu_{0.25(1-x)}Zn_{0.25(1-x)})O$ [80], $Ba(ZrSnTiHfNb)O_3$ [81], (MgNiCoCuZnSc)O [82], and (MgCoNiCuZn)O ceramic films [83]. Nanoporous HE oxide films were also prepared by the PLD strategy [84].

Besides the vapor phase method, strategies for depositing HEM films and coatings from liquid precursors are attractive. Among them, electrochemical deposition is the most cost-effective method since it usually conducts at room temperature and does not require complicated equipment [85]. The slurry coating is also a cost-effective method for the preparation of HEC coatings [86] such as $(Yb_{0.2}Y_{0.2}Lu_{0.2}Sc_{0.2}Gd_{0.2})_2Si_2O_7$ environmental barrier coating (EBC) on ceramic matrix composites.

LC, in which the pre-selected coating material and the thin surface layer of the substrate are simultaneously melted and solidified rapidly under the heat source of laser, is a technique for adding one material to the surface of another. FeCoNiCrCu [87] and FeCoCrBNiSi [88] coatings were prepared by LC. The effect processing parameters on the microstructure and mechanical properties of HEA coatings were summarized in the recent review of Arif et al. [89]. Atmospheric plasma spraying (APS) is a conventional technique to prepare HEM coatings. High-entropy $AlSiTiCrFeCoNiMo_{0.5}$ and $AlSiTiCrFeNiMo_{0.5}$ [90], AlCoCrFeNi and MnCoCrFeNi alloys coatings [91] and $(La_{0.2}Nd_{0.2}Sm_{0.2}Eu_{0.2}Gd_{0.2})_2Zr_2O_7$/YSZ double-ceramic-layer (DCL) coating [92] were prepared through APS. It has come to light that the presence of pores and oxides is the main drawback of PS. To overcome this obstacle, high-velocity oxy-fuel (HVOF) spraying was used to prepare HEM coatings

[93,94]. Compared to the PS $FeCoCrNiMo_{0.2}$ coating, Li et al. [95] found that the HVOF coating had lower porosity and less content of oxides. Cold spray is a newly developed solid-state deposition method, which allows for the avoidance of oxidation and phase transformation in the final product. FeCoNiCrMn HEA coatings [96] were prepared using this method.

3.4 STRATEGIES FOR THE SYNTHESIS OF DENSE AND POROUS BULK HEMs

Strategies for the synthesis of dense HEMs bulks include arc melting (AM), induction melting (IM), spark plasma sintering (SPS), hot-press sintering, flash sintering, pressureless sintering, HPT, and equal-channel angular pressing (ECAP). For the synthesis of porous HEMs bulk, strategies such as chemical dealloying, liquid metal dealloying, in-situ reaction/partial sintering, and sol-gel methods, etc. have been used.

AM and IM are the most common methods for the preparation of HEA bulks. It is worth mentioning that the pioneering work on high-entropy alloy ingots was prepared by Yeh et al. [66] and Cantor et al. [97] using AM and IM, respectively. The advantages of these methods include direct melting and solidification, and large-scale production. AM was also used to prepare dense HECs such as HE transitions metal diboride [98].

The *Bridgman* method uses a crucible containing the melt, which is moved in a vertical temperature gradient toward the cooler regions of the furnace. This process was used [99] to prepare HEAs bulk and single-crystal $CoCrFeNiAl_{0.3}$ alloy [100]. HEMs single crystals can also be grown by optical floating zone [101] and micro-pulling-down [102] method. SPS is a pressure-assisted rapid sintering technique to produce high-entropy alloys and ceramics. Bulk CoCrFeNiTiAl, AlCoCrCuFe, NiCoCrCuFe, $Co_{0.5}FeNiCrTi_{0.5}$, and CoCrFeNiMn, $Co_{25}Ni_{25}Fe_{25}Al_{7.5}Cu_{17.5}$ [103–106] alloys and HECs like $(HfZrTaNbTi)B_2$, $(HfZrMoNbTi)B_2$ and $(HfMoTaNbTi)B_2$ [107] were prepared using the SPS process. Hot pressing (HP) is also a well-established pressure-assisted sintering method and was used to prepare high-entropy borides, nitride-carbides, oxides, and fluorides [108–110]. In addition, oscillatory pressure sintering, a variant of HP, was used in the sintering of high-entropy borides [111].

Although pressure-assisted sintering can promote sintering and accelerate the densification process, for the preparation of large-size and complex shape components, pressureless sintering is the most suitable method. In addition to near fully dense bulk (MgCoNiCuZn)O [112], $(CaSrBa)ZrO_3$ [113], $(La_{0.2}Nd_{0.2}Gd_{0.2}Sm_{0.2}Pr_{0.2})MgAl_{11}O_{19}$, $(La_{0.2}Nd_{0.2}Gd_{0.2}Sm_{0.2}Dy_{0.2})Mg\ Al_{11}O_{19}$, and $La(Mg_{0.2}Fe_{0.2}Co_{0.2}Ni_{0.2}Zn_{0.2})Al_{11}O_{19}$ [114] prepared by pressureless sintering, transparent $(Lu_{0.2}Y_{0.395}Gd_{0.2}Yb_{0.2}Tm_{0.005})_2O_3$ was also prepared through vacuum sintering.

Severe plastic deformation (SDP) processes, such as HPT, were also used to synthesize bulk HEAs such as CoCrFeMnNi [115,116], $Al_{0.3}CoCrFeNi$ [117], $Al_{0.3}Cu_{0.5}CoCrFeNi$ [118], TiZrNbHfTa [119], CoCrFeNi [120], and TiAlFeCoNi [121]. The HEAs processed by this method show significantly small grain size and high hardness. ECAP is also a well-developed SPD processing technique, in which a rod-like sample is pressed through a die constrained within a channel that is bent at an abrupt angle. HE AlFeMgTiZn [122] and CoCrFeNiMn [123] were consolidated via ECAP.

For the preparation of porous HEAs, dealloying was widely used. Through dealloying, the precursor alloys in a NaOH solution, nanoporous AlNiCuPtPdAu, AlNiCuPtPdAuCoFe, and AlNiCuMoCoFe [124] were prepared. Besides acid and base solution, liquid metal was also used as a dealloying agent. Nanoporous TiVNbMoTa and TaMoNbVNi HEAs were prepared by removing the Ni elements from $(TiVNbMoTa)_{25}Ni_{75}$ and $(TaMoNbV)_{25}Ni_{75}$ precursor alloys in the Mg melt [125]. In addition, additive manufacturing was also used for the processing of porous HEAs [126] (Table 3.3).

To prepare porous bulk ceramics, a novel *in-situ* reaction/partial sintering method was developed [127–129] because carbides and borides are synthesized through carbothermal, borothermal, and boro/carbothermal reductions. The gases released during these reactions can not only increase the

TABLE 3.3
Strategies for the Synthesis of Dense and Porous Bulk HEMs

Strategy	Advantages	Suitable Materials
Arc melting	Direct melting and solidification, large-scale production	Dense HEAs and HECs
Induction melting	Direct melting and solidification, large-scale production	Dense HEAs and HECs
Bridgman solidification	Simplicity and ease of implementation, small and equiaxed gain size	Dense and single crystal HEAs and HECs
Optical floating zone	Light heating, crucible-free, atmosphere adjustable	Single crystal HECs
Micro-pulling-down	Use of microchannel or micro-outlets at the bottom of the melt reservoir	Single crystal HECs
Spark plasma sintering	Rapid heating and cooling, pulsed electric current and pressure-assisted sintering, fine microstructure	Dense HEAs and HECs
Flash sintering	Rapid heating and cooling, nanometer particle size, uniform elemental distribution	Dense and porous HEAs and HECs
Microwave heating	Rapid heating and cooling, nanometer particle size and uniform elemental distribution	Dense and porous HEAs and HECs
Hot pressing	Pressure-assisted sintering, large-scale production	Dense HEAs and HECs
Pressureless sintering	Low-cost, large-scale production, no limit to the size and shape of product	Dense and porous HEAs and HECs
High-pressure torsion	Severe plastic deformation, low-temperature processing, small grain size and enhanced mechanical properties	Dense HEAs
Equal-channel angular pressing	Severe plastic deformation, low-temperature processing, small grain size, high hardness	Dense HEAs
Chemical dealloying	Simple process, easy precursor design and composition control	Porous HEAs
Liquid metal dealloying	Simple process, precursor designable	Porous HEAs
Additive manufacturing	Customized geometries, multi-structure, high porosity	Porous HEAs and HECs
In-situ reaction/partial sintering	Integrated synthesis and sintering, no pore-forming agent, high porosity, small grain size	Porous HECs
Sol-gel process	Low temperature, homogeneous composition, controllable shape and size of pores	Porous HECs

porosity but also block the large shrinkage during the reactions. After the reactions, the porous body was partially sintered at higher temperatures. Thus, no pore-forming agent is needed, and the synthesis/partial sintering can be finished in a single cycle. High porosity $(Zr_{0.2}Hf_{0.2}Ti_{0.2}Nb_{0.2}Ta_{0.2})C$ [130], $(Zr_{0.2}Hf_{0.2}Ti_{0.2}Nb_{0.2}Ta_{0.2})B_2$ [131], and $(Y_{0.2}Yb_{0.2}Sm_{0.2}Nd_{0.2}Eu_{0.2})B_6$ [132] were prepared using this novel method. Sol-gel process was used for the synthesis of spherical mesoporous HECs [133,134].

3.5 CONCLUSION REMARKS

In this chapter, strategies for the synthesis of different forms of HEMs are reviewed. Using various examples, we have demonstrated that rapid heating/cooling strategies are suitable for the synthesis of entropy-stabilized materials. Cost-efficient strategies are also introduced. Finally, strategies for preparing porous HEMs are summarized.

REFERENCES

[1] W. Łoński, M. Spilka, M. Kądziołka-Gaweł, P. Gębara, A. Radoń, T. Warski, K. Młynarek-Żak, R. Babilas, The effect of cooling rate on the structure and selected properties of AlCoCrFeNiSix (x=0; 0.25; 0.5; 0.75) high entropy alloys, *J Alloy Compd* 905 (2022) 164074.

[2] H. Xiang, F. Dai, Y. Zhou, *High-entropy materials: From basics to applications*, Weinheim, Germany: Wiley-VCH GmbH (2023).

[3] N. Kaushik, A. Meena, H. S. Mali, High entropy alloy synthesis, characterization, manufacturing & potential applications: A review, *Mater Manuf Process* 37 (2022) 1085–1109.

[4] J. M. Torralba, P. Alvaredo, A. García-Junceda, High-entropy alloys fabricated via powder metallurgy. A critical review, *Powder Metall* 62 (2019) 84–114.

[5] H. M. Xiang, Y. Xing, F. Z. Dai, H. J. Wang, L. Su, L. Miao, G. J. Zhang, Y. G. Wang, X. W. Qi, L. Yao, H. L. Wang, B. Zhao, J. Q. Li, Y. C. Zhou, High-entropy ceramics: Present status, challenges, and a look forward, *J Adv Ceram* 10 (2021) 385–441.

[6] S. Akrami, P. Edalati, M. Fuji, K. Edalati, High-entropy ceramics: Review of principles, production and applications, *Mater Sci Eng R* 146 (2021) 100644.

[7] C. Suryanarayana, Mechanical alloying and milling, *Prog Mater Sci* 46 (2001) 1–184.

[8] I. Kunce, M. Polanski, K. Karczewski, T. Plocinski, K. J. Kurzydlowski, Microstructural characterisation of high-entropy alloy AlCoCrFeNi fabricated by laser engineered net shaping, *J Alloy Compd* 648 (2015) 751–758.

[9] H. Qiao, M. T. Saray, X. Wang, S. Xu, G. Chen, Z. Huang, C. Chen, G. Zhong, Q. Dong, M. Hong, H. Xie, R. S. Yassar, L. B. Hu, Scalable synthesis of high entropy alloy nanoparticles by microwave heating, *ACS Nano* 15 (2021) 14928–14937.

[10] M. Bondesgaard, N. L. N. Broge, A. Mamakhel, M. Bremholm, B. B. Iversen, General solvothermal synthesis method for complete solubility range bimetallic and high-entropy alloy nanocatalysts, *Adv Funct Mater* 29 (2019) 1905933.

[11] S. Gao, S. Hao, Z. Huang, Y. Yuan, S. Han, L. Lei, X. Zhang, S. R. Yassar, J. Lu, Synthesis of high-entropy alloy nanoparticles on supports by the fast moving bed pyrolysis, *Nat Commun* 11 (2020) 2016.

[12] C. F. Tsai, P. W. Wu, P. Lin, C. G. Chao, K. Y. Yeh, Sputter deposition of multi-element nanoparticles as electrocatalysts for methanol oxidation, *Jpn J Appl Phys* 47 (2008) 5755–5761.

[13] Y. Yao, Z. Huang, P. Xie, S. D. Lacey, R. J. Jacob, H. Xie, F. Chen, A. Nie, T. Pu, M. Rehwoldt, D. Yu, M. R. Zachariah, C. Wang, S. R. Yassar, J. Li, L. B. Hu, Carbothermal shock synthesis of high-entropy-alloy nanoparticles, *Science* 359 (2018) 1489–1494.

[14] S. Rajendrachari, An overview of high-entropy alloys prepared by mechanical alloying followed by the characterization of their microstructure and various properties, *Alloys* 1 (2022) 116–134.

[15] S. Varalakshmi, M. Kamaraj, B. S. Murty, Synthesis and characterization of nanocrystalline AlFeTiCrZnCu high entropy solid solution by mechanical alloying, *J Alloy Compd* 460 (2008) 253–257

[16] S. Varalakshmi, M. Kamaraj, B. S. Murty, Formation and stability of equiatomic and nonequiatomic nanocrystalline CuNiCoZnAlTi high-entropy alloys by mechanical alloying, *Metall Mater Trans A* 41 (2010) 2703–2709.

[17] Y. Chen, Y. Hu, C. Tsai, C. A. Hsieh, S. W. Kao, J. W. Yeh, T. S. Chin, S. K. Chen, Alloying behavior of binary to octonary alloys based on Cu-Ni-Al-Co-Cr-Fe-Ti-Mo during mechanical alloying, *J Alloy Compd* 477 (2009) 696–705.

[18] H. Wang, T. Gao, J. Niu, P. Shi, J. Xu, Y. Wang, Microstructure, thermal properties, and corrosion behaviors of FeSiBAlNi alloy fabricated by mechanical alloying and spark plasma sintering, *Int J Miner Metall Mater* 23 (2016) 77.

[19] X. Tan, G. Zhang, Q. Zhi, Z. X. Liu, Effects of milling on the microstructure and hardness of $Al_2NbTi_3V_2Zr$ high-entropy alloy, *Mater Des* 109 (2016) 27–36.

[20] M. Vaidya, A. Prasad, A. Parakh, B. S. Murty, Influence of sequence of elemental addition on phase evolution in nanocrystalline AlCoCrFeNi: Novel approach to alloy synthesis using mechanical alloying, *Mater Des* 126 (2017) 37–46.

[21] M. Vaidya, G. Muralikrishna, B. S. Murty, High-entropy alloys by mechanical alloying: A review, *J Mater Res*, 34(5) (2019) 664–686.

[22] S. Rajendrachari, An overview of high-entropy alloys prepared by mechanical alloying followed by the characterization of their microstructure and various properties, *Alloys* 1 (2022) 116–134.

[23] D. Moskovskikh, S. Vorotilo, A. Sedegov, K. V. Kuskov, K. V. Bardasova, Ph. V. Kiryukhantsev-korneev, M. Zhukovsky, A. S. Mukasyan, High-entropy (HfTaTiNbZr)C and (HfTaTiNbMo)C carbides fabricated through reactive high-energy ball milling and spark plasma sintering, *Ceram Int* 46 (2020) 19008–19014.

[24] J. Guan, D. Li, Z. Yang, B. Wang, D. Cai, X. Duan, P. He, D. Jia, Y. Zhou, Synthesis and thermal stability of novel high-entropy metal boron carbonitride ceramic powders, *Ceram Int* 46 (2020) 26581–26589.

[25] T. Jin, X. Sang, R. Unocic, R. Kinch, X. Liu, J. Hu, H. Liu, S. Dai, Mechanochemical-assisted synthesis of high-entropy metal nitride via a soft urea strategy, *Adv Mater* 30 (2018)1707512.

[26] H. Chen, W. Lin, Z. Zhang, K. Jie, D. R. Mullins, X. Sang, S. Yang, C. J. Jafta, C. A. Bridges, X. Hu, R. R. Unocic, F. Jie, P. Zhang, S. Dai, Mechanochemical synthesis of high entropy oxide materials under ambient conditions: Dispersion of catalysts via entropy maximization, *ACS Mater Lett* 1 (2019) 83–88.

[27] Y. Liu, J. Wang, Q. Fang, B. Liu, Y. Wu, S. Chen, Preparation of superfine-grained high entropy alloy by spark plasma sintering gas atomized powder, *Intermetallics* 68 (2016) 16–22.

[28] P. Ding, A. Mao, X. Zhang, X. Jin, B. Wang, M. Liu, X. Gu, Preparation, characterization and properties of multicomponent $AlCoCrFeNi_{2.1}$ powder by gas atomization method, *J Alloy Compd* 721 (2017) 609–614.

[29] C. Li, K. Chang, A. Yeh, On the microstructure and properties of an advanced cemented carbide system processed by selective laser melting, *J Alloy Compd* 782 (2019) 440–450.

[30] P. Zhou, D. Xiao, Z. Wu, Q. Qu, $Al_{0.5}FeCoCrNi$ high entropy alloy prepared by selective laser melting with gas-atomized pre-alloy powders, *Mater Sci Eng A* 739 (2019) 86–89.

[31] W. Lee, K. Park, K. Yi, S. Lee, K. Park, T. Lee, T. Na, H. Park, Synthesis of spherical V-Nb-Mo-Ta-W high-entropy alloy powder using hydrogen embrittlement and spheroidization by thermal plasma, *Metals* 9 (2019) 1296.

[32] P. Edalati, Q. Wang, H. Razavi-Khosroshahi, M. Fuji, T. Ishihara, K. Edalati, Photocatalytic hydrogen evolution on a high-entropy oxide, *J Mater Chem A* 8 (2020) 3814–3821.

[33] P. Edalati, X. Shen, M. Watanabe, T. Ishihara, M. Arita, M. Fuji, K. Edalati, High-entropy oxynitride as a low-bandgap and stable photocatalyst for hydrogen production, *J Mater Chem A* 9 (2021) 15076.

[34] C. M. Rost, E. Sachet, T. Borman, A. Moballegh, E. Dichkey, D. Hou, J. Jones, S. Curtarolo, J Maria, Entropy-stabilized oxides, *Nat Commun* 6 (2015) 8485.

[35] J. Dąbrowa, M. Stygar, A. Mikula, A. Knapik, K. Mroczka, W. Tejchman, M. Danielewski, M. Martin, Synthesis and microstructure of the $(Co, Cr, Fe, Mn, Ni)_3O_4$ high entropy oxide characterized by spinel structure, *Mater Lett* 216 (2018) 32–36.

[36] Y. Pu, Q. Zhang, R. Li, M. Chen, X. Du, S. Zhou, Dielectric properties and electrocaloric effect of high-entropy $(Na_{0.2}Bi_{0.2}Ba_{0.2}Sr_{0.2}Ca_{0.2})TiO_3$ ceramic, *Appl Phys Lett* 115 (2019) 223901.

[37] H. Chen, Z. Zhao, H. Xiang, F. Dai, W. Xu, K. Sun, J. Liu, Y. Zhou, High entropy $(Y_{0.2}Yb_{0.2}Lu_{0.2}Eu_{0.2}Er_{0.2})_3A_{15}O_{12}$: A novel high temperature stable thermal barrier material. *J Mater Sci Technol* 48 (2020) 57–62.

[38] L. Sun, Y. Luo, X. Ren, Z. Gao, T. Du, J. Wang, A multicomponent γ-type $(Gd_{1/6}Tb_{1/6}Dy_{1/6}Tm_{1/6}Yb_{1/6}Lu_{1/6})_2Si_2O_7$ disilicate with outstanding thermal stability, *Mater Res Lett* 8 (2020) 424–430.

[39] G. Tallarita, R. Licheri, S. Garroni, R. Orrù, G. Cao, Novel processing rout for the fabrication of bulk high-entropy metal diborides, *Scrip Mater* 158 (2019) 100104.

[40] D. Liu, T. Wen, B. Ye, Y. Chu, Synthesis of superfine high-entropy metal diboride powders, *Scr Mater* 167 (2019) 110–114.

[41] D. Liu, H. Liu, S. Ning, B. Ye, Y. Chu, Synthesis of high-purity high-entropy metal diboride powders by boro/carbothermal reduction, *J Am Ceram Soc* 102 (2019) 7071–7076.

[42] W. Zhang, B. Zhao, H. Xiang, F. Dai, S. Wu, Y. Zhou, One-step synthesis and electromagnetic absorption properties of high entropy rare earth hexaborides (HE REB_6) and high entropy rare earth hexaborides/borates (HE REB_6/HE $REBO_3$) composite powders, *J Adv Ceram* 10 (2020) 62–77.

[43] W. M. Zhang, B. Zhao, N. Ni, H. M. Xiang, F. Z. Dai, S. J. Wu, Y. C. Zhou, High entropy rare earth hexaborides/tetraborides (HE REB_6/HE REB_4) composite powders with enhanced electromagnetica wave abosorption performance, *J Mater Sci Tech* 87 (2021) 155–166.

[44] C. Peng, X. Gao, M. Wang, L. Wu, H. Tang, X. Li, Q. Zhang, Y. Ren, F. Zhang, Y. Wang, B. Zhang, B. Gao, Q. Zou, Y. Zhao, Q. Yang, D. Tian, H. Xiao, H. Gou, W. Yang, X. Bai, W. Mao, K. Mao, Diffusion-controlled alloying of single-phase multi-principal transition metal carbides with high toughness and low thermal diffusivity, *Appl Phys Lett* 114 (2019) 011905.

[45] J. Song, G. Chen, H. Xiang, F. Dai, S. Dong, W. Han, X. Zhang, Y. Zhou, Regulating the formation ability and mechanical properties of high-entropy transition metal carbides by carbon stoichiometry, *J Mater Sci Technol* 121 (2022) 181–189.

[46] Y. He, C. Peng, S. Xin, K. Li, S. Liang, X. Liu, N. Kang, H. Xue, X. Shen, T. Shen, M. Wang, Vacancy effect on the preparation of high-entropy carbides, *J Mater Sci* 55 (2020) 6754–6760.

[47] H. Chen, B. Zhao, Z. Zhao, H. Xiang, F. Dai, J. Liu, Y. Zhou, Achieving strong microwave absorption capability and wide absorption bandwidth through a combination of high entropy rare earth silicide carbides/rare earth oxides, *J Mater Sci Technol* 47 (2020) 216–222.

[48] D. Liu, X. Peng, J. Liu, Y. Chen, Ultrafast synthesis of entropy stabilized oxide at room temperature, *J Eur Ceram Soc* 40 (2020) 2504–2508.
[49] B. Yoon, R. Avila, R. Raj, L. M. Jesus, Reactive flash sintering of the entropy-stabilized oxide $(Mg_{0.2}Ni_{0.2}Co_{0.2}Cu_{0.2}Zn_{0.2})O$, *Scr Mater* 181 (2020) 48–52.
[50] K. W. Wang, B. S. Ma, T. Li, C. Xie, Z. Sun, D. Liu, J. Liu, L. An, Fabrication of high-entropy perovskite oxide by reactive flash sintering, *Ceram Inter* 46 (2020) 18358–18361.
[51] M. Singh, C. Srivastava, Synthesis and electron microscopy of high entropy alloy nanoparticles, *Mater Lett* 160 (2015) 419–422.
[52] B. Niu, F. Zhang, H. Ping, N. Li, J. Zhou, L. Lei, J. Xie, J. Zhang, W. Wang, Z. Fu, Sol-gel autocombustion synthesis of nanocrystalline high-entropy alloys, *Sci Rep* 7 (2017) 3421.
[53] F. Okejiri, Z. Zhang, J. Liu, M. Liu, S. Yang, S. Dai, Room-temperature synthesis of high-entropy perovskite oxide nanoparticle catalysts via ultrasonication-based method, *Chem Sus Chem* 13 (2020) 111–115.
[54] A. Sarkar, R. Djenadic, N. Usharani, K. Sanghvi, V. S. K. Chakravadhanula, A. S. Gandhi, H. Hahn, S. S. Bhattacharya, Nanocrystalline multicomponent entropy stabilised transition metal oxides, *J Eur Ceram Soc* 37 (2016) 747–754.
[55] Z. Zhao, H. Xiang, F. Dai, Z. Peng, Y. Zhou, $(La_{0.2}Ce_{0.2}Nd_{0.2}Sm_{0.2}Eu_{0.2})_2Zr_2O_7$: A novel high-entropy ceramic with low thermal conductivity and sluggish grain growth rate, *J Mater Sci Technol* 35 (2019) 2647–2651.
[56] Y. Xu, X. Xu, L Bi, A high-entropy spinel ceramic oxide as the cathode for proton-conducting solid oxide fuel cells, *J Adv Ceram* 11 (2022) 794–804.
[57] Z. Zhao, H. Xiang, H. Chen, F. Dai, X. Wang, Z. Peng, Y. Zhou, High-entropy $(Nd_{0.2}Sm_{0.2}Eu_{0.2}Y_{0.2}Yb_{0.2})_4Al_2O_9$ with good high temperature stability, low thermal conductivity, and anisotropic thermal expansivity, *J Adv Ceram* 9 (2020) 595–605.
[58] Z. Zhao, H. Chen, H. Xiang, F. Dai, X. Wang, Z. Peng, Y. Zhou, $(La_{0.2}Ce_{0.2}Nd_{0.2}Sm_{0.2}Eu_{0.2})$PO4: A high-entropy rare-earth phosphate monazite ceramic with low thermal conductivity and good compatibility with Al_2O_3, *J Mater Sci Technol* 35 (2019) 2892–2896.
[59] A. Mao, H. Xiang, Z. Zhang, K. Kuramoto, H. Yu, S. Ran, Solution combustion synthesis and magnetic property of rock-salt $(Co_{0.2}Cu_{0.2}Mg_{0.2}Ni_{0.2}Zn_{0.2})O$ high-entropy oxide nanocrystalline powder, *J Mag Mag Mater* 484 (2019) 245–252.
[60] D. Wang, Z. Liu, S. Du, Y. Zhang, H. Li, Z. Xiao, W. Chen, R. Chen, Y. Wangm Y. Zou, S. Wang, Low-temperature synthesis of small-sized high-entropy oxides for water oxidation, *J Mater Chem A* 7 (2019) 24211.
[61] A. Sarkar, R. Djenadic, D. Wang, C. Hein, R. Kautenburger, O. Clemens, H. Hahn, Rare earth and transition metal based entropy stabilised perovskite type oxides, *J Eur Ceram Soc* 38 (2018) 2318–2327.
[62] R. Djenadic, A. Sarkar, O. Clemens, Multicomponent equiatomic rare earth oxides, *Mater Res Lett* 5 (2017) 102–109.
[63] A. Abdelhafiz, B. M. Wang, A. R. Harutyunyan, J. Li, Carbothermal shock synthesis of high entropy oxide catalysts: Dynamic structural and chemical reconstruction boosting the catalytic activity and stability toward oxygen ovolution reaction, *Adv Energy Mater* 12 (2022) 2200742.
[64] G. W. Wang, Laser ablation in liquids: Applications in the synthesis of nanocrystals, *Prog Mater Sci* 52 (2007) 648–698.
[65] B. Wang, C. Wang, X. Yu, Y. Cao, L. Gao, C. Wu, Y. Yao, Z. Lin, Z. Zou, General synthesis of high-entropy alloy and ceramic nanoparticles in nanoseconds, *Nat Synth* 1 (2022) 138–146.
[66] J. Yeh, S. Chen, S. J. Lin, J. Gan, T. Chin, T. Sun, C. Tsau, S. Chang. Nanostructured high-entropy alloys with multiple principal elements: Novel alloy design concepts and outcomes, *Adv Eng Mater* 6 (2004) 299–303.
[67] M. Tsai, J. Yeh, J. Gan, Diffusion barrier properties of AlMoNbSiTaTiVZr high-entropy alloy layer between copper and silicon, *Thin Solid Films* 516 (2008) 5527–5530.
[68] N. Tüten, D. Canadinc, A. Motallebzadeh, B. Bai, Microstructure and tribological properties of TiTaHfNbZr high entropy alloy coatings deposited on Ti-6Al-4V substrates, *Intermetallics* 105 (2019) 99–106.
[69] N. A. Khan, B. Akhavan, H. Zhou, L. Chang, Y. Wang, L. Sun, M. Bilek, Z. Liu, High entropy alloy thin films of $AlCoCrCu_{0.5}FeNi$ with controlled microstructure, *Appl Surf Sci* 495 (2019) 143560.
[70] W. Liao, S. Lan, L. Gao, H. Zhang, S. Xu, J. Song, X. Wang, Y. Lu, Nanocrystalline high-entropy alloy $(CoCrFeNiAl_{0.3})$ thin-film coating by magnetron sputtering, *Thin Solid Films* 638 (2017) 383–388.
[71] B. R. Braeckman, F. Boydens, H. Hidalgo, P. Dutheil, M. Jullien, A. L. Thomann, D. Depla, High entropy alloy thin films deposited by magnetron sputtering of powder targets, *Thin Solid Films* 580 (2015) 71–76.

[72] N. A. Khan, B. Akhavan, C. Zhou, H. Zhou, L. Chang, Y. Wang, Y. Liu, L. Fu, M. M. Bilek, Z. Liu, RF magnetron sputtered $AlCoCrCu_{0.5}FeNi$ high entropy alloy (HEA) thin films with tuned microstructure and chemical composition, *J Alloy Compd* 836 (2020) 155348.

[73] L. Z. Medina, L. Riekehr, U. Jansson, Phase formation in magnetron sputtered CrMnFeCoNi high entropy alloy, *Surf Coat Technol* 403 (2020) 126323.

[74] S. Fritze, C. M. Koller, L. Fieandt, P. Malinofskis, K. Johansson, E. Lewin, P. Mayrhofer, U. Jansson, Influence of deposition temperature on the phase evolution of HfNbTiVZr high-entropy thin films, *Mater* 12 (2019) 587.

[75] P. H. Mayrhofer, A. Kirnbauer, P. Ertelthaler, C. M. Koller, High-entropy ceramic thin films; A case study on transition metal diborides, *Scr Mater* 149 (2018) 93–97.

[76] T. K. Chen, T. T. Shun, J. Yeh, M. S. Wong, Nanostructured nitride films of multi-element high-entropy alloys by reactive DC sputtering, *Surf Coat Technol* 188–189 (2004) 193–200.

[77] H. Chang, P. Huang, A. Davison, J. Yeh, C. Tsau, C. Yang, Nitride films deposited from an equimolar Al-Cr-Mo-Si-Ti alloy target by reactive direct current magnetron sputtering, *Thin Solid Films* 516 (2008) 6402–6408.

[78] M. D. Cropper, Thin films of AlCrFeCoNiCu high-entropy alloy by pulsed laser deposition, *Appl Surf Sci* 455 (2018) 153–159.

[79] T. Liu, C. Feng, Z. Wang,K. Liao, Z. Liu, Y. Xie, J. Hu, W. Liao, Microstructures and mechanical properties of $CoCrFeNiAl_{0.3}$ high-entropy alloy thin films by pulsed laser deposition, *Appl Surf Sci* 494 (2019) 72–79.

[80] P. B. Meisenheimer, T. J. Kratofl, J. T. Heron, Giant enhancement of exchange coupling in entropy-stabilized oxide heterostructures, *Sci Rep* 7 (2017) 13344.

[81] Y. Sharma, B. Musico, X. Gao, C. Hua, A. F. May, A. Herklotz, A. Rastogi, D. Mandrus, J. Yan, H. N. Lee, M. F. Chisholm, V. Keppens, T. Z. Ward, Single-crystal high entropy perovskite oxide epitaxial films, *Phys Rev Mater* 2 (2018) 060404(R).

[82] G. N. Kotsonis, C. M. Rost, D. T. Harris, J. Maria, Epitaxial entropy-stabilized oxides: Growth of chemically diverse phases via kinetic bombardment, *MRS Comm* 8 (2018) 1371–1377.

[83] V. Jacobson, D. Diercks, B. To, A. Zakutayev, G. Brennecka, Thin film growth effects on electrical conductivity in entropy stabilized oxides, *J Eur Ceram Soc* 41 (2021) 2617–2624.

[84] H. Guo, X. Wang, A. D. Dupuy, J. Schoenung, W. Bowman, Growth of nanoporous high-entropy oxide thin films by pulsed laser deposition, *J Mater Res* 37 (2022) 124–135.

[85] Z. Shojaei, G. R. Khayati, E. Darezereshki, Review of electrodeposition methods for the preparation of high-entropy alloys, *Int J Miner Metall Mater* 29 (2022) 1683–1696.

[86] Y. Dong, K. Ren, Y. Lu, Q. Wang, J. Liu, Y. Wang, High-entropy environmental barrier coating for the ceramic matrix composites, *J Eur Ceram Soc* 39 (2019) 2574–2579.

[87] H. Zhang, Y. Pan, Y. He, Synthesis and characterization of FeCoNiCrCu high-entropy alloy coating by laser cladding, *Mater Des* 32 (2011) 1910–1915.

[88] F. Shu, B. Yang, S. Dong,H. Zhao, B. Xu, B. Liu, P. He, J. Feng, Effects of Fe-to-Co ratio on microstructure and mechanical properties of laser cladded FeCoCrBNiSi high-entropy alloy coatings, *Appl Surf Sci* 450 (2018) 538–544.

[89] Z. U. Arif, M. Y. Khalid, E. Rehman, S. Ullah, M. Atif, A. Tariq, A review on laser cladding of high-entropy alloys, their recent trends and potential applications, *J Manuf Proc* 68 (2021) 225–273.

[90] P. Huang, J. Yeh, T. Shun, S. Chen, Multi-principal-element alloys with improved oxidation and wear resistance for thermal spray coating, *Adv Eng Mater* 6 (2004) 74–78.

[91] A. S. M. Ang, C. C. Berndt, M. L. Sesso, A. Anupam, P. S. Ravi, S Kottada, B. S. Murty, Plasma-sprayed high entropy alloys: Microstructure and properties of AlCoCrFeNi and MnCoCrFeNi, *Metal Mater Trans A* 46A (2015) 791–800.

[92] L. Zhou, F. Lia, J. X. Liu, Q. Hu, W. Bao, Y. Wu, X. Cao, F. F. Xu, G. J. Zhang, High-entropy thermal barrier coating of rare-earth zirconate: A case study on $(La_{0.2}Nd_{0.2}Sm_{0.2}Eu_{0.2}Gd_{0.2})_2Zr_2O_7$ prepared by atmospheric plasma spraying, *J Eur Ceram Soc* 40 (2020) 5731–5739.

[93] M. Löbel, T. Lindner, T. Mehner,T. Lampke, Microstructure and wear resistance of AlCoCrFeNiTi high-entropy alloy coatings produced by HVOF, *Coat* 7 (2017) 144.

[94] M. Srivastava, M. Jadhav, Chethan, R. P. S. Chakradhar, M. Muniprakash, S. Singh, Synthesis and properties of high velocity oxy-fuel sprayed $FeCoCrNi_2Al$ high entropy alloy coating, *Surf Coat Technol* 378 (2019) 124950.

[95] T. Li, Y. Liu, B. Liu,W. Guo, L. Xu, Microstructure and wear behavior of $FeCoCrNiMo_{0.2}$ high entropy coatings prepared by air plasma spray and the high velocity oxy-fuel spray processes, *Coat* 7 (2017) 151.

[96] S. Yin, W. Li, B. Song, X. Yan, M. Kuang, Y. Xu, K. Wen, R. Lupoi, Deposition of FeCoNiCrMn high entropy alloy (HEA) coating via cold spraying, *J Mater Sci Technol* 35 (2019) 1003–1007.

[97] B. Cantor, I. T. H. Chang, P. Knight, A. J. B. Vincent, Microstructural development in equiatomic multicomponent alloys, *Mater Sci Eng A* 375–377 (2004) 213–218.

[98] S. Failla, P. Galizia, S. Fu, et al. Formation of high entropy metal diborides using arc-melting and combinatorial approach to study quinary and quaternary solid solutions, *J Eur Ceram Soc* 40 (2020) 588–593.

[99] Y. Zhang, S. Ma, J. Qiao, Morphology transition from dendrites to equiaxed grains for AlCoCrFeNi high-entropy alloys by copper mold casting and Bridgman solidification, *Metal Mater Trans A* 43 (2012) 2625–2630.

[100] S. Ma, S. Zhang, J. Qiao, Z. Wang, M. Gao, Z. Jiao, H. Yang, Y. Zhang, Superior high tensile elongation of a single-crystal $CoCrFeNiAl_{0.3}$ high-entropy alloy by Bridgman solidification, *Intermetallics* 54 (2014) 104–109.

[101] C. Kinsler-Fedon, Q. Zheng, Q. Huang, E. S. Choi, J. Yan, H. Zhou, D. Mandrus, V. Keppens, Synthesis, characterization, and single-crystal growth of a high-entropy rare-earth pyrochlore oxide, *Phys Rev Mater* 4 (2020) 104411.

[102] M. Pianassola, M. Loveday, B. C. Chakoumakos, M. Koschan, C. L. Melcher, M. Zhuravleva, Crystal growth and elemental homogeneity of the multicomponent rare-earth garnet $(Lu_{1/6}Y_{1/6}Ho_{1/6}Dy_{1/6}Tb_{1/6}Gd_{1/6})_3Al_5O_{12}$, *Cryst Growth Des* 10 (2020) 6769–6776.

[103] K. Zhang, Z. Fu, J. Zhang, W. Wang, S. Lee, K. Niihara, Characterization of nanocrystalline CoCrFeNiTiAl high-entropy solid solution processed by mechanical alloying, *J Alloy Compd* 495 (2010) 33–38.

[104] Z. Fu, W. Chen, H. Xiao, L. Zhou, D. Zhu, S. Yang, Fabrication and properties of nanocrystalline Co0.5FeNiCrTi0.5 high entropy alloy by MA-SPS technique, *Mater Des* 44 (2013) 535–539.

[105] W. Ji, W. Wang, H. Wang, J. Zhang, Y. Wang, F. Zhang, Z. Fu, Alloying behavior and novel properties of CoCrFeNiMn high-entropy alloy fabricated by mechanical alloying and spark plasma sintering, *Intermetallics* 56 (2015) 24–27.

[106] Z. Fu, W. Chen, H. Wen, D. Zhang, Z. Chen, B. Zheng, Y. Zhou, E. J. Lavernia, Microstructure and strengthening mechanisms in an FCC structured single-phase nanocrystalline $Co_{25}Ni_{25}Fe_{25}A_{17.5}Cu_{17.5}$ high-entropy alloy, *Acta Mater* 107 (2016) 59–71.

[107] Y. Zhang, Z. Jiang, S. Sun, W. Guo, Q. Chen, J. Qiu, K. Plucknett, H. Lin, Microstructure and mechanical properties of high-entropy borides derived from boro/carbothermal reduction, *J Eur Ceram Soc* 39 (2019) 3920–3924.

[108] J. Liu, X. Shen, Y. Wu,F. Li, Y. Liang, G. Zhang, Mechanical properties of hot-pressed high-entropy diboride-based ceramics, *J Adv Ceram* 9 (2020) 503–510.

[109] T. Wen, B. Ye, M. C. Nguyen, M. Ma, Y. Chu, Thermophysical and mechanical properties of novel high-entropy metal nitride-carbides, *J Am Ceram Soc* 103 (2020) 6475–6489.

[110] X. Chen, Y. Wu, High-entropy transparent fluoride laser ceramics, *J Am Ceram Soc* 103 (2020) 750–756.

[111] M. Li, X. Zhao, G. Shao, H. Wang, J. Zhu, W. Liu, B. Fan, H. Xu, H. Lu, Y. Zhou, R. Zhang, Oscillatory pressure sintering of high entropy $(Zr_{0.2}Ta_{0.2}Nb_{0.2}Hf_{0.2}Mo_{0.2})$B2 ceramic, *Ceram Int* 47 (2021) 8707–8710.

[112] M. Biesuz, L. Spiridigliozzi, G. Dell'Agli, M. Bortolotti, V. Sglavo, Synthesis and sintering of (Mg, Co, Ni, Cu, Zn)O entropy-stabilized oxides obtained by wet chemical methods, *J Mater Sci* 53 (2018) 8074–8085.

[113] S. Qiu, M. Li, G. Shao, H. Wang, J. Zhu, W. Liu, B. Fan, H. Xu, H. Lu, Y. Zhou, R. Zhang, (Ca, Sr, Ba)ZrO_3: A promising entropy-stabilized ceramic for titanium alloys smelting, *J Mater Sci Technol* 65 (2021) 82–88.

[114] H. L. Zhu, L. Liu, H. M. Xiang, F, Z. Dai, X. H. Wang, Z. Ma, Y. B. Liu, Y. C. Zhou, Improved thermal stability and infrared emissivity of high-entropy $REMgAl_{11}O_{19}$ and $LaMAl_{11}O_{19}$ (RE=La, Nd, Gd, Sm, Pr, Dy; M=Mg, Fe, Co, Ni, Zn), *Mater Sci Technol* 104 (2022) 131–144.

[115] B. Schuh, F. Mendez-Martin, B. Völker, E. George, H. Clemens, R. Pippan, A. Hohenwarter, Mechanical properties, microstructure and thermal stability of a nanocrystalline CoCrFeMnNi high-entropy alloy after severe plastic deformation, *Acta Mater* 96 (2015) 258–268.

[116] H. Shahmir, J. He, Z. Lu, M. Kawasaki, T. Langdonet, Effect of annealing on mechanical properties of a nanocrystalline CoCrFeNiMn high-entropy alloy processed by high-pressure torsion, *Mater Sci Eng A* 676 (2016) 294–303.

[117] Q. Tang, Y. Huang, Y. Huang, X. Liao, T. Langdon, P. Dai, Hardening of an $Al_{0.3}CoCrFeNi$ high entropy alloy via high-pressure torsion and thermal annealing, *Mater Lett* 151 (2015) 126–129.

[118] H. Yuan, M. Tsai, G. Sha, F. Liu, Z. Horita, Y. Zhu, J. Wang, Atomic-scale homogenization in an fcc-based high-entropy alloy via severe plastic deformation, *J Alloy Compd* 686 (2016) 15–23.

[119] B. Schuh, B. Völker, J. TodtN. Schell, L. Perriere, J. Li, J. P. Couzinie, A. Hohenwarter, Thermodynamic instability of a nanocrystalline, single-phase TiZrNbHfTa alloy and its impact on the mechanical properties, *Acta Mater* 142 (2018) 201–212.

[120] J. Gubicza, P. T. Hung, M. Kawasaki, J. Han, Y. Zhao, Y. Xue, J. L. Lábár, Influence of severe plastic deformation on the microstructure and hardness of a CoCrFeNi high-entropy alloy: A comparison with CoCrFeNiMn, *Mater Chara* 154 (2019) 304–314.

[121] P. Edalati, R. Floriano, Y. Tang, A. Mohammadi, K. Pereora, A. Luchessi, K. Edalati, Ultrahigh hardness and biocompatibility of high-entropy alloy TiAlFeCoNi processed by high-pressure torsion, *Mater Sci Eng C* 112 (2020) 110908.

[122] V. H. Hammond, M. A. Atwater, K. A. Darling, H. Nguyen, L. Kecskes, Equal-channel angular extrusion of a low-density high-entropy alloy produced by high-energy cryogenic mechanical alloying, *JOM* 66 (2014) 2021–2029.

[123] H. Shahmir, T. Mousavi, J. He, Z. Lu, M. Kawasaki, T. Landon, Microstructure and properties of a CoCrFeNiMn high-entropy alloy processed by equal-channel angular pressing, *Mater Sci Eng A* 705 (2017) 411–419.

[124] H. Qiu, G. Fang, Y. Wen,P. Liu, G. Xie, X. Liu, S. Sun, Nanoporous high-entropy alloys for highly stable and efficient catalysis, *J Mater Chem A* 7 (2019) 6499–6506.

[125] S. Joo, J. Bae, W. Park, Y. Shimada, T. Wada, H. Kim, A. Takeuchi, T. Konno, H. Kato, I. Okulov, Beating thermal coarsening in nanoporous materials via high-entropy design, *Adv Mater* 32 (2019) 1906160.

[126] Z. Xu, Z. Zhu, P. Wang, G. K. Meenashisundaram, S. Nai, J. Wei, Fabrication of porous CoCrFeMnNi high entropy alloy using binder jetting additive manufacturing, *Addit Manuf* 35 (2020) 101441.

[127] Q. Guo, H. Xiang, X. Sun,X. Wang, Y. Zhou, Preparation of porous YB_4 ceramics using a combination of in situ borothermal reaction and high temperature partial sintering, *J Eur Ceram Soc* 35 (2015) 3411–3418.

[128] H. Chen, H. Xiang, F. Dai, J. Liu, Y. Zhou, Low thermal conductivity and high porosity ZrC and HfC ceramics prepared by in situ reduction reaction/partial sintering method for ultrahigh temperature applications, *J Mater Sci Technol* 35 (2019) 2778–2784.

[129] H. Chen, H. Xiang, F. Dai, J. Liu, Y. Zhou, High strength and high porosity YB_2C_2 ceramics prepared by a new high temperature reaction/ partial sintering process, *J Mater Sci Technol* 35 (2019) 2883–2891.

[130] H. Chen, H. M. Xiang, F. Z. Dai, J. C. Liu, Y. M. Lei, J. Zhang, High porosity and low thermal conductivity high entropy $(Zr_{0.2}Hf_{0.2}Ti_{0.2}Nb_{0.2}Ta_{0.2})C$, *J Mater Sci Tech* 35 (2019) 1700–1705.

[131] H. Chen, H. M. Xiang, F. Z. Dai, J. C. Liu, Y. C. Zhou, Porous high entropy $(Zr_{0.2}Hf_{0.2}Ti_{0.2}Nb_{0.2}Ta_{0.2})$ B2: A novel strategy towards making ultrahigh temperature ceramics thermal insulating, *J Mater Sci Technol* 35 (2019) 2404–2408.

[132] H. Chen, Z. Zhao, H. Xiang, F. Dai, S. Wang, J. Zhang, Y. Zhou, Effect of reaction routes on the porosity and permeability of porous high entropy $(Y_{0.2}Yb_{0.2}Sm_{0.2}Nd_{0.2}Eu_{0.2})$B6 for transpiration cooling, *J Mater Sci Technol* 38 (2020) 80–85.

[133] G. Wang, J. Qin, Y. Feng, B. Feng, S. Yang, Z. Wang, Y. Zhao, J. Wei, Sol-gel synthesis of spherical mesoporous high-entropy oxides, *ACS Appl Mater Inter* 12 (2020) 45155–45164.

[134] X. Zhang, L. Xue, F. Yang, Z. Shao, H. Zhang, Z. Zhao, K. Wang, $(La_{0.2}Y_{0.2}Nd_{0.2}Gd_{0.2}Sr_{0.2})CrO_3$: A novel conductive porous high-entropy ceramic synthesized by the sol-gel method, *J Alloy Compd* 863 (2021) 158763.

4 High-Entropy Materials
Composition and Structural Aspects

L.J. Jing, J.Y. Zhang, T. Yang, and Y.L. Zhao

4.1 INTRODUCTION

The composition and microstructure of high-entropy materials (HEMs) play a crucial role in determining their properties. Unlike traditional alloy design strategy, the innovative design paradigm of HEMs, based on multi-principal elements [1], has profoundly enlarged the landscape of materials engineering by expanding the compositional space and the microstructural diversity, holding promise for achieving unprecedented properties [2]. However, such complex and diverse nature of the constitutions in HEMs makes the criteria for component selection and controllable microstructural modulation challenging.

It is within this context that various approaches have been proposed to predict the phase formation behavior of the HEMs. In this chapter, we will mainly focus on the metallic alloy systems and outline the widely used approaches for the phase formation criterion in high-entropy alloys (HEAs), including empirical criteria, CALculation of PHAse Diagrams (CALPHAD) based approach, and Machine Learning (ML) aided approach. Those approaches can reasonably be used to predict the phase constitutions in the HEAs, such as single-phase solid solution (SS) [1,3], metastable dual-phase HEAs [4], nanoprecipitation or interstitial strengthening HEAs [5,6], chemical complex intermetallic alloys (CCIMAs) [7]. Furthermore, we will concentrate on the microstructural heterogeneities arising from both atomic and mesoscales and discuss their impact on the mechanical properties of the HEAs. Finally, some outstanding issues, prospects, and challenges will be briefly addressed.

4.2 PHASE FORMATION CRITERION

4.2.1 Empirical Models

In the first decade since the advent of HEAs, investigations have been extensively focused on uncovering the mystery of why SS prevails over intermetallic compounds in the HEAs. Yeh et al. attributed the stabilization of SS to the high configurational entropy of mixing (ΔS_{mix}) in comparison to traditional alloys [1]. In addition to the ΔS_{mix}, various other parameters have been proposed to understand the formation of disordered SS in HEAs, such as the mean atomic radius (a), the δ-parameter (atomic size differences between the elements), the enthalpy of mixing (ΔH_{mix}), the Ω-parameter ($T_m \Delta S_{mix}/|\Delta H_{mix}|$), valence electron concentration (VEC), electronegativity ($\Delta\chi$). Those parameters have achieved different degrees of success in understanding the formation of disordered SS in HEAs [8–12]. Works by Zhang et al. suggested that higher ΔS_{mix} (~ 12–17.5 J/ K·mol), lower δ (~ 1–6), and not too negative and positive ΔH_{mix} (~ –15 to 5 KJ/mol) are beneficial for the formation of the SS (Figure 4.1) [12].

To simplify the parameter selection, Anandh Subramaniam et al. proposed a geometrical parameter Λ, combining the effect of ΔS_{mix} and δ (following $\Lambda = \Delta S_{mix}/\delta^2$) to predict the formation of

DOI: 10.1201/9781003391388-4

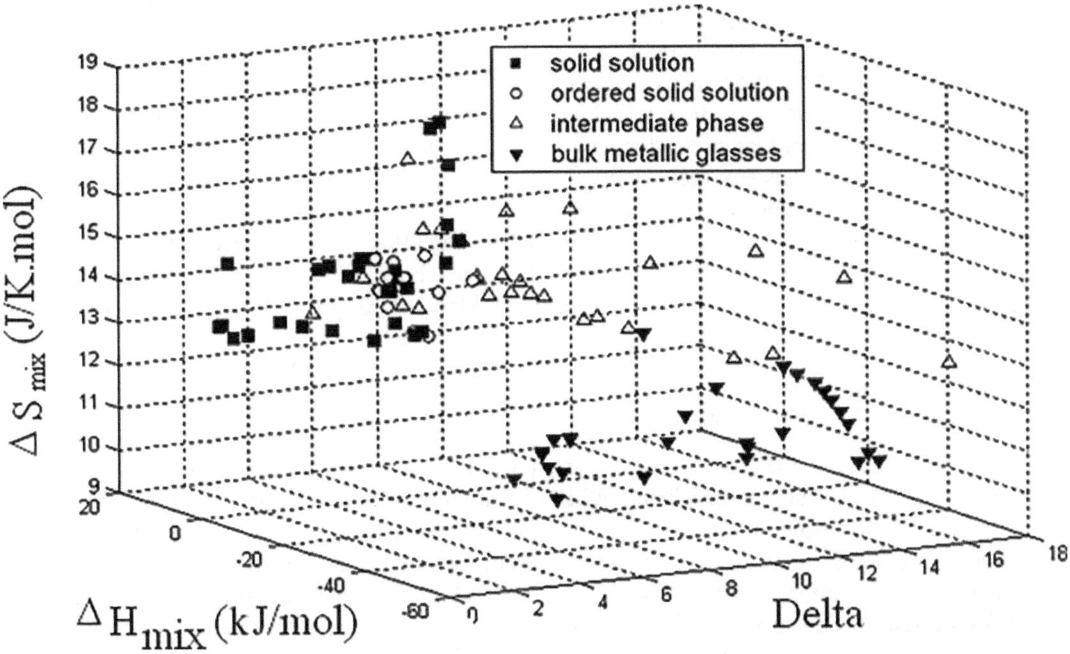

FIGURE 4.1 The effect of ΔS_{mix}, ΔH_{mix}, and δ on the phase formation of HEAs. Adapted with permission [12]. Copyright 2008, Wiley.

disordered SS [10]. By analyzing the phase formation of various reported multicomponent alloys, values of the parameter are prescribed for the formation of single-phase disordered SS ($\Lambda > 0.96$), two-phase mixtures ($0.24 < \Lambda < 0.96$) and mixtures involving intermetallics (IMs) ($\Lambda < 0.24$). Later, based on the hard-sphere model, Yang et al. set an entropy-based criterion for the formation of a single-phase SS in the as-cast HEAs [11]. Two dimensionless thermodynamic parameters are involved in their study: one is related to the entropic departure of the alloy from the ideal solution ($|S_E|/S_c$) and the other to the average heat of mixing scaled by the "ideal" configurational entropy of mixing ($|H_m|/TS_c$). They demonstrated that the boundary separating the single-phase SS from the multi-phase region is $|S_E|/S_c \sim 0.04$, and $4 \times 10^{-3} < |H_m|/TS_c < 2 \times 10^{-1}$. Such an empirical rule is in accord with the theoretical prediction that a single-phase SS is formed when $|S_E|/S_c \ll 1 - |H_m|/TS_c$. Independent of existing experimental or modeled thermodynamic data, King et al. proposed a new parameter to forecast the SS formation in the HEAs [8]. The parameter ϕ is defined as the ratio between the Gibbs free energy of a completely disordered SS to that of the most likely IM or segregated binary system. The results showed that the formation of random SS is favored when $\phi > 1$, demonstrating a reasonable forecasting proficiency when compared with 185 experimentally reported HEAs. Some deviations, i.e., 16 out of 185 alloy systems, come from the partial ordering present in these SS.

In regard to the performance of the HEAs, researchers gradually realized that the SS does not necessarily exhibit superior properties. Research focuses have gradually shifted to precipitation-strengthened HEAs [5], high-entropy ceramics [13], high-entropy intermetallics (HEIs) [7,14], etc. Among them, HEIs, also referred to as CCIMAs or multicomponent IM alloys, have attracted a lot of attention. This kind of novel alloy possesses crystal structures (such as $L1_2$, B_2, $D0_{22}$, etc.) that are like the conventional binary IM phases. Yet they are composed of multiple principal elements in their sublattice, leading to an increased configurational ΔS_{mix} in each sublattice. O.N. Senkov and D.B. Miracle defined a parameter k_1^{cr} and proposed a simple thermodynamic criterion, $k_1^{cr}(T)$ vs.

$\Delta H_{IM}/\Delta H_{mix}$, to predict what types of equilibrium HEIs are present in a HEA at a given temperature [9]. It is demonstrated that HEIs appear below the critical line corresponding to the condition $k_1^{cr}(T)=\Delta H_{IM}/\Delta H_{mix}$. The rule was evidenced using 45 available HEAs and demonstrated a good correlation with the experiments. Besides, electronic and geometric factors, such as VEC, $\Delta\chi$, δ and packing efficiency, also influence the formation of HEIs. Liu et al. found that with the decreasing VEC, phase transformations from tetragonal (VEC > 8.75) to ordered hexagonal (7.89 < VEC < 8.75) crystal structure, and further to ordered cubic structures (VEC < 7.89) in the NiCoFeV alloy system [15]. Similarly, based on the calculated overall VEC of the alloy, He et al. successfully introduced $D0_{22}$ superlattice (γ'' phase) as a strengthening phase in a $Ni_2CoCrFeNb_{0.15}$ HEA [16]. They also claimed that a high VEC value (>8.4) and adding the elements in group VB tends to promote the precipitation of γ'' phase.

4.2.2 CALPHAD-Based Approach

Although the empirical rules can be helpful in predicting phase formation in HEAs, their performances are not always satisfactory. The CALPHAD method allows for a more detailed investigation of phase compositions, fractions, and phase stabilities as a function of constituent elements, pressures, and temperatures. A reliable thermodynamic database is a prerequisite for CALPHAD modeling. Currently, there are two main commercial databases that are specific to HEAs. The TCHEA database was developed to include all the binary and most ternary systems associated with 26 elements [17], and the PanHEA database contains all the binary and ternary systems related to HEAs consisting of Al, Co, Cr, Cu, Fe, Mn, and Ni [18].

He et al. used the TCHEA database to evaluate the nucleation competition between $L1_2$ and $L2_1$ Heusler phases in a $(FeCoNiCr)_{100-x-y}Ti_xAl_y$ (x=1–3 and y=4–9 at.%) HEA system [19]. Compared to the prediction results for HEAs alloy with different Ti and Al contents, they found that $L1_2$ and $L2_1$ phases can be only formed when Ti and Al are simultaneously added. The calculation results suggested that increasing Ti and Al contents could increase the volume fraction of the $L1_2$ phase but be accompanied by a sacrifice of its usage temperature range, while it will promote the nucleation and stability of the Heusler phase. These calculation results agree highly with the experimental observations in terms of the competing behaviors of two main precipitates. MacDonald et al. conducted a comprehensive analysis by comparing the CALPHAD-predicted phases with experimental observations in the CoCuFeMnNi HEA using the TCHEA database [20]. The calculated diagram indicated that triple phases (i.e., Fe–Co B2 phase, Cu-rich phase, face-centered cubic FCC matrix) co-exist at temperatures below 559°C. They further compared the chemistry of the three phases from the experimental and predicted data at 500°C. Despite the difference in the forming temperature, a relatively good agreement can be observed.

Nevertheless, the extrapolation from common binary and ternary systems to multicomponent systems may lead to inadequate accuracy. Some studies indicated CALPHAD approach cannot be used to predict the formation of an unknown new phase or σ phase in the high-order HEA systems [21]. Zhao et al. evaluated the phase stability of the complex $(CoCrNi)_{97-x}Ti_3$–Al_x alloy system by the CALPHAD method with the TTNI8 database (Figure 4.2) [22]. It can be found that a pure $\gamma+\gamma'$ two-phase region only exists in a narrow temperature range of 940°C–980°C when Al=3 at. % (Figure 4.2a). However, their experimental results suggested that such a clean two-phase microstructure can be also observed in the HEA aging at 800°C, which contradicts the prediction results where a small mole fraction of σ phase is supposed to appear (Figure 4.2b). The researchers identified two factors that could lead to the inconsistency: the insufficient thermodynamic database currently available for HEAs, and the inadequate aging time due to the sluggish kinetics observed in HEA systems.

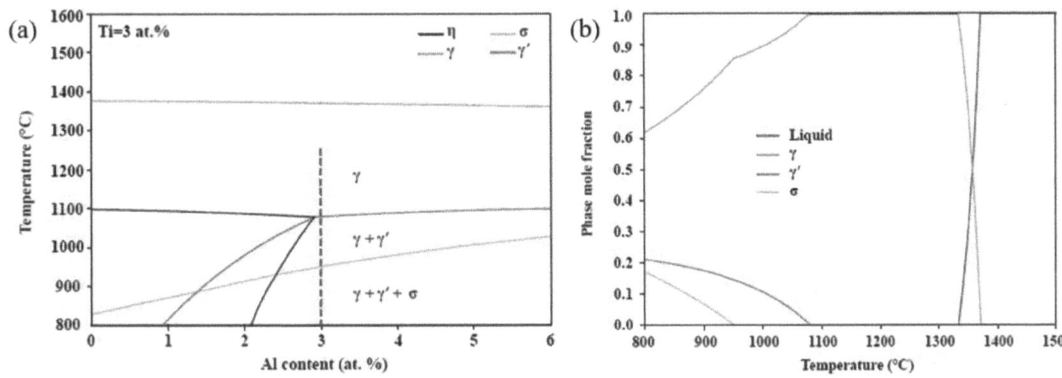

FIGURE 4.2 (a) Phase diagram and (b) the mole fraction variation of the equilibrium phase of the $(CoCrNi)_{97-x}Ti_3\text{-}Al_x$ alloy system using the TTNI8 database. Adapted with permission [22]. Copyright 2017, Elsevier.

4.2.3 Machine-Learning-Aided Approach

In contrast to the CALPHAD approach, data-based ML shows great potential to improve both the accuracy and efficiency of predictions for HEAs. One of the main challenges is the limited availability and quality of training data, especially for newer and more complex HEAs. This problem can usually be somewhat mitigated with the assistance of theoretical calculations, such as density functional theory (DFT) calculation, molecular dynamics (MD) simulation, and CALPHAD modeling.

Zhuang et al. used three different ML algorithms: K-nearest neighbors (KNN), support vector machine (SVM), and artificial neural network (ANN) to determine the phase selection using a dataset of 401 HEAs [23]. They found that in addition to the hyperparameters, the suitability of the algorithm is also of great importance for the predicting performance. Their results showed that the trained ANN model performs the best accuracy (up to 94.3% for SS+IM and IM phase) among the three ML algorithms, which is efficient to explore phase selection rules and to assist in the design of new HEAs. The authors also suggested that fine-tuning the number of hidden layers and neurons, as well as adjusting the internal parameters like the training function, is also expected to enhance the testing accuracy.

Zhou et al. explored the use of additional new parameters in the ML model for the phase design rules for the HEAs and examined the sensitivity of these parameters for the phase formation [24]. A dataset collected from 601 as-cast alloys with 131 SS alloys, 248 IM alloys, 165 amorphous (AM) alloys, 6 mixed IM and AM alloys, and 51 mixed SS and IM alloys were trained for the ML model. In addition, an expanded descriptor space containing 13 parameters was used, including older parameters like a, δ, H_{mix}, S_{id}, χ, $\Delta\chi$, VEC, as well as new parameters like standard deviation of VEC (σ_{VEC}), standard deviation of mixing enthalpy (σ_H), an average of the melting temperatures of constituent elements (T_m), the standard deviation of melting temperature (σ_T), mean bulk modulus (K), and standard deviation of bulk modulus (σ_K). This study examined the magnitude of the correlation of individual features and obtained the sensitivity matrices for the three phases, i.e., AM, IM, and SS (Figure 4.3). Clearly, δ plays a very important role in the formation of SS, whereas ΔH_{mix} has almost the same degree of influence as S_{id}. Based on their findings, it appears that the accuracy of the ML model based on the multi-descriptors has been substantially improved as compared to the general ML model based solely on a single parameter.

Furthermore, ML modeling is also capable of distinguishing the crystal structures of the phases in HEAs. Based on an experimental dataset containing 322 as-cast HEAs, Li et al. constructed

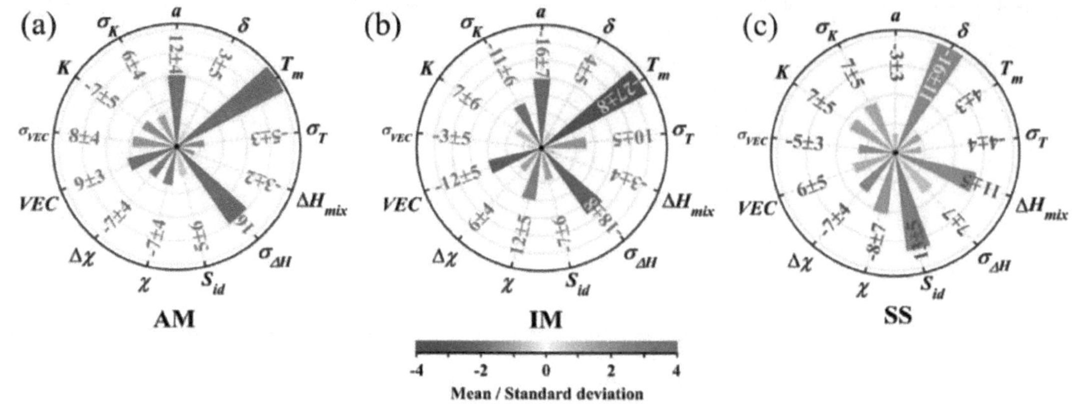

FIGURE 4.3 The sensitivity measures of the design parameters for (a) AM, (b) IM, and (c) SS phases. Adapted with permission [24].

a SVM model to identify the FCC, BCC single phases, and the remaining phases in those alloy systems, which achieves a cross-validation accuracy of over 90% after training and testing [25]. The model predicted numerous composition constitutions for the 16 metallic elements, including 267 BCC and 369 FCC equimolar HEAs, many of which were found to be consistent with both experiments and first-principles calculations.

The application of ML techniques can also facilitate the design of precipitation-strengthened HEAs. Zheng et al. designed an ultra-strong HEA with nanoprecipitation strengthening via the combination of ANN-based ML model and prestrain aging [26]. In this approach, the volume fraction of the nanoprecipitates is optimized by an ML strategy based on the database of nickel-based superalloy, while the morphology of precipitates is kinetically regulated by the prestrain aging treatment. The output is highly consistent with the target model for the prediction of γ' phase volume fraction and yield strength.

However, a direct prediction of the composition-property relationships by ML modeling remains elusive and challenging due to the limited valid databases, and the human bias in feature and algorithms selection. Active learning with improved accuracy involves selecting and labeling the most relevant and informative data points, which can be particularly useful in cases where data is scarce or where labeling data is expensive or time-consuming. However, the active learning approach relies on simple surrogate models and Bayesian optimization methods, making their usage limited to low-dimensional data. Additionally, the improvement in model accuracy can be slow and requires many iterations. Recently, Dierk Raabe et al. successfully overcame such a critical issue of sparse datasets by developing an active learning framework integrating ML with DFT, thermodynamic calculations, and experiments feedback [27]. Specifically, the study involved training a learning algorithm to generate candidate compositions for targeted alloys. The selected candidates were further selected by a two-stage ensemble regression framework with additional inputs from DFT and thermodynamic databases. To validate and refine the model, the top three candidate alloys were experimentally measured, and their results were fed back to the database. The iteration was repeated until a targeted alloy was discovered. The entire workflow shows a high proficiency in processing and requires only a few months to accomplish, which is far less than the conventional alloy design approaches. Using this approach, Raabe and his co-workers screened two high-entropy Invar alloys with extremely low thermal expansion coefficients out of millions of possible compositions [27], demonstrating the power of this approach for the accelerated development of HEAs with desirable properties.

The above-mentioned different methods, i.e., empirical models, CALPHAD-based approach, and ML-aided approach, have their advantages and shortcomings. The empirical models for the prediction of the HEAs proposed so far can only indicate a general direction. They are usually relatively simple and are appropriate for alloy systems with simple phase constituents. Also, their accuracy of predictions is usually inadequate. The CALPHAD-based approach liberates labor and saves time, which is a powerful tool for predicting the HEA structure under different conditions, providing rational guidance for experimental synthesis. Yet, this method is not always efficient for a large search space of the HEAs and generally provides equilibrium-phase diagrams. The prediction accuracy depends highly on the accuracy of the adopted databases. Many data of HEAs inherent with multicomponent features need to be further verified by experiments. Accordingly, parameters in the related databases should also be optimized. In contrast to empirical and CALPHAD methods, the ML-aided approach can greatly improve both the accuracy and efficiency of phase predictions for HEAs. However, apart from relatively small experimental databases, the lack of transparency and interpretability of the physical mechanisms of the ML models hampers its further applications. Fortunately, those drawbacks could be compensated by integrating ML modeling with other methods, such as DFT, MD, CALPHAD, experiments, and so on. Therefore, the synergy of multiple methods is more conducive to the design of the targeted HEAs.

4.3 STRUCTURAL MODULATION AND PROPERTIES

4.3.1 Atomic Structural Tuning

As mentioned in the previous section, in the early stage of the development, HEAs were usually considered as random SS with the single-phase FCC or BCC structure due to their high configurational entropies. And the solid-solution strengthening brought about by their multicomponent characteristics was considered as a key strengthening mechanism of single-phase HEAs [2,28]. With the rapid development of HEAs, there is increasing evidence that chemical-order heterogeneities at the nanoscale exist widely in HEAs and profoundly affect their macroscale mechanical properties [29]. Compared to the low yield strength contributed by single solid-solution strengthening and the loss of ductility caused by grain-refinement strengthening or precipitation strengthening, atomic structural tuning (containing the short-range order (SRO), local chemical order (LCO), and long-range order (LRO)) offer a promising avenue to push the envelope of mechanical properties into a previously inaccessible territory. Therefore, a deep insight into the atomic structure characteristics and properties of HEAs is essential for understanding and designing high-performance HEAs.

Figure 4.4 presents a schematic illustration of five possible scenarios of atomic stacking for a ternary alloy, namely disordered solid solution, SRO, LCO, intermetallic precipitates, and LRO superlattice. In fact, due to the different atomic radii and the complex interactions between constituent elements, HEAs are more inclined to form energetically favorable atomic configurations, i.e., the SRO or LCO, rather than the ideal random solid solution. SRO and LCO are used to describe the thermally induced infinitesimal compositional fluctuations, or in other words, chemical ordering in adjacent atomic shells and local domains. Specifically, the SRO usually represents the chemical ordering on the short-range (~1 nm in real space), confined to the nearest and the second nearest-neighbor atomic shells, while the LCO is no longer confined to the second nearest-neighbor atomic shell and can be extended to medium-range or long-range domains in some dimensions. The LCO contains the SRO, and in most cases, no distinction is deliberately made between the two chemical ordering as the area of the LCO or SRO is much smaller than that of the intermetallic precipitates with full-blown chemical ordering. The key point in distinguishing between SRO and intermetallic precipitates is that the long-range ordered structure in the precipitates gives sharp spots in reciprocal space (diffraction pattern), while the diffraction signal from SRO is weak and diffuse. When the chemical ordering extends infinitely to the entire real space, the atomic packing then becomes an LRO superlattice structure, which means that each site inside the unit cell is occupied by one or

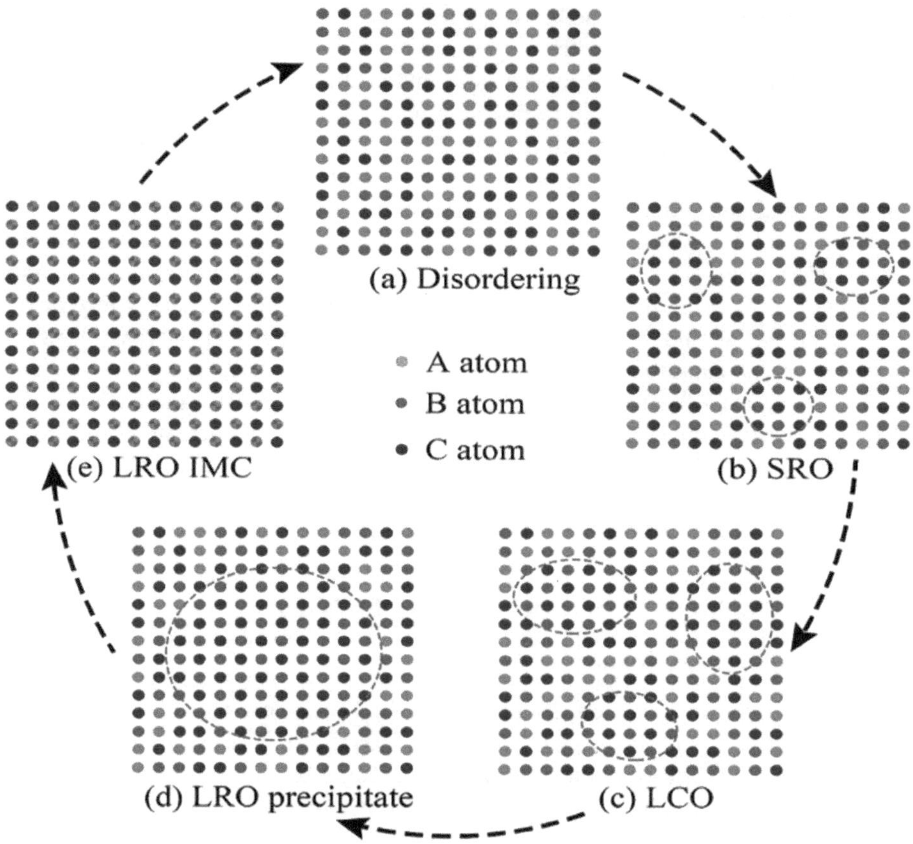

FIGURE 4.4 Schematic illustration of the atomic packing in a ternary alloy.

more specific atoms. Similarly, the local chemical disordering may also arise in the LRO superlattice structure. Tuning chemical SRO in disordered solid solution or tuning disorder in LRO superlattice can be considered as a powerful means to further optimize the macroscopic mechanical and physical properties of HEAs.

Since SROs are usually confined to chemically ordered domains around 1 nm, direct observation of SROs is very difficult and challenging. Early studies on SRO were limited to verifying the existence of SRO in HEAs through Monte-Carlo simulation and DFT [30]. Zhang et al. have experimentally confirmed the existence of SRO in the NiCoCr alloy by extended X-ray absorption fine structure (EXAFS), where the Cr tends to form atomic pairs with Ni and Co in some localized nano-regions, respectively [31]. In fact, SRO is not rare in traditional alloys and has been characterized and quantitatively analyzed through energy-filtered transmission electron microscopy (TEM) imaging [32]. Such a method enables the detailed intensity distribution and shape of the diffraction contrast resulting from chemical SRO to be obtained by equipping the TEM with an omega-type energy filter to attenuate the background noise from inelastic scattering [32]. Based on this method, Zhang et al. observed SRO-induced streaks along the reciprocal lattice vectors that are normal to the {111} planes in the aged NiCoCr alloy, as shown in Figure 4.5 [33]. Compared with the weak streaking along the {111} directions in the recrystallized NiCoCr alloy (Figure 4.5a), the aged counterpart (Figure 4.5b) shows a marked increase in streaking strength, implying an enhanced degree of SRO. In addition, the dark-field image obtained by placing the objective aperture in the streaked region between the two {111} planes further proves that the SRO is negligible or undetectably low in the recrystallized NiCoCr alloy (Figure 4.5c), but is clearly observed in the aged alloy

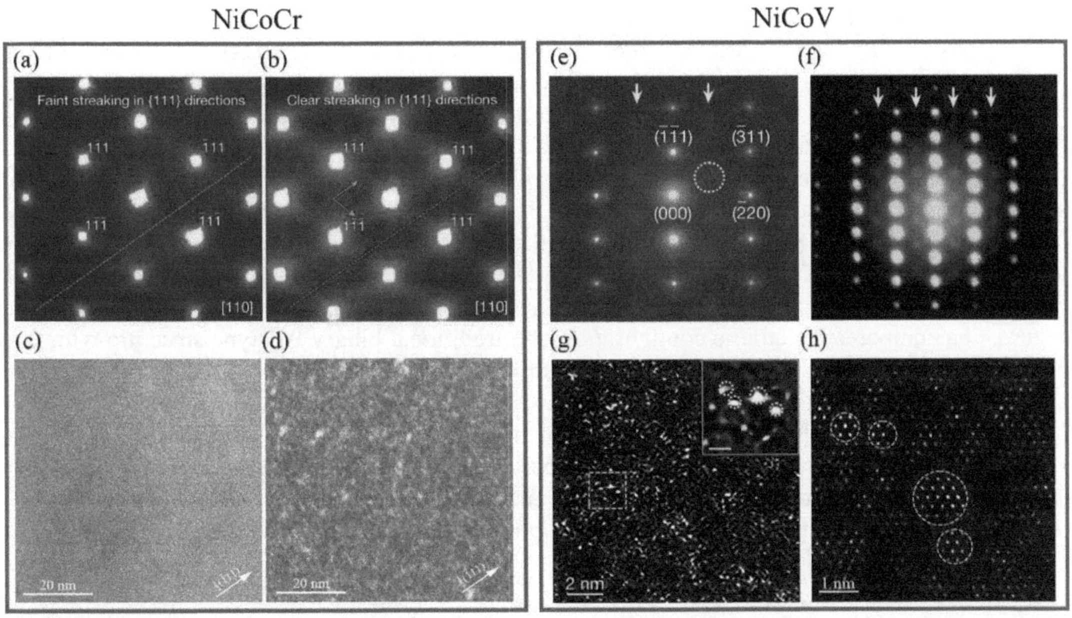

FIGURE 4.5 TEM evidence of chemical SRO in the FCC structured NiCoCr (a–d) and NiCoV alloy (c–h). (a–d) Adapted with permission [33]. Copyright (2020), Springer Nature. (c–h) Adapted with permission [34]. Copyright 2021, Springer Nature.

(Figure 4.5d), as reflected in the contrast due to chemical inhomogeneities on the nanoscale. The more pronounced and intuitive SRO was observed in the deformed NiCoV alloy, where Chen et al. found that the superlattice spots (indicated by the yellow dashed circle) induced by SRO appeared on the $\frac{1}{2}\{\bar{3}11\}$ positions under a specific zone axis, i.e., the [112] axis (Figure 4.5e) [34]. The use of the nano-beam electron diffraction pattern (EDP) makes these extra reflections more discernible (Figure 4.5f). In addition, both the energy-filtered dark-field TEM image taken using the extra reflections (Figure 4.5g) and the inverse fast Fourier transform (FFT) map (Figure 4.5h) clearly show the SRO domains.

Admittedly, both the SRO or LCO arising from chemical-order heterogeneities on the nanoscale would inevitably provide obstacles to the dislocation movement and thus affect the mechanical behavior of the HEAs. Many studies have also been based on this to investigate the effect of SRO on the mechanical behaviors of HEAs through hybrid Molecular-dynamics/Monte-Carlo simulations. For example, the SFE of NiCoCr alloy can be tailored by changing the degree of SRO, which in turn affects the formation and growth of deformation-accommodating defects [35]. In addition, higher degrees of SRO have been shown to not only heighten the rugged energy landscape and thus increase the activation barrier for dislocation movement [36], but they also increase the strain required for Shockley partial dislocations nucleation, thereby increasing the yield strength of NiCoCr alloy [37]. Recently, SRO has also been shown to alter the nucleation and propagation of dislocations in refractory HEAs, promoting the dominance of edge dislocations upon tensile deformation [38]. Although numerous simulation results have shown that SRO changes the dislocation activities and thus affects the mechanical behaviors of the HEAs, Yin et al. demonstrated experimentally that the SRO has a negligible effect on the strength of NiCoCr alloy under typical processing conditions (annealed at 700°C and 600°C for 384 h, respectively, water quenched) [39]. Similar findings were also confirmed by Zhang et al., who found that varying degrees of SRO had essentially no effect on the tensile strength and ductility of NiCoCr alloy [40]. However, the results of the nanoindentation test show that stress for the onset of plasticity is closely related to the SRO, and the different degrees of

SRO affect the obstacle for dislocation nucleation, resulting in different nanoindentation strengths [40]. Up to now, although the SRO has been confirmed to exist widely in HEAs, its influence on the macroscopic mechanical behavior remains elusive and needs to be further studied.

In addition to the SRO or LCO, another structural feature that has received extensive attention is the LRO superlattice structure. By combining the alloy design concept of HEAs and LRO superlattice, a number of multicomponent ordered alloys with superior mechanical properties and thermal stability than traditional binary ordered alloys have been designed, such as the $L1_2$-structured $Ni_{43.9}Co_{22.4}Fe_{8.8}Al_{10.7}Ti_{11.7}B_{2.5}$ alloy (at. %) and the B2-structured $Ni_{32.3}Fe_{12.1}Cr_{8.8}Co_{16.9}Al_{29.9}$ alloy (at. %), also known as CCIMAs [7,41]. Here we take the alloy with an $L1_2$ structure as an example to demonstrate the unique structural characteristics of CCIMAs. The schematic diagram in Figure 4.6a compares the atomic configuration of a traditional binary $L1_2$-type structure with that of a quinary $L1_2$-type CCIMA structure. In contrast to the single occupancy of elements in the simple binary structure, the quinary $L1_2$-type CCIMA exhibits a chemically complex site occupancy, i.e., the A site or B site can be occupied by multiple elements at the same time. The STEM–HAADF image and corresponding energy dispersive X-ray (EDX) maps in Figure 4.6a clearly show that Ni and Co atoms tend to occupy the A site (the face-center of the $L1_2$unit cell), while the Al and Ti atoms with relatively large atomic radii mainly occupy the B site (vertices of the

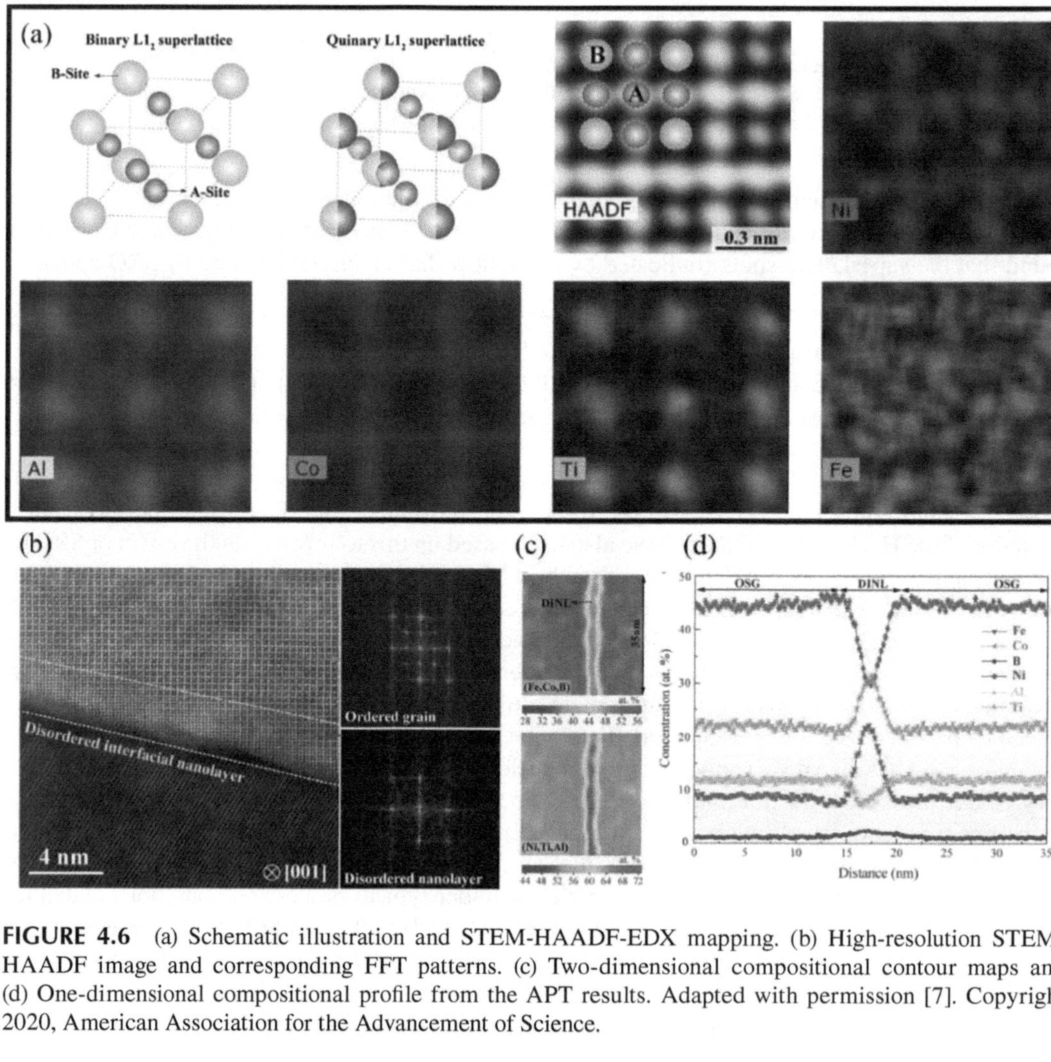

FIGURE 4.6 (a) Schematic illustration and STEM-HAADF-EDX mapping. (b) High-resolution STEM-HAADF image and corresponding FFT patterns. (c) Two-dimensional compositional contour maps and (d) One-dimensional compositional profile from the APT results. Adapted with permission [7]. Copyright 2020, American Association for the Advancement of Science.

$L1_2$ unit cell). It is noteworthy that the Fe atoms have no evident occupancy tendency, suggesting their simultaneous presence in both sublattice sites, so as to maintain the (Ni, Co, $Fe)_3$(Al, Ti, Fe)-type $L1_2$ structure. Similar to the appearance of SRO in disordered solid solutions, nanoscale disordered domains appear adjacent to grain boundaries in ordered CCIMAs, attributed to the multicomponent alloying that significantly alters the localized microstructures near grain boundaries, as shown in Figure 4.6b. The two-dimensional compositional contour maps (Figure 4.6c) and the one-dimensional compositional profile (Figure 4.6d) obtained by APT show that the co-segregation of Co, Fe, and B elements near the grain boundaries destroys the ordered superlattice structure, enabling the occurrence of order-to-disorder phase transformation (from $L1_2$ to disordered FCC) at the nanoscale. The multicomponent ordered superlattice structure combined with the disordered interfacial nanolayer (DINL) allows the CCIMAs to increase both strength and ductility compared to conventional binary alloys [7].

4.3.2 Heterogeneous Grain Structure

In addition to chemical-order heterogeneities on the nanoscale, another structural feature that has received widespread attention in HEAs is microstructural heterogeneities, also called heterogeneous grain structures (HGS). As early as 2002, Wang et al. revealed the importance of HGS (bimodal microstructure) in the mechanical properties of materials, which allows for a good combination of strength and ductility of pure copper [42]. Subsequently, Wu et al. constructed an HGS (soft micro-grained lamellae embedded in hard ultrafine-grained lamella matrix) on pure titanium by partial recrystallization, giving it both the strength of ultrafine-grained metals and the ductility of coarse-grained metals [43]. In recent years, HGS has also gained considerable interest in HEAs, and various types of heterogeneous-grain-structured HEAs (multi-modal grain structures [44], harmonic grain structures [45], and gradient grain structures [46]) have been successfully fabricated and shown superior mechanical and physical properties than conventional homogeneous counterparts.

A typical example to show the characteristics and unique mechanical behavior of HEAs with HGS is the study on the partial recrystallized NiCoCr alloy containing multi-modal grain structures, including micrometer-sized, ultrafine-sized, and nanosized grains [44]. Noticeably, the nanosized grains are mainly distributed on the grain boundaries and triple junctions of the ultrafine-sized grains. The HGS NiCoCr alloy exhibits excellent room-temperature and low-temperature mechanical properties, with a tensile yield strength of 1150 MPa at room temperature (well above that of the homogeneous counterpart) and good tensile ductility (~30%). The strength difference between the hard region (ultrafine-sized and nanosized grains) and soft region (micrometer-sized grains) makes them exhibit inconsistent elastoplastic transitions during tensile straining, resulting in a unique strain distribution that allows a combination of high strength and good ductility to be maintained.

Interestingly, this unique HGS can also effectively alleviate intermediate-temperature intergranular embrittlement, a headache for equiaxed microstructures. Figure 4.7a and b shows the corresponding fully recrystallized and partially recrystallized microstructures of the $Ni_{29.9}Co_{30}Fe_{13}Cr_{15}Al_6Ti_6B_{0.1}$ (at.%) HEA, respectively [47]. In contrast to the equiaxed grains in the fully recrystallized alloys, the partially recrystallized alloys consist of unrecrystallized deformed grains (banded area) and fine recrystallized grains. The tensile curves (Figure 4.7c) and corresponding SEM fractographs (Figure 4.7d) tested at 800°C indicate that the heterostructured HEAs effectively avoid the occurrence of intergranular embrittlement that occurs in the equiaxed-grain-structured counterpart and exhibit a tensile ductility of up to 20%. Similar HGS morphology is also observed in the partially recrystallized $Co_{40}Cr_{20}Ni_{30}Al_5Ti_5$ (at.%) HEA (Figure 4.7e and f) [48]. The deformed grains (uncrystallized regions) are divided by shear bands (dashed lines in Figure 4.7e) and occupy 30%, while fine recrystallized grains occupy 70%. This heterostructured HEA exhibits not only ultra-high room-temperature and low-temperature yield strength and considerable ductility, but also more excellent tensile ductility at 500°C, 600°C, and 700°C (Figure 4.7g). In addition, it can be seen from Figure 4.7h and i that the fracture morphologies after tensile deformation at 500°C and 700°C are dominated by

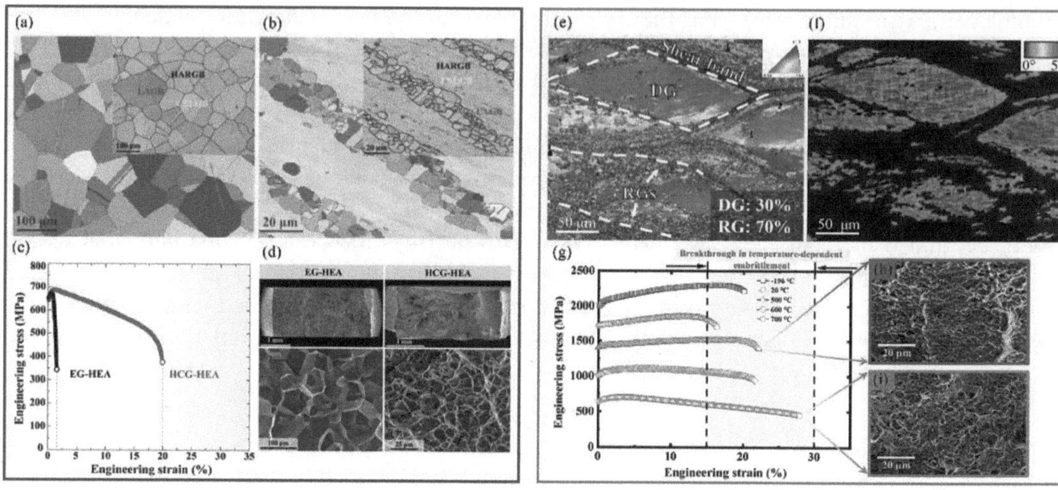

FIGURE 4.7 Microstructure, mechanical property and fractographs of heterogeneous structured $Ni_{29.9}Co_{30}Fe_{13}Cr_{15}Al_6Ti_6B_{0.1}$ (a–d) and $Co_{40}Cr_{20}Ni_{30}Al_5Ti_5$ HEA (e–i). (a–d) Adapted with permission [47]. Copyright 2021, Elsevier. (e–i) Adapted with permission [48]. Copyright 2023, Elsevier.

dimples. These all indicate that the HGS can help to maintain high strength while avoiding the occurrence of intergranular embrittlement, in which the fine recrystallized grains provide good ductility, and the hard deformed grains are expected to hinder crack propagation or nucleation.

Apart from the common HGS caused by grain size differences, grain-boundary segregation is also considered a specific type of HGS. Improving the mechanical or physical properties of alloys by segregating or depleting certain elements on the grain boundaries is called grain-boundary segregation engineering. Li et al. found that increasing the Fe content in the FeCoNiCr quaternary system can promote the segregation of Cr element on the grain boundaries [49]. These chemical heterogeneities at the grain boundaries enhance the cohesive strength of the grain boundaries and hinder the local hydrogen segregation at the grain boundaries, resulting in excellent hydrogen-embrittlement resistance. However, the chemical heterogeneities of grain boundaries are not always beneficial. For example, Ming et al. found that the gradual decrease in ductility of Cantor alloys with increasing temperatures was due to the nano-segregation of Ni, Cr, and Mn elements on the grain boundaries, which reduced the cohesion of the grain boundaries and thus promoted crack propagation along the grain boundaries [50]. They followed this up with the design of the FCC+σ dual-phase structured HEA ($Cr_{26}Mn_{20}Fe_{20}Co_{20}Ni_{14}$), allowing the principal elements that diffuse rapidly to grain boundaries to move into the pre-existed Cr-rich σ phase, thus avoiding the loss of ductility. In short, it is worthy of further exploration to rationally tailor the chemical and microstructural heterogeneities (including intragranular regions and grain boundaries) of the HEAs to achieve optimum mechanical properties.

4.4 FUTURE PROSPECTS

1. Compared to the empirical models and the CALPHAD-based approach, ML modeling exhibits higher capability in guiding the phase selection and accelerating the discovery of new HEAs. However, its accuracy and efficiency are highly correlated to the reliability and size of the training databases. Additionally, the lack of transparency and the poor interpretability of the physical mechanisms of the ML modeling seriously hamper its further applications. Thus, efforts on constructing frameworks that integrate ML modeling with physical-based approaches like DFT, thermodynamic calculations, and experiments are highly needed.

2. Simulators and experiments have shown that chemical SRO widely exists in HEAs and can affect the macroscopic mechanical behavior through the interaction with dislocations. However, there are still many controversies about the influence of SRO on mechanical behaviors, and more experiments are required to verify it.
3. Tailoring disorder in LRO superlattice structures deserves further investigation, where the interaction of dislocations with disordered nanolayers and the deformation mechanisms of the chemically complex ordered structures need to be further clarified.
4. The HGS solves the issue of temperature-dependent brittleness encountered by many structural materials with equiaxed grain structures, especially the intermediate-temperature brittleness. However, apart from good mechanical properties, excellent thermal stability is also required to guarantee the secure and long-term use of heterostructured HEAs in elevated-temperature environments, which still needs more detailed research.

ACKNOWLEDGMENT

The authors from the Harbin Institute of Technology (Shenzhen) thank the Shenzhen Science and Technology Program (Grant No. RCBS20210609103202012, JCYJ20220531095217039) and the financial support from the National Natural Science Foundation of China (Grant No. 52101135). The authors from the City University of Hong Kong greatly acknowledge the financial supports from the Shenzhen Science and Technology Program (Grant No. SGDX20210823104002016).

REFERENCES

[1] J.-W. Yeh, S.-K. Chen, S.-J. Lin, J.-Y. Gan, T.-S. Chin, T.-T. Shun, C.-H. Tsau, S.-Y. Chang, Nanostructured high-entropy alloys with multiple principal elements: Novel alloy design concepts and outcomes, *Adv. Eng. Mater.* 6(5) (2004) 299–303.
[2] D.B. Miracle, O.N. Senkov, A critical review of high entropy alloys and related concepts, *Acta Mater.* 122 (2017) 448–511.
[3] B. Cantor, I.T.H. Chang, P. Knight, A.J.B. Vincent, Microstructural development in equiatomic multicomponent alloys, *Mater. Sci. Eng. A* 375–377 (2004) 213–218.
[4] Z. Li, K.G. Pradeep, Y. Deng, D. Raabe, C.C. Tasan, Metastable high-entropy dual-phase alloys overcome the strength-ductility trade-off, *Nature* 534 (7606) (2016) 227–230.
[5] T. Yang, Y.L. Zhao, Y. Tong, Z.B. Jiao, J. Wei, J.X. Cai, X.D. Han, D. Chen, A. Hu, J.J. Kai, K. Lu, Y. Liu, C.T. Liu, Multicomponent intermetallic nanoparticles and superb mechanical behaviors of complex alloys, *Science* 362(6417) (2018) 933–937.
[6] J. Su, X. Wu, D. Raabe, Z. Li, Deformation-driven bidirectional transformation promotes bulk nanostructure formation in a metastable interstitial high entropy alloy, *Acta Mater.* 167 (2019) 23–39.
[7] T. Yang, Y.L. Zhao, W.P. Li, C.Y. Yu, J.H. Luan, D.Y. Lin, L. Fan, Z.B. Jiao, W.H. Liu, X.J. Liu, J.J. Kai, J.C. Huang, C.T. Liu, Ultrahigh-strength and ductile superlattice alloys with nanoscale disordered interfaces, *Science* 369(6502) (2020) 427–432.
[8] D.J.M. King, S.C. Middleburgh, A.G. McGregor, M.B. Cortie, Predicting the formation and stability of single phase high-entropy alloys, *Acta Mater.* 104 (2016) 172–179.
[9] O.N. Senkov, D.B. Miracle, A new thermodynamic parameter to predict formation of solid solution or intermetallic phases in high entropy alloys, *J. Alloy. Compd.* 658 (2016) 603–607.
[10] A.K. Singh, N. Kumar, A. Dwivedi, A. Subramaniam, A geometrical parameter for the formation of disordered solid solutions in multi-component alloys, *Intermetallics* 53 (2014) 112–119.
[11] Y.F. Ye, Q. Wang, J. Lu, C.T. Liu, Y. Yang, The generalized thermodynamic rule for phase selection in multicomponent alloys, *Intermetallics* 59 (2015) 75–80.
[12] Y. Zhang, Y.J. Zhou, J.P. Lin, G.L. Chen, P.K. Liaw, Solid-Solution Phase Formation Rules for Multi-component Alloys, Adv. *Eng. Mater.* 10(6) (2008) 534–538.
[13] C. Oses, C. Toher, S. Curtarolo, High-entropy ceramics, *Nat. Rev. Mater.* 5(4) (2020) 295–309.
[14] H. Wang, Q.-F. He, Y. Yang, High-entropy intermetallics: From alloy design to structural and functional properties, *Rare Metals* 41(6) (2022) 1989–2001.
[15] C.T. Liu, J.O. Stiegler, Ductile ordered intermetallic alloys, *Science* 226(4675) (1984) 636–642.

[16] F. He, D. Chen, B. Han, Q. Wu, Z. Wang, S. Wei, D. Wei, J. Wang, C.T. Liu, J.-j. Kai, Design of D022 superlattice with superior strengthening effect in high entropy alloys, *Acta Mater.* 167 (2019) 275–286.
[17] J.O. Andersson, T. Helander, L. Höglund, P. Shi, B. Sundman, Thermo-Calc & DICTRA, computational tools for materials science, *Calphad* 26(2) (2002) 273–312.
[18] S.L. Chen, S. Daniel, F. Zhang, Y.A. Chang, X.Y. Yan, F.Y. Xie, R. Schmid-Fetzer, W.A. Oates, The PANDAT software package and its applications, *Calphad* 26(2) (2002) 175–188.
[19] J.Y. He, H. Wang, Y. Wu, X.J. Liu, H.H. Mao, T.G. Nieh, Z.P. Lu, Precipitation behavior and its effects on tensile properties of FeCoNiCr high-entropy alloys, *Intermetallics* 79 (2016) 41–52.
[20] B.E. MacDonald, Z. Fu, X. Wang, Z. Li, W. Chen, Y. Zhou, D. Raabe, J. Schoenung, H. Hahn, E.J. Lavernia, Influence of phase decomposition on mechanical behavior of an equiatomic CoCuFeMnNi high entropy alloy, *Acta Mater.* 181 (2019) 25–35.
[21] F. Zhang, C. Zhang, S.L. Chen, J. Zhu, W.S. Cao, U.R. Kattner, An understanding of high entropy alloys from phase diagram calculations, *Calphad* 45 (2014) 1–10.
[22] Y.L. Zhao, T. Yang, Y. Tong, J. Wang, J.H. Luan, Z.B. Jiao, D. Chen, Y. Yang, A. Hu, C.T. Liu, J.J. Kai, Heterogeneous precipitation behavior and stacking-fault-mediated deformation in a CoCrNi-based medium-entropy alloy, *Acta Mater.* 138 (2017) 72–82.
[23] W. Huang, P. Martin, H.L. Zhuang, Machine-learning phase prediction of high-entropy alloys, *Acta Mater.* 169 (2019) 225–236.
[24] Z. Zhou, Y. Zhou, Q. He, Z. Ding, F. Li, Y. Yang, Machine learning guided appraisal and exploration of phase design for high entropy alloys, *NPJ Comput. Mater.* 5(1) (2019) 128.
[25] Y. Li, W. Guo, Machine-learning model for predicting phase formations of high-entropy alloys, *Phys. Rev. Mater.* 3(9) (2019) 095005.
[26] T. Zheng, X. Hu, F. He, Q. Wu, B. Han, D. Chen, J. Li, Z. Wang, J. Wang, J.-j. Kai, Z. Xia, C.T. Liu, Tailoring nanoprecipitates for ultra-strong high-entropy alloys via machine learning and prestrain aging, *J. Mater. Sci. Technol.* 69 (2021) 156–167.
[27] Z. Rao, P.-Y. Tung, R. Xie, Y. Wei, H. Zhang, A. Ferrari, T.P.C. Klaver, F. Körmann, P.T. Sukumar, A. Kwiatkowski da Silva, Y. Chen, Z. Li, D. Ponge, J. Neugebauer, O. Gutfleisch, S. Bauer, D. Raabe, Machine learning-enabled high-entropy alloy discovery, *Science* 378(6615) (2022) 78–85.
[28] E.P. George, W.A. Curtin, C.C. Tasan, High entropy alloys: A focused review of mechanical properties and deformation mechanisms, *Acta Mater.* 188 (2020) 435–474.
[29] Y. Wu, F. Zhang, X. Yuan, H. Huang, X. Wen, Y. Wang, M. Zhang, H. Wu, X. Liu, H. Wang, S. Jiang, Z. Lu, Short-range ordering and its effects on mechanical properties of high-entropy alloys, *J. Mater. Sci. Technol.* 62 (2021) 214–220.
[30] S. Chen, Z.H. Aitken, S. Pattamatta, Z. Wu, Z.G. Yu, R. Banerjee, D.J. Srolovitz, P.K. Liaw, Y.-W. Zhang, Chemical-affinity disparity and exclusivity drive atomic segregation, short-range ordering, and cluster formation in high-entropy alloys, *Acta Mater.* 206 (2021) 116638.
[31] F.X. Zhang, S. Zhao, K. Jin, H. Xue, G. Velisa, H. Bei, R. Huang, J.Y.P. Ko, D.C. Pagan, J.C. Neuefeind, W.J. Weber, Y. Zhang, Local structure and short-range order in a NiCoCr solid solution alloy, *Phys. Rev. Lett.* 118(20) (2017) 205501.
[32] D. Shindo, A. Gomyo, J. Zuo, J.C.H. Spence, Short-range ordered structure of Ga0.47In0.53 as studied by energy-filtered electron diffraction and HREM, *Microscopy* 45(1) (1996) 99–104.
[33] R. Zhang, S. Zhao, J. Ding, Y. Chong, T. Jia, C. Ophus, M. Asta, R.O. Ritchie, A.M. Minor, Short-range order and its impact on the CrCoNi medium-entropy alloy, *Nature* 581(7808) (2020) 283–287.
[34] X. Chen, Q. Wang, Z. Cheng, M. Zhu, H. Zhou, P. Jiang, L. Zhou, Q. Xue, F. Yuan, J. Zhu, X. Wu, E. Ma, Direct observation of chemical short-range order in a medium-entropy alloy, *Nature* 592(7856) (2021) 712–716.
[35] J. Ding, Q. Yu, M. Asta, R.O. Ritchie, Tunable stacking fault energies by tailoring local chemical order in CrCoNi medium-entropy alloys, Proc. *Natl. Acad. Sci.* 115(36) (2018) 8919–8924.
[36] Q.-J. Li, H. Sheng, E. Ma, Strengthening in multi-principal element alloys with local-chemical-order roughened dislocation pathways, *Nat. Commun.* 10(1) (2019) 3563.
[37] W.-R. Jian, Z. Xie, S. Xu, Y. Su, X. Yao, I.J. Beyerlein, Effects of lattice distortion and chemical short-range order on the mechanisms of deformation in medium entropy alloy CoCrNi, *Acta Mater.* 199 (2020) 352–369.
[38] S. Chen, Z.H. Aitken, S. Pattamatta, Z. Wu, Z.G. Yu, D.J. Srolovitz, P.K. Liaw, Y.-W. Zhang, Short-range ordering alters the dislocation nucleation and propagation in refractory high-entropy alloys, *Mater. Today* 65 (2023) 14–25.
[39] B. Yin, S. Yoshida, N. Tsuji, W.A. Curtin, Yield strength and misfit volumes of NiCoCr and implications for short-range-order, *Nat. Commun.* 11(1) (2020) 2507.

[40] M. Zhang, Q. Yu, C. Frey, F. Walsh, M.I. Payne, P. Kumar, D. Liu, T.M. Pollock, M.D. Asta, R.O. Ritchie, A.M. Minor, Determination of peak ordering in the CrCoNi medium-entropy alloy via nanoindentation, *Acta Mater.* 241 (2022) 118380.

[41] R. Feng, Y. Rao, C. Liu, X. Xie, D. Yu, Y. Chen, M. Ghazisaeidi, T. Ungar, H. Wang, K. An, P.K. Liaw, Enhancing fatigue life by ductile-transformable multicomponent B2 precipitates in a high-entropy alloy, *Nat. Commun.* 12(1) (2021) 3588.

[42] Y. Wang, M. Chen, F. Zhou, E. Ma, High tensile ductility in a nanostructured metal, *Nature* 419(6910) (2002) 912–915.

[43] X. Wu, M. Yang, F. Yuan, G. Wu, Y. Wei, X. Huang, Y. Zhu, Heterogeneous lamella structure unites ultrafine-grain strength with coarse-grain ductility, *Proc. Natl. Acad. Sci.* 112(47) (2015) 14501–14505.

[44] M. Yang, D. Yan, F. Yuan, P. Jiang, E. Ma, X. Wu, Dynamically reinforced heterogeneous grain structure prolongs ductility in a medium-entropy alloy with gigapascal yield strength, *Proc. Natl. Acad. Sci.* 115(28) (2018) 7224–7229.

[45] Z. Zhang, X. Zhai, G. Chen, X. Chen, K. Ameyama, Enhanced synergy of strength-ductility and low-cycle fatigue resistance of high-entropy alloy through harmonic structure design, *Scr. Mater.* 213 (2022) 114591.

[46] M.N. Hasan, Y.F. Liu, X.H. An, J. Gu, M. Song, Y. Cao, Y.S. Li, Y.T. Zhu, X.Z. Liao, Simultaneously enhancing strength and ductility of a high-entropy alloy via gradient hierarchical microstructures, *Int. J. Plasticity* 123 (2019) 178–195.

[47] B.X. Cao, H.J. Kong, L. Fan, J.H. Luan, Z.B. Jiao, J.J. Kai, T. Yang, C.T. Liu, Heterogenous columnar-grained high-entropy alloys produce exceptional resistance to intermediate-temperature intergranular embrittlement, *Scr. Mater.* 194 (2021) 113622.

[48] T.H. Chou, W.P. Li, H.W. Chang, B.X. Cao, J.H. Luan, J.C. Huang, T. Yang, Suppressing temperature-dependent embrittlement in high-strength medium-entropy alloy via hetero-grain/precipitation engineering, *Scr. Mater.* 229 (2023) 115377.

[49] Q. Li, J.W. Mo, S.H. Ma, F.H. Duan, Y.L. Zhao, S.F. Liu, W.H. Liu, S.J. Zhao, C.T. Liu, P.K. Liaw, T. Yang, Defeating hydrogen-induced grain-boundary embrittlement via triggering unusual interfacial segregation in FeCrCoNi-type high-entropy alloys, *Acta Mater.* 241 (2022) 118410.

[50] K. Ming, L. Li, Z. Li, X. Bi, J. Wang, Grain boundary decohesion by nanoclustering Ni and Cr separately in CrMnFeCoNi high-entropy alloys, *Sci. Adv.* 5(12) (2019) eaay0639.

5 Mathematical Modeling for High Entropy Materials

Babak Sokouti

5.1 INTRODUCTION

Metal alloys are characterized by their solid solution or secondary phases formed when they are alloyed with secondary elements, which determine their material properties. High entropy alloys (HEAs) are brittle intermetallic compounds created using a unique alloying procedure that involves introducing several primary elements in the equiatomic or near-equiatomic composition as solid solutions [1].

With HEAs, introduced in 2004, a new paradigm for alloying has emerged. HEAs contain at least five primary elements in near-equimolar concentrations [2,3]. Due to the lack of discovery of new base materials in recent decades, traditional alloys have seen a very limited amount of advancement [4]. New technologies have also led to the limits of traditional alloys. It is, however, very promising to use HEAs for a wide range of applications, including aerospace structural elements at high temperatures, steam turbines, and nuclear energy infrastructure including molten salt reactors, fusion reactors, and cryogenics. Additionally, they show outstanding mechanical properties and high-temperature strength, excellent oxidation resistance, and outstanding corrosion resistance [4].

Scientists have been fascinated by HEAs because of their remarkable mechanical properties. In addition to its excellent mechanical strength, this material is extremely wear-resistant, extremely strain-sensitive, extremely damage-tolerant, and extremely thermally stable. A major challenge has been to develop a HEA with better mechanical properties. Experimental data supports the root mean square atomic displacement (RMSAD), atomic volume differential, and local atomic strain. Several atomic distortion parameters are linearly related to the critical resolved shear stress (CRSS) [5].

According to Yeh and colleagues [6], a relatively new type of alloy has been named "high entropy alloy" because of the equal or nearly equal atomic concentrations of more than five elements. The entropy of their configurations is quite high, HEAs have been recognized as stable single-phase solid solutions, with a crystal structure that is simpler than that found in classical metallurgical systems [6]. A number of researchers have questioned the significance of configurational entropy, and the necessity of equimolar compositions and component numbers, over the past few years. Due to their complex compositions, these alloys may possess unique properties based on the rapidly expanding body of research [7].

Recent years have seen HEAs emerge as one of the most important areas of materials science research due to their seemingly endless potential for countless applications. According to two recent reviews highlighting experiments and theoretical progress, this field must continue to grow by combining theory, experiment, and modeling. Predicting compound composition and attributes has been simplified by many theoretical and computational techniques. There is a particular maturity in the area of HEA with King et al.'s work involving the physical characteristics, phase structure, order/disorder transitions, and segregation of various HEAs, as well as the development of a complete software tool to screen compositions for multicomponent system [8]. HEAs have now become a testing ground for a number of approaches and studies because of the extensive experimental work currently available [9].

DOI: 10.1201/9781003391388-5

Chemical species with low entropy have recently been studied as well as precipitation hardening as an approach to enhancing alloy performance (such as medium entropy alloys and composite alloys). There aren't many stable HEAs with four, five, or more components, according to CALculation of PHAse Diagrams (CALPHAD) surveys of a variety of chemistries [10]. When considering available elements, CALPHAD approaches must use extrapolation to calculate phase stability for more complex equiatomic alloys. As of now, experiments and theory indicate that ternary and quaternary alloys are the most suitable materials for the study of single-phase solid-solution alloys. Complex concentrated alloys (CCAs) have been found to be strengthened by even the slightest deviation from average atomic sizes [11].

In recent years, corrosion has made it difficult to achieve many engineering grand challenges. As technology applications and security develop, novel corrosion-resistant alloys (CRAs) must survive strong corrosion assaults in severe service environments and also have improved inherent corrosion resistance to maintain technical operations safe and secure [12].

Complex nonequimolar HEA corrosion investigations are uncommon since most CRAs are produced using empirical approaches, which are time-consuming, expensive, and incremental. As a result of the multidimensional compositional space within which HEAs are designed and synthesized, design and synthesis are extremely complex problems. Experimental approaches alone cannot handle the complexity. A corrosion-resistant HEA or corrosion-resistant CRA clearly requires an advanced approach. In integrated computational materials engineering (ICME), models are linked across multiple length scales, so that specific performance criteria are met when designing materials. Many alloy properties can be predicted using ICME's suite of tools, which includes Phase Diagrams (CALPHAD) and density functional theory (DFT), provided that accurate databases describing materials' behavior are available. Therefore, ICME is the best way to search for and design CR-HEAs with desired characteristics efficiently and exhaustively. Although alloy composition-based corrosion structures can be developed at a fundamental research stage, they are not yet at a stage where they can be incorporated into ICME. It is feasible to find potential corrosion resistance-HEAs by combining empirical corrosion resistance concepts with mathematical tools based on previous information [12].

If HEAs are used at potential use temperatures, it is unclear if they establish thermodynamically stable solid solutions. In spite of the ease of casting HEAs, evidence suggests long-term annealing can lead to their decomposition into more conventional, multiphase microstructures. Decomposition has been observed to result from long-term annealing at elevated temperatures, raising questions about the suitability of HEA-based components at such temperatures [13]. Understanding the long-term stability of HEA at elevated temperatures is crucial for developing components based on HEA for extreme environments. Molecular dynamics (MD) and DFT can be used to predict the stability of HEA phases. It is, however, fundamentally difficult to use such methods because solid solutions are disordered, temperature changes involve complex physics, and multiple phases compete for stability with HEA. Since it uses thermodynamic calculations of protein folding, CALPHAD is ideal for predicting the long-term stability of HEAs at arbitrary temperatures. Since CALPHAD has a great deal of success in predicting traditional alloy phase equilibrium, it can be easily modeled for HEA compositions. Solution thermodynamic models are used to describe multicomponent complex systems in the CALPHAD approach [13].

Fu et al. emphasize the importance of empirical results over experimental results [14]. Multiscale computational modeling tools are useful for not only predicting the microstructures of different HEAs but also validating the experimental phases of these HEAs before they are applied. Gao and coworkers have discovered many important computational approaches that may supplement experimental phase diagram examination, including CALPHAD modeling, empirical parameter criteria, DFT, and MD simulations [15]. These approaches supplement the observation of experimental phase diagrams. As indicated in the aforementioned literature, these methods have had considerable application with various HEA systems. HEAs, however, can be extremely costly, complex, and sometimes do not perform sufficiently well when using this computational tool.

It is important to use experimental data to verify the reliability of these computational tools when attempting to develop new HEAs. In an effort to simplify the usage of multiscale computational tools, current materials research has concentrated on exploring a number of machine learning (ML) techniques. As a result, they are able to examine the effects of these algorithms on parameters as well as a broader variety of alloy compositions [16] and their effects on parameters.

Several reports have focused on high-speed experiments, high-speed calculations, and material database development in order to develop new materials more rapidly and cost-effectively using materials genome engineering (MGE) [17]. Materials informatics provides insights into the underlying physics, guides experiments, and generates new data by analyzing and predicting new materials' properties. Materials discovery has been accelerated by this field as a result of the need to accelerate it. For example, ML models can be used to predict crystal structures, physical properties, and thermodynamic stability, and optimization algorithms are used by designing new experiments and calculations in vast search spaces to find global optima efficiently [18].

A ML approach typically builds a surrogate model (*f*) based on a set of material descriptors (X) in order to determine the relationship between the target property (Y) and the descriptors. A robust ML prediction is essential to designing new materials based on material characteristics (X) and models (*f*). As a result, *f* and X should be combined to identify the optimal combination. It is possible to combine all of the models $p \times (2n-1)$, where p denotes the number of materials descriptors, p refers to the total number of mathematical models, whereas n represents the total number of permutations. There have been many developments in materials science that have used kernels, trees, neural networks, etc. These models can be used to obtain large numbers of *p*, providing a solution to materials science problems. By introducing a wide range of materials descriptors, a large pool of n can be obtained. This provides an explanation for the influence of thermodynamics and kinetics on material properties, including composition, microstructure, processing, etc. [19].

It is difficult to understand the physical mechanisms that govern HEAs (e.g., dilute alloys) in comparison to traditional metal alloys. The mechanical properties of HEAs have been explained by large lattice distortions and slow diffusion. Experimental literature rarely describes lattice structure, defect structures, or deformation mechanisms at the atomic level in detail. Considering the structure-property hypothesis, it is possible to enhance HEAs' physical characteristics, in particular their plastic deformation processes, by learning more about their internal structures. In order to resolve defects at an atomistic level, first-principles simulations and MD can be used to obtain this information. State-of-the-art techniques have not been reviewed or assessed comprehensively. A comprehensive review and systematic assessment of the results of atomistic studies have been conducted on HEAs. By using existing atomic simulations, we can understand the mechanical behavior of existing HEAs, which we can then use to design new ones. Due to the ongoing development of novel simulation methodologies, atomistic simulations of HEAs will inevitably be used for much broader applications in the future. This chapter's objective is to offer a general summary of the many mathematical simulation approaches that are currently in use. By discussing the current challenges associated with such simulations, ways of explaining HEAs' unparalleled mechanical properties were proposed.

5.2 MATHEMATICAL MODELING

A mathematical model is an idealized scientific model, one that uses mathematics to describe an object of study as well as to conduct research using those methods. It is called the mathematical model when the object is described in mathematical terms. Mathematical models are mathematical operators A that map inputs X of models to their outputs Y. They are mathematical operators A with their inputs X becoming the outputs Y of the model. Mathematics modeling techniques are generally classified as follows.

5.2.1 Atomistic Modeling

According to Debais et al. [9], a study using a modeling environment analyzed the fine details of the atomic distribution in AlNbTaTiZr and VNbTaTiZr alloys with high entropy. The two solid solutions form specific domains (regions where different elements mix together) in atomic modeling because they evolve differently. For alloys with a specific concentration of the fifth element (Al or V), the Bozzolo–Ferrante–Smith (BFS) method can determine the entropy using a multicomponent system [20]. This algorithm can eliminate or minimize the need for extensive experimental work in determining the concentrations of various elements for high entropy alloys.

Pasianot et al. tested a new model of dislocations using quinary model interatomic potentials [21]. The results indicate that the potential model can be used to simulate quinary random alloys with a high degree of entropy. Using atomistic simulations of a face-centered cubic (FCC) alloy with a high degree of entropy, dissociated ½⟨1 1 0⟩ dislocations were investigated. To simulate an equiatomic alloy with random dislocations, researchers use embedded atom method (EAM) potentials and an average atom potential. Stacking fault energies and stable and unstable fault energies depend locally on the alloy composition, as do stacking fault energies and stacking fault energies for complex alloys. Edge dislocations are more prone to this effect than screw dislocations. Peierls stresses for dislocations in alloys are significantly higher than those calculated by using an average atom potential or pure components.

The Peierls–Nabarro framework was used to develop an atomistic model of bilayer HEAs that incorporates atomic level randomness of interlayer dislocations [22]. The atomistic model was transformed into a continuum model by the use of asymptotic analysis and the limit theorem. To simulate the nonlinear interactions that occur between layers, the stochastic misfit energy is used. This energy takes into account the stochastic elastic energy that exists in both layers. A stochastic generalization of the classical Peierls–Nabarro model may be seen as an explanation of how the derived continuum model corresponds to deterministic Peierls–Nabarro models. This can be done by seeing the model as a stochastic generalization.

The analysis of HEAs including five elements (Mo, Ta, V, W, and Zr) necessitated a technique for large atomistic simulations; therefore, the researchers turned to the BFS approach for energetics to complement the experimental findings [23]. The calculated lattice parameter and measured density of the body-centered cubic (BCC) solid solution agree well. Because of the limitations imposed by the calculations, the results of the calculations cannot be directly compared with those of the experiments. The examination of the annealed sample, together with the computational cell properties, gives a more complete picture of atomic behavior than the experimental description.

5.2.2 CALPHAD

Kivy experimentally and computationally investigated phase equilibria and $Al_2FeCoNiCu$ HEA microstructures [16]. The phase diagram calculation (CALPHAD) and experimental analysis showed that the body-centered cubic phase precipitated first, followed by the FCC phase precipitating as a cross-dendritic phase at lower temperatures from the BCC phase. Additionally, interdendritic regions were observed to contain highly directional Widmanstastten-like precipitates. As a result, it can be concluded that equiatomic AlFeCoNiCu alloys with similar microstructures with higher Al enable Widmanstätten-like precipitates to form. By removing Cr from $Al_2FeCoNiCu$ HEA, its mechanical and chemical properties will be improved, since Cr is hard (brittle) and corrosion-resistant.

For the purpose of obtaining isopleth phase diagrams for the $Co_xCr_xFe_xNi_xTi_{2x}$ systems [24], researchers developed CALPHAD formulae. All systems were analyzed for temperature and composition variations. By using equiatomic alloys for the analysis, the thermodynamic stability of single-phase FCC solid solutions were explained. There are extensive sections in the phase diagrams of Ni, Co, and Fe where single-phase FCC structures may be stabilized, demonstrating that these molecules have a tendency to stabilize FCC phase structures. Despite its reputation as a bcc stabilizer, Cr is really somewhat FCC-rich. The Gibbs phase rule is a universally observed property of phase diagrams, and it has been shown that BCC, FCC, and ETA (η) phases all exhibit dominating three-phase areas.

The ordered B2 phase was isolated from the Fe/Cr-rich BCC solid solution using Al- and Ti-rich alloys [25]. Grain/dendrite borders are often where the Fe/Cr-rich BCC solid solution phase may be found in the ordered B2 phase, which is richer in Al, Co, Ni, and Ti. Fe/Cr-rich BCC crystalline phases are more likely to occur when Fe and Cr are omitted from ordered B2. A Gibbs free energy model that takes structural ordering and thermal influences into account is excellently matched with CALPHAD predictions. By constructing vertical sections of multicomponent temperature and composition, the model can be used to produce phase boundaries from BCC+B2 to BCC to BCC. CALPHAD predictions show the model's usefulness for developing multicomponent, multiphase alloys with good agreement with CALPHAD predictions [25].

The corrosion behavior and composition of a series of single-phase, processable CR-HEAs have been systematically modified by the application of the ICME method to their design [12]. Therefore, localized corrosion susceptibility can be evaluated in the lab, and different compositional choices were made from the original HEA (which included Mo, W, and Ru) to preserve the single-phase ETA structure. CALPHAD's predictions of homogeneous single-phase microstructures in all four nonequimolar HEA compositions were confirmed, as were the anticipated trends in corrosion behavior when Cr content was varied.

There has been considerable interest in HEAs containing vanadium (V), which will exhibit large hardening effects in solid solutions. Due to its importance, sigma phases have been evaluated primarily for predicting mechanical properties in V-added HEAs [26]. The prediction of V-added HEA systems still requires a strong thermodynamic database, even if phase structures are predicted by CALPHAD. Understanding the thermodynamic stability of the sigma phase is necessary for Cr–Fe–Ni–V HEA. System models with multiple sublattices of Cr–NiV and Fe–NiV can be modeled using a parameterization method, thus simplifying the extension to higher-order systems. In order to verify the validity of the thermodynamic description developed, several CrFe–Ni–V HEAs have been designed, fabricated, and analyzed.

A mixture of such compositions has shown extraordinarily good corrosion resistance, and it may be selected using CALPHAD's combination of computational techniques and empirical information [12]. Even complex and nonequimolar corrosion-resistant HEAs were successfully validated in the simulations by ICME, even if they were complex and nonequimolar.

The CALPHAD method cannot be used before the genetic algorithm is run in another approach to genetic algorithm-based optimization [27]. After the optimization routine was completed, the CALPHAD method was combined with the four databases in order to perform an additional a posteriori verification (in order to determine whether a single disordered solid solution would be predicted to form at a given temperature) and adjust the alloy's temperature during heat treatment. By comparing CALPHAD predictions with observations of long-term heat treatment systems, phase equilibria of HEA systems can be accurately predicted [13]. To eliminate metastable inhomogeneities reported previously through multiple FCC phases, it is necessary to compare CALPHAD predictions and experimental references. To determine whether homogenization is needed, a Scheil simulation can be used to predict the segregation process during solidification. In addition, providing information on appropriate phase equilibrium in HEAs can improve the quantitative accuracy of CALPHAD databases. The design of alloys requires accurate CALPHAD databases, and the optimization of alloy parameters within HEA composition spaces requires these computational tools as well. Based on CALPHAD modeling with the TCNI8 database, an as-cast MoNbTaTiV HEA was fabricated by arc-melting with only one BCC phase [28]. To characterize the material's mechanical properties, an XRD analysis as well as an SEM analysis were conducted. The compression and

Vickers hardness of the material was tested. The DFT approach was used in order to arrive at a result that is both fair and accurate about the elastic characteristics of the MoNbTaTiV alloy and the constituents that comprise it.

5.2.3 Cluster-Plus-Glue-Atom Model

A novel structural model based on clusters and glue atoms has recently emerged [29]. The local order of the distribution of the solute atoms in a solid solution alloy may be characterized by the "cluster-plus-connected-atom" structural model, which accounts for both the structural unit and atomic occupancy of the solvent and solute atoms [30]. Therefore, HEA alloying will be used instead of conventional alloying, saving both time and effort, and the bridge design will be greatly improved. High-strength HEA coatings on 904L stainless steel were successfully fabricated by Ding and coworkers using this model [31]. Also, rutile-structured multicomponent oxides (Me=Ga, Fe, Cr) were successfully designed and synthesized based on the existing Al–Cr–Nb–Ta–Ti–O high entropy materials [32].

It is more likely that cluster formation will be improved if the cluster center replaces the target CCO atom with an atom with a similar structure, but glue atoms can come from any atom, regardless of its structure. The cationic radius can be calculated for components that fit the cluster-plus-glue-atom model so that a single-phase CCO can be formed. A CCO composition can also be modified and different elements can be added to control the lattice variation. In addition to using the design method to design compounds with compositional complexity such as carbides, nitrides, silicides, phosphides, fluorides, etc., it is also capable of taking into account crystal structure, electronic data, and geometrical information [33]. A cluster-plus-glue-atoms model has been used to analyze HEAs produced by Al-TMs with BCC and FCC structures. Due to its slower softening process, Al–$Ti_3V_3Nb_4Mo_4$Al alloy has greater strength at room temperature and greater plastic deformation than other alloys. Ta replaced some Mo in [Al–$Ti_3V_3Nb_4Mo_3$Ta]Al, resulting in a dramatic reduction in strength at room temperature and high temperature. The degradation of Al-TM properties may be caused by Ta. Additional investigation was also conducted on high entropy oxides (HEOs) with successful outcomes [34,35].

5.2.4 ML and MD Simulations

Using semi-empirical CALPHAD calculations or ML augmented by CALPHAD calculations is one way of unraveling the thermodynamics of eutectic high entropy alloys (EHEAs), which are the most robust methods for predicting eutectic composition. The pseudo-binary phase diagram of HEAs is commonly used in CALPHAD calculations to pinpoint their composition, which is heavily impacted by the element grouping strategy and the reliability of thermodynamic databases. As HEA compositions are typically eutectic plateaus, calculating EHEA compositions is time-consuming due to the large composition space [36].

Ruestes et al. looked at an equiatomic quinary FeNiCrCoCu alloy at both ambient temperature and cryogenic temperatures (77 K) [37]. By comparing the HEA alloy with standard FCC substrates made of nickel and average atom FCC crystals, it may better be understood what sets the HEA alloy different. Their results show that random composition has a significant effect on the plasticity, deformation processes, and microstructure of multicomponent alloys.

Mechanical behavior and microstructure development during indentation were shown to be distinct for FeCoCrNiCu HEAs, as predicted by a microstructure-based constitutive model and verified by MD simulations [38]. FeCoCrNiCu experiences intense work hardening as a result of dislocations interacting with the solution. FeCoCrNiCu exhibits good plasticity due to abundant slip systems, resulting in good work hardening. When it comes to the plastic deformation of FeCoCrNiCu, dislocation loop motion is a key factor. FeCoCrNiCu HEAs are very strong and robust because of the lattice distortion strengthening and the greater number of activable slip systems compared to Cu.

The deformation behavior of a quinary alloy with high entropy was studied in simulation [38]. Using empirical interatomic potentials, a massively parallel MD simulation was conducted at the atomistic level. Due to grain boundaries, random alloys have a harder time gliding dislocations emitted from grain boundaries than average atom alloys. As a result, the complex alloy is stronger and less plastic-deformable.

The MD model was created to replicate the mechanical and tribological characteristics of both HEA-coated metal and pure aluminum [39]. Nanoindentation and tribological characteristics of AlCoCrFes generated by additive methods on aluminum substrates were analyzed. During the additive manufacturing process of HEA coating, it seems that more Al is melted in the substrate to react with other elements in the coating layer as the laser heating temperature rises. The resulting HEA coating and Al substrate contact area agree qualitatively with tests reported in the literature.

The researchers studied the near-surface composition profile of an equimolar alloy composed of CoCrFeMnNi using two approaches [40]. Computer simulations of atoms are carried out using Monte Carlo, MD, and molecular statics techniques, and analytical calculations are carried out using a multilayer model and nearest neighbor calculations. Based on the results obtained, the two approaches appear to be qualitatively similar. Using an analytical model on a laptop is faster than using a computer simulation that requires 100 parallel processors. The atomic-scale picture provided by computer simulations may imply massive computation, in order to optimize the composition of a multicomponent alloy, despite the presumably more reliable results. Using the analytical model can be an effective method of narrowing the composition space in complex multinary systems, as with the CALPHAD method for rapid investigation of phase diagrams.

A HEA must be designed with phases that are thermally stable and selected carefully. A study was conducted to determine phase formation in cast AlFeCoNiCu HEAs [41]. To design HEAs, phase selection and thermal stability must be carefully considered. The MD simulation of AlFeCoNiCu at different temperatures was conducted using *ab-initio* molecular dynamics (AIMD). To determine the crystal structure of phases at various temperatures, an AIMD simulation of AlFeCoNiCu was performed with an equiatomic composition.

The thermodynamic behavior of interstitial oxygen was investigated beyond dilute limits in Mo–Nb–Ta–W alloys with high entropy. Within an isothermal–isobaric ensemble [4], Monte Carlo simulations were conducted on HEA and HEA–oxygen configurations. The interstitial oxygen at tetrahedral and octahedral sites is comparable in stability. A correlation has been found between the metallic environment composition and oxygen stability in interstitial environments. A decrease in solubility of the interstitial oxygen in the alloy with an increase in temperature was also observed as oxygen clusters formed at the O-sites and bulk moduli increased with an increase in oxygen content.

5.3 CONCLUSIONS

In this chapter, we have covered the topic modeling of nanoscale plasticity in HEAs and atomistic simulations of HEAs. We concluded by reviewing HEA atomistic simulation studies, including those areas where atomistic simulation could provide insight into the physical mechanisms underlying their mechanical properties. HEAs are difficult to model accurately using DFT due to length scale limitations. The chemical disorder can be statistically averaged using first-principles methods (mean-field coherent potential approximation (CPA) and supercell techniques) but has both advantages and disadvantages. In addition to the local lattice distortions associated with mean-field applications, HEAs cannot also capture the local heterogeneity associated with supercells. The local heterogeneity is not captured effectively by mean-field methods that are computationally efficient. As MD simulations of HEAs continue to be a challenging issue due to the lack of accurate interatomic potentials, techniques that can address this computational cost will be beneficial in the future. HEAs can be analyzed using existing DFT techniques to calculate the vacancies' formation and migration energies. In HEAs, Suzuki locking is a noteworthy issue concerning point defect diffusion. Interstitials and interstitial diffusion have also been calculated using these techniques.

This mechanism has been suggested to strengthen dislocations in HEAs. Atomistic simulations may shed light on the possible role that strain fields of dislocations play in bolstering point defects in HEAs by modeling their interactions with point defects. In both FCC and BCC HEAs, atomistic dislocation configurations may be obtained by the use of atomistic modeling methods. Additionally, MD simulations of dislocation shearing have been performed to capture the variability in dislocation dissociation distances reported experimentally.

Dislocation processes, like cross-slip, may be simulated using MD, and the technique can also be used to analyze equilibrium dislocation structures and Peierls stresses. While transmission electron microscopy (TEM) evidence indicates that dislocation structures are highly planar post-deformation, MD simulations of deformed HEAs can demonstrate spontaneous cross-slip at 300 K. Atomic simulations can help clarify these findings. It will also be useful to understand dislocation–dislocation interactions in deformed HEAs to identify dislocation structures and strengthening mechanisms. Through simulations of atomistic defects like stacking faults, twin boundaries, grain boundaries, and phase boundaries, planar defects can be understood more thoroughly. It has not been widely studied via atomistic simulations how grain and phase boundaries work, but simulating them offers several opportunities to investigate their structure and dynamics. In HEA applications that involve point and line defects, boundary interactions are particularly important. In the future, synergizing experiments with atomistic simulations will also prove very beneficial. And, experimental studies of HEAs at the atomic level will provide valuable insights. Simulations using atomistic models will be stimulated as crystallographic information about active dislocations and twin systems becomes available. Because the system contains atomic heterogeneity, it can be studied using atomistic simulations. There are few insights into the atomistic nature of HEA plasticity, despite experimental findings that describe it fairly accurately. During the next few years, developments in computational techniques will allow atomistic simulations to reveal more of these details. Simulations can certainly give us a better understanding of HEA materials and lead us to design new HEAs based on these simulations.

REFERENCES

1. Aitken ZH, Sorkin V, Zhang Y-W. Atomistic modeling of nanoscale plasticity in high-entropy alloys. *Journal of Materials Research* 2019; 34: 1509–1532.
2. Ikeda Y, Grabowski B, Körmann F. Ab initio phase stabilities and mechanical properties of multicomponent alloys: A comprehensive review for high entropy alloys and compositionally complex alloys. *Materials Characterization* 2019; 147: 464–511.
3. Miracle DB, Senkov ON. A critical review of high entropy alloys and related concepts. *Acta Materialia* 2017; 122: 448–511.
4. Samin AJ. A computational investigation of the interstitial oxidation thermodynamics of a Mo-Nb-Ta-W high entropy alloy beyond the dilute regime. *Journal of Applied Physics* 2020; 128: 215101.
5. Ali ML, Shinzato S, Wang V, Shen Z, Du J-p, Ogata S. An atomistic modeling study of the relationship between critical resolved shear stress and atomic structure distortion in FCC high entropy alloys-relationship in random solid solution and chemical-short-range-order alloys-. *Materials Transactions* 2020; 61: 605–609.
6. Yeh J-W, Chen S-K, Lin S-J, Gan J-Y, Chin T-S, Shun T-T, Tsau C-H, Chang S-Y. Nanostructured high-entropy alloys with multiple principal elements: Novel alloy design concepts and outcomes. *Advanced Engineering Materials* 2004; 6: 299–303.
7. Smith TM, Hooshmand MS, Esser BD, Otto F, McComb DW, George EP, Ghazisaeidi M, Mills MJ. Atomic-scale characterization and modeling of 60 dislocations in a high-entropy alloy. *Acta Materialia* 2016; 110: 352–363.
8. King DJM, Middleburgh SC, McGregor AG, Cortie MB. Predicting the formation and stability of single phase high-entropy alloys. *Acta Materialia* 2016; 104: 172–179.
9. Debais G, Mosca HO, Bozzolo G. Atomistic modeling of AlNbTaTiZr and VNbTaTiZr high entropy alloys. *Computational Materials Science* 2020; 177: 109557.
10. Miracle DB, Miller JD, Senkov ON, Woodward C, Uchic MD, Tiley J. Exploration and development of high entropy alloys for structural applications. *Entropy* 2014; 16: 494–525.

11. Antillon E, Woodward C, Rao S, Akdim B, Parthasarathy T. Chemical short range order strengthening in a model FCC high entropy alloy. *Acta Materialia* 2020; 190: 29–42.
12. Lu P, Saal JE, Olson GB, Li T, Swanson OJ, Frankel G, Gerard AY, Quiambao KF, Scully JR. harsh environments. *Scripta Materialia* 2018; 153: 19–22.
13. Saal JE, Berglund IS, Sebastian JT, Liaw PK, Olson GB. Equilibrium high entropy alloy phase stability from experiments and thermodynamic modeling. *Scripta Materialia* 2018; 146: 5–8.
14. Fu X, Schuh CA, Olivetti EA. Materials selection considerations for high entropy alloys. *Scripta Materialia* 2017; 138: 145–150.
15. Gao MC. Design of high-entropy alloys. *High-entropy alloys: Fundamentals and applications* 2016, 369–398.
16. Kivy MB, Mahata AK, Thompson R, Palominos Jr J, Kestenbaum M, Hunter L. Experimental and computational study of microstructure of Al2FeCoNiCu high-entropy alloy. *Journal of Phase Equilibria and Diffusion* 2023, 1–10.
17. Rickman JM, Lookman T, Kalinin SV. Materials informatics: From the atomic-level to the continuum. *Acta Materialia* 2019; 168: 473–510.
18. Zhang Y, Wen C, Wang C, Antonov S, Xue D, Bai Y, Su Y. Phase prediction in high entropy alloys with a rational selection of materials descriptors and machine learning models. *Acta Materialia* 2020; 185: 528–539.
19. Ouyang R, Curtarolo S, Ahmetcik E, Scheffler M, Ghiringhelli LM. SISSO: A compressed-sensing method for identifying the best low-dimensional descriptor in an immensity of offered candidates. *Physical Review Materials* 2018; 2: 083802.
20. Bozzolo G, Ferrante J, Smith JR. Method for calculating alloy energetics. *Physical Review B* 1992; 45: 493–496.
21. Pasianot R, Farkas D. Atomistic modeling of dislocations in a random quinary high-entropy alloy. *Computational Materials Science* 2020; 173: 109366.
22. Jiang T, Xiang Y, Zhang L. Stochastic Peierls–Nabarro model for dislocations in high entropy alloys. *SIAM Journal on Applied Mathematics* 2020; 80: 2496–2517.
23. Anzorena MS, Bertolo A, Gagetti L, Kreiner AJ, Mosca HO, Bozzolo G, del Grosso MF. Characterization and modeling of a MoTaVWZr high entropy alloy. *Materials & Design* 2016; 111: 382–388.
24. Anis G, Attallah MM, Youssef M, Salem H. Temperature-dependent enthalpy and entropy stabilization of solid solution phases in non-equiatomic CoCrFeNiTi high entropy alloys: Computational phase diagrams and thermodynamics. *Modelling and Simulation in Materials Science and Engineering* 2022; 30: 045013.
25. Conway PL, Golay D, Bassman L, Ferry M, Laws KJ. Thermodynamic modelling to predict phase stability in BCC+ B2 Al-Ti-Co-Ni-Fe-Cr high entropy alloys. *Materials Chemistry and Physics* 2022; 276: 125395.
26. Choi W-M, Jo YH, Kim DG, Sohn SS, Lee S, Lee B-J. A thermodynamic modelling of the stability of sigma phase in the Cr-Fe-Ni-V high-entropy alloy system. *Journal of Phase Equilibria and Diffusion* 2018; 39: 694–701.
27. Menou E, Toda-Caraballo I, Rivera-Díaz-del PEJ, Pineau C, Bertrand E, Ramstein G, Tancret F. Evolutionary design of strong and stable high entropy alloys using multi-objective optimisation based on physical models, statistics and thermodynamics. *Materials & Design* 2018; 143: 185–195.
28. Yao H, Qiao J, Hawk J, Zhou H, Chen M, Gao M. Mechanical properties of refractory high-entropy alloys: Experiments and modeling. *Journal of Alloys and Compounds* 2017; 696: 1139–1150.
29. Chuan-Pu H, Qing W, Ren-Tao M, Ying-Min W, Jian-Bing Q, Dong C. Cluster-plus-glue-atom model in bcc solid solution alloys. *Acta Physica Sinica* 2011; 60: 116101.
30. Pang C, Jiang B, Shi Y, Wang Q, Dong C. Cluster-plus-glue-atom model and universal composition formulas [cluster](glue atom) x for BCC solid solution alloys. *Journal of Alloys and Compounds* 2015; 652: 63–69.
31. Ding K, Wang Z, Wu X, Shang X, Liu Q. A novel compositional design of [Cr-Fe4Co4Ni4] Cr2. 6− x Ti x Mo0. 4 high-entropy alloy coating based on the cluster-plus-glue-atom model and its mechanism of resisting strong acid erosion. *Advanced Engineering Materials* 2023 25: 2201483.
32. Yu Y, Zhang S, Liu S, Ren Y, Xia Z, Wang N, Wang H, Jiang W, Liu C, Ding W. Cluster-model-based design and synthesis for rutile-type Ti-Sn-Nb-Ta-Me-O (Me= Ga, Fe, Cr) high/medium entropy oxides. *Journal of Alloys and Compounds* 2022; 926: 166771.
33. Liu T-y, Shi K, Zhao J, Liu S-b, Liu H-y, Liu B-l, Mei X-m, Ning Z-s, Chen X-m. Composition formula of Al-TMs high-entropy alloys derived by cluster-plus-glue-atoms model and its experimental verification. *China Foundry* 2022; 19: 544–550.

34. Guo Y, Li X, Liu Q. A novel biomedical high-entropy alloy and its laser-clad coating designed by a cluster-plus-glue-atom model. *Materials & Design* 2020; 196: 109085.
35. Spiridigliozzi L, Ferone C, Cioffi R, Dell'Agli G. Compositional design of single-phase rare-earth based high-entropy oxides (HEOs) by using the cluster-plus-glue atom model. *Ceramics International* 2023; 49: 7662–7669.
36. Bai K, Ng CK, Lin M, Wei F, Li S, Teo SL, Tan DCC, Wang P, Wuu D, Lee JJ. Short-range ordering heredity in eutectic high entropy alloys: A new model based on pseudo-ternary eutectics. *Acta Materialia* 2023; 243: 118512.
37. Ruestes CJ, Farkas D. Dislocation emission and propagation under a nano-indenter in a model high entropy alloy. *Computational Materials Science* 2022; 205: 111218.
38. Luo G, Li L, Fang Q, Li J, Tian Y, Liu Y, Liu B, Peng J, Liaw P. Microstructural evolution and mechanical properties of FeCoCrNiCu high entropy alloys: A microstructure-based constitutive model and a molecular dynamics simulation study. *Applied Mathematics and Mechanics* 2021; 42: 1109–1122.
39. Yang X, Zhang J, Sagar S, Dube T, Kim B-G, Jung Y-G, Koo DD, Jones A, Zhang J. Molecular dynamics modeling of mechanical and tribological properties of additively manufactured AlCoCrFe high entropy alloy coating on aluminum substrate. *Materials Chemistry and Physics* 2021; 263: 124341.
40. Chatain D, Wynblatt P. Surface segregation in multicomponent high entropy alloys: Atomistic simulations versus a multilayer analytical model. *Computational Materials Science* 2021; 187: 110101.
41. Kivy MB, Zaeem MA, Lekakh S. Investigating phase formations in cast AlFeCoNiCu high entropy alloys by combination of computational modeling and experiments. *Materials & Design* 2017; 127: 224–232.

6 Characterizations of High Entropy Materials

Yixuan Hu, Yumeng Zhang, Xiaodong Wang, and Kolan Madhav Reddy

6.1 INTRODUCTION

Since the birth of high entropy alloy (HEA) in 2004, the field has been widely explored, leading to a large family of high entropy material (HEM), including alloys, oxides, carbides, borides, and others. With the continuous exploration of the vast composition space, HEM has demonstrated highly enhanced and unexpected properties, which continue to attract the attention of materials science. The understanding of the unique properties of HEM relies on various advanced characterization techniques, which helps the researchers gain insights into the intrinsic composition-structure-property relationship. In this chapter, we will mainly focus on the characterization of HEM and the unique features revealed by various characterization methods that distinguish HEM from other materials.

6.2 STRUCTURAL CHARACTERIZATION

The key idea behind the formation thermodynamics in HEM is to achieve a high degree of disorder and mixing entropy, which can compensate for the enthalpy of mixing and the elastic strain energy associated with the formation of a solid solution. To achieve this, all elements are usually distributed in equimolar or near-equimolar ratios on the same lattice sites to reach a maximum mixing entropy, and fabrication is typically conducted at elevated temperatures to thermodynamically increase the lowered Gibbs free energy by the entropy term. With the different properties of constituent elements, the resulting HEM could possess different structural information such as crystal structure and single solution formability. This structural information can be tracked by techniques such as X-ray diffraction and state-of-the-art transmission electron microscopy.

6.2.1 X-RAY DIFFRACTION

X-ray diffraction (XRD) is an exceptionally powerful tool for the structural characterization of materials. Based on Bragg's law, the X-rays will only be diffracted by the crystal planes when the incident beam strikes the crystal at a specific angle, resulting in a diffraction pattern. It is a non-destructive technique that provides valuable information on the crystal structure, lattice parameters, and phase composition of materials. In the case of HEM, XRD proves to be an indispensable tool that can identify the formation of solid solutions and potential crystallographic phase transitions that may occur because of changes in temperature or composition. Rost et al. [1] showed an entropy-stabilized phenomenon in a (MgNiZnCuCo)O system (Figure 6.1a), where they heated the equimolar mixture of MgO, NiO, ZnO, CuO, and CoO at different temperatures and found that the feature peaks representing the impurity phases fade with increasing temperature. Moreover, the entropy-stabilized feature is revealed by bringing the single rocksalt structure high entropy oxides (HEO) reversely equilibrated at 1000°C and 750°C, the transition between single-phase and multiphase system is found to occur reversibly. This leads to a fabrication concept of HEM that the

 DOI: 10.1201/9781003391388-6

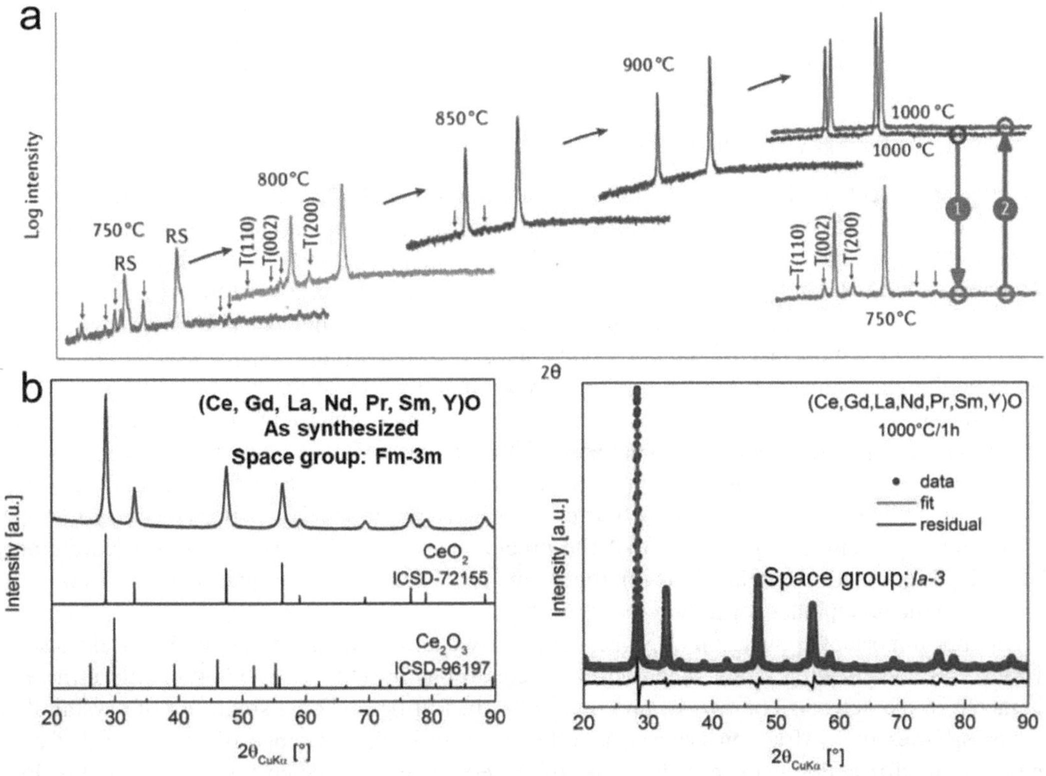

FIGURE 6.1 (**a**) XRD patterns of heating equimolar mixture of MgO, CoO, NiO, CuO and ZnO from 750°C to 1000°C and back down to 750°C. Peaks indexed with (T) and with (RS) correspond to tenorite and rocksalt phases, respectively. Adapted with permission [1]. Copyright The Authors, some rights reserved; exclusive licensee Springer Nature. Distributed under a Creative Commons Attribution License 4.0 (CC BY). (**b**) XRD profile of (Ce, Gd, La, Nd, Pr, Sm, Y)O in as-synthesized state and after 1000°C annealing for 1h. The space group of the sample transformed from Fm-3m to Ia-3. Adapted with permission [4]. Copyright The Authors, some rights reserved; exclusive licensee Taylor & Francis. Distributed under a Creative Commons Attribution License 4.0 (CC BY).

single-phase HEM forms at elevated temperatures due to high entropy, the same structure can be retained with quenching and remain stable at room temperature due to the slow kinetics [2].

However, as HEMs are a newly developed family of materials, they cannot be precisely matched with existing Powder Diffraction Files in the database. Therefore, the Rietveld refinement method [3] is commonly used to accurately extract important structural information, such as lattice parameters, crystal size, and atomic occupation information. This method is based on the least-squares fitting of a theoretical model to the experimental XRD data, allowing for the determination of the crystal structure with high precision and accuracy.

Many researchers have utilized the Rietveld refinement method to determine the crystal structure and lattice parameters, particularly when distinguishing structures with similar symmetry. For example, Djenadic et al. [4] and Sarkar [5] demonstrated a symmetry degradation with annealing at 1000°C in rare earth HEO, as shown in Figure 6.1b. As structures with similar symmetry could yield similar diffractions, it is difficult to determine the symmetry from an intuitive analysis of the XRD pattern. With Rietveld refinement of the experimental XRD patterns based on input structures with different possible symmetries, they could select the most probable structure with the lowest Rietveld weighted profile (R_{wp}) value. Assisted by this method, they demonstrated that the HEOs synthesized with Fm-3m space group will degrade into Ia-3 space group after annealing at 1000°C for 1 hour.

6.2.2 Selected Area Electron Diffraction

Selected area electron diffraction (SAED) is a diffraction technique based on Bragg's law, like XRD but uses electrons instead of X-rays as the diffraction source. The used incident electron beam in SAED is highly focused, which renders the advantage of SAED for analysis at micro-scale. Moreover, SAED is complemented within the transmission electron microscopy (TEM) imaging, indicating a clearer structure-morphology correlation could be revealed. Lv et al. [6] utilized SAED to demonstrate the existence of stacking faults and nano twins (Figure 6.2a), which dominate the strengthening mechanism in $(CuZnMnNi)_{100-x}Sn_x$ high entropy brass alloy. Besides using SAED to determine the crystal structure of single crystals, one of the essential applications of SAED is to determine the orientation relationship between different phases. In Figure 6.2b and c, Liu et al. [12] showed the morphology of round and lath-shaped precipitates and the SAED pattern of the coexistence of precipitates and matrix. Furthermore, from analyzing the SAED pattern of the lath-shaped precipitates and matrix, it is known that the precipitates with FCC structure share Kurdjumov–Sachs (K–S) orientation relationship with the B2 matrix. The deduced orientation relationship helps understand the precipitates formation mechanism.

Another important application of SAED is directly characterizing defects in materials, such as dislocations and stacking faults. The two-beam conditions in SAED can be efficiently utilized to characterize the defects. The two-beam conditions are fulfilled when a single set of crystal plane groups satisfies the diffraction criteria, ideally resulting in the presence of only two diffracted beams in the diffraction pattern. When one diffracted beam is collected for imaging, the dislocations would be invisible or visible under central Dark-field Imaging and Weak-beam Dark-field Imaging, respectively. Therefore, the dislocation contrast would be clear, providing abundant microscopic information behind the macroscopic strengthening mechanism. Zhang et al. [7] used the two-beam condition to characterize the distribution of dislocations (Figure 6.2d and e) under different heat treatments. They found the randomly distributed dislocations in a water-quenched sample while a more regular distribution for the sample aged at 1000°C.

Due to the intrinsic complexity of HEM, the chemical short-range order (SRO) may occur. It may strongly affect the mechanical properties of HEM, which has been proved by various molecular dynamic simulation studies [8]. Proving the existence of SRO experimentally can be a challenging task. Several attempts have been made to determine the SRO from SAED patterns [7,9]. Zhang et al. [7] demonstrated that the SRO changes the SAED patterns (Figure 6.2f and g) along <110> zone axis in CrCoNi alloy, where the sample without SRO shows faint streaks and the sample with SRO shows a clear streaking feature. Chen et al. [9] demonstrated that additional spots appear in CrCoNi alloys along <112> zone axis, which indicates the existence of SRO as shown in Figure 6.2i and h.

6.2.3 High-Resolution TEM (HRTEM) and Scanning TEM (STEM)

In addition to SAED in TEM, other important tools for structural characterization are HRTEM and STEM. For HRTEM imaging at high magnifications, when the electron beam interacts with the atomic columns of a crystalline specimen, it produces various contrast mechanisms known as phase contrast. The resolution can be up to the sub-angstrom level, allowing for probing atomic arrangements by analyzing interferometric fringes and observing local structural features such as defects and interfaces. An et al. [11] used HRTEM (Figure 6.3a) to show the partial dislocations of CoCrFeNiMn HEA, which contribute to the overall strengthening mechanism. Wang et al. [12]

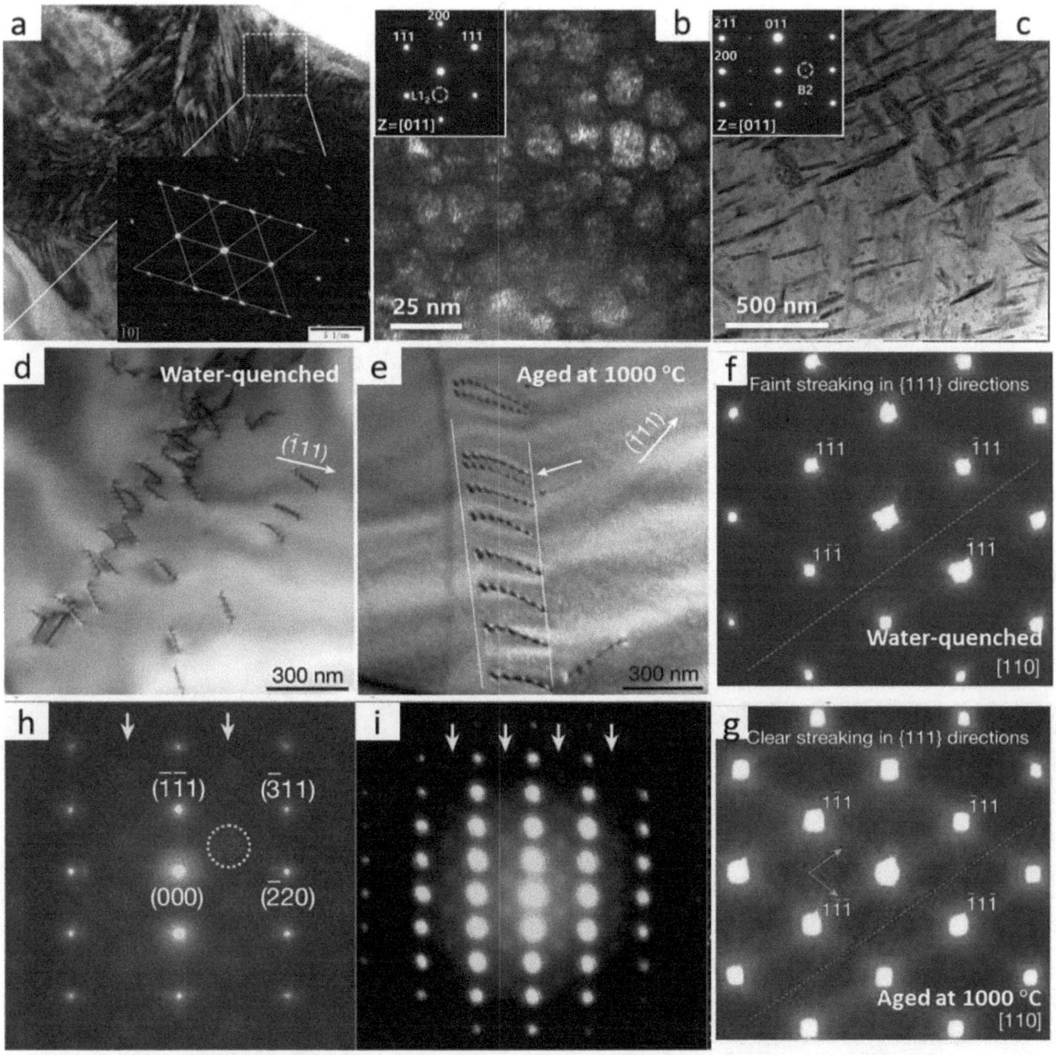

FIGURE 6.2 **(a)** Dark-field imaging showing the existence of twins and stacking faults. Inset is the corresponding SAED pattern. Adapted with permission [6]. Copyright (2023), Elsevier. **(b and c)** Morphology of precipitates. Insets are corresponding SAED patterns. Adapted with permission [10]. Copyright (2023), Elsevier. **(d and e)** Dislocation morphology and **(f and g)** SAED patterns of CrCoNi in water-quenched and 1000°C aged states. Adapted with permission [7]. Copyright (2020), Springer Nature. **(h and i)** SAED and Nano-beam Electron Diffraction Pattern along [112] zone axis. Arrows indicate superlattice reflection. Adapted with permission [9]. Copyright (2021), Springer Nature.

presented a clear interface relationship (Figure 6.3b) of the hcp lamella and bcc matrix utilizing the Inverse Fourier filtered transform (IFFT) HRTEM images, which shows an interlayer of 3–5 nm in width smoothly connecting the bcc and hcp lattice. Hu et al. [13] characterized the phase evolution after annealing the (CeGdLa–Zr/Hf)O_x at elevated temperatures. A detailed XRD study confirmed that the single phase (CeGdLaHf)O_x split into two phases with the same fluorite structure but different lattice parameters. Using HRTEM, they confirmed two types of crystals with the lattice parameters of 5.460 Å and 5.232 Å, as shown in Figure 6.3c, which STEM-energy dispersive X-ray spectroscopy (EDS) further proved to be rich in oxygen and lack oxygen separately.

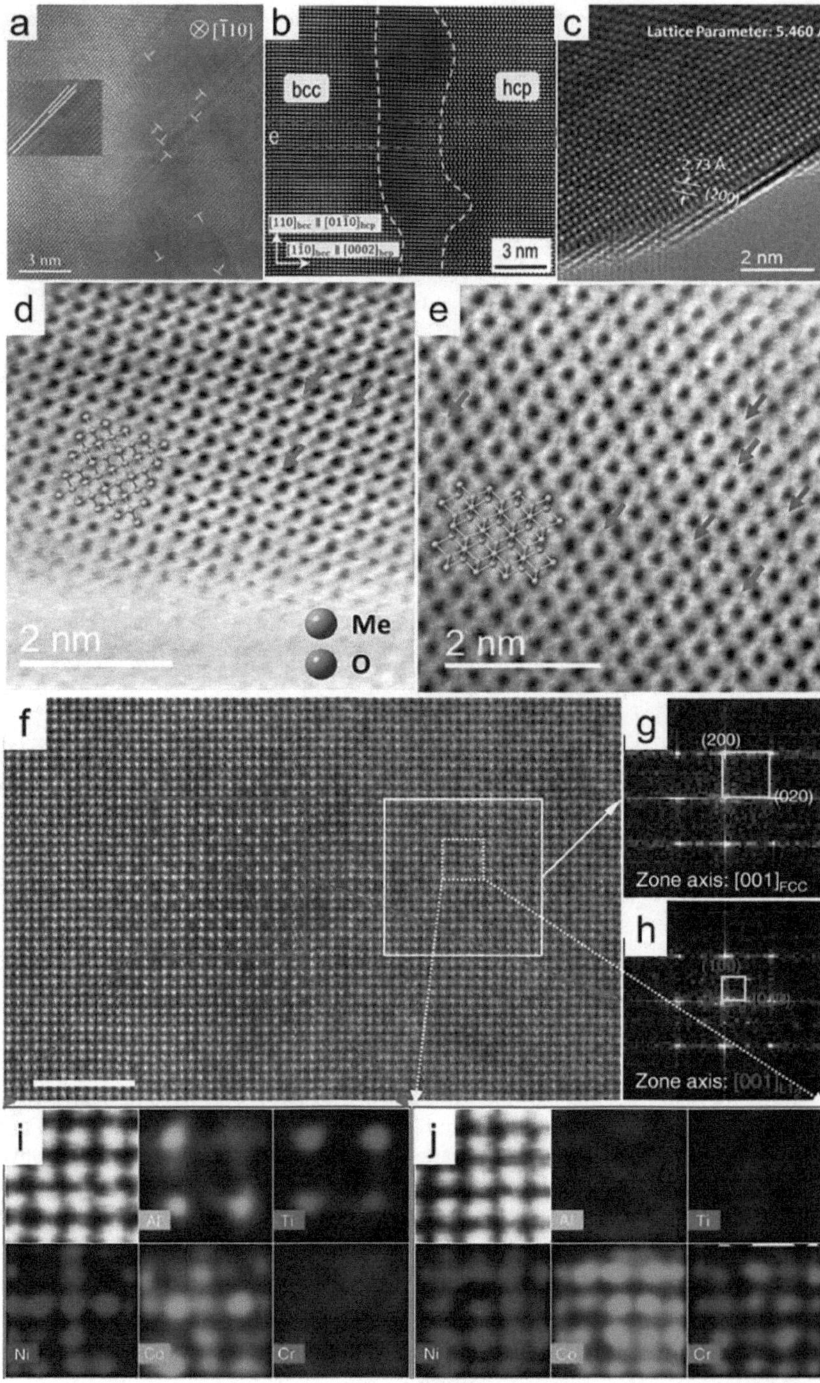

FIGURE 6.3 **(a)** HRTEM image of CoCrFeNiMn. T marks the dislocations. Adapted with permission [11]. Copyright (2023), Elsevier. **(b)** IFFT HRTEM showing transition area between bcc and hcp phase in $Ta_{0.5}HfZrTi$ HEA. Adapted with permission [12]. Copyright (2020), Elsevier. **(c)** HRTEM image of fluorite type $(CeGdLaHf)O_x$ with marked lattice plane spacing and corresponding lattice parameter. **(d and e)** ABF-STEM image of oxygen-rich and oxygen-poor region. **(f and j)** The $L1_2$ AlTiNiCoCr precipitates in NiCoCr matrix. **(f)** HAADF-STEM image. **(g and h)** corresponding FFT. **(i-j)** corresponding atomic-resolution EDS result of precipitate and matrix. Adapted with permission [14]. Copyright The Authors, some rights reserved; exclusive licensee Springer Nature. Distributed under a Creative Commons Attribution License 4.0 (CC BY).

STEM uses a highly focused electron beam to scan across a thin sample, detecting the electrons that pass through to form an image. In contrast to HRTEM's phase contrast imaging, STEM directly images the atomic columns in a material, making it a powerful tool for structural characterization. There are two types of imaging modes in STEM: Bright Field (BF) imaging and Dark Field (DF). Due to the different imaging principles, BF imaging generally provides stronger signals for light elements than DF, while DF imaging provides stronger signals for heavier elements.

Utilizing the Annular Bright Field (ABF) imaging, Hu et al. [13] could characterize the atomic arrangement of light element O, as shown in Figure 6.3d and e. Regarding the two phases with different oxygen vacancies concentration in the annealed $(CeGdLaHf)O_x$ system, the difference in oxygen concentration leads to a different oxygen contrast. In the region lacking oxygen vacancies (Figure 6.3d), it is clear that all possible O occupation sites are homogeneously filled with weak contrast. While in the region with abundant oxygen vacancies (Figure 6.3e, elemental quantitative results show the O contents lower than 10 at. %), the contrast of O columns appears very low or even invisible as those O columns are filled with low amount of O atoms or even empty.

While using DF imaging, the High-Angle Annular Dark-Field (HAADF) imaging is usually adopted. In a HAADF-STEM image, the atomic column intensity is proportional to the square of atomic number (Z) of the element in that column. Thus, it is possible to distinguish the superstructure of the coherent precipitates possessing $L1_2$ superlattice. Figure 6.3f and j shows the coherent relationship between $L1_2$ superlattice precipitates and the FCC matrix. From the HAADF-STEM image in Figure 6.3f, the atomic columns occupied with elements of lower atomic numbers (Al, Ti) should express darker contrast, which makes the $L1_2$ superlattice distinguishable from HAADF-STEM with ordered dark-bright contrast. Moreover, this is further proved by FFT (Figure 6.3g and h) and atomic-resolution EDS results (Figure 6.3i and j). STEM can also be used for mapping out the chemical composition and electronic states of materials with the help of EDS and electron energy loss spectroscopy (EELS), respectively (will be discussed in detail in later sections).

6.3 ELEMENTAL DISTRIBUTION CHARACTERIZATION

Given the complexity of HEMs, examining the distribution of elements from a macroscopic to atomic level is crucial. For the rational design of HEM, macroscopic elemental distribution is required to confirm the synthesized material presenting a uniform elemental distribution without macroscopic segregation to reach the designed property. A characterization technique such as SEM-EDS is then required. For specific designs such as heterogeneous structures, an elemental distribution at higher scale such as TEM-EDS, and 3D-atom probe tomography (APT) is then required. While studying SRO or superstructure, atomic-resolution characterization technique such as atomic-resolution EDS is then required. These elemental distribution characterization techniques complement each other to provide a full-scale understanding of the composition-structure-property relationship.

6.3.1 Energy-Dispersive X-ray Spectroscopy

EDS is a technique for probing the chemical composition of a material by detecting the characteristic X-rays emitted from the sample when it is bombarded with an electron beam. EDS is widely used in scanning electron microscope (SEM) to complement the morphology study of secondary electron (SE) or backscattering electron (BSE) imaging. It provides accurate information about the elements' uniformity that researchers aiming for single-phase HEM production widely use SEM-EDS to prove the successful single solution production. Otto et al. [15] studied the decomposition of the classical CrMnFeCoNi HEA by annealing at elevated temperatures for ultra long time. They utilized SEM-EDS to show that the CrMnFeCoNi alloy presented a uniform elemental distribution in the as-cast state (Figure 6.4a), while part of the alloy decomposed into NiMn, Cr, and FeCo after 500 days annealing at 500°C (Figure 6.4b).

However, when the phases to be identified are in the fine scale, i.e., nano- and micrometer level, the resolution of SEM may not be sufficient to fully characterize phases, particularly if the phases

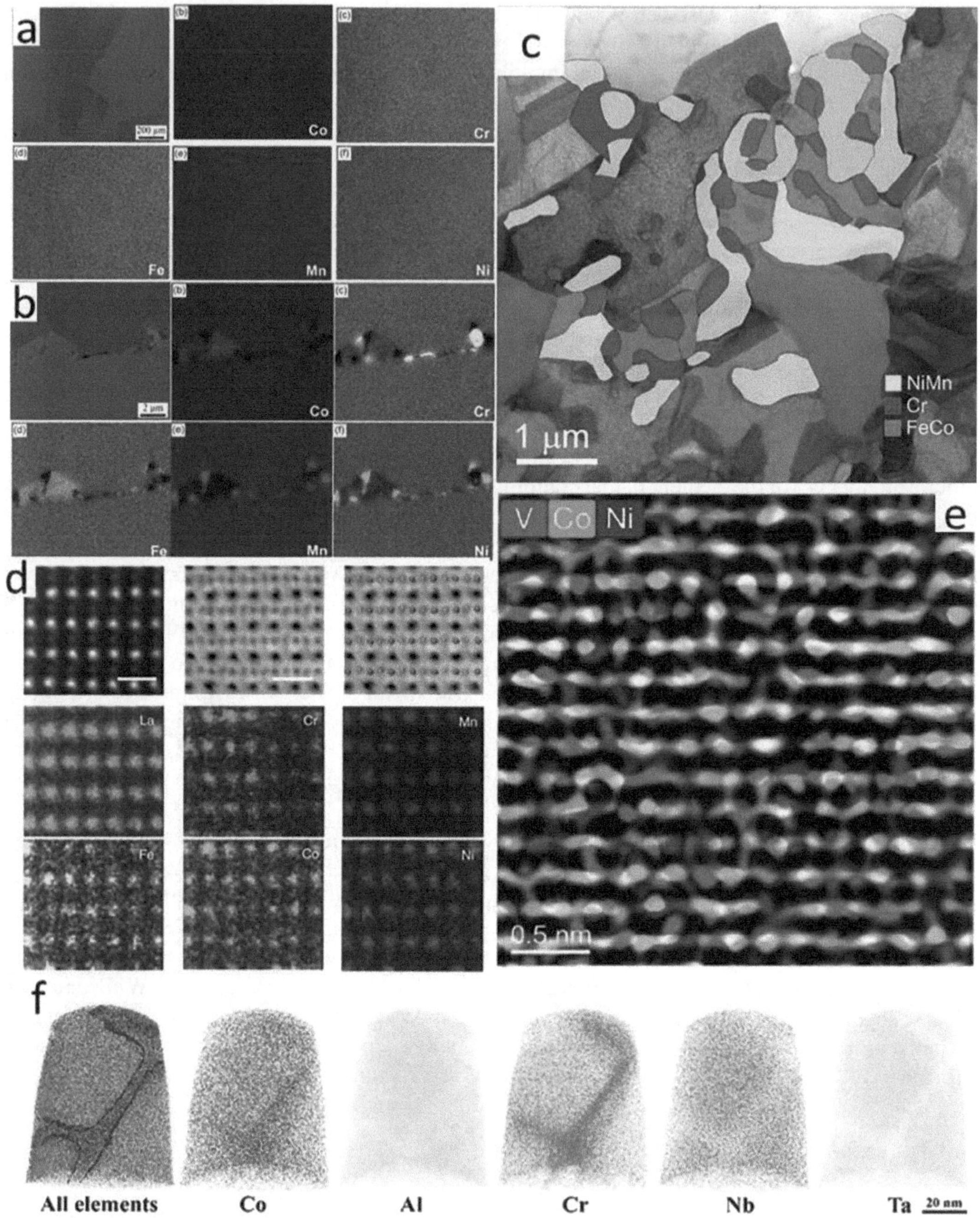

FIGURE 6.4 **(a-c)** SEM-EDS results of CrCoFeNiMn in **(a)** as-prepared state and **(b)** after annealed at 500°C for 500 h. **(c)** decomposed phases identified from TEM-EDS point scanning results. Adapted with permission [15]. Copyright (2016), Elsevier. **(d)** atomic-resolution STEM-EDS results of $La(Fe_{0.2}Co_{0.2}Cr_{0.2}Ni_{0.2}Mn_{0.2})O_3$. Adapted with permission [16]. Copyright The Authors, some rights reserved; exclusive licensee Springer Nature. Distributed under a Creative Commons Attribution License 4.0 (CC BY) **(e)** overlapped atomic-resolution STEM-EDS results of VCoNi medium entropy alloy (MEA). Adapted with permission [9]. Copyright (2021), Springer Nature. **(f)** Atom maps reconstructed using 3D-APT of $Co_{43.1}Ni_{28.8}Cr_{10.2}Al_{10.0}Ti_{2.1}Mo_{2.2}Ta_{2.1}Nb_{1.5}$. Adapted with permission [17]. Copyright (2021), Elsevier.

are closely intermixed or have fine microstructures. In such cases, TEM-EDS can be a powerful tool for phase identification and characterization. TEM-EDS can provide higher resolution and sensitivity than SEM-EDS due to the smaller probe size and higher electron dose. In the decomposition of CrMnFeCoNi alloy [15], the decomposed products are in nanometers scale, which is beyond the capability of SEM-EDS. While in TEM, a mixture of three types of decomposed products (NiMn, CoFe and Cr) ranging from tens to hundreds of nanometers can be distinguished, as shown in Figure 6.4c, by analyzing the point spectrum of grains with different morphologies.

The state-of-art aberration-corrected STEM combined with a high-resolution EDS detector in TEM can even reach the atomic level resolution, which could be particularly useful for analyzing HEM. Su et al. [16] synthesized a ABO_3 perovskite structured HEO with La distributed in the A site while Cr, Mn, Fe, Co, and Ni are distributed in the B site. In the presented atomic-resolution EDS as shown in Figure 6.4d, the designed distribution of atoms is reached, as no interlayer elemental disorder could be observed from the EDS result. Du et al. [14] used the atomic resolution to present the elemental distribution of the $L1_2$ superlattice NiCoCrAlTi precipitate and NiCoCr matrix. It is clearly shown in Figure 6.3i and j that in the matrix, the Ni, Co, and Cr are overall uniformly distributed, while in the $L1_2$ superlattice precipitate, the Al and Ti occupy the A sites and the Ni, Co, and Cr occupy the other B sites. The atomic-resolution EDS results strongly validated the successful rational design. Moreover, the atomic-resolution EDS is often used to study the SRO in HEM. Chen et al. [9] claimed the direct observation of SRO in a VCoNi alloy, as shown in Figure 6.4e. The atomic-resolution EDS along with the SAED pattern together prove the existence of SRO.

6.3.2 Three-Dimensional Atom Probe Tomography

While EDS can provide elemental distribution information in a 2D projection, a 3D elemental distribution, such as in the case of nano precipitate and layered structure, is necessary to demonstrate the intrinsic interaction of the heterostructure. 3D-APT is a powerful analytical technique used to determine the 3D atomic-scale composition of materials. It is a type of mass spectrometry that uses a sharp, pointed tip to extract atoms from tip surface, one at a time. The extracted atoms are then ionized, and their mass-to-charge ratio is measured to determine their identity, which can be reconstructed to yield the spatial elemental distribution. Cao et al. [17] demonstrated that a $L1_2$-strengthened multicomponent Co-rich HEA $Co_{43.1}Ni_{28.8}Cr_{10.2}Al_{10.0}Ti_{2.1}Mo_{2.2}Ta_{2.1}Nb_{1.5}$ with superb high-temperature tensile strength (755 MPa at 800°C). Using 3D-APT, the existence and distribution of $L1_2$ precipitates could be revealed as shown in Figure 6.4f. The APT results show that the Ta and Nb elements are important for stabilizing the $L1_2$ ordered precipitates, which provided a solution for the stabilization of $L1_2$ precipitates at elevated temperatures.

6.4 SPECTROSCOPIC CHARACTERIZATION

Besides structural and elemental analysis, spectroscopic characterization techniques play a critical role in understanding the HEM in local chemical environment, coordination, and vibrational modes et al. These techniques can provide valuable insights into materials' physical and chemical properties. Raman spectroscopy, for example, is useful for studying the vibrational modes of materials, while X-ray absorption spectroscopy (XAS) and X-ray photoelectron spectroscopy (XPS), can provide information on the electronic and chemical environments around specific atoms. EELS is also a powerful tool for studying materials' local electronic properties. Spectroscopic techniques play an essential role in the comprehensive characterization of materials, providing information that is not accessible through other methods.

6.4.1 Raman Spectroscopy

As a non-destructive vibrational spectroscopic technique, Raman spectroscopy can be useful to probe specific vibration modes, such as in HEO that can effectively identify the structural symmetry change. Raman spectroscopy is based on the inelastic scattering of light by materials, a process known as Raman scattering, which results in the deviation of scattered radiation from its incident frequency due to activated vibrations. This technique has been widely used to characterize structural and defect information, especially in HEO. For instance, in the widely studied fluorite type HEO, there are three vibration modes $F_{2g} \sim 465\,cm^{-1}$, $T_{1u}TO \sim 300\,cm^{-1}$, and $T_{1u}LO \sim 605\,cm^{-1}$. Among these F_{2g} vibration mode represents the eightfold metal-oxygen bond vibration in fluorite structure, while $T_{1u}TO$ vibration mode usually evolves when oxygen vacancies occur. Sarkar et al. [5] synthesized a series of rare earth oxides with fluorite structure and recorded the Raman spectra as shown in Figure 6.5a. Considering the ceria (CeO_2) as a parent fluorite structure, it is found that with increasing number of constituent elements, the eightfold CeO_2 symmetry breaks and leads to the drop of the intensity of F_{2g} symmetric vibration mode. While at the same time, the peaks at ~570 cm^{-1} evolve due to the necessary existence of O vacancies for charge compensation. The ratio of the integral intensity of the $T_{1u}LO$ vibration mode and the integral intensity of F_{2g} vibration mode portrays the relative oxygen vacancies concentration. This allows a deeper understanding of evolution of oxygen vacancies with the composition variation in rare earth HEOs.

6.4.2 Electron Energy Loss Spectroscopy

EELS is an analytical technique used in TEM to measure the energy loss of electrons passing through the observed sample. The energy losses are the consequence of the interaction of high-energy electrons with the sample for excitation of core electrons. These energy losses are measured, and the resulting spectrum provides information about the sample's chemical composition and electronic structure. Zhang et al. [18] synthesized $NiFeXO_4$ (X=Fe, Ni, Al, Mo, Co, Cr), which promoted oxygen evolution reaction (OER) higher than the $NiFeYO_4$(Y=Fe, Ni, Al, Mo) and $NiFe_2O_4$. The EELS spectra shown in Figure 6.5b and c could give an explanation about the promoted performance. The K-edge of O split into two peaks as a result of crystal field splitting due to electronic correlation, which further reflects the metal-oxygen bonding nature and lattice distortion. Furthermore, from $NiFe_2O_4$ to $NiFeXO_4$, the L2 peak of Fe shifts to 721.39 eV from 721.11 eV and the L3 peak shifts to 708.45 eV from the original 708.14 eV. Moreover, the ratio of integrated L3 and L2 peaks decreased from 1.82 to 1.24. The peaks shift along with the L3 to L2 ratio change both indicate the formation of Fe with higher valence state, which promotes the formation of cation vacancies, further enhancing the OER performance.

6.4.3 UV–Vis Diffuse Reflectance Spectroscopy

Bandgap engineering is a fundamental aspect of material science, as it allows for the control and modification of HEM's electronic and optical properties. The most used technique for experimentally determining the bandgap is UV–Vis Diffuse Reflectance Spectroscopy (UV–Vis DRS). By measuring the amount of light transmitted or reflected from a material as a function of wavelength, the material's optical absorption properties can be recorded. Further post-processing the recorded data using the Tauc equation could obtain the bandgap value. In Hu's work [13], they discovered oxygen vacancy generation process in fluorite $(CeGdLaZr/Hf)O_x$ multi-component oxide (MCO), where the $(CeGdLaZr)O_x$ and $(CeGdLaHf)O_x$ show oxygen losses at elevated temperatures to different extents. $(CeGdLaZr)O_x$ shows an oxygen loss about 21 at%, and UV–Vis DRS (Figure 6.5d and e) further shows this oxygen loss leads to a bandgap narrowing of 0.08 eV. While $(CeGdLaHf)O_x$ split into two phases, with one containing poor oxygen, the UV–Vis DRS shows that the phase with deficient oxygen finally leads to a zero-bandgap feature in 1500°C annealed $(CeGdLaHf)O_x$.

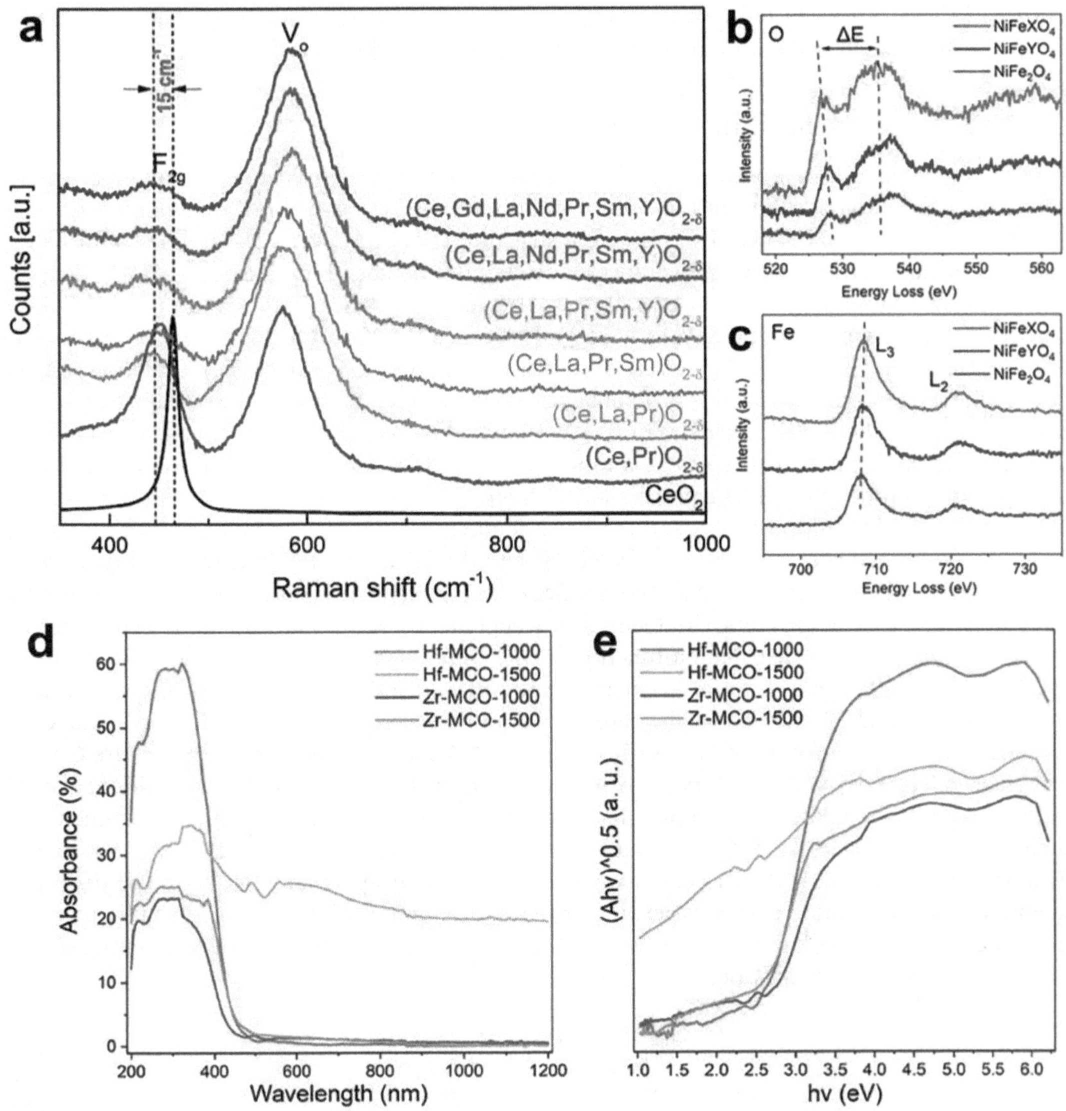

FIGURE 6.5 (**a**) Raman spectra of rare earth HEOs. Adapted with permission [5]. Copyright (2017), Royal Society of Chemistry. (**b-c**) EELS spectra of (**b**) O K edge and (**c**) Fe L edge for $NiFe_2O_4$-based entropy-stabilized oxides. Adapted with permission [18]. Copyright (2023), American Chemical Society. (**d**) UV–Vis spectra and corresponding (**e**) Tauc plots of (CeGdLa-Hf/Zr)O_x.

6.4.4 X-RAY PHOTOELECTRON SPECTROSCOPY

XPS is a surface-sensitive analytical technique that is widely used to study the overall elemental composition, chemical bonding states, and electronic properties of HEM. In XPS, X-rays are directed onto the surface of a sample, causing the emission of photoelectrons. These emitted photoelectrons are then collected and analyzed to obtain information about the sample's elemental composition and chemical bonding states. XPS has been widely used for characterize the valency state of constituent elements of HEMs, especially those with polyvalent nature. Such as in rare earth fluorite HEO [5], the participation of Ce is usually necessary to keep a stable fluorite structure. However, Ce can exist in oxide as Ce^{4+} or Ce^{3+}. By carefully comparing the 3d spectra of Ce with the 3d spectra of Ce^{4+} in CeO_2, the existence of Ce^{4+} in the rare earth HEOs is confirmed.

Further, Ce^{4+} is proved to be the key factor stabilizing fluorite structure in the rare earth HEOs system, as the system without Ce participating cannot form a single fluorite phase.

6.4.5 X-RAY ABSORPTION SPECTROSCOPY

XAS is a technique that probes the local electronic and geometric structure of atoms in materials. X-ray absorption spectra are typically divided into two regions: X-ray absorption near edge structure (XANES) and X-ray absorption fine structure (EXAFS). XANES provides information about the local electronic structure and oxidation state of the absorbing atom. The XANES spectrum consists of a series of peaks and valleys associated with the excitation of electrons from the core level to unoccupied orbitals. Wang et al. [19] recorded the XANES change (Figure 6.6a–d) of $Mg_{0.2}Co_{0.2}Ni_{0.2}Cu_{0.2}Zn_{0.2}O$ HEO working as anode in a lithium-ion battery. The oxidation state of each element can then be tracked at different electrochemical reaction states. For instance, after the first discharging, the XANES spectra of Co, Ni, Cu, and Zn match well with each corresponding metallic reference, proving these cations reduced to metallic state after this step. This helps the understanding of the intermediate process of the reaction.

EXAFS provides information about the local geometric structure around the absorbing atom, including bond distances, coordination numbers, and disorder. The EXAFS spectrum is a series of oscillations that arise from the interference between the photoelectron wave and the waves scattered from the surrounding atoms. Mathematically applying a Fourier transformation of the EXAFS spectrum yields the pair distribution function (PDF). While peaks in PDF are indicative of interatomic distances within a material, element-specific PDFs can be obtained when the spectrum is measured at energies slightly above the core-level binding energy of a particular element. Tan et al. [20] measured the element-specific EXAFS (Figure 6.6e) and corresponding PDF (Figure 6.6f) of Zr-center in TiZrHfNbTa refractory HEA before and after the tensile test. By carefully fitting the PDF spectra, they could obtain the pair distance of Zr/Nb with neighboring other constituent elements. Figure 6.6g shows the Zr/Nb-center pair distances before and after the tensile test. Ta and Ti atoms are located near the Zr center, while Zr, Nb, and Hf atoms are far away. The distortion magnitude becomes more evident after the tensile test, as indicated by the solid square markers. Utilizing the pair distance results, it is possible to calculate the element-specified local lattice distortion using the following equation:

$$\chi_i = \sqrt{\Sigma_j^n C_j \left(1 - \frac{\gamma_1 i - j}{\overline{\gamma}}\right)} \times 100\% \quad (1)$$

where χ_i denotes the local lattice distortion centered around the i-th atom, $r_{i\text{-}j}$ represents the distance between the i-th and j-th atoms, r indicates the averaged atom pair distance calculated from lattice parameter a, and C_j represents the atomic content of j-element atoms. This gives the local lattice distortion around the Zr-center about 5.2±0.3% in an as-cast state (Figure 6.6h). This distortion value rises to 6.8% after the tensile test, probably because of the dislocation and twinning activity.

6.5 MICROSTRUCTURE CHARACTERIZATION (SEM, EBSD, TEM)

The microstructure of material comprises various phases that exhibit diverse morphological characteristics, including but not limited to grains, precipitates, dendrites, spherulites, lamellae, and pores, which exhibit varying forms, sizes, and distributions. To comprehend the material's behavior, it is imperative to establish correlations between the macroscopic properties and the underlying microstructural phenomena. In HEMs, which possess a complex and unique composition, understanding the microstructure is paramount to reveal their properties' origin. SEM, EBSD, and TEM are important techniques used to probe the microstructure of materials. SEM and EBSD provide information about

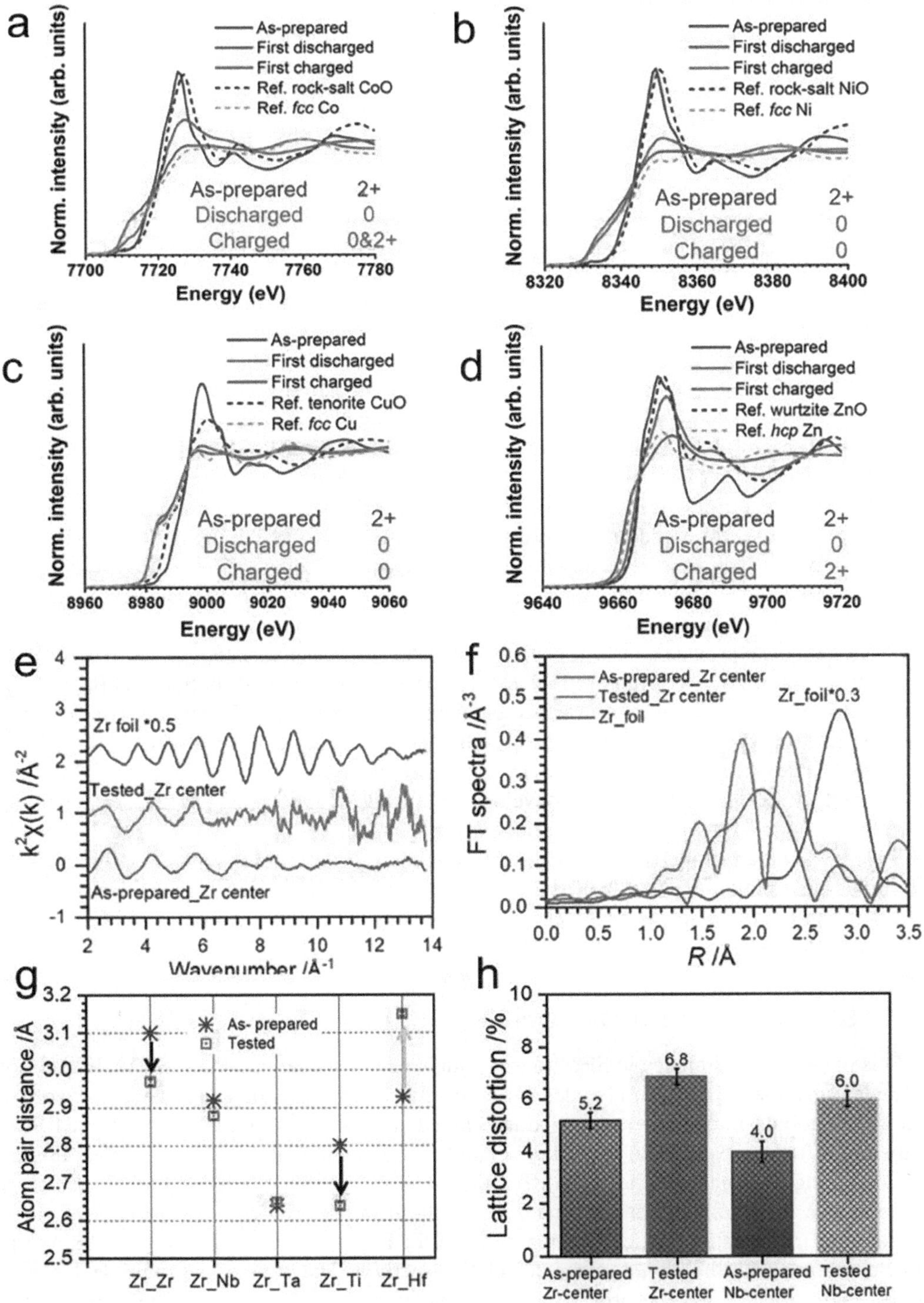

FIGURE 6.6 **(a-d)** XANES spectra of Co, Ni, Cu, and Zn in $Mg_{0.2}Co_{0.2}Ni_{0.2}Cu_{0.2}Zn_{0.2}O$, reference metal, and oxides. Adapted with permission [19]. Copyright The Authors, some rights reserved; exclusive licensee Springer Nature. Distributed under a Creative Commons Attribution License 4.0 (CC BY). **(e-h): (e-f)** Extracted k^2-weight and Fourier-transformed EXAFS spectra of Zr K-edge for as-prepared and tensile TiZrHfNbTa, respectively. **(g)** atom pair evolution around Zr-center and **h**. lattice distortion of Zr-center in as-prepared and tensile TiZrHfNbTa. Adapted with permission [20]. Copyright (2023), Elsevier.

the surface topography and crystallographic orientation of grains, while TEM can reveal information about the internal structure of materials with sub-nanometer resolution. By utilizing these techniques, researchers can better understand the microstructural features that govern the properties of HEM.

6.5.1 Scanning Electron Microscopy

SEM is a powerful imaging technique used to investigate the surface morphology and microstructure of HEM. It works by scanning a focused beam of electrons across the sample surface and detecting the secondary or backscattered electrons produced by the interaction of the beam with the sample. Among which the SE imaging provides high-resolution topographical information with larger depth of field, while BSE imaging advances in providing additional contrast regarding the atomic number of the elements. The two imaging modes differ in contrast mechanisms and are often used complementarily to provide comprehensive information about the sample.

For HEM to be used as catalysts to achieve high catalytic activity, it is essential to achieve intricate nanostructures with high specific surface area by rational design and engineering. SE imaging, with a large depth of field, is widely adopted to image the morphology of synthesized HEM. Jin et al. [21] synthesized a fourteen-component high entropy alloy@oxide bifunctional electrocatalyst which showed a record-low ΔE of 0.61 V for Zn–air batteries. The key to excellent performance is the nanoporous nanowire structure, proved by the combination of SE imaging and TEM study. SE imaging showed a macroscopic nanowires morphology at micrometers level (Figure 6.7a), while TEM complementarily showed the nanoporous nature and loaded HEA nanoclusters at nanometer level (Figure 6.7b), which together achieved the excellent performance.

BSE imaging can provide contrast related to the atomic number variation of the different phases in a sample, making it a convenient tool for demonstrating their distribution. While SEM-EDS can also provide information about surface elemental distribution, BSE imaging is advantageous regarding imaging speed and resolution. Therefore, BSE imaging is a valuable technique for characterizing HEM, where multiple phases with different atomic numbers can coexist. Meghwal et al. [22] utilized a CALPHAD approach to develop a HEA (AlCoCrFeNi) - MEA (CoCrFeNi) composite coating material. For composite materials, an even mixing of components is crucial to achieving optimized overall properties. The BSE image (Figure 6.7c) provided by the authors showed a well-mixed distribution of the AlCoCrFeNi and CoCrFeNi phases. The compositional difference in Al between the two alloys enabled clear contrast between the two phases, making them easily distinguishable.

6.5.2 Electron Backscatter Diffraction

EBSD is one of the techniques in SEM that is widely used for characterizing HEM. In addition to phase identification, EBSD can also determine crystal orientations of grains and analyze crystal defects in materials, including grain boundaries and dislocations. The basic principle of EBSD is to obtain the diffraction patterns of the Kikuchi bands corresponding to the different crystallographic planes utilizing backscattered electron diffraction. The standard crystallographic database is used as a reference for automated pattern indexing of the different phases using dedicated software. Ye et al. [23] investigated the deformation behavior of $L1_2$-structure FeCoNiV HEA using in situ EBSD combined with the digital image correlation method. They found that multiple sub-regions with different orientations appeared within individual grains with increasing deformation strain, indicating inhomogeneous deformation within the grain (Figure 6.7e). In addition, the evolution of the local strain distribution during the tensile deformation was analyzed with the help of the kernel average misorientation (KAM) maps of EBSD. The EBSD results demonstrate that the formation of deformation substructures can coordinate intergranular deformation and ensure the stability of the grain boundaries.

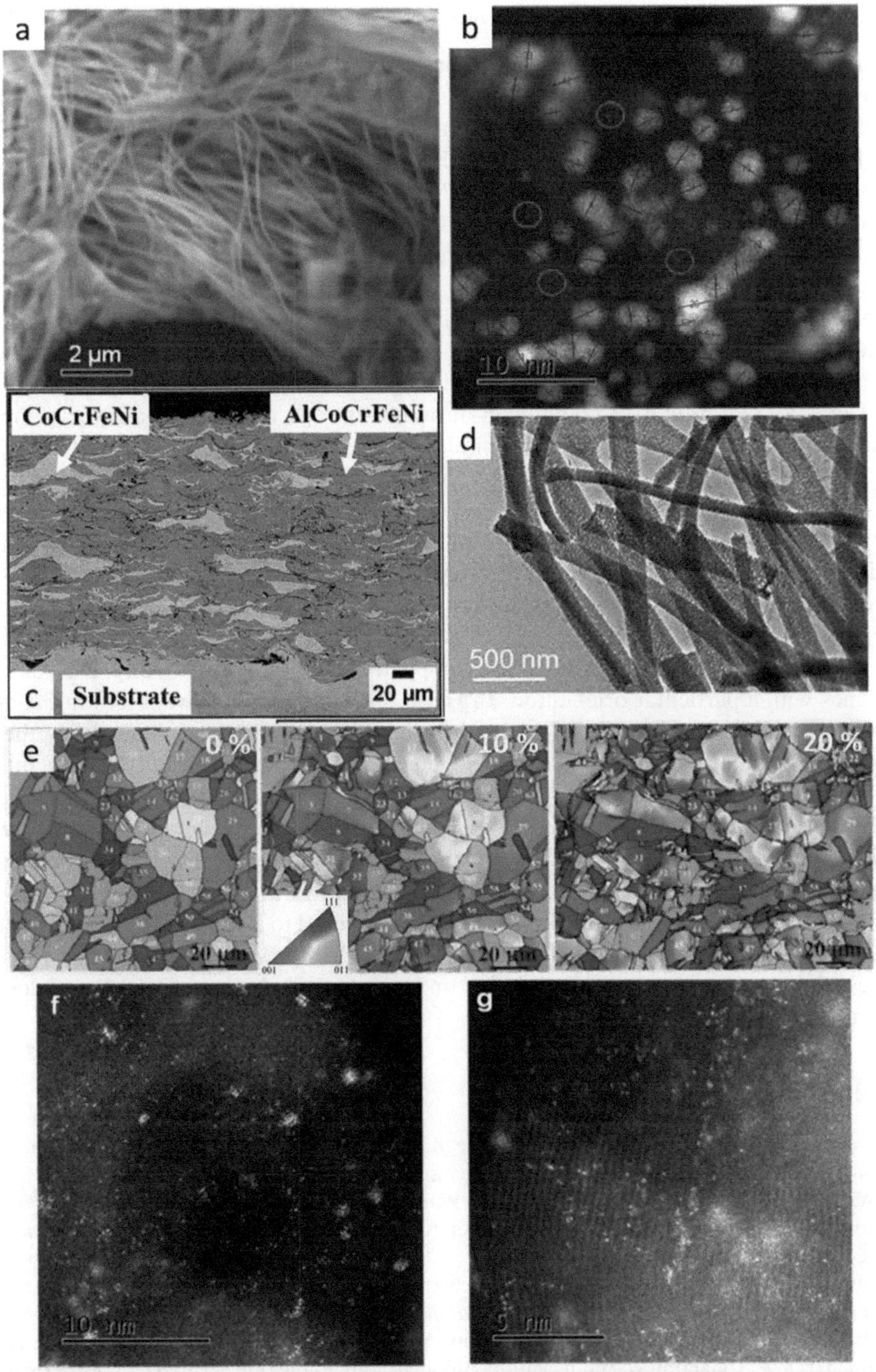

FIGURE 6.7 **(a-b): (a)** SE imaging the morphology of nanowires. **(b)** STEM imaging of nanoclusters on the nanowire substrate. Adapted with permission [21]. Copyright (2022) The Royal Society of Chemistry. **(c)** BSE imaging of CoCrFeNi and AlCoCrFeNi composite coating. Adapted with permission [22]. Copyright (2023), Elsevier. **(d)** TEM imaging of nanoporous nanowires. Adapted with permission [18]. Copyright (2023), American Chemical Society. **(e)** In-situ EBSD inverse pole figure (IPF) maps in the deformation process of 0%, 10%, and 20%. Adapted with permission [23]. Copyright (2022), Elsevier. **(f-g)**: HAADF-STEM imaging of single atoms and nanoclusters on the graphene substrate. Adapted with permission [24]. Copyright (2023), American Chemical Society.

6.5.3 Transmission Electron Microscopy

TEM is an advanced microscopy technique that uses a focused beam of electrons to transmit through a thin sample to obtain high-resolution images at the atomic scale. Various analytical techniques such as SAED, EELS, Precession Electron Diffraction (PED), et al. are also applied in TEM using the high-energy electron beam. Thus, in HEM, TEM can provide valuable information on the microstructure, morphology, crystallographic orientation, and defect structures of materials.

With its much higher resolution compared to SEM, TEM can observe samples up to the atomic scale level, enabling the imaging of microscopic features that may not be visible in SEM. For example, nanopores in nanowires with a diameter of ~200 nm may not be observable in SEM, but the resolution of TEM imaging can easily reach this level, allowing for the identification and characterization of such nanopores shown in Figure 6.7d [18]. Moreover, STEM possesses the ability to probe real atoms, making the observations of single-atom catalysts and nanocluster's catalysts possible. Lin et al. [24] used a 12-component ultrahigh entropy nanoporous alloy as template, acting as inhibited surface diffusion under even elevated temperature range (800–1000°C), to grow small-pore-size nanoporous graphene. When removing the HEA template, the metal single atoms and nanoclusters are naturally left and anchored on the graphene. The STEM images of Figure 6.7f and g clearly show a uniform distribution of single atoms and nanoclusters.

Additionally, DF imaging in TEM is advantageous to probe microstructure details that are difficult to see in BF imaging. In DF imaging, only scattered or diffracted light is collected and detected by the objective lens, whereas in BF imaging, scattered and transmitted light is detected. This allows for the selective highlighting of regions of the sample that are highly diffracting, such as defects or crystal planes with a particular orientation. In HEA, DF imaging has advantages in the characterization of nanoprecipitates. By collecting the diffracted light of the precipitates, only the contrast of the round-shaped nano precipitates is visible, enabling their identification. This technique also distinguishes the contrast of layered precipitates, as shown in Figure 6.2b and c. In addition, when the two-beam condition is achieved, central DF and weak-beam DF imaging can be selectively collected from different diffracted light, allowing for the clear characterization of defects such as dislocations, as demonstrated in Figure 6.2d and e.

In conclusion, various characterization methods can be used to study HEMs. These methods can provide valuable information on the microstructural features, crystal structure, and properties of HEMs. Researchers can use these techniques to understand the behavior of HEMs and develop new applications for this class of materials.

REFERENCES

[1] C.M. Rost, E. Sachet, T. Borman, A. Moballegh, E.C. Dickey, D. Hou, J.L. Jones, S. Curtarolo, J.-P. Maria, Entropy-stabilized oxides, *Nature Communications*, 6 (2015) 8485.

[2] W. Wang, W. Zhou, S. Song, K.M. Reddy, X. Wang, Effect of deformation induced B2 precipitates on the microstructure and mechanical property of Al0.3CoCrFeNi high-entropy alloy, *Journal of Alloys and Compounds*, 821 (2020) 153445.

[3] L. Lutterotti, R. Vasin, H.-R. Wenk, Rietveld texture analysis from synchrotron diffraction images. I. *Calibration and basic analysis, Powder Diffraction*, 29 (2014) 76–84.

[4] R. Djenadic, A. Sarkar, O. Clemens, C. Loho, M. Botros, V.S.K. Chakravadhanula, C. Kübel, S.S. Bhattacharya, A.S. Gandhi, H. Hahn, Multicomponent equiatomic rare earth oxides, *Materials Research Letters*, 5 (2016) 102–109.

[5] A. Sarkar, C. Loho, L. Velasco, T. Thomas, S.S. Bhattacharya, H. Hahn, R. Djenadic, Multicomponent equiatomic rare earth oxides with a narrow band gap and associated praseodymium multivalency, *Dalton Transactions*, 46 (2017) 12167–12176.

[6] Y. Lv, X. Lang, C. Su, L. Cao, L. Wang, Stacking fault and nano-twins dominating strengthening mechanism of (CuZnMnNi)100-xSnx high entropy brass alloy prepared by mechanical alloying and fast hot pressing sintering, *Materials Letters*, 312 (2022) 131614.

[7] R. Zhang, S. Zhao, J. Ding, Y. Chong, T. Jia, C. Ophus, M. Asta, R.O. Ritchie, A.M. Minor, Short-range order and its impact on the CrCoNi medium-entropy alloy, *Nature*, 581 (2020) 283–287.

[8] S. Yin, Y. Zuo, A. Abu-Odeh, H. Zheng, X.G. Li, J. Ding, S.P. Ong, M. Asta, R.O. Ritchie, Atomistic simulations of dislocation mobility in refractory high-entropy alloys and the effect of chemical short-range order, *Nature Communications*, 12 (2021) 4873.

[9] X. Chen, Q. Wang, Z. Cheng, M. Zhu, H. Zhou, P. Jiang, L. Zhou, Q. Xue, F. Yuan, J. Zhu, X. Wu, E. Ma, Direct observation of chemical short-range order in a medium-entropy alloy, *Nature*, 592 (2021) 712–716.

[10] L. Liu, Z. Wang, Q. Wu, Y. Jia, Q. Xu, F. He, J. Li, J. Wang, A hypoeutectic high-entropy alloy with hierarchical microstructure for high-temperature application, *Scripta Materialia*, 232 (2023) 115502.

[11] Z. An, S. Mao, Y. Liu, L. Yang, A. Vayyala, X. Wei, C. Liu, C. Shi, H. Jin, C. Liu, J. Zhang, Z. Zhang, X. Han, Inherent and multiple strain hardening imparting synergistic ultrahigh strength and ductility in a low stacking faulted heterogeneous high-entropy alloy, *Acta Materialia*, 243 (2023) 118516.

[12] P. Wang, Y. Bu, J. Liu, Q. Li, H. Wang, W. Yang, Atomic deformation mechanism and interface toughening in metastable high entropy alloy, *Materials Today*, 37 (2020) 64–73.

[13] Y. Hu, M. Anandkumar, J. Joardar, X. Wang, A.S. Deshpande, K.M. Reddy, Effective band gap engineering in multi-principal oxides (CeGdLa-Zr/Hf)O(x) by temperature-induced oxygen vacancies, *Scientific Report*, 13 (2023) 2362.

[14] X.H. Du, W.P. Li, H.T. Chang, T. Yang, G.S. Duan, B.L. Wu, J.C. Huang, F.R. Chen, C.T. Liu, W.S. Chuang, Y. Lu, M.L. Sui, E.W. Huang, Dual heterogeneous structures lead to ultrahigh strength and uniform ductility in a Co-Cr-Ni medium-entropy alloy, *Nature Communications*, 11 (2020) 2390.

[15] F. Otto, A. Dlouhý, K.G. Pradeep, M. Kuběnová, D. Raabe, G. Eggeler, E.P. George, Decomposition of the single-phase high-entropy alloy CrMnFeCoNi after prolonged anneals at intermediate temperatures, *Acta Materialia*, 112 (2016) 40–52.

[16] L. Su, H. Huyan, A. Sarkar, W. Gao, X. Yan, C. Addiego, R. Kruk, H. Hahn, X. Pan, Direct observation of elemental fluctuation and oxygen octahedral distortion-dependent charge distribution in high entropy oxides, *Nature Communications*, 13 (2022) 2358.

[17] B.X. Cao, H.J. Kong, Z.Y. Ding, S.W. Wu, J.H. Luan, Z.B. Jiao, J. Lu, C.T. Liu, T. Yang, A novel L12-strengthened multicomponent Co-rich high-entropy alloy with both high γ'-solvus temperature and superior high-temperature strength, *Scripta Materialia*, 199 (2021) 113826.

[18] Q. Zhang, Y. Hu, H. Wu, X. Zhao, M. Wang, S. Wang, R. Feng, Q. Chen, F. Song, M. Chen, P. Liu, Entropy-Stabilized Multicomponent Porous Spinel Nanowires of NiFeXO(4) (X=Fe, Ni, Al, Mo, Co, Cr) for Efficient and Durable Electrocatalytic Oxygen Evolution Reaction in Alkaline Medium, *ACS Nano*, 17 (2023) 1485–1494.

[19] K. Wang, W. Hua, X. Huang, D. Stenzel, J. Wang, Z. Ding, Y. Cui, Q. Wang, H. Ehrenberg, B. Breitung, C. Kübel, X. Mu, Synergy of cations in high entropy oxide lithium ion battery anode, *Nature Communications*, 14 (2023) 1487.

[20] Y.-Y. Tan, T. Li, Y. Chen, Z.-J. Chen, M.-Y. Su, J. Zhang, Y. Gong, T. Wu, H.-Y. Wang, L.-H. Dai, Uncovering heterogeneity of local lattice distortion in TiZrHfNbTa refractory high entropy alloy by SR-XRD and EXAFS, *Scripta Materialia*, 223 (2023) 115079.

[21] Z. Jin, X. Zhou, Y. Hu, X. Tang, K. Hu, K.M. Reddy, X. Lin, H.J. Qiu, A fourteen-component high-entropy alloy@oxide bifunctional electrocatalyst with a record-low DeltaE of 0.61 V for highly reversible Zn-air batteries, *Chemical Science*, 13 (2022) 12056–12064.

[22] A. Meghwal, S. Singh, S. Sridar, W. Xiong, C. Hall, P. Munroe, C.C. Berndt, A.S.M. Ang, Development of composite high entropy-medium entropy alloy coating, *Scripta Materialia*, 222 (2023) 115044.

[23] Z. Ye, C. Li, M. Zheng, X. Zhang, X. Yang, J. Gu, In situ EBSD/DIC-based investigation of deformation and fracture mechanism in FCC- and L12-structured FeCoNiV high-entropy alloys, *International Journal of Plasticity*, 152 (2022) 103247.

[24] X. Lin, Y. Hu, K. Hu, X. Lin, G. Xie, X. Liu, K.M. Reddy, H.-J. Qiu, Inhibited surface diffusion of high-entropy nano-alloys for the preparation of 3D nanoporous graphene with high amounts of single atom dopants, *ACS Materials Letters*, 4 (2022) 978–986.

7 Stability Landscape and Charge Compensation Mechanism for Isovalent and Aliovalent Substitution in High Entropy Oxides

Ashwani Gautam and Md. Imteyaz Ahmad

7.1 INTRODUCTION

In this era of material development, there is a perpetual hunt for novel, technologically advanced materials with tailorable features. The traditional approach for developing materials introduces a limited quantity of elements or compounds such as alloying or dopants into base materials or mixed component materials to form a composite. Until recently, enthalpy was the main thermodynamic knob utilized for material design, and the entropy term was not much explored [1]. In 2004, Canter et al. [2] and Yeh et al. [3] independently proposed a new approach by introducing more than four elements in near-equiatomic ratio to produce a single-phase solid solution alloy. Yeh et al. [3] suggested that the tendency to form a single-phase solid solution is mainly caused owing to the increased configurational entropy of mixing (ΔS_{mix}) resulting from the presence of several components in near-equiatomic proportions, and consequentially introduced the term called the high entropy alloys (HEA). The advent of HEA opened the exploration of central regions of multicomponent phase diagrams, which remained largely untouched, where potentially unique properties can result from the synergistic effects of the component elements. In contrast to traditional alloys, HEAs feature distinctive compositions, microstructures, and adaptive properties that have earned them a high level of recognition. The concept of HEA had recently been extended to compounds where, in 2015, Rost et al. [4] demonstrated a single-phase rocksalt structure in a five-component equiatomic transition metal oxide. Since then, numerous non-metallic chemical groups, including oxides [4,5], borides [6], nitrides [7], and carbides [8], have been reported. Since the first report, several different compositions, and structures of high entropy oxides (HEO) have been reported, including rocksalt [5], spinel [4], fluorite [9], pyrochlore [10], and perovskite [11].

7.2 THERMODYNAMIC BASIS

The stability of a solution is assessed in terms of free energy change due to mixing (ΔG^{mix}). The solution has a lower free energy i.e $\Delta G^{\text{mix}} < 0$. The ΔG^{mix} can be expressed as:

$$\Delta G^{\text{mix}} = \Delta H^{\text{mix}} - T\Delta S^{\text{mix}} \quad (7.1)$$

where ΔH^{mix}, ΔS^{mix}, and T stand for enthalpy of mixing, the entropy of mixing, and temperature, respectively. When several components are mixed to form a solid solution, the interaction between the components plays an important role. When the interaction between the components is such that

DOI: 10.1201/9781003391388-7

the deviation from the ideal solution is positive, i.e., $\Delta H^{mix} > 0$, the components mixing will not be spontaneous. In such cases, the entropy of mixing plays a crucial role in stabilizing the solid solution. The solution can be achieved when the $T\Delta S^{mix} > \Delta H^{mix}$. In other words, even if ΔH^{mix} is positive, the solid solution can be stable when the $T\Delta S^{mix}$ term more than compensates for the positive enthalpy term. The configurational entropy of mixing can be expressed as:

$$\Delta S_{config} = -R\left(\sum_{i=1}^{n} x_i \ln x_i\right)_{cation} + -R\left(\sum_{i=1}^{n} y_i \ln y_i\right)_{anion} \qquad (7.2)$$

where R, x_i and y_i stands for universal gas constant, the mole fraction of components at cation and anion sites, respectively. In HEOs, the anion being oxygen, the anion sublattice is fixed while several cations are incorporated in the second term of the equation (7.2). Thus, for HEO systems ΔS_{config} is given by:

$$\Delta S_{config} = -R\left(\sum_{i=1}^{n} x_i \ln x_i\right)_{cation}$$

According to the hypothesis presented by Yeh et al. [3] entropy stabilization occurs when both the entropy and enthalpy of mixing are positive and, as a result ΔG^{mix} becomes negative above a certain temperature (T_c). The solution will become unstable below this critical temperature and will either produce an intermetallic compound or produce a low-temperature multiphase mixture. The configurational entropy (S_{conf} or ΔS_{mix}) of the system can be increased by increasing the number of elements in the system. Further, the maximum configurational entropy is achieved when all components are present in an equiatomic composition (Figure 7.1a) [5]. For a given ΔS_{conf}, the solution becomes stable at temperatures greater than T_c [12, 13]. Murty et al. [14] classified alloy materials based on their configurational entropy. When configuration entropy is greater than 1.5R the materials are called high entropy, while the materials having ΔS_{conf} in between 1.5 R and R are called medium entropy and low entropy materials, respectively. A similar nomenclature is also adopted to the oxides system [15].

7.3 SELECTION CRITERIA

It was earlier shown that it is possible to achieve multicomponent (>4) solid solution over the entire composition range, even when the atomic radii of some component cation size differences are greater than 14% and have different electronegativity or anion coordination, as long as the valency of all the constituent cations is same. However, later studies have shown that the initial oxidation state of the starting materials may not be the primary attribute for the formation of the solid solution phase, as the equilibrium can be attained at high temperatures with oxygen vacancy formation, which can be maintained at room temperature upon quenching. This increases the possibilities of different structure formations with aliovalent substitutions. At the same time, vacancy thus obtained would also contribute toward increasing the entropic term. The possibility of aliovalent substitution adds another handle for tailoring the structure and widening the properties.

7.4 ENTROPY-DRIVEN STABILIZATION

The multicomponent solid solution phase can be stabilized when the enthalpy is negative, or the entropic term is greater than the positive enthalpy term. However, in the cases where the enthalpy of mixing is already negative, the entropy does not play any role in the stabilization of the solid solution. On the other hand, when the enthalpy of mixing is positive at lower temperatures, the negative entropic term ($T\Delta S^{mix}$) may not be large enough to overcome the positive entropy term. However, at higher temperatures above T_c, the entropic more than compensates for the positive enthalpy term. Therefore, the reversibility with respect to temperature is a characteristic of entropy-stabilized materials. This was demonstrated by performing a cyclic heat treatment on transition metal (TM)-based rocksalt HEO $(Co_{0.2}Cu_{0.2}Mg_{0.2}Ni_{0.2}Zn_{0.2})O$. In $(Co_{0.2}Cu_{0.2}Mg_{0.2}Ni_{0.2}Zn_{0.2})O$, reversible phase transition was noticed when single-phase solid solution was obtained at high temperature, and the segregation/decomposition into multiple phases (rocksalt, tenorite, and wurtzite) at lower temperature was observed as shown in Figure 7.1b. The reversibility has been observed in medium entropy oxides $(Ni_{0.2}Zn_{0.2}Cu_{0.2})O$ and low entropy oxides $(Ni_{0.2}Zn_{0.2})O$ as well. However, for the medium entropy oxides (MEOs) and low entropy oxides systems, a high processing temperature is necessary for the single-phase formation [16].

7.5 PARAMETER AFFECTING HEOS FORMATION

In general, the same thermodynamic parameter that governs the phase stability of HEAs also governs HEOs [17]. Schematic indicating important factors influencing HEO single-phase formation is depicted in Figure 7.1c. In this section, we have discussed the general guidelines for selecting suitable elements for the formation of single-phase HEOs based on previous studies.

7.5.1 Enthalpy of Mixing

For HEOs, the effect of mixing enthalpy is relatively identical to that of HEAs, which is close to zero. The high-entropy effect may not compensate for a large positive ΔH^{mix} of different-structured metal oxides and requires a considerably higher temperature for phase transition and mixing. In contrast, a large negative enthalpy of formation (ΔH^{mix}) ensures spontaneous mixing where configuration entropy does not play any role in stabilizing solid solution. One of the conditions for the entropy-driven stabilization of solid solution is the different crystal structure or coordination number of some of the constituents. Rost et al. synthesized an HEO with a phase pure rocksalt crystal structure mixing the five elements with varying crystal structures: rocksalt (MgO, CoO, and NiO), tenorite (CuO), and wurtzite (ZnO). Tenorite and wurtzite have a positive enthalpy associated with their structural transition to the rocksalt structure (Figure 7.1d). Therefore, in this case, the formation of a single-phase rocksalt structure is attributed to entropy-driven stabilization. It is postulated that there should be at least one different crystal structure to demonstrate entropy-driven stabilization.

7.5.2 Ionic Radius

Similar to the difference in atomic size radii for HEAs, the difference in ionic radii of cations influences lattice distortion and single-phase HEO stabilization. Metal cations with similar ionic radii are considered for the cubic rocksalt and fluorite crystal structure. However, owing to some geometrical and electrical charge compensation mechanisms (discussed in sections) come into play when cation size differences are more than 14%. For multi-cationic structures, other structure-specific factors also need to be considered. For instance, in the case of perovskite (ABO_3 type), Goldschmidt's tolerance factor (t) is an important parameter for the structure prediction [19]. It is the dimensionless expressed as:

$$t = \frac{\overline{r}_A + r_O}{\sqrt{2}(\overline{r}_B + r_O)}$$

where $\overline{r}_A$ and $\overline{r}_B$ stand for the average ionic radius of cation in sites A and B, respectively while r_O the radius for oxygen. It has been shown that a hexagonal or tetragonal type perovskite phase would form when $t > 1.0$, an orthorhombic or rhombohedral perovskite forms when $t < 0.9$, while a cubic perovskite is most likely when $0.9 \leq t \leq 1.0$ [11].

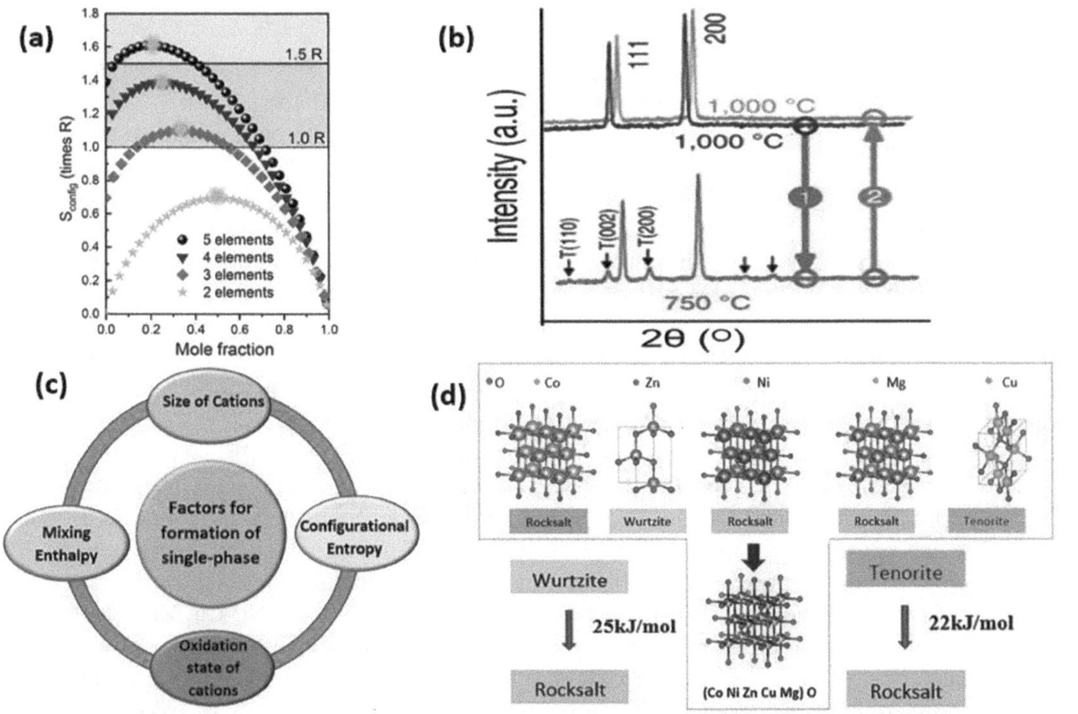

FIGURE 7.1 (a) Correlation between configurational entropy (S_{conf}) and the number of elements and their respective mole fractions. Adapted with permission [18]. Copyright (2023), Oxford Academic Press. (b) Depicts reversibility which is an essential factor for entropy-stabilized oxides. Adapted with permission [5]. Copyright (2015), Nature Communications. (c) Schematic highlights the essential factors for designing single-phase HEOs. (d) Schematic of structural transition in single phase ($Co_{0.2}Cu_{0.2}Mg_{0.2}Ni_{0.2}Zn_{0.2}$)O HEOs.

7.5.3 Oxidation State

For single-phase HEOs formation, the oxidation state of constituent ions plays an important role. In the case of the cubic rocksalt phase, it is assumed that all the cations possess a similar oxidation state of +2 and form MO-type oxides, whereas, in the case of fluorite structure, all the cations should have an average oxidation state of +4 and crystallize in MO_2-type oxides. However, later studies showed that the mutual charge compensation mechanism permits cations with different oxidation states to stabilize in single-phase oxides. In spinel and perovskite-type HEOs, cation with mixed oxygen state interacts. Other factors, such as the coordination number of the cations, also play an important role [17].

7.6 STABILITY LANDSCAPE FOR DIFFERENT CRYSTAL STRUCTURE HEOS

This section will briefly discuss the stability mechanisms for different crystal structures. The uniformity in the size of cations and maintaining electrical charge neutrality in the oxide system is an essential factor that minimizes the enthalpy term. As mentioned earlier, increasing the number of cations in a specific lattice site increases the configurational entropy and thus overcomes the enthalpy penalty. The configurational entropy maximum at equimolar composition for the particular number of cations, and when its value is > 1.5 R, it is referred to as high entropy. However, exploring a single phase with varying compositions would give more freedom in tailoring the properties of these materials. Therefore, exploring the stability landscape of HEO with varying sizes and valency of the component cations is important.

7.6.1 Rocksalt

As discussed earlier, the basic criteria for selecting these cations were based on Hume-Rothery and Paulling's rules that involve uniformity in ionic radii, coordination number, and oxidation state [20]. The initial research on HEOs was concentrated primarily on rocksalt systems with a single Wyckoff site for the cations. $(Co_{0.2}Cu_{0.2}Mg_{0.2}Ni_{0.2}Zn_{0.2})O$ rocksalt HEO was synthesised by mixing five binary oxides and calcined at a temperature > 850°C followed by quenching. Reversible single-phase formation on heating above T_c is the main feature of this type of structure. On cyclic heat treatment, phase transition from single phase to multiphase and vice-versa (reversibility) is evident from the high-temperature XRD pattern recorded at varying temperatures (Figure 7.2b). It has also been observed that when any constituent element is excluded from $(Co_{0.2}Cu_{0.2}Mg_{0.2}Ni_{0.2}Zn_{0.2})O$, multiple secondary phases form on processing in similar conditions (Figure 7.2a). Further similar to the valence electron count tuning by doping in HEAs [21], partial substitution either by aliovalent element in $(Co_{0.2}Cu_{0.2}Mg_{0.2}Ni_{0.2}Zn_{0.2})O$ facilitates tuning of its properties. Ionic charge balance and geometrical compatibility must be considered when designing HEOs [22]. The rocksalt phase formed in the case of $(Co_{0.2}Cu_{0.2}Mg_{0.2}Ni_{0.2}Zn_{0.2})O$ is sturdy to substitutions owing to the charge compensation mechanisms that maintain the charge neutrality between the cationic and anionic sublattices. These compensation mechanisms are often accompanied by oxygen vacancy formation. It has been demonstrated that the partial substitution of Li^{+1} (~0.76 Å) up to $x = 0.3$ can be made to form $(Co, Mg, Ni, Cu, Zn)_{1-x}Li_xO$ while still achieving the single-phase rocksalt structure [23]. The ionic size of Li^{+1} is slightly larger (in octahedral geometry) than the average ionic radius of five divalent cations. To compensate for the Li^{+1} in lattice some of the divalent M^{+2} cations should oxidize to M^{+3} state or oxygen vacancies should form [23,24]. The oxidation of some of the cations from M^{+2} to M^{+3} would also reduce the size of these cations which helps in accommodating the large size of Li^{+1} ions. Figure 7.2b shows two distinct regions with the linear dependency of lattice parameter on the concentration of Li^{+1}. Two different slopes indicate two distinct charge compensation mechanisms involved. Since Co^{+3} is more stable in an octahedral environment than Ni^{+3}, when Li concentration is low ($x < 0.2$), charges are compensated mainly through the oxidation of Co^{+2} to oxidation of Co^{+3} with only traces of Ni^{+3} observed in XPS shown in Figure 7.2c and d. However, when the concentration of Li^{+1} increases ($x > 0.2$), a second mechanism of the creation of oxygen vacancies in the lattice comes into play [23]. The charge compensation can also be achieved by doping Li^{+1} with Ga^{+3} called self-compensation which was shown by forming single-phase phase rocksalt $(Co, Mg, Ni, Cu, Zn)_{0.8}(Li\ Ga)_{0.2}O$ phase [36]. The average ionic radius of Li^{+1} (~0.76 Å) and Ga^{+3} (~0.62 Å) is smaller than the average ionic size of cations in $(Co_{0.2}Cu_{0.2}Mg_{0.2}Ni_{0.2}Zn_{0.2})O$ and a significant reduction in lattice parameter compare to pristine HEOs is noticed. However, when the Ga^{+3} cation (smaller ionic size compared to the average ionic radius of the element in pristine HEOs) is substituted alone, it leads to the evolution of secondary phases [25]. This indicates that the charge compensation is important for the formation of a single-phase solid solution. Some studies were performed to further test the size tolerance by substituting even larger Na^{+1} cation. The ionic radius of Na^{+1} is around 1.02 Å. The size difference is even greater than Hume-Rothery ±14% criterion compared to the average size of the cation present in $(Co_{0.2}Cu_{0.2}Mg_{0.2}Ni_{0.2}Zn_{0.2})O$. Even with such a large cation size difference, a single-phase rocksalt (Co, Mg, Ni, Cu, Zn, Na)O could be achieved on annealing at a somewhat high temperature (1000°C) compared to that of pristine rocksalt HEOs [26]. The lattice parameter after substitution decreased when compared to $(Co_{0.2}Cu_{0.2}Mg_{0.2}Ni_{0.2}Zn_{0.2})O$. From the XPS study (Figure 7.3), it is obvious that only Co^{+2} oxidation to Co^{+3} state and the creation of oxygen vacancies is not sufficient to compensate both the charge and size imbalance in the case Na^{+1} substitution, but some of the Ni^{+2} also took the higher oxidation Ni^{+3}. In addition, Cu was present in both +1 and +2 oxidation states. When a

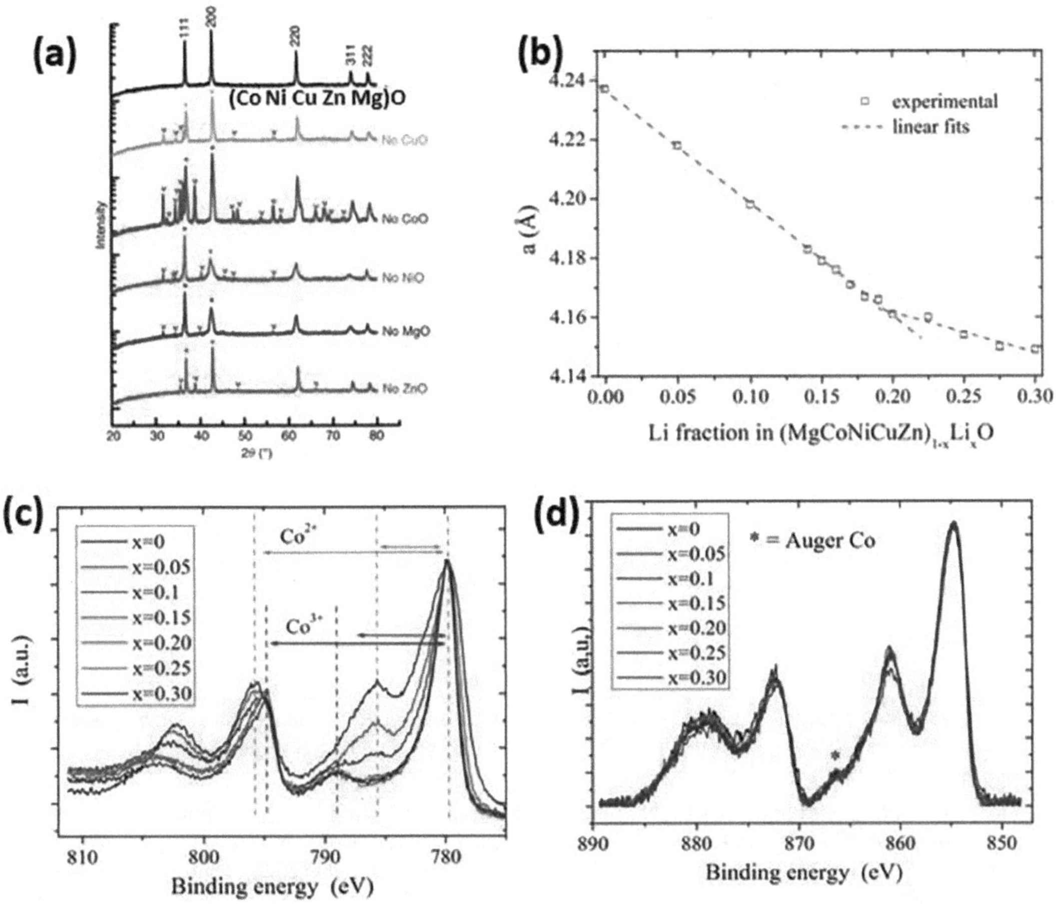

FIGURE 7.2 (a) XRD pattern showing the role of individual cations in single-phase formation. Adapted with permission [21]. Copyright (2015), Nature Communications. (b) Depicts the linear relation between the lattice parameter of (Co, Mg, Ni, Cu, Zn)$_{1-x}$Li$_x$O HEOs and the lithium concentration. Two distinct slopes indicate two compensation mechanisms involved for single-phase HEOs formation (c) XPS spectra of Co 2p in (Co, Mg, Ni, Cu, Zn)$_{1-x}$Li$_x$O showing that all the Co present in divalent state converted to Co^{+3} at (x=0.2) Lithium concentration. (d) XPS spectra of Ni 2p in (Co, Mg, Ni, Cu, Zn)$_{1-x}$Li$_x$O show only a marginal change in spectra upon increasing lithium doping concentration. Adapted with permission [23]. Copyright (2019), Journal of the American Ceramic Society.

combination of Na^{+1}, and Ga^{+3} is substituted in $(Co_{0.2}Cu_{0.2}Mg_{0.2}Ni_{0.2}Zn_{0.2})O$, similar to Li^{+1} and Ga^{+3} co-substitution, the self-compensating mechanism was thought to work for the formation of single phase (Co, Mg, Ni, Cu, Zn)$_{0.8}$(Na Ga)$_{0.2}$O. However, the average ionic radius of Na^{+1} and Ga^{+3} is larger than the average ionic radius of cations in $(Co_{0.2}Cu_{0.2}Mg_{0.2}Ni_{0.2}Zn_{0.2})O$. Similar to these, a wide range of compositions can be designed, enabling tailoring of desired properties.

Isovalent Ca^{+2} (ionic radii ~1 Å) was partially substituted to get insight into the effect of the size difference separately. The large size of the Ca^{+2} results in lattice distortion. To reduce the positive energy contribution associated with the lattice strain, some of the host cations would reduce the atomic radii by going into +3 state or creating oxygen vacancies [27–29]. The large size of Ca^{+2} hinders the complete solubility in the lattice and about ~7% CaO is formed as a

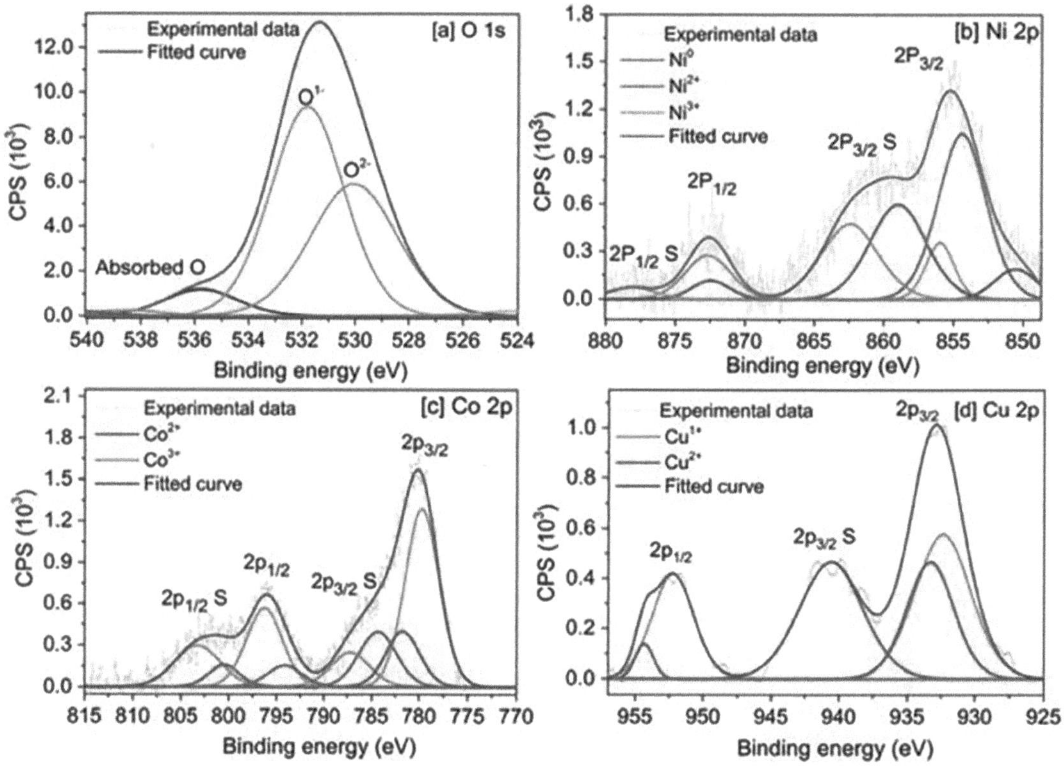

FIGURE 7.3 XPS spectra of (a) O1s (b) Ni2p (c) Co 2p and (d) Cu 2p indicate charge compensation mechanism in single phase (Co, Mg, Ni, Cu, Zn, Na)O HEOs which include oxidation of divalent cation as well as the formation of oxygen vacancies. Adapted with permission [26]. Copyright (2020), Dalton Transactions.

secondary phase [26]. However, (Co, Mg, Ni, Cu, $Zn)_{1-x}Ca_xO$ non-equimolar HEOs (NE–HEOs) with Ca doping ($x \leq 0.1$) were synthesized, and properties were explored. The possible mechanism for their successful fabrication includes partial oxidation of Co^{+2} and Ni^{+2} as well as the creation of oxygen vacancies.

7.6.2 Fluorite

In recent years, the concept of entropy-driven stabilization has been widened to other oxides systems such as rare earth (RE) and TM oxides with different crystal structures including cubic fluorite structure. In addition to the previous criteria for selecting elements we discussed in the earlier section, some other criteria are also important for the formation of a single-phase fluorite structure. It is postulated that Cerium (Ce) and Zirconium (Zr) element are crucial elements for any composition to crystallize in a single-phase fluorite structure [9,30]. RE-based systems without Ce or Zr always result in multiple secondary phases, regardless of the number of incorporated elements or synthesis parameters, as shown in Figure 7.4a and b, where Ce^{+4} is demonstrated as an element for stabilization of solid solution fluorite structure. This might be due to the fact that +4 is their most stable oxidation state. A more detailed study is required to gain insight into the other competing stability mechanisms for high entropy fluorite structures. The most explored fluorite HEOs include $(CeRE)O_{2-\delta}$ and $(HfZrCeM)O_{2-\delta}$, where RE stands for rear earth elements such as La, Pr, Sm, Y, etc., M represents for other metals, while δ depends on the average cation valency. The other explored systems

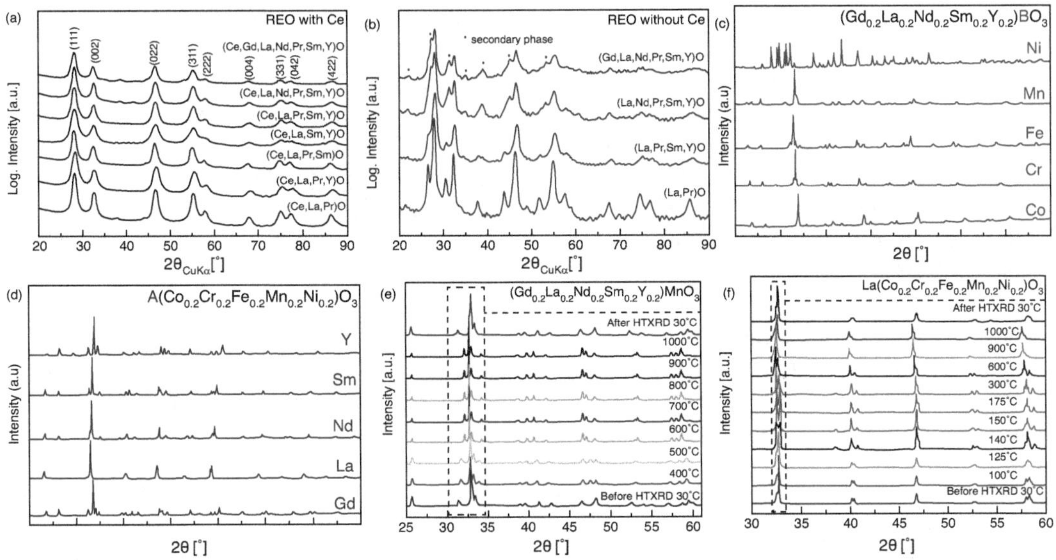

FIGURE 7.4 (a and b) XRD pattern showing the role of Ce^{+4} in the design of single-phase fluorite HEOs. Without Ce^{+4}, system crystallizes in secondary multiphase oxides. Adapted with permission [30]. Copyright (2017), *Material Research Letters*. (c) XRD pattern of $(5A_{0.2})BO_3$ type HEPO, where B=Co, Cr, Fe, Mn, Ni. (d) XRD pattern of A $(5B_{0.2})O_3$ type HEPO, where A site cation comprised of Gd, La, Nd, Sm, Y. (e & f) XRD pattern depicts entropy-driven stabilization of$(Gd_{0.2}La_{0.2}Nd_{0.2}Sm_{0.2}Y_{0.2})MnO_3$, and La $(Co_{0.2}Cr_{0.2}Fe_{0.2}Mn_{0.2}Ni_{0.2})O_3$ at a wide range of temperatures. Adapted with permission [19]. Copyright (2018), Journal of European ceramic society.

are based on the general formula $RE_2TM_2O_7$, where TM stands for transition metals. In this case, a single-phase high entropy fluorite structure forms when the ratio of the ionic radius of the RE cation to that of the TM cation is close to 1.46 [30].

7.6.3 Perovskite

One of the popular high entropy material groups includes high entropy perovskite oxide (HEPO). The general formula for perovskite is ABO_3, in which two cation sublattices A and B sit at two types of cation sites. The perovskite structure is comprised up of twofold coordinated A-type cations, sixfold coordinated B-type cations, and octahedral oxygen anions. These perovskites may exist in cubic, orthorhombic, or hexagonal crystal structures. The Goldschmidt's tolerance factor predicts the stability of different types of perovskite structures, which has been discussed in an earlier section [11]. Ionic radii, oxidation state, and synergistic effect of multiple cations are the other important factors that strongly impact structural parameters and can lead to unusual phase compositions in HEOs. For instance, $(Gd_{0.2}La_{0.2}Nd_{0.2}Sm_{0.2}Y_{0.2})MnO_3$ has a tolerance factor similar to $NdMnO_3$ and $SmMnO_3$, which favors orthorhombic structure; however, at ambient temperature, HEO–$(Gd_{0.2}La_{0.2}Nd_{0.2}Sm_{0.2}Y_{0.2})MnO_3$ exhibits a dispersion of eight orthorhombic perovskites phases [31]. High entropy perovskite can be broadly represented as general formulae (Sr Ba)$(M)O_3$ and $(RE)(TM)O_3$, where M stands for metals, RE represent rare earth elements, and TM refers to transition metals.

Jiang et al. [11] synthesized six single-phase HEPO having general formulae $(Sr)(M)O_3$ and $(Ba)(M)O_3$. In their study, only the B cation site was substituted by five different elements in equiatomic ratio, and the A site was occupied by one or at most two elements. They correlated

Goldschmidt's tolerance factor with the single-phase formation and thermal stability of multi-cation HEPO. Jiang et al. [11] Goldschmidt's tolerance factor close to unity ($t \approx 1$) is probably a necessary but insufficient condition to form a single-phase HEPO. On the other hand, Sarkar et al. [31] studied a more complex HEPO system with a completely different composition in which cation site A is populated with Gd, La, Nd, Sm, Y, and cation site B with Co, Cr, Fe, Mn, and Ni in equimolar ratios. The cations were selected such that all the cations on specific cation sites A and B should have the same ionic radii, specific oxidation state, and coordination number and at least one should exhibit a different prototype structure (e.g. $YMnO_3$). These five elements in cation site A or Site B increase the configurational entropy and thus highlights the possible role in entropic stabilization. Five {($5A0.2)CoO_3$, $(5A_{0.2})FeO_3$, $Gd(5B_{0.2})O_3$, $La(5B_{0.2})O_3$, $Nd(5B_{0.2})O_3$} system out of ten senary system (six cationic) crystalized in orthorhombic single-phase HEPO, whereas the other five compositions were multiphase mixture even at high temperature. It was believed that in these systems, the positive enthalpy term was quite high, which couldn't be compensated by the entropic term even at high temperatures (up to 1500°C, Figure 7.4c and d) The same research group also reported an equiatomic ten cation system (five cations on A site and five cations on B site) single-phase orthorhombic $(Gd_{0.2}La_{0.2}Nd_{0.2}Sm_{0.2}Y_{0.2})$ $(Co_{0.2}Cr_{0.2}Fe_{0.2}Mn_{0.2}Ni_{0.2})O_3$ HEPO. Out of all compositions, $(Gd_{0.2}La_{0.2}Nd_{0.2}Sm_{0.2}Y_{0.2})MnO_3$, $La(Co_{0.2},Cr_{0.2},Fe_{0.2}Mn_{0.2}Ni_{0.2})O_3$ was found to be reversible (Figure 7.4e and f) on thermal cycling, showing that solid solution phases with general formula $(5A_{0.2})MnO_3$, La $(5B_{0.2})O_3$ are indeed stabilized due to high configurational entropy [19].

7.6.4 Spinel

Spinels (AB_2O_4) have two distinct cation sites, just like perovskites. Spinels can also be represented by general formula M_3O_4, where M assumes both +2 and +3 states. In this structure, divalent (+2) cations should ideally occupy the tetrahedral A-sites while the trivalent (+3) cations should occupy the octahedral B-sites. $(Co_{0.2}Cr_{0.2}\ Fe_{0.2}\ Mn_{0.2}\ Ni_{0.2})_3O_4$ was first reported spinel type high entropy oxide (S-HEO) using basic criteria for the selecting cation such as similar ionic radius of constituting cation and element possess good solubility in a binary and ternary subsystem. For the formation of single-phase S-HEOs, some elements (TM) afford multiple valence states. For instance, in the case of $(Co_{0.2}Cr_{0.2}\ Fe_{0.2}\ Mn_{0.2}\ Ni_{0.2})_3O_4$, Co, Ni, Fe, exists in both +2 as well as +3 oxidation states, whereas Mn exists in +4 and +3 oxidation states [4] (Figure 7.5). Further aliovalent doping, such as Li^{+1}, results in a change of oxidation of constituent cation or oxygen vacancy formation. Since S-HEOs contain TM that can afford multiple oxidation states and thus charge compensation can be expected. As a result of charge compensation, a structural transition was observed from S-HEOs to R-HEOs [34]. According to Fracchia et al. [35], new ferrites can be developed from spinel HEOs by efficiently synthesizing an eight-cation spinel $(Co,Mg,Mn,Ni,Zn)(Al,Co,Cr,Fe,Mn)_2O_4$. This study revealed that Al, Cr, and Ni occupied the octahedra site, whereas Mg and Zn were present in the tetrahedral site, while mixed site occupancy is reported for Mn, Co, and Fe. For this arrangement of cations, configurational entropy turns out to be maximum (~3.3R) (Table 7.1).

7.6.5 Pyrochlore

These oxides are a significant structural/functional ceramics class that takes the general formula $A_2B_2O_7$ and contains two different types of A and B cations. The general formula of the pyrochlores can be written as $(RE)_2(TM)_2O_7$ wherein the A cations are typically RE elements and the B cations are TM [10]. In addition to ionic radii and valence state of the elements, size disorder adds to the main driving force for the single-phase high entropy pyrochlore. The average radius ratio of A and B cations $\left({r_a}/{r_b} \right)$ is an important factor for pyrochlore–fluorite phase transition in HEO. In the case of $(RE)_2(TM)_2O_7$ type pyrochlore structure, it is postulated that if the ratio of the ionic size of RE cations (r_a) to that of TM cation (r_b) is in the range of 1.46–1.78, the composition crystallizes as

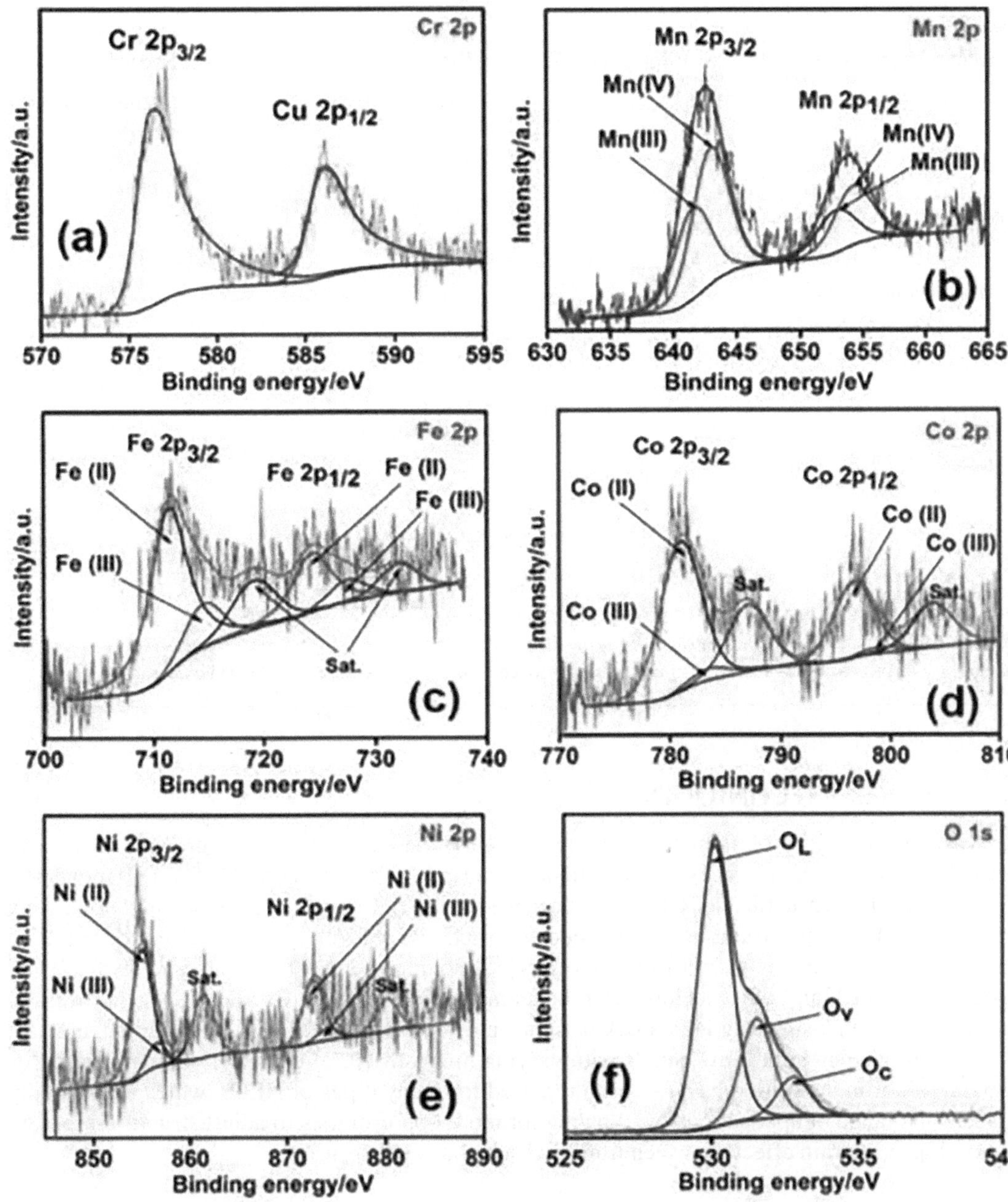

FIGURE 7.5 XPS spectra of (a) Cr 2p (b) Mn 2p (C) Fe 2p (d) Co 2p (e) Ni 2p (f) Ols in $(Co_{0.2}Cr_{0.2}\ Fe_{0.2}\ Mn_{0.2}\ Ni_{0.2})_3O_4$ showing cations present in multiple oxidation states in the lattice. Adapted with permission [36]. Copyright (2021), *Journal of Energy Storage.*

single-phase pyrochlore HEOs, whereas when this ratio ranges between 1.38 and 1.45 high entropy phase shows fluorite structure. The multiple complex phases appear when the ratio $\left(r_a/r_b\right)$ lies between 1.78 and 1.80. For instance, rear earth zirconates $((RE)_2Zr_2O_7)$, where the ratio between the ionic radius of rear earth cation (in +3 oxidation state) and that of Zr^{+4} lies in the range 1.46–1.78, are shown to crystallize in pyrochlore structure [37].

TABLE 7.1
Showing Various Published Single-Phase Compositions in Different Crystal Structures

HEOs Compositions	Entropy stabilization	Literature reference
Rocksalt Type HEOs		
$(Co_{0.2}Cu_{0.2}Mg_{0.2}Ni_{0.2}Zn_{0.2})O$ (several derivatives of this system contain additional elements like Li, Na, Ga, and K)	Yes	[5 25–27,32]
Fluorite Type HEOs		
$(Ce_{0.2}La_{0.2}Pr_{0.2}Sm_{0.2}Y_{0.2})O_{2-\delta}$	No	[30]
$(Nd_{0.16}Ce_{0.16}La_{0.16}Pr_{0.16}Sm_{0.16}Y_{0.16})O_{2-\delta}$	No	[33]
$(Ce_{0.2}\ Zr_{0.2}\ Hf_{0.2}\ Sn_{0.2}\ Ti_{0.2})O_{2-\delta}$	Yes	[9]
Perovskite-Type HEOs		
$Ba(Zr_{0.2}\ Hf_{0.2}\ Sn_{0.2}\ Ti_{0.2}\ Nb_{0.2})O_3$	No[a]	[11]
$Sr(Zr_{0.2}\ Hf_{0.2}\ Sn_{0.2}\ Ti_{0.2}\ Nb_{0.2})O_3$	No	[11]
$(Gd_{0.2}La_{0.2}Nd_{0.2}Sm_{0.2}Y_{0.2})MnO_3$	Yes	[19]
Spinel Type HEOs		
$(Co_{0.2}Cr_{0.2}\ Fe_{0.2}\ Mn_{0.2}\ Ni_{0.2})_3O_4$	No[a]	[4]

Source: Adapted with permission [12]. Copyright (2019) *Advanced Materials*.

[a] Cyclic heat (reversibility) treatment not studied in the literature.

The table also depicts a clear distinction between entropy-driven oxides and non-entropy-stabilized oxides.

7.7 SYNTHESIS TECHNIQUES

The ease and flexibility of synthesis are one of the primary reasons for the growing interest in HEOs. There are various processing techniques reported for synthesizing HEOs. Different morphologies and microstructures of HEOs can be achieved depending on the synthesis and processing techniques. For example, traditional solid-state approaches enable the synthesis of microcrystalline HEOs, whereas techniques such as spray pyrolysis [38], solution combustion synthesis (SCS) [39,40], sonochemistry [41], and other techniques produce nanocrystalline HEOs [12]. Various processes, such as conventional sintering [42], spark plasma sintering, flash sintering [43], and so on, are also known for producing bulk HEO pellets with varying microstructural characteristics. Additionally, epitaxial thin film deposition has also been reported for many types of HEOs, which undoubtedly broadens the range of applications by allowing for more opportunities to adjust structure-property relationships via strain effects between films and substrates [12,15,16].

7.8 CONCLUSIONS

This chapter highlights the enthalpy-driven stabilization as entropy can be manipulated in terms of both compositional complexity and structural disorder thus widening the access to functional properties and applications. High ΔS_{mix} associated with multiple components or ions in near-equimolar ratios, thermodynamically favors the synthesis of a single-phase solid solution. Subsequent research has revealed that various thermodynamic and structural parameters and kinetic factors play a key role in designing single-phase solid solutions. When considering the charge compensation mechanism to accommodate larger cation or aliovalent cation substitution, any composition should contain multivalent cations because the presence of multivalent cation in any composition decides the level of charge neutrality compensation mechanism. Various studies revealed that accommodating large-size cation oxygen stoichiometry is essential. In some studies, it has been observed that Hume-Rothery and Paulling's rule regarding size limitation and the same oxidation states of cation are not essential criteria for the formation of phase pure crystal structure. From a fundamental standpoint, these various phase stability mechanisms make HEOs special, which encourages more investigation to better understand the causes of phase stability in HEOs.

REFERENCES

[1] S. GUO, C.T. LIU, Phase stability in high entropy alloys: Formation of solid-solution phase or amorphous phase, *Prog. Nat. Sci: Mater. Int.* 21 (2011) 433–446.

[2] B. Cantor, I.T.H. Chang, P. Knight, A.J.B. Vincent, Microstructural development in equiatomic multicomponent alloys, *Mater. Sci. and Eng. A.* 375–377 (2004) 213–218.

[3] J.W. Yeh, S.K. Chen, S.J. Lin, J.Y. Gan, T.S. Chin, T.T. Shun, C.H. Tsau, S.Y. Chang, Nanostructured high-entropy alloys with multiple principal elements: Novel alloy design concepts and outcomes, *Adv. Eng. Mater.* 6 (2004) 299–303. https://doi.org/10.1002/adem.200300567.

[4] A. Mao, F. Quan, H.Z. Xiang, Z.G. Zhang, K. Kuramoto, A.L. Xia, Facile synthesis and ferrimagnetic property of spinel $(CoCrFeMnNi)_3O_4$ high-entropy oxide nanocrystalline powder, *J. Mol. Struct.* 1194 (2019) 11–18.

[5] C.M. Rost, E. Sachet, T. Borman, A. Moballegh, E.C. Dickey, D. Hou, J.L. Jones, S. Curtarolo, J.P. Maria, Entropy-stabilized oxides, *Nat. Commun.* 6 (2015) 9485.

[6] J. Gild, Y. Zhang, T. Harrington, S. Jiang, T. Hu, M.C. Quinn, W.M. Mellor, N. Zhou, K. Vecchio, J. Luo, High-entropy metal diborides: A new class of high-entropy materials and a new type of ultrahigh temperature ceramics, *Sci. Rep.* 6 (2016) 37946.

[7] T.K. Chen, T.T. Shun, J.W. Yeh, M.S. Wong, Nanostructured nitride films of multi-element high-entropy alloys by reactive DC sputtering, *Surf. Coat. Technol.* 188–189 (2004) 193–200.

[8] M. Braic, V. Braic, M. Balaceanu, C.N. Zoita, A. Vladescu, E. Grigore, Characteristics of (TiAlCrNbY)C films deposited by reactive magnetron sputtering, *Surf. Coat. Technol.* 204 (2010) 2010–2014.

[9] J. Gild, M. Samiee, J.L. Braun, T. Harrington, H. Vega, P.E. Hopkins, K. Vecchio, J. Luo, High-entropy fluorite oxides, *J. Eur. Ceram. Soc.* 38 (2018) 3578–3584.

[10] Z. Teng, L. Zhu, Y. Tan, S. Zeng, Y. Xia, Y. Wang, H. Zhang, Synthesis and structures of high-entropy pyrochlore oxides, *J. Eur. Ceram. Soc.* 40 (2020) 1639–1643.

[11] S. Jiang, T. Hu, J. Gild, N. Zhou, J. Nie, M. Qin, T. Harrington, K. Vecchio, J. Luo, A new class of high-entropy perovskite oxides, *Scr. Mater.* 142 (2018) 116–120.

[12] A. Sarkar, Q. Wang, A. Schiele, M.R. Chellali, S.S. Bhattacharya, D. Wang, T. Brezesinski, H. Hahn, L. Velasco, B. Breitung, High-entropy oxides: Fundamental aspects and electrochemical properties, *Adv. Mater.* 31 (2019) 1806236.

[13] B.L. Musicó, D. Gilbert, T.Z. Ward, K. Page, E. George, J. Yan, D. Mandrus, V. Keppens, The emergent field of high entropy oxides: Design, prospects, challenges, and opportunities for tailoring material properties, *APL Mater.* 8 (2020) 040912.

[14] B.S. Murty, J.-W. Yeh, S. Ranganathan, P.P. Bhattacharjee, *High-Entropy Alloys*, Elsevier, 2019.

[15] C. Oses, C. Toher, S. Curtarolo, High-entropy ceramics, *Nat. Rev. Mater.* 5 (2020) 295–309.

[16] S. Akrami, P. Edalati, M. Fuji, K. Edalati, High-entropy ceramics: Review of principles, production and applications, *Mater. Sci. Eng. R.* 146 (2021) 100644.

[17] A. Amiri, R. Shahbazian-Yassar, Recent progress of high-entropy materials for energy storage and conversion, *J. Mater. Chem. A.* 9 (2021) 782–823.

[18] A. Sarkar, H. Hahn, R. Kruk, High Entropy Oxides, in: A. S. M. A. Haseeb (Ed.), *Encyclopedia of Materials: Electronics*, Academic Press, Oxford, 2023: pp. 536–545.

[19] A. Sarkar, R. Djenadic, D. Wang, C. Hein, R. Kautenburger, O. Clemens, H. Hahn, Rare earth and transition metal based entropy stabilized perovskite type oxides, *J. Eur. Ceram. Soc.* 38 (2018) 2318–2327.

[20] A. Salian, S. Mandal, Entropy stabilized multicomponent oxides with diverse functionality-a review, *Crit Rev. Solid State Mater. Sci.* 47(2022) 142–193.

[21] D.B. Miracle, O.N. Senkov, A critical review of high entropy alloys and related concepts, *Acta Mater.* 122 (2017) 448–511.

[22] A. Sarkar, B. Breitung, H. Hahn, High entropy oxides: The role of entropy, enthalpy and synergy, *Scr. Mater.* 187 (2020) 43–48.

[23] N. Osenciat, D. Bérardan, D. Dragoe, B. Léridon, S. Holé, A.K. Meena, S. Franger, N. Dragoe, Charge compensation mechanisms in Li-substituted high-entropy oxides and influence on Li superionic conductivity, *J. Am. Ceram. Soc.* 102 (2019) 6156–6162.

[24] J. Wang, Y. Cui, Q. Wang, K. Wang, X. Huang, D. Stenzel, A. Sarkar, R. Azmi, T. Bergfeldt, S.S. Bhattacharya, R. Kruk, H. Hahn, S. Schweidler, T. Brezesinski, B. Breitung, Lithium containing layered high entropy oxide structures, *Sci. Rep.* 10 (2020) 18430.

[25] D. Bérardan, S. Franger, D. Dragoe, A.K. Meena, N. Dragoe, Colossal dielectric constant in high entropy oxides, *Phys. Status Solidi - Rapid Res. Lett.* . 10 (2016) 328–333.

[26] N.J. Usharani, R. Shringi, H. Sanghavi, S. Subramanian, S.S. Bhattacharya, Role of size, alio-/multi-valency and non-stoichiometry in the synthesis of phase-pure high entropy oxide (Co,Cu,Mg,Na,Ni,Zn)O, *Dalton Trans.* 49 (2020) 7123–7132.

[27] G. Anand, A.P. Wynn, C.M. Handley, C.L. Freeman, Phase stability and distortion in high-entropy oxides, *Acta Mater.* 146 (2018) 119–125.

[28] A. Kretschmer, D. Holec, K. Yalamanchili, H. Rudigier, M. Hans, J.M. Schneider, P.H. Mayrhofer, Strain-stabilized Al-containing high-entropy sublattice nitrides, *Acta Mater.* 224 (2022).

[29] S.T. Murphy, H. Lu, R.W. Grimes, General relationships for isovalent cation substitution into oxides with the rocksalt structure, *J. Phy. Chem. Solids.* 71 (2010) 735–738.

[30] R. Djenadic, A. Sarkar, O. Clemens, C. Loho, M. Botros, V.S.K. Chakravadhanula, C. Kübel, S.S. Bhattacharya, A.S. Gandhi, H. Hahn, Multicomponent equiatomic rare earth oxides, *Mater. Res. Lett.* 5 (2017) 102–109.

[31] A. Sarkar, R. Djenadic, D. Wang, C. Hein, R. Kautenburger, O. Clemens, H. Hahn, Rare earth and transition metal based entropy stabilized perovskite type oxides, *J. Eur. Ceram. Soc.* 38 (2018) 2318–2327.

[32] N. Qiu, H. Chen, Z. Yang, S. Sun, Y. Wang, Y. Cui, A high entropy oxide ($Mg_{0.2}Co_{0.2}Ni_{0.2}Cu_{0.2}Zn_{0.2}O$) with superior lithium storage performance, *J. Alloys. Compd.* 777 (2019) 767–774.

[33] A. Sarkar, C. Loho, L. Velasco, T. Thomas, S.S. Bhattacharya, H. Hahn, R. Djenadic, Multicomponent equiatomic rare earth oxides with a narrow band gap and associated praseodymium multivalency, *Dalton Trans.* 46 (2017) 12167–12176.

[34] J. Wang, D. Stenzel, R. Azmi, S. Najib, K. Wang, J. Jeong, A. Sarkar, Q. Wang, P.A. Sukkurji, T. Bergfeldt, M. Botros, J. Maibach, H. Hahn, T. Brezesinski, B. Breitung, Spinel to rock-salt transformation in high entropy oxides with Li incorporation, *Electrochem.* 1 (2020) 60–74.

[35] M. Fracchia, M. Manzoli, U. Anselmi-Tamburini, P. Ghigna, A new eight-cation inverse high entropy spinel with large configurational entropy in both tetrahedral and octahedral sites: Synthesis and cation distribution by X-ray absorption spectroscopy, *Scr. Mater.* 188 (2020) 26–31.

[36] B. Talluri, M.L. Aparna, N. Sreenivasulu, S.S. Bhattacharya, T. Thomas, High entropy spinel metal oxide $(CoCrFeMnNi)_3O_4$ nanoparticles as a high-performance supercapacitor electrode material, *J. Energy Storage.* 42 (2021) 103004.

[37] J. Wu, X. Wei, N.P. Padture, P.G. Klemens, M. Gell, E. García, P. Miranzo, M.I. Osendi, Low-thermal-conductivity rare-earth zirconates for potential thermal-barrier-coating applications, *J. Am. Ceram. Soc.* 85 (2002) 3031–3035.

[38] N.J. Usharani, P. Arivazhagan, T. Thomas, S.S. Bhattacharya, Factors determining the band gap of a nanocrystalline multicomponent equimolar transition metal based high entropy oxide (Co,Cu,Mg,Ni,Zn)O, *Mater. Sci. Eng. B Solid State Mater. Adv. Technol.* 283 (2022) 115847.

[39] A. Gautam, M.I. Ahmad, Low-temperature synthesis of five component single phase high entropy oxide, *Ceram. Int.* 47 (2021) 22225–22228.

[40] J. Sushil, A. Kumar, A. Gautam, Md. I. Ahmad, High entropy phase evolution and fine structure of five component oxide, *Mater. Chem. Phys.* 259 (2021) 124014

[41] H. Chen, W. Lin, Z. Zhang, K. Jie, D.R. Mullins, X. Sang, S.Z. Yang, C.J. Jafta, C.A. Bridges, X. Hu, R.R. Unocic, J. Fu, P. Zhang, S. Dai, Mechanochemical synthesis of high entropy oxide materials under ambient conditions: Dispersion of catalysts via entropy maximization, *ACS Mater. Lett.* 1 (2019) 83–88.

[42] X.C. Yang, W. Riehemann, M. Dubiel, H. Hofmeister, Nanoscaled ceramic powders produced by laser ablation, *Mater. Sci. Eng. B Solid State Mater. Adv. Technol.*, 95 (2002) 299–307

[43] A. Kumar, G. Sharma, A. Aftab, M.I. Ahmad, Flash assisted synthesis and densification of five component high entropy oxide (Mg, Co, Cu, Ni, Zn)O at 350°C in 3 min, *J. Eur. Ceram. Soc.* 40 (2020) 3358–3362.

8 Mechanical and Electrical Properties of High Entropy Materials

Sheetal Kumar Dewangan, Cheenepalli Nagarjuna, and Vinod Kumar

8.1 INTRODUCTION

The process of transforming pure metal into alloys and superalloys has played a vital role in the creation of new metallic materials since ancient times. This development has progressed in tandem with the advancement of civilization, from the Stone Age to the Bronze Age, the Iron Age, and modern time. Figure 8.1 depicts a schematic diagram illustrating the various stages in the evolution of metallic materials. In the past few years, there has been a significant interest in high entropy alloys (HEAs) owing to their exceptional physical and mechanical features, including high strength, ductility, and corrosion resistance [1,2]. Unlike conventional materials, HEAs are composed of at least five primary elements, with each element having nearly the same concentration ranging between 5 and 35 at. %, which results in the formation of simple solid solution phases such as face-centered cubic (FCC), body-centered cubic (BCC), and hexagonal close-packed (HCP) structures. To date, most of the research on HEAs has concentrated on the development of phase, microstructure, and mechanical properties [3].

Apart from superior mechanical performance, HEAs possess electrical properties such as electrical resistivity, hall coefficient, thermoelectric properties, catalysts, and superconductivity [5]. These properties can be influenced by several factors including alloy composition, crystal structure, valance electron concentration, and lattice distortion effects. Since the presence of multiple elements in HEAs, complex electronic structures can be formed, including localized and delocalized states. So far, there are no extensive works that have been made to understand the electrical properties of HEAs due to their vast compositional space. Therefore, the understanding of the fundamental electronic behavior of HEAs has become active research in recent years to explore them in potential functional applications such as energy storage and catalysis.

HEAs are a new class of alloys that are composed of multiple elements in roughly equal proportions, typically consisting of three or more principal elements. Due to their unique composition and microstructure, HEAs display exceptional mechanical characteristics, such as high strength, high ductility, and excellent resistance to fatigue and wear.

8.1.1 Different Mechanical Characteristics of HEAs

8.1.1.1 The Hardness Behavior

The hardness of HEAs depends on several aspects such as the configuration of the alloy, the processing route, and the examination used. Generally, HEAs have high hardness values because of their complex microstructures, which often comprise multiple phases and crystal structures. One of the main challenges in studying the hardness of HEAs is the lack of a clear relationship between

DOI: 10.1201/9781003391388-8

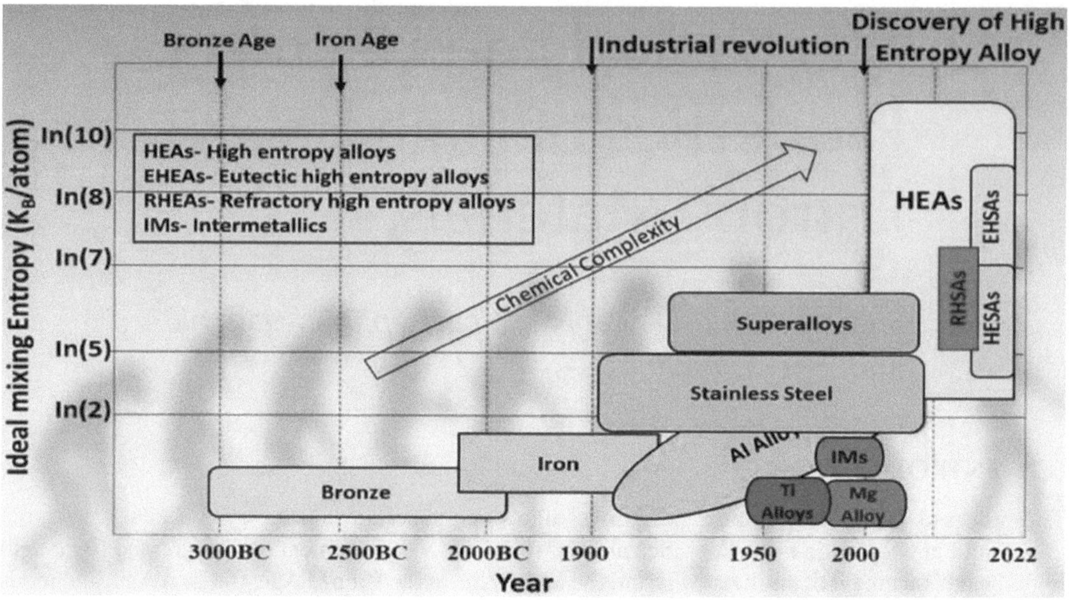

FIGURE 8.1 Development of material relating to increases in chemical complexity. Adapted with permission [4]. Copyright (2021), Elsevier.

microstructure and mechanical properties. This is because HEAs have a high degree of unpredictability and disorder, making it difficult to predict their behavior based on conventional material science principles [1, 4, 6].

Another challenge is the processing of HEAs, which can affect their hardness and further mechanical properties. For example, the use of different heat treatment methods, deformation rates, and strain rates can all impact the microstructure and therefore the hardness of HEAs. Despite these challenges, researchers and scientists continue to study the hardness of HEAs by means of a variety of techniques, including microhardness testing, nanoindentation, and in-situ mechanical testing. These studies have discovered that HEAs can have hardness values ranging from relatively low to enormously high, dependent on the specific composition and synthesis parameters. Overall, the hardness of HEAs is a complex and multifaceted subject demands further research. However, these materials demonstrate promise as high-performance materials for an extensive range of usage due to their unique properties and potential for tailoring their microstructure and mechanical properties [1].

8.1.1.2 Different Strengthening Mechanisms in HEAs

HEA has unique mechanical properties that are attributed to its complex microstructure and a high degree of randomness [7–9]. The hardness of HEAs is influenced by several mechanisms, which are described below:

i. **Solid solution strengthening**: HEAs contain multiple principal elements in nearly equal atomic percentages, resulting in a random solid solution microstructure. This microstructure can contribute to the hardness of HEAs by improving the resistance dislocation movement, known as solid solution strengthening [10].
ii. **Grain size refinement**: HEAs often contain multiple phases with different crystal structures, resulting in a complex grain structure. Reducing the grain size can improve the hardness of HEAs by enhancing the number of grain boundaries, which act as barriers to dislocation movement [11].

iii. **Precipitation hardening**: Some HEAs can form precipitates due to the presence of multiple elements. These precipitates can impede dislocation movement, resulting in increased hardness [12,13].
iv. Transformation-induced plasticity (TRIP): HEAs can undergo phase transformations during deformation, which can enhance the hardness and other mechanical properties. For example, TRIP can occur when a soft phase transforms to a hard phase during deformation, increasing the hardness of the material [14,15].
v. **Work hardening**: HEAs can undergo work hardening during deformation, which occurs when dislocations become trapped and accumulate in the material, increasing hardness.
vi. **Stacking fault energy (SFE)**: HEAs with low SFE values have higher hardness due to the presence of stacking faults, which can act as blockades to dislocation movement [16].

8.1.1.3 Tensile Properties of HEAs

Overall, the hardness mechanisms in HEAs are complex and multifaceted and depend on a variety of factors including the alloy composition, microstructure, processing conditions, and testing methods used. By understanding these mechanisms, researchers can tailor HEAs' microstructure and mechanical properties for specific applications [12,17,18]. In terms of tensile properties, HEAs generally exhibit high strength and good ductility. The strength of HEAs can be attributed to several factors, including the solid solution strengthening effect from the multiple principal elements, the development of nanocrystalline and/or amorphous phases, and the presence of coherent precipitates. The ductility of HEAs is often related to their ability to undergo deformation via multiple slip systems as well as the suppression of strain localization and crack initiation.

Studies have shown that the tensile properties of HEAs can be influenced by various factors, including the composition and processing conditions of the alloy, in addition to the testing temperature and strain rate. For example, increasing the content of certain elements, such as aluminum or titanium, can improve the tensile strength of HEAs while reducing the testing temperature can improve their ductility. Overall, the unique composition and microstructure of HEAs make them promising candidates for an extensive range of structural applications, where high strength and ductility are critical properties. However, further research is needed to fully understand the fundamental mechanisms that govern the mechanical behavior of these materials and to optimize their properties for specific applications [19,20].

8.1.1.4 Compression Properties of HEAs

HEAs exhibit a unique microstructure that is characterized by the development of solid solution phases, nanocrystalline phases, and/or intermetallic compounds, resulting in exceptional mechanical properties. In terms of compression properties, HEAs typically exhibit high strength, good ductility, and excellent resistance to deformation and damage. The strength of HEAs under compression is typically attributed to several factors, including the presence of multiple principal elements that increase the number of possible deformation mechanisms, along with the formation of nanocrystalline and/or amorphous phases that can impede dislocation motion. In addition, the presence of precipitates and other microstructural features can enhance the strength of HEAs under compression [21].

The ductility of HEAs under compression is generally good, although it can vary depending on the specific alloy composition and processing conditions. Studies have shown that increasing the content of certain elements, such as aluminum or titanium, can improve the ductility of HEAs under compression while reducing the testing temperature can also enhance their ductility. Furthermore, HEAs have been shown to exhibit excellent resistance to deformation and damage under compression, which is attributed to their unique microstructure and deformation mechanisms. For example, HEAs can undergo deformation via multiple slip systems and exhibit a high degree of work hardening, which helps to suppress localized deformation and improve the overall ductility of the material. In general, the compression properties of HEAs are promising for a wide range of structural applications, where high strength and ductility are critical properties. However, more research is needed to fully understand the fundamental mechanisms that govern the mechanical behavior of these materials and to optimize their properties for specific applications [1,22].

8.2 STRENGTHENING AND TOUGHENING MECHANISMS OF HEAS

Certain shortcomings in the features have come to light due to the development of HEAs. For instance, refractory HEAs, such as the Re–Mo–Ta–W HEA with a maximum strength of 1451 MPa and a failure strain of 5.69% [23], exhibit low plasticity and toughness despite having high strength, hardness, and strong wear resistance. Although HEAs can be strengthened and more challenging using conventional ways, a set of mechanisms for these processes has not yet been developed [24].

The strengthening of solid solutions is primarily a result of alterations in atomic size. When solute atoms are introduced, they deform the lattice and create a stress field, which increases the resistance to dislocation movement. This type of strengthening can be further categorized into displacement solution strengthening and interstitial solution strengthening, as shown in Figure 8.2. In the case of HEAs, solid solution strengthening can be achieved through a combination of these mechanisms. The elastic interaction between dislocations and solute atoms, which is influenced by the size effect and modulus relationship, is the primary cause of solid solution strengthening. Yeh et al. noted that the lack of a concept of the matrix and the fact that all atoms are solute atoms provide a higher saturation degree of solid solution as the cause of the high hardness of HEAs.

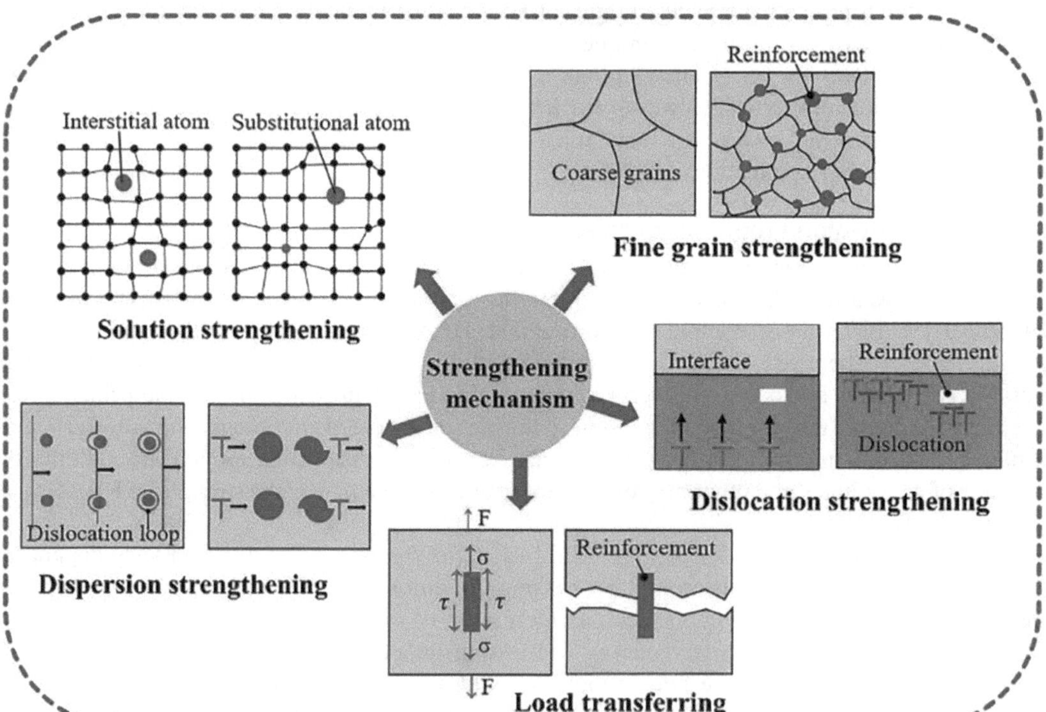

FIGURE 8.2 Representation of mechanism of strengthening of HEAs. Adapted with permission [24]. Copyright (2021), De Gruyter.

8.3 MECHANICAL PROPERTIES AT DIFFERENT TEMPERATURE ENVIRONMENTS

8.3.1 Room Temperature Properties of HEAs

In HEAs, several peer-reviewed papers have been published, and the main focus of the studies is on room temperature characteristics. Therefore, the chapter on HEA characteristics at room temperatures could prove beneficial to research groups investigating applications for HEA. The current assessment also summarizes key information from the literature on phase formation, microstructure analysis, thermal stability, and various high-temperature features. Additionally, the evaluation provides a brief overview of HEAs for a specific application according to their qualities. New research directions are also provided by understanding the procedure for creating a HEA and the many instruments used. In the end, the chapter would be helpful for the development and comprehension of HEA material features at high temperatures by the metallic material research communities. Thus, a graphical presentation of the properties of HEAs is shown in Figure 8.3.

8.3.2 High-Temperature Properties of HEAs

HEAs have attracted significant interest in recent years due to their exceptional mechanical properties at both ambient and elevated temperatures. In addition to their excellent mechanical properties, HEAs have also been found to exhibit unique high-temperature properties such as high-temperature strength, oxidation resistance, and thermal stability.

One of the key advantages of HEAs at high temperatures is their high melting point, which is typically higher than that of conventional alloys. This allows HEAs to retain their mechanical properties even at high temperatures, where conventional alloys would start to deform and lose strength. The high melting point of HEAs is attributed to their unique crystal structure, which

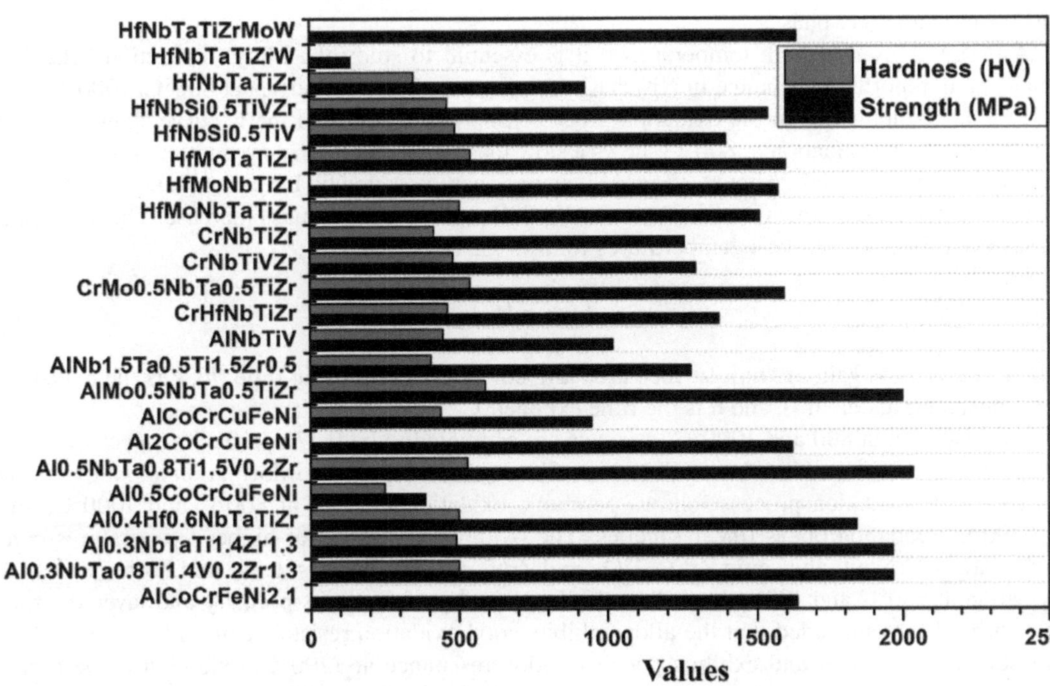

FIGURE 8.3 Room temperature properties (hardness and strength) of different high entropy alloys. Adapted with permission [4]. Copyright (2021), Elsevier.

is typically characterized by an FCC or BCC structure. HEAs have also been found to exhibit good high-temperature strength, which is important for applications such as gas turbines and jet engines. This is because HEAs typically have a high density of dislocations, which impede the movement of defects and dislocations in the material, leading to improved strength and hardness at high temperatures.

Microstructural stability is very crucial for the HEA at higher temperatures to offer exceptional properties. The stability of a nanocrystalline MoNbTaTiV refractory HEA at high temperatures has been studied by Wang et al. [25]. The alloy was prepared by mechanical alloying. XRD and SEM analysis were utilized to study the phases present and grain size as a function of milling time (0–40 hours). XRD pattern shows an amalgamation of all the elements into a BCC phase at 20 hours, as the peaks related to the elements disappeared at 20 hours. SEM analysis depicted that the lamellar spacing and holes present in the alloy decreased with increased milling time. Annealing treatment estimated the thermal stability of the alloy. The nanoscale grain size and bcc phase were maintained after the annealing treatment at 800°C to 1200°C. These represent the excellent thermal stability of the MoNbTaTiV HEA at high temperatures.

In another study done by Yao et al. [26], the phase stability of a single ductile phase BCC $Hf_{0.5}Nb_{0.5}Ta_{0.5}Ti_{1.5}Zr$ HEA at various temperatures has been reported. The alloy was prepared by suction casting, and its structure was single-phase, having a BCC structure. For the investigation of the thermal stability of phases, annealing (for 14 days) at various temperatures of 500, 700, 800, and 900°C was done. At 900°C, the bcc phase remains stable. Upon examining samples at 800°C, tiny second-phase particles were discovered at the grain boundaries of the HEA. The HEA had a single-phase structure, similar to that at 900°C. Subsequent SEM analysis of the HEA annealed at 800°C revealed NbTa-rich precipitates mainly at the grain boundaries and in some intragranular regions. Conversely, after annealing at 500°C, a three-phase microstructure was observed, consisting of an HCP phase enriched with Ti, a BCC1 phase enriched with Hf and Zr, and a BCC2 phase in Nb and Ta. This study provides evidence of the phase stability of the $Hf_{0.5}Nb_{0.5}Ta_{0.5}Ti_{1.5}Zr$ HEA and suggests that the HEA is capable of maintaining its stability at high temperatures without disintegrating into other phases.

As HEA is used at high temperatures, it is essential to study the alloy's oxidation behavior. Zhang et al. reported the change in NbZrTiCrAl refractory HEA properties at 800°C, 1000°C, and 1200°C for 50 hours due to the oxidation process [27]. The oxidation kinetic curve shows that the kinetics curve for oxidation at 800°C shows a slow mass increment during oxidation. For oxidation at 1000°C, in the initial 25 hours, the mass change increases slowly, but after that, it increases a little bit faster. At the same time, oxidation at 1200°C showed a sharp mass increase. The mass gain behavior of the alloys at these temperatures follows the Power law.

$$\left(\Delta m = kt^{n}\right) \tag{8.1}$$

where Δm is mass gain per unit surface area (mg cm^{-2}), t is the holding time (*h*), ***k*** is the oxidation rate constant (mg cm^{-2}h^{-1}), and *n* is the time exponent).

For oxidation at 800 and 1000°C, n values are estimated to be 0.57 and 0.64, respectively, and at 1200°C the estimated *n* value is close to 1. These n values provide information about oxidation kinetics. The oxidation process follows parabolic oxidation kinetics at 800°C and 1000°C, and at 1200°C, oxidation obeys linear kinetics. The oxide layer formed at higher temperatures contains mainly $CrNbO_4$ and some ZrO_2, TiO_2, and $ZrNb_2O_7$. The layer is dense and homogenous for oxidation at 800°C and 1000°C, while at 1200°C, the layer possesses porosity and layered structure. It has been concluded that the alloy exhibits good oxidation resistance at 800°C and 1000°C but degrades after that and exhibits poor oxidation resistance at 1200°C. oxidation curve represented that mass gain is higher for oxidation at 1200°C than that at 800°C and 1000°C. Thus, in summary, it is suggested that HfMoTaTiZr HEA has shown an excellent yield strength at 1000°C while HfNbTaTiZr HEA presents a similar (~ 780 MPa) at 1400°C. Accordingly, these HEAs can work in higher temperatures than the other HEAs.

8.4 APPLICATION OF HIGH ENTROPY ALLOYS

HEAs have shown promising mechanical properties such as high strength, hardness, and ductility, which make them suitable for various applications. Some of the applications of HEAs based on their mechanical properties are:

- **Structural materials:** HEAs are considered potential structural materials for various applications due to their high strength, toughness, and corrosion resistance. They can be used in the construction of bridges, buildings, and other structures that require high strength and durability.
- **Wear-resistant materials:** HEAs exhibit excellent wear resistance due to their high hardness and toughness. They can be used in the production of cutting tools, wear-resistant coatings, and other wear-resistant components.
- **Aerospace applications:** HEAs have excellent mechanical properties at high temperatures, which makes them suitable for use in aerospace applications. They can be used in the production of high-temperature components such as jet engine parts, turbine blades, and heat exchangers.
- **Medical implants**: HEAs have biocompatibility properties, which make them suitable for use in the production of medical implants such as dental implants, artificial joints, and bone plates.
- **Energy applications**: HEAs have been explored for energy applications such as the production of hydrogen storage materials, solid oxide fuel cells, and thermoelectric materials.

In conclusion, HEAs have shown great potential for various applications based on their excellent mechanical properties which are shown in Table 8.1 However, more research is needed to optimize their properties for specific applications and to scale up their production.

TABLE 8.1
Processing Routes, Phase Formation, and Proposed Applications of HEAs [4]

Alloy Composition	Processing Techniques	Phase	Proposed Application
Al–Ti–Cr–Mn–V	Arc melting	BCC + FCC	Transportation and energy industries
$Al_{0.5}$CoCrFeNi	Arc melting	BCC + FCC	Aviation and turbine manufacturing, marine, and ship vessels
FeCoNiCrMn	Cold spraying	BCC + FCC	Coatings in the automobile Sector
AlCOCrFeNi	Electron beam Melting	BCC + FCC	Tools and high-temperature applications
$MgMoNbFeTi_2Y_x$	Laser cladding	BCC + FCC	The lightweight structural material, automotive, and electronic sector
CoCrFeMnNi	MA+ hot Isostatic sintering	FCC	Welding
	Vacuum introduction melting	FCC	Cryogenic application
CoCrFeNi	Gas atomization+spark Plasma sintering	FCC	The structural component In a nuclear reactor
$FeCoNi_{1.5}CuY_{0.2}Bx$	Cold isostatic Pressure+Microwave sintering	FCC	Structural application and high load applications

8.5 ELECTRICAL PROPERTIES OF HEAs

Apart from superior mechanical performance, HEAs possess electrical properties that can be found in various fields such as thermoelectric, catalysts, and superconductivity [28]. For instance, the electrical conductivity of HEAs is significant for their use in electrical contacts and conductive coatings. Similarly, the thermoelectric property is crucial for thermoelectric generators, which turn waste heat into electrical energy. Recently, HEAs have been considered promising high-temperature thermoelectric materials due to their low lattice thermal conductivity induced by severe lattice distortion effect [29]. Also, HEA catalysts have recently drawn great attention because of their promising catalytic performance and are widely used for various catalytic applications such as oxygen evolution reactions, oxygen reduction reactions, hydrogen evolution reactions, carbon dioxide reduction reactions, and ammonia reduction reactions [30]. Furthermore, HEA superconductors are a new class of functional materials that can be used in extreme environmental conditions due to their superior hardness and irradiation resistance [31]. Since the number of applications regarding the electronic properties, understanding the fundamental electronic behavior of HEAs emerged as active research in recent years to explore them in potential functional applications such as energy storage and catalysis.

8.5.1 Electrical Resistivity

Electrical resistivity is one of the fundamental electronic properties of materials. Compared to pure metals, the electrical resistivity of multicomponent HEAs is considerably enhanced due to disorder-induced electron scattering, which results in energy dissipation in the form of Joule heat. The electrical resistivity of HEAs can vary widely depending on the specific alloy composition and can be tuned by altering the concentration of constituent elements. For instance, Huang et al. [32] reported the increase of Cu content altered the electronic structure of the $AlCu_xNiTiZr_{0.75}$ HEA film, thus decreasing the resistivity obviously from 360 to 66 μΩ·cm. Furthermore, Table 8.2 shows the comparison of the electrical resistivities of some representative pure metals, traditional alloys, bulk metallic glasses (BMGs), and HEAs. It is found that the multicomponent HEAs exhibit higher resistivity, which is higher than that of traditional and pure elements due to their distorted lattice that enhances the carrier scattering [33]. The electrical resistivities of CoCrNi, CoCrFeNi, CoCrFeMnNi, and CoCrFeNiPd are marginally higher than that of 304 stainless steels, also comparable to the Inconel 718 alloy, and lesser than that of Ni-based BMGs. Some HEAs have been shown to have higher electrical resistivity than traditional alloys, while others have lower resistivity. It is concluded that the actual tendency of electrical resistivity is not straightforward, so it is further required to understand the electronic structure of specific HEAs.

In multicomponent HEAs, the mixing entropy can influence electrical resistivity by affecting the arrangement of atoms in the alloy. When the atoms are arranged in a more disordered manner caused by higher mixing entropy, the electrical resistivity can be decreased. More clearly, the electrical resistivity of alloys is dependent on the number of elements that exist and their electronic configuration. If the alloy system has a smaller number of elements, the scattering would be less, while the increased number of elements in the alloy system enables higher disorder and increases the electron scattering. For instance, Jin et al. [42] studied the series of Ni-based equiatomic FCC alloys such as Ni, NiCo, NiFe, NiCoFe, NiCoCr, NiCoFeCr, NiCoFeCrMn, and NiCoFeCrPd were chosen to study the electrical properties at 300 K. The resultant electrical resistivity of those alloys is shown in Figure 8.4. It shows the increased electrical resistivity, whereas an increasing number of elements in the system due to increased randomness higher the electron scattering. Since HEAs have complex chemical compositions and microstructural characteristics, their electrical properties are lesser than pure metals. The typical increase in electrical resistivity by adding several elements is mainly attributed to the following factors including the significant difference in atomic size, valance electron concentration, and chemical properties when compared to the pure elements.

TABLE 8.2
Comparison of Electrical Resistivities of Some Pure Elements and Conventional Alloys with HEAs

Category	Composition	Electrical Resistivity (μΩ·cm)	Reference
Pure element	Al	3	[34]
	Fe	10	
	Ni	7	
	Ti	42	
	Cu	2	
Conventional alloy	7075 AI alloy	6	[35]
	Low Carbon steel	17	
	304 Stainless Steel	69	
	Inconel 718	125	
	Ti-6Al-4V	168	
Bulk metallic glass	$Zr_{41}Ti_{14}Cu_{12.5}Ni_{10}Be_{22.5}$	171	[36]
	$Fe_{78}Si_9B_{13}$	137	[37]
	$Co_{63}Fe_9Zr_8B_{20}$	188	[37]
High entropy alloys	CoCrFeNi	142	[38]
	AlCoCrFeNi	221	[38]
	$Al_2CoCrFeNi$	211	[38]
	NbMoTaW	168	[39]
	AlCoCrFeNiCu	290	[40]
	AlCoCrFeNiTi	360	[40]
	FeCoNi(AlSi)0.2	69.5	[41]
	CoCrFeMnNi	110	[42]
	CoFeNiCrPd	134	[42]
	FeCoNiCuMn	287	[43]
	CoCrFeNi	135.1	[44]

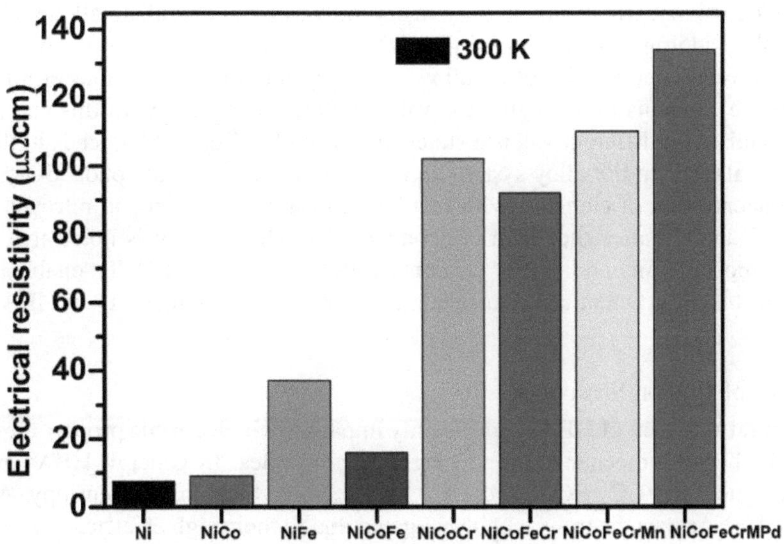

FIGURE 8.4 Temperature dependence of electrical resistivity with an increasing number of elements at 300 K. Adapted with permission [42]. Copyright (2016), Nature.

It is noteworthy to mention that in some cases, there is a decrease in electrical resistivity by the addition of more elements into the alloy. In particular, the mobility of electrons can be improved if the added elements possess similar atomic size, electronic configuration and crystal structure with the base elements. Therefore, the careful evolution of electronic properties by varying the chemical compositions is still necessary for various functional applications. Some of the major factors that influence the electrical properties of HEAs can be summarized as follows.

8.5.2 Effect of Mixing Entropy

Since HEAs contain multiple elements, mixing entropy plays a crucial role in determining their crystal structure and mechanical properties, including electrical properties. By the definition of entropy, the mixing of multiple elements can increase the randomness of atomic positions within the crystal lattice of HEAs. Owing to the disorder in the atomic positions, many lattice defects including vacancies, dislocations, and point defects can be formed, which influence the electronic properties of materials. Furthermore, a complex electronic band structure is formed with a high density of states (DOS), which can result in unique electronic properties like high electrical conductivity for specific alloy compositions. Therefore, understanding the electronic band structure of various HEAs still requires significant research [38].

8.5.1.2 Effect of Severe Lattice Distortion

In contrast to conventional alloys, the severe lattice distortion effect is a promising effect in HEAs that can increase the scattering of charge carriers, thereby reducing electrical conductivity. This is due to the presence of multiple elements with different atomic radii and crystal structures. For instance, Zhang et al. [45] reported increased electrical resistivity by the addition of Si content in FeCoNi(AlSi)x HEA due to increased lattice distortion.

8.5.1.3 Effect of Valance Electron Concentration

Further, VEC is an effective parameter to tune the electrical properties of HEAs. VEC usually determines how many electrons are present in an atom of a metal or alloy, and it is known to have a significant impact on their electrical properties. Thus, controlling VEC has a considerable effect on electrical conductivity, which can be done in several ways. In general, an alloy with a high VEC can be prepared by adding elements with a high VEC. For example, the addition of Al, Si, or P can increase the electrical conductivity of an alloy by raising the VEC of the master alloy. The other approach is to use elements by changing the valence states in the alloy. In this case, the addition of transition metals with different valence states, including Fe, Co, and Ni, leads to the creation of local charge imbalances in the alloy system and increases the electrical conductivity. Besides, the decrease in concentration of elements with low VEC, including carbon and nitrogen can improve the overall VEC and enhance the electrical conductivity. Therefore, it is noted that the selection of the alloy composition and addition of elements with high or low VEC that enable the change of the valence states is an efficient route to regulate the electrical conductivity of HEAs for specific applications.

8.5.1.4 Effect of Crystal Structure

Finally, the crystal structure of HEAs significantly impacts their electronic properties, which in turn influences their thermal, mechanical, and magnetic properties. In general, HEAs exhibit simple solid solutions, such as FCC, BCC, and HCP, due to their high mixing entropy. Among these, FCC-structured HEAs have been widely investigated due to their high electrical conductivity when compared to BCC structures. The reason for this high electrical conductivity in FCC structure is due to the highly symmetrical arrangement of atoms leading to the movement of a larger number of free electrons through the material. However, BCC HEAs show a more complex electronic structure compared to the FCC HEAs with multiple band crossings and a lower DOS at the Fermi level. Therefore, BCC HEAs demonstrate a lower electrical conductivity, but attractive magnetic properties such as high susceptibility and magnetic ordering. On the other hand, HCP HEAs are the least studied crystal structures; however, they exhibit unique electronic properties due to the high DOS near the Fermi level, leading to show high electrical conductivity at elevated temperatures that make them useful for high-temperature applications. Therefore, high symmetrical crystal structures of HEAs are likely to achieve high convergence of the valance bands near the Fermi level, thereby obtaining a high Seebeck coefficient [46]. Therefore, HEAs are potential candidates to tune the thermoelectric performance by adjusting the phase composition.

8.5.3 Thermoelectric Properties

With the increasing energy crisis and environmental concern, green thermoelectric technology (TE) has drawn emerging research due to its ability to convert a temperature difference into electricity and vice versa based on the Seebeck and Peltier effects, respectively [47]. The performance of TE materials is usually scalable by a dimensionless figure of merit, ZT= $(S^2\sigma/\kappa)\ T$, where S, σ, κ, and T are the Seebeck coefficient, electrical conductivity, thermal conductivity, and absolute temperature, respectively. To improve TE performance, a TEM should always possess a high-power factor ($S^2\sigma$) and low κ; however, it's difficult to optimize electronic and thermal transport properties together due to their strong interdependency. Thus, the new concept of HEAs has recently drawn great attention for TE applications due to the higher mixing entropy effect tends to form simple solid solution phases with high crystalline symmetry [45]. It is noted that the high symmetry of crystal structures facilitates the convergence of the Fermi level bands, and hence a high Seebeck coefficient was achieved [46]. While the extreme disorder in terms of atomic, crystallographic, and chemical behavior suggests reduced thermal conductivity by enhanced phonon scattering. In recent years, tuning configurational entropy (ΔS) is promising to improve the TE performance of HEAs by increasing the number of elements in a HEA system. In general, the configurational entropy of an alloy system can be estimated by using the following equation.

$$\Delta S = k_B \sum_{i=1}^{n} x_i \ln(x_i) \tag{8.2}$$

where K_B is the Boltzmann constant, x_i is the atomic percentage of the ith element. As mentioned above, the increased ΔS with an increasing number of elements tends to optimize the electrical and thermal properties of TE materials. For instance, Jiang et al. [48] introduced different amounts of Sb, Sn, Te, and S into the PbSe system, thereby increasing the ΔS, and forming a single-phase $Pb_{0.89}Sb_{0.012}Sn_ySe_{1-2x}Te_xS_x$ HEAs. As a result, they found the lowest lattice thermal conductivity of ~0.3 W/mK and improved ZT up to 1.8 at 900 K in $Pb_{0.89}Sb_{0.012}Sn_{0.1}Se_{0.5}$ $Te_{0.25}S_{0.25}$ sample. In addition, Wang et al. [49] prepared PbTe, GeTe, and MnTe co-alloying with SnTe to form a single-phase solid solution. Owing to the inclusion of different elements at the cationic (Sn^{2+}) site, the ΔS increases and enhances the Seebeck coefficient due to the band modification via co-alloying. Figure 8.5a shows the measured Seebeck coefficient of Ga_x $(Sn_{0.25}Pb_{0.25}Mn_{0.25}Ge_{0.25})_{1-x}Te$ (x=0, 0.015, 0.02, 0.025, and 0.03) HEAs with an increasing Ga content. It shows that the Seebeck coefficient is increased up to 0.025 at. % and reduced for the higher content of 0.3 at. %. The measured Seebeck coefficient is ~160 μV/K at 323 K and increases to the peak value ~of 246 μV/K at 673 K for the sample with 0.025 at. % of Ga. Meanwhile, Figure 8.5b shows the total thermal conductivity measurement over the range of temperature from 323 to 823 K. It shows the decrease in thermal conductivity from x=0 to x=0.025 and then increases when the content of Ga increases to x=0.03 throughout the measurement temperature. The lowest thermal conductivity of ~0.7 W/mK was obtained for the sample with 0.025 at.% of Ga content. As the temperature increases, the thermal conductivity decreases and reaches its minimum value of 0.6 W/mK at 573 K.

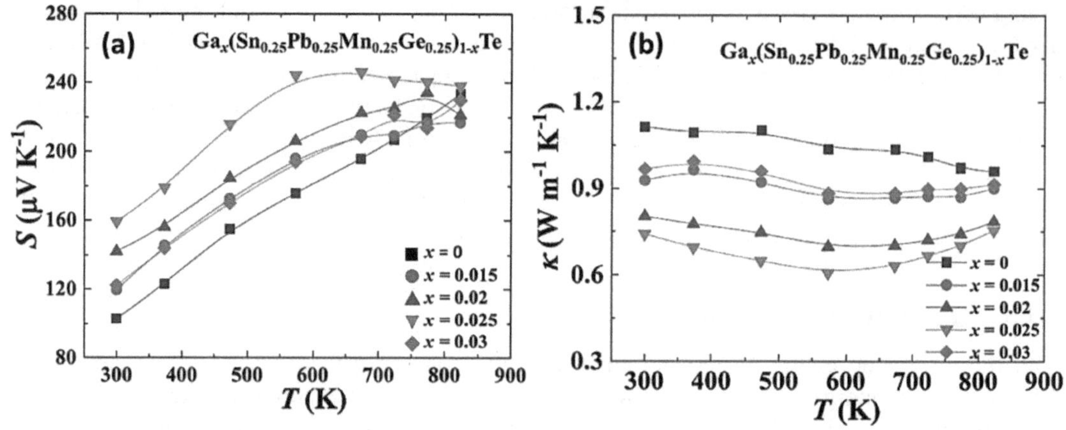

FIGURE 8.5 (a) Measurement of Seebeck coefficient (S) and thermal conductivity (κ) of $Ga_x(Sn_{0.25}Pb_{0.25}Mn_{0.25}Ge_{0.25})_{1-x}Te$ (x=0, 0.015, 0.02, 0.025, and 0.03) alloy with increasing entropy in the system. Adapted with permission [49]. Copyright (2021), Elsevier.

8.5.4 Catalytic Applications

The development of highly active and stable catalysts is a substantial challenge for clean energy conversion. In general, the catalytic performance of metals is highly dependent on their crystalline structure, such as their phase structure and composition. Binary compounds usually show a large immiscible gap with the formation of intermetallic phases, which inhibits the catalytic activity. In contrast, HEAs exhibit stable solid solution phases, induced by higher mixing entropy. Therefore, HEAs have attracted great attention in the development of electro or thermocatalytic clean energy conversion because of their unique phase, impressive thermal stability, and excellent catalytic activities [30]. Figure 8.6 shows the schematic representation of HEAs for catalysis applications, indicating that stable solid solutions with highly crystalline materials show attractive electrical properties, which play a significant role in their catalytic performance, and increase the resistance to degradation during catalytic reactions [50]. For many catalytic reactions, conductive materials can make it easier for electrons to move between reactants and catalyst surfaces. Fortunately, the electronic properties of HEAs can be varied by changing the elemental composition for specific reactions. The severe lattice distortion effect caused by the difference in atomic radii induces a thermodynamically non-equilibrium state, thus decreasing the energy barrier during catalytic reactions, which can enhance the catalytic performance. In addition, the tunability of HEAs electronic structures, like their d-band centers, and lattice strain is advantageous for further improving catalytic performance. From the perspective of catalytic stability, high entropy, and slow diffusion significantly increase thermal stability, enhancing the longevity of catalysts under varied reaction circumstances. Overall, the unique combination of properties is advantageous for the enhancement of catalytic activity, making HEAs highly effective catalysts for catalytic applications in a wide range of fields, including energy conversion, environmental remediation, and chemical synthesis.

8.5.5 Superconductivity of HEAs

Since the first discovery of superconductivity in Ti–Zr–Nb–Ta–Hf refractory HEA, the research has received considerable attention because of their superconducting property under extreme pressure conditions [52]. Till now, most of the HEA superconductors are found in BCC and HCP crystal structures rather than FCC [53]. According to Matthias et al. [54], superconductivity occurs empirically based on TC variation with the number of outer electrons per atom (e/a). The optimum conditions for the existence of superconductivity are found to occur for 5 and 7 valance electrons per atom. But the reason behind the lack of superconductivity in FCC structures is due to their larger

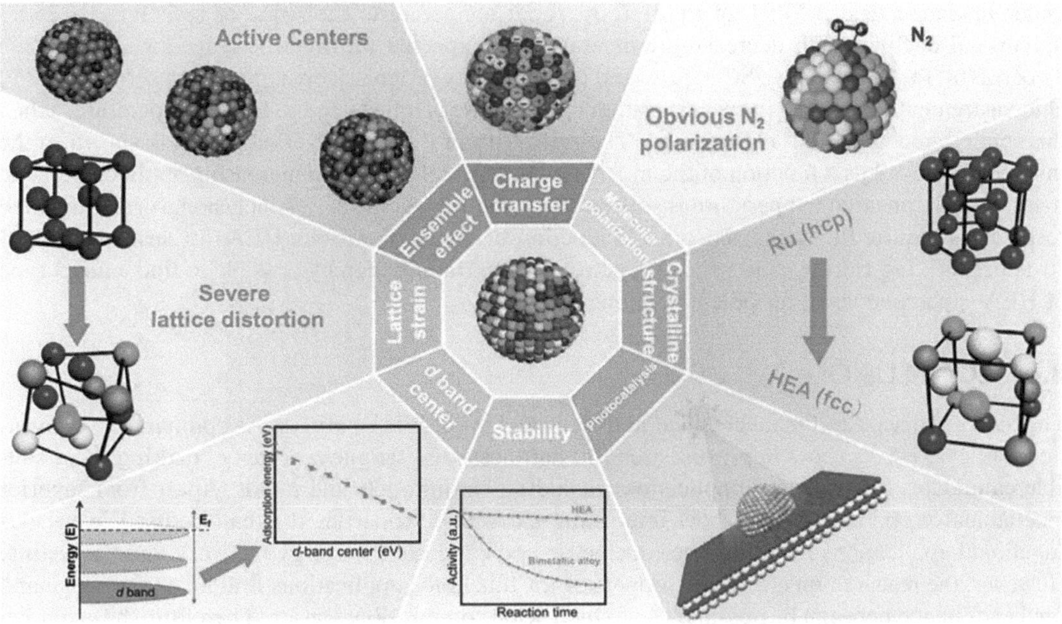

FIGURE 8.6 Representation diagram of the proposed future development directions for HEA-based catalysis. Adapted with permission [51]. Copyright (2020), American Chemical Society.

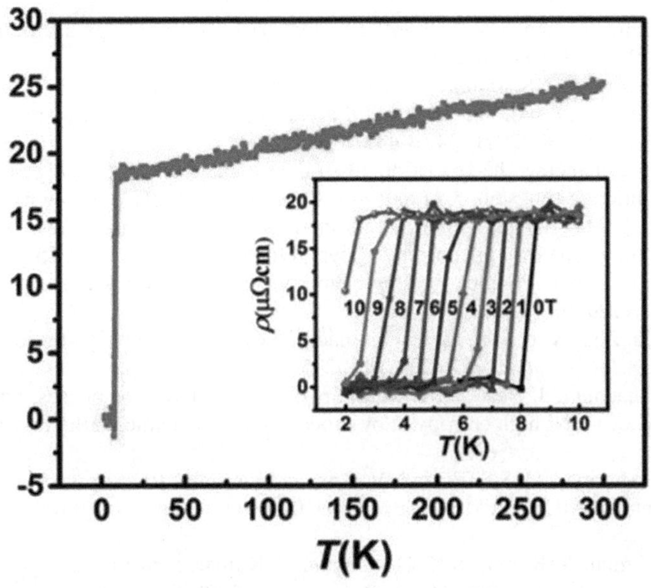

FIGURE 8.7 Electrical resistivity of $Ti_{15}Zr_{15}Nb_{35}Ta_{35}$ HEA in zero fields as a function of temperature. The inset figure illustrates the change in resistivity with temperature under a magnetic field up to 10 T. Adapted with permission [56]. Copyright (2018), Frontiers.

VEC (>8). For instance, Guo et al. [55] reported the BCC-structured Ta–Nb–Hf–Zr–Ti HEA shows extraordinarily robust zero-resistance superconductivity under pressure up to 190.6 GPa, making it a promising candidate for new applications under extreme pressure. In another study, Yuan et al. [56] reported the superconducting behavior of BCC-structured $Ti_{15}Zr_{15}Nb_{35}Ta_{34}$ HEA by varying the chemical composition. The temperature dependence of electrical resistivity ρ (T) in zero fields is shown in Figure 8.7, and the inset shows the variation of resistivity as a function of temperature under magnetic field ρ (T)H up to 10 T. At room temperature, the value of resistivity is 25±1 µΩcm and declines with decreasing temperature as expected for metallic behavior. At 8 K, the $Ti_{15}Zr_{15}Nb_{35}Ta_{35}$ HEA shows SC nature, and the resistivity progressively approaches zero. The ρ (T) H measurements show that the SC transition temperature is transferred to lower temperatures when the applied magnetic field is increased. The resistivity in the SC transition region is shown in the inset of Figure 8.7 as a function of the magnetic field up to 10 T. As an increasing applied magnetic field, the SC transition temperature is shifted to a lower temperature. In conclusion, further detailed research is required to study the superconducting behavior of various HEAs in terms of crystal structure, mixing entropy, and valance electron concentration, thereby possible to find a novel type of HEA superconductors for potential applications.

8.6 CONCLUSION

This chapter discusses the mechanical and electrical properties of HEAs. As compared to conventional alloys, HEAs show improved strength, hardness, and fatigue resistance, making them suitable candidates for structural applications including cutting tools and molds. Apart from superior mechanical properties, HEAs show interesting electronic properties that can be used in various functional applications including thermoelectric and catalysts for energy recovery and harvesting. However, the research on electronic properties for functional applications is at an earlier stage and further enhancement can be possible due to their wide compositional space. Therefore, the evolution of electronic properties for various HEAs is emerging research for expanding their applications.

REFERENCES

[1] B.S. Murty, S. Ranganathan, J.W. Yeh, P.P. Bhattacharjee, High-entropy alloys, *High-Entropy Alloys*, (2019) 1–363. ISBN: 978-0-12-800251-3

[2] S.K. Dewangan, D. Kumar, S. Samal, V. Kumar, Microstructure and mechanical properties of nanocrystalline AlCrFeMnNiW*x* (*x*=0, 005, 01, 05) high-entropy alloys prepared by powder metallurgy route, *J. Mater. Eng. Perform.*, 30 (2021) 4421–4431.

[3] S.K. Dewangan, Enhancing the oxidation resistance of nanocrystalline high-entropy AlCuCrFeMn alloys by the addition of tungsten. *J. Mater. Res. Technol.*, 21 (2021) 4960–4968.

[4] S.K. Dewangan, A. Mangish, S. Kumar, A. Sharma, B. Ahn, V. Kumar, A review on high-temperature applicability: A milestone for high entropy alloys, *Eng. Sci. Technol. Int. J*, 35 (2022) 101211.

[5] X. Wang, W. Guo, Y. Fu, High-entropy alloys: Emerging materials for advanced functional applications, *J. Mater. Chem. A Mater.*, 9 (2021) 663–701.

[6] L. Qiao, Y. Liu, J. Zhu, A focused review on machine learning aided high-throughput methods in high entropy alloy, *J. Alloys Compd.*, 877 (2021) 160295.

[7] V. Shivam, Y. Shadangi, J. Basu, N.K. Mukhopadhyay, Evolution of phases, hardness and magnetic properties of AlCoCrFeNi high entropy alloy processed by mechanical alloying, *J. Alloys Compd.*, 832 (2020) 154826.

[8] D. Kumar, V.K. Sharma, Y.V.S.S. Prasad, V. Kumar, Materials-structure-property correlation study of spark plasma sintered AlCuCrFeMnW*x* (*x*=0, 005, 01, 05) high-entropy alloys, *J. Mater. Res.*, 34 (2019) 767–776.

[9] O. Maulik, D. Kumar, S. Kumar, S.K. Dewangan, V. Kumar, Structure and properties of lightweight high entropy alloys: A brief review, *Mater. Res. Express*, 5 (2018) 052001.

[10] Y. Zhang, Y.J. Zhou, J.P. Lin, G.L. Chen, P.K. Liaw, Solid-solution phase formation rules for multi-component alloys, *Adv. Eng. Mater.*, 10 (2008) 534–538.

[11] G. Qin, W. Xue, R. Chen, H. Zheng, L. Wang, Y. Su, H. Ding, J. Guo, H. Fu, Grain refinement and FCC phase formation in AlCoCrFeNi high entropy alloys by the addition of carbon, *Materialia*, 6 (2019) 100259.

[12] J.Y. He, H. Wang, H.L. Huang, X.D. Xu, M.W. Chen, Y. Wu, X.J. Liu, T.G. Nieh, K. An, Z.P. Lu, A precipitation-hardened high-entropy alloy with outstanding tensile properties, *Acta Mater.*, 102 (2016) 187–196.

[13] E. Leroy, B.M. Laurent-brocq, X. Sauvage, A. Akhatova, Precipitation and hardness of carbonitrides in a CrMnFeCoNi high entropy alloy, *Adv. Eng. Mater.*, 20 (2017) 1–4.

[14] X. Wu, D. Mayweg, D. Ponge, Z. Li, Microstructure and deformation behavior of two TWIP/TRIP high entropy alloys upon grain refinement, *Mater. Sci. Eng., A*, 802 (2021).
[15] S. Chen, H. Seok Oh, B. Gludovatz, S. Jun Kim, E. Soo Park, Z. Zhang, R.O. Ritchie, Q. Yu, Real-time observations of TRIP-induced ultrahigh strain hardening in a dual-phase CrMnFeCoNi high-entropy alloy, *Nat. Comm.*, 826 (2020) 1–8.
[16] A.J. Zaddach, C. Niu, C.C. Koch, D.L. Irving, Mechanical properties and stacking fault energies of NiFeCrCoMn high-entropy alloy, *JOM*, 65 (2013) 1780–1789.
[17] L. Zhang, Y. Zhang, Tensile properties and impact toughness of AlCoxCrFeNi_{31-x} (x=04, 1) high-entropy alloys, *Front Mater.*, 7 (2020) 1–8.
[18] B. Kang, T. Kong, H.J. Ryu, S.H. Hong, The outstanding tensile strength of Ni-rich high entropy superalloy fabricated by powder metallurgical process, *Mater Chem. Phys.*, 235 (2019) 121749.
[19] H. Jiang, D. Qiao, W. Jiao, K. Han, L. Yiping, P.K. Liaw, Tensile deformation behavior and mechanical properties of a bulk cast Al09CoFeNi2 eutectic high-entropy alloy, *J. Mater Sci. Technol.*, 61 (2021) 119–124.
[20] Y. Zhang, X. Wang, J. Li, Y. Huang, Y. Lu, X. Sun, Deformation mechanism during high-temperature tensile test in an eutectic high-entropy alloy $AlCoCrFeNi_{21}$, *Mater. Sci. Eng., A*, 724 (2018) 148–155.
[21] C. Nagarjuna, K. Yong Jeong, Y. Lee, S. Min Woo, S. Ig Hong, H. Seop Kim, S.J. Hong, Strengthening the mechanical properties and wear resistance of CoCrFeMnNi high entropy alloy fabricated by powder metallurgy, *Adv. Powder Technol.*, 33 (2022).
[22] C. Nagarjuna, A. Sharma, K. Lee, S.J. Hong, B. Ahn, Microstructure, mechanical and tribological properties of oxide dispersion strengthened CoCrFeMnNi high-entropy alloys fabricated by powder metallurgy, *J. Mater. Res. Technol.*, 22 (2023) 1708–1722.
[23] Q. Wei, Q. Shen, J. Zhang, B. Chen, G. Luo, L. Zhang, Microstructure and mechanical property of a novel ReMoTaW high-entropy alloy with high density, *Int. J. Refract Metals Hard Mater.*, 77 (2018) 8–11.
[24] J. Chen, X. Jiang, H. Sun, Z. Shao, Y. Fang, R. Shu, Phase transformation and strengthening mechanisms of nanostructured high-entropy alloys, *Nanotechnol. Rev.*, 10 (2021) 1116–1139.
[25] H. Materials, G. Wang, Q. Liu, J. Yang, X. Li, X. Sui, Y. Gu, Y. Liu, International journal of refractory metals synthesis and thermal stability of a nanocrystalline MoNbTaTiV refractory high-entropy alloy via mechanical alloying, *Int. J. Refract Metals Hard Mater.*, 84 (2019) 104988.
[26] J.Q. Yao, X.W. Liu, N. Gao, Q.H. Jiang, N. Li, G. Liu, W.B. Zhang, Z.T. Fan, Intermetallics refractory high-entropy alloy, *Intermetallics* 98 (2018) 79–88.
[27] P. Zhang, Y. Li, Z. Chen, J. Zhang, B. Shen, Oxidation response of a vacuum arc melted NbZrTiCrAl refractory high entropy alloy at 800-1200°C, *Vacuum*, 162 (2019) 20–27.
[28] M.H. Tsai, Physical properties of high entropy alloys, *Entropy*, 15 (2013) 5338–5345.
[29] M.A. Al Hasan, J. Wang, S. Shin, D.A. Gilbert, P.K. Liaw, N. Tang, W.L.N.C. Liyanage, L. Santodonato, L. DeBeer-Schmitt, N.P. Butch, Effects of aluminum content on thermoelectric performance of AlxCoCrFeNi high-entropy alloys, *J. Alloys Compd.*, 883 (2021) 160811.
[30] Y. Xin, S. Li, Y. Qian, W. Zhu, H. Yuan, P. Jiang, R. Guo, L. Wang, High-entropy alloys as a platform for catalysis: Progress, challenges, and opportunities, *ACS Catal.*, 10 (2020) 11280–11306.
[31] S.G. Jung, Y. Han, J.H. Kim, R. Hidayati, J.S. Rhyee, J.M. Lee, W.N. Kang, W.S. Choi, H.R. Jeon, J. Suk, T. Park, High critical current density and high-tolerance superconductivity in high-entropy alloy thin films, *Nat. Commun.*, 13 (2022) 3373.
[32] K. Huang, G. Wang, H. Qing, Y. Chen, H. Guo, Effect of Cu content on electrical resistivity, mechanical properties and corrosion resistance of $AlCu_xNiTiZr_{0.75}$ high entropy alloy films, *Vacuum*, 195 (2022).
[33] J.-W. Yeh, Recent progress in high-entropy alloys, *Annales de Chimie Science Des Matériaux*, 31 (2006) 633–648.
[34] D.R. Lide, B. Moraes, *Handbook of Chemistry and Physics 84th*, 72nd edition, CRC Press, Academia (n.d.). ISBN-13: 978-0849304729.
[35] C.J. Smithells, W.F. Gale, T.C. Totemeier, *Smithells metals reference book*, Elsevier 8 (2004).
[36] Y. Li, H.Y. Bai, Superconductivity in a representative Zr-based bulk metallic glass, *J. Non Cryst. Solids*, 351 (2005) 2378–2382.
[37] A. Inoue, Stabilization of metallic supercooled liquid and bulk amorphous alloys, *Acta Mater.*, 48 (2000) 279–306.
[38] Y.F. Kao, S.K. Chen, T.J. Chen, P.C. Chu, J.W. Yeh, S.J. Lin, Electrical, magnetic, and Hall properties of AlxCoCrFeNi high-entropy alloys, *J. Alloys Compd.*, 509 (2011) 1607–1614.
[39] H. Kim, S. Nam, A. Roh, M. Son, M.H. Ham, J.H. Kim, H. Choi, Mechanical and electrical properties of NbMoTaW refractory high-entropy alloy thin films, *Int. J. Refract Metals Hard Mater.*, 80 (2019) 286–291.

[40] M. Dada, P. Popoola, N. Mathe, S. Pityana, S. Adeosun, Electrical resistivity and oxidation behavior of Cu and Ti doped laser deposited high entropy alloys, *World J. Eng.* 20 (2022) 868–876.

[41] Y. Zhang, T. Zuo, Y. Cheng, P.K. Liaw, High-entropy alloys with high saturation magnetization, electrical resistivity, and malleability, *Sci. Rep.*, 3 (2013) 1455.

[42] K. Jin, B.C. Sales, G.M. Stocks, G.D. Samolyuk, M. Daene, W.J. Weber, Y. Zhang, H. Bei, Tailoring the physical properties of Ni-based single-phase equiatomic alloys by modifying the chemical complexity, *Sci. Rep.*, 6 (2016) 20159.

[43] M. Harivandi, M. Malekan, S.A. Seyyed Ebrahimi, Soft magnetic high entropy FeCoNiCuMn alloy with excellent ductility and high electrical resistance, *Met. Mater. Int.*, 27 (2022) 556–564.

[44] W. Huo, X. Liu, S. Tan, F. Fang, Z. Xie, J. Shang, J. Jiang, Ultrahigh hardness and high electrical resistivity in nano-twinned, nanocrystalline high-entropy alloy films, *Appl. Surf. Sci.*, 439 (2018) 222–225.

[45] Y. Zhang, T. Zuo, Y. Cheng, P.K. Liaw, High-entropy alloys with high saturation magnetization, electrical resistivity and malleability, *Sci. Rep.*, 3 (2013) 1–7.

[46] Y. Pei, X. Shi, A. Lalonde, H. Wang, L. Chen, G.J. Snyder, Convergence of electronic bands for high performance bulk thermoelectrics, *Nature*, 473 (2011) 66–69.

[47] C. Nagarjuna, P. Dharmaiah, J.H. Lee, K.B. Kim, G. Song, J.K. Lee, S.J. Hong, Formation of inhomogeneous micro-scale pores attributed ultralow κlat and concurrent enhancement of thermoelectric performance in p-type $Bi_{0.5}Sb_{15}Te_3$ alloys, *J. Alloys Compd.*, 881 (2021) 160499.

[48] B. Jiang, Y. Yu, J. Cui, X. Liu, L. Xie, J. Liao, Q. Zhang, Y. Huang, S. Ning, B. Jia, B. Zhu, S. Bai, L. Chen, S.J. Pennycook, J. He, High-entropy-stabilized chalcogenides with high thermoelectric performance, *Science*, (1979), 371 (2021) 830–834.

[49] X. Wang, H. Yao, Z. Zhang, X. Li, C. Chen, L. Yin, K. Hu, Y. Yan, Z. Li, B. Yu, F. Cao, X. Liu, X. Lin, Q. Zhang, Enhanced thermoelectric performance in high entropy alloys $Sn_{0.2}5Pb_{0.25}Mn_{0.25}Ge_{0.25}Te$, *ACS Appl. Mater. Interf.*, 13 (2021) 18638–18647.

[50] T. Löffler, A. Ludwig, J. Rossmeisl, W. Schuhmann, What makes high-entropy alloys exceptional electrocatalysts? *Angew. Chem. Int. Ed.*, 60 (2021) 26894–26903.

[51] Y. Xin, S. Li, Y. Qian, W. Zhu, H. Yuan, P. Jiang, R. Guo, L. Wang, High-entropy alloys as a platform for catalysis: Progress, challenges, and opportunities, *ACS Catal.*, 10 (2020) 11280–11306.

[52] P. Koželj, S. Vrtnik, A. Jelen, S. Jazbec, Z. Jagličić, S. Maiti, M. Feuerbacher, W. Steurer, J. Dolinšek, Discovery of a superconducting high-entropy alloy, *Phys. Rev. Lett.*, 113 (2014) 1–5.

[53] L. Sun, R.J. Cava, High-entropy alloy superconductors: Status, opportunities, and challenges, *Phys. Rev. Mater.*, 3 (2019) 090301.

[54] B.T. Matthias, Empirical relation between superconductivity and the number of valence electrons per atom, *Phy. Rev.*, 97 (1955) 74.

[55] J. Guo, H. Wang, F. Von Rohr, Z. Wang, S. Cai, Y. Zhou, K. Yang, A. Li, S. Jiang, Q. Wu, R.J. Cava, L. Sun, Robust zero resistance in a superconducting high-entropy alloy at pressures up to 190GPa, *Proc. Natl. Acad. Sci. U. S. A.*, 114 (2017) 13144–13147.

[56] Y. Yuan, Y. Wu, H. Luo, Z. Wang, X. Liang, Z. Yang, H. Wang, X. Liu, Z. Lu, Superconducting $Ti_{15}Zr_{15}Nb_{35}Ta_{35}$ high-entropy alloy with intermediate electron-phonon coupling, *Front Mater.*, 5 (2018) 72.

9 High-Entropy Materials for Methanol Oxidation Reactions

Yayun Zhao and Yichao Lin

9.1 INTRODUCTION

Fuel cells, as one of the representatives of new clean energy, are considered the fourth generation of power generation technology after hydroelectric, thermal, and nuclear power [1]. It can directly convert the chemical energy of fuel into electrical energy, with a theoretical efficiency of 85%–90%. Compared to thermodynamic processes, the use of fuel cells can reduce carbon dioxide emissions by over 40%, thereby attracting the attention of both academia and the industry. As an important member of fuel cells, direct methanol fuel cells (DMFCs) have multiple advantages, including high energy density, efficient energy conversion efficiency, ease of recharging, quick start-up at low temperature, nonpolluting using methanol as fuel directly without hydrogen reforming and portability [1–3].

DMFC device involves the oxidation of CH_3OH into CO_2 and H_2O, thus converting chemical energy into electrical energy. Unlike other alcohol-based fuels (e.g., ethanol), the kinetics of its anode reaction is easier and faster, since there is no need to break the strong C–C bond commonly present in other alcohol fuels [4,5]. The main structure of DMFC consists of three parts: the fuel electrode as anode, the oxidant electrode as cathode, and the proton (or alkaline anion)-exchange membrane sandwiched between them [6]. In acidic media, CH_3OH reacts with H_2O in the anode region to form CO_2, H^+ and e^-. Meanwhile, CO_2 is discharged from the anode outlet, and H^+ and e^- are transferred to the cathode region through the proton-exchange membrane and external circuit, respectively. H^+ and e^- react with O_2 in the cathode region to form H_2O, which is discharged from the cathode outlet [3,7]. In alkaline media, H_2O, CO_2 and e^- are produced by the reaction of CH_3OH and OH^- in the anode region, and CO_2 and H_2O are discharged from the anode outlet. Besides, e^- is transferred to the cathode region and reacts with O_2 and H_2O to form OH^-, which is transferred to the anode through the alkaline anion-exchange membrane [6]. The total reactions for DMFC with an acidic or alkaline electrolyte are described in Table 9.1.

Compared to the cathodic reaction, the methanol oxidation reaction (MOR) is a very slow process, and how to effectively improve the methanol oxidation activity and resistance to CO poisoning of anode catalysts are the technical requirements to achieve large-scale commercialization of DMFC.

TABLE 9.1
The Fuel Cell Reactions for DMFC

	Acidic Electrolyte	Alkaline Electrolyte
Anode	$CH_3OH + H_2O \rightarrow CO_2 + 6H^+ + 6e^-$ $\left(E^\theta = 0.02\ V\right)$	$CH_3OH + 6OH^- \rightarrow CO_2 + 5H_2O + 6e^-$ $\left(E^\theta = -0.81\ V\right)$
Cathode	$3/2O_2 + 6H^+ + 6e^- \rightarrow 3H_2O$ $\left(E^\theta = 1.23\ V\right)$	$3/2O_2 + 3H_2O + 6e^- \rightarrow 6OH^-$ $\left(E^\theta = 0.40\ V\right)$
Overall	$CH_3OH + 3/2O_2 \rightarrow CO_2 + 2H_2O$ $\left(E^\theta = 1.21\ V\right)$	$CH_3OH + 3/2O_2 \rightarrow CO_2 + 2H_2O$ $\left(E^\theta = 1.21\ V\right)$

DOI: 10.1201/9781003391388-9

The emergence of high-entropy materials (HEMs) may provide a viable technological pathway to these problems. In this chapter, the mechanism of MOR and the responding catalysts will be introduced, and the recent progress on the HEMs is emphatically introduced, including noble metal-based high-entropy alloys (HEAs) and high-entropy intermetallic (HEI), as well as non-noble metal-based HEAs. Meanwhile, the possible reasons for the enhanced performance are also introduced.

9.2 MECHANISM AND CATALYSTS OF MOR

9.2.1 Mechanism of MOR

MOR is the anode reaction of DMFCs, involving the transfer of six protons and six electrons. In acidic electrolytes, CH_3OH is adsorbed on the specific catalyst surface, followed by the dissociation of CH_3OH with the activation of the C–H bond to form CO intermediates, then H_2O adsorbed on the catalysts surface are activated and OH_{ads} is formed; subsequently, CO intermediates are oxidized by OH_{ads} to form CO_2 (Figure 9.1a). Take Pt catalysts as an example, the possible reaction mechanism of MOR in acidic media can be written as follows [4,8]

$$CH_3OH + Pt(s) \rightarrow Pt - CH_2OH + H^+ + e^- \quad (9.1)$$

$$Pt - CH_2OH + Pt(s) \rightarrow Pt_2 - CHOH + H^+ + e^- \quad (9.2)$$

$$Pt_2 - CHOH + Pt(s) \rightarrow Pt_3 - COH + H^+ + e^- \quad (9.3)$$

$$Pt_3 - COH \rightarrow Pt - CO + 2Pt + H^+ + e^- \quad (9.4)$$

$$H_2O + Pt(s) \rightarrow Pt - OH_{ads} + H^+ + e^- \quad (9.5)$$

$$Pt - CO + Pt - OH_{ads} \rightarrow Pt(s) + CO_2 + H^+ + e^- \quad (9.6)$$

In alkaline electrolyte, CH_3OH and OH^- are adsorbed on the catalysts' surface, then a series of carbonaceous intermediates are formed through the oxidation reaction steps involving electron

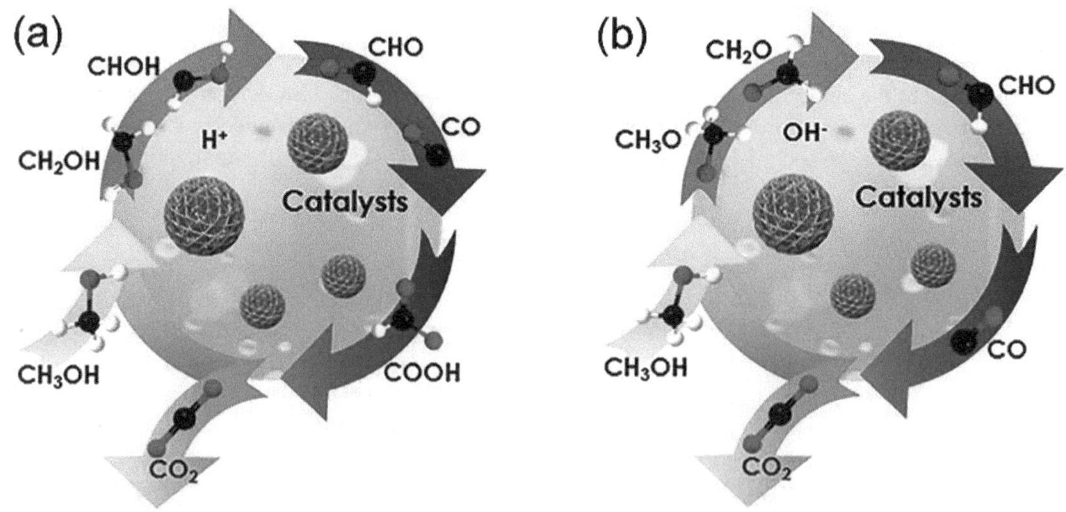

FIGURE 9.1 Mechanisms of MOR in (a) acidic and (b) alkaline electrolyte. Adapted with permission [4]. Copyright 2019, John Wiley and Sons.

transfer; these species eventually react with OH_{ads} to generate CO_2 (Figure 9.1b). MOR process on Pt catalysts in alkaline electrolyte is proposed to take place through the bellowing steps [4,9]

$$OH^- + Pt(s) \rightarrow Pt - OH_{ads} + e^- \tag{9.7}$$

$$CH_3OH + Pt(s) \rightarrow Pt - CH_3OH_{ads} + H_2O + e^- \tag{9.8}$$

$$Pt - CH_3OH_{ads} + OH^- \rightarrow Pt - CH_3O + H_2O + e^- \tag{9.9}$$

$$Pt - CH_3O + OH^- \rightarrow Pt - CH_2O + H_2O + e^- \tag{9.10}$$

$$Pt - CH_2O + OH^- \rightarrow Pt - CHO + H_2O + e^- \tag{9.11}$$

$$\begin{aligned} &Pt - CHO + OH^- \rightarrow Pt - CO + H_2O + e^- \\ &\text{or } Pt - CHO + Pt - OH_{ads} + 2OH^- \rightarrow Pt(s) + CO_2 + 2H_2O + 2e^- \end{aligned} \tag{9.12}$$

$$Pt - CO + Pt - OH_{ads} + OH^- \rightarrow Pt(s) + CO_2 + H_2O + e^- \tag{9.13}$$

9.2.2 The Design of MOR Catalysts

Pt-based nanomaterials have been considered as the most widely used catalysts for MOR. However, Pt catalysts are easily poisoned by surface-absorbed carbonaceous intermediates (especially CO_{ads}) generated during the electrooxidation of CH_3OH, which leads to a drastic decrease in catalyst activity [10]. To solve these problems, researchers have devoted enormous efforts to improve and develop MOR electrocatalyst. In summary, there are three main factors that are closely related to the MOR performance for a catalyst, including its morphology, exposed crystalline facets, and chemical compositions.

9.2.2.1 Morphology

Catalyst morphology-based studies currently include nanoparticles (NPs), nanowires (NWs) and nanoflowers (NFs). Different morphological catalysts can not only regulate their specific area, but also increase their contact area with the reactants. For instance, Huang and coworkers [11] reported an ultrafine (2 nm) Pt NWs with dense twin-crystal defects (Pt-MTNN) using a biopeptide-oriented bionanosynthesis method, and the cyclic voltammetry (CV) results showed the ultrafine Pt-MTNN exhibited excellent catalytic performance for MOR in 0.1 M H_2SO_4+0.5 M CH_3OH electrolyte. Considering the fact that MOR mainly occurs on the catalyst' surface, core-shell structures have been developed to fully utilize Pt atoms, in which the core is non-Pt metal atoms and an ultra-thin PtM alloy shell is formed on the surface of the core [12]. This special structure can not only reduce the Pt dosage but also greatly increase the utilization of Pt. Wu et al. [13] recently reported a PtCuNi core-shell alloy with a Cu-rich core and a Pt-rich shell for MOR, and the core-shell catalyst exhibits efficient resisting of CO poisoning with a high I_f/I_b ratio (I_f: forward current density, I_b: backward current density), which is an important factor to evaluate the tolerance of the catalyst to the poisoning of intermediate carbonaceous species in the anode oxidation reaction. It also shows an excellent mass activity of 1.93 A/mg_{Pt} and specific activity of 3.04 mA/cm^2, 5.7 and 5.1 times higher than those of Pt/C, respectively.

9.2.2.2 Exposed Crystal Facets

During the electrooxidation of CH_3OH, the generated CO molecules would occupy the active site of catalysts and lead to the poisoning and deactivation of catalytic sites when there is not enough active OH produced to oxidize it to CO_2. This is closely related to the exposed crystal facets. Jarvi et al. [14] have studied the charge passed, the extent of poisoning degree, and CO_2 yield of Pt(111) and Pt(100) catalysts. They found that only CO was produced on Pt(100) when the potential is below 0.50 V, while small quantities of CO_2 on Pt(111) were generated slowly but sustainably. Pt(111) exhibits higher resistance to CO toxicity, although a greater tendency to dehydrogenate methanol occurred on Pt(100). Pt(111) and Pt(100) are low index facets and stable, however, it has been found that Pt catalysts with high index facets were more active for catalytic electrooxidation than those with low index facets [15]. However, the surface energy of the high index facets is relatively high, and the growth rate of high index factets is faster during the preparation process, resulting in difficulties in the synthesis of catalysts with high index facets.

9.2.2.3 Chemical Compositions

The conflict between catalytic activity and resistance to CO poisoning of Pt catalysts suggests that monometallic Pt is not an ideal choice. To solve this problem, Pt-based binary alloy has been widely studied by introducing a second element. By far, extensive endeavor has been dedicated toward the design and synthesis of bimetallic alloy catalysts including PtRu, PtAu, PtPd, and PtNi [16–19]. For example, Rana et al. [20] prepared $Pd_{73}Pt_{27}$ NWs with less than 30nm using Te NWs as sacrificial templates and applied the $Pd_{73}Pt_{27}$ NWs for MOR catalysis. The electrochemically active area of $Pd_{73}Pt_{27}$ NWs was more than twice that of Pt/C and the mass activity was nearly ten times that of Pt/C, suggesting the good MOR performance of PdPt alloy nanocatalysts. It is widely accepted that the bifunctional mechanism and electronic effects can significantly increase MOR activity modulating the binding strength between Pt atoms and adsorbed oxygen-containing carbon atom intermediates [21].

However, although binary catalysts have some advantages in terms of catalytic activity, the variable dimensions of metal atoms in alloys and their different crystal structures and electron densities have constrained the composition of binary alloys and affected the formation conditions, leading to undesired electronic structures. Researchers are gradually focusing on multi-alloy materials, especially the recently emerged HEMs.

9.3 HEMS ELECTROCATALYSTS FOR MOR

HEMs generally comprise five or more elements with equimolar or near-equimolar concentrations, which are distributed uniformly in a single-phase or unevenly in two-phase solid solutions [22,23]. Currently, the HEMs used as electrocatalysts for MOR are mainly HEAs. The reported HEM electrocatalysts toward the MOR is summarized in Table 9.2.

9.3.1 Noble Metal-Based HEA NPs

Nobel metal-based catalysts have excellent catalytic performance for the electrooxidation of small organic molecules and may serve as the ideal catalyst for complex reactions owing to their multiple adsorption sites on the surface [24]. For MOR, Pt-based catalysts are still considered to be the most popular and efficient electrocatalysts, and the performance of Pt-based HEM is probably higher than that of binary and ternary alloys due to multiatom synergy [25]. As early as 2008, Tsai and coworkers, inspired by the concept of HEA, fabricated a series of PtFeCoNiCuAg HEA NPs with adjustable metal ratios using the sputter deposition method [26]. According to Boltzmann hypothesis, the configuration entropy of $Pt_{52}Fe_{11}Co_{10}Ni_{11}Cu_{10}Ag_8$, whose molar ratio of Pt:Fe:Co:Ni:Cu:Ag is simplified as 5:1:1:1:1:1, could be calculated as below:

$$\Delta S_{\text{conf}} = -R\left[\frac{1}{2}\ln\frac{1}{2}+5\times\frac{1}{10}\ln\frac{1}{10}\right] = 1.498\ R\ (\text{J/K}\cdot\text{mol}) \quad (9.14)$$

This value was a little lower than the configuration entropy of the corresponding HEA system with an equal atomic ratio, which could be expressed as follows:

$$\Delta S_{\text{conf,equal}} = -R\left[6\times\frac{1}{6}\ln\frac{1}{6}\right] = 1.792\ R\ (\text{J/K}\cdot\text{mol}) \quad (9.15)$$

TABLE 9.2
Summary of the Features, Synthesis Methods and the Electrocatalytic MOR Performance of the Reported HEMs

Catalyst Composition	Morphology	Structures	Synthetic Method	Electrolyte	I_f/I_b	Mass Activity	Durability	Ref.
$Pt_{52}Fe_{11}Co_{10}Ni_{11}Cu_{10}Ag_8$	Nanoparticle	*fcc*	Sputter deposition	0.5 M H_2SO_4 1 M CH_3OH	1.09	0.462–0.504 A/mg_{Pt}	-	[26]
HEA-700 (PtFeCoNiCu)	Nanoparticle	*fcc*	Thermal decomposition	0.1 M $HClO_4$ 1 M CH_3OH	1.21	1.4 A/mg_{Pt}	72.5% after 1000 cycles	[30]
$Pt_{28}Cr_{20}Ta_{20}V_{11}Fe_{11}Al_{10}$	Nanoparticle	*fcc*	Hydrogen cold plasma	0.1 M $HClO_4$ 0.5 M CH_3OH	-	0.204 A/mg_{Pt}	-	[44]
AlNiCuPtPdAu AlNiCuPtPdAuCoFe	Nanoporous Nanoporous	*fcc* *fcc*	Dealloying	0.5 M H_2SO_4 1 M CH_3OH	-		-	[45]
PtCoCuRuNiFe	Nanoparticle	*fcc*	SD-TDR	0.5 M H_2SO_4 0.5 M CH_3OH	1.3	1.511 A/mg_{Pt}	-	[25]
PtRuCuOsIr	Nanoporous	*fcc*	Mechanical alloying & dealloying	0.5 M H_2SO_4 0.5 M CH_3OH	1.26	0.86 A/mg_{Pt}	-	[34]
$Ir_{0.19}Os_{0.22}Re_{0.21}Rh_{0.20}Ru_{0.19}$	Nanoparticle	*hcp*	Thermal decomposition	1 M H_2SO_4 1 M CH_3OH	-	-	-	[27]
CuAgAuPtPd	Nanoparticle	*fcc*	Arc-melting & mechanical alloying	0.5 M KNO_3 500 μL CH_3OH	4	-	-	[46]
$Pt_{18}Ni_{26}Fe_{15}Co_{14}Cu_{27}$	Nanoparticle	*fcc*	Colloidal synthesis	1 M KOH 1 M CH_3OH	3.26	15.04 A/mg_{Pt}	93.6% after 1000 cycles	[28]
PdNiCoCuFe	Nanotube	-	Electrodeposition method	0.5 M NaOH 0.5 M CH_3OH	-	~95 A/mg_{Pd}	-	[47]
PtRhBiSnSb	Nanoplate	*hcp*	Wet-chemistry method	1 M KOH 1 M CH_3OH	-	19.529 A/mg_{Pt+Rh}	70.2% after 5000 cycles	[40]
PdPtCuAgAu	Nanowire	*fcc*	One-pot solution-phase	1 M KOH 1 M CH_3OH	-	~2.1 A/mg_{Pd+Pt}	-	[37]
PtRuRhCoNi	Nanowire	*fcc*	Colloidal synthesis	1 M KOH 1 M CH_3OH	-	8.20 A/mg_{Pt} 6.65 A/mg_{PtRuRh}	90.2% after 2000 cycles	[38]
$Ni_2Fe_{0.5}Co_{0.5}$-BP	Nanoparticle	-	Aqueous sol-gel	1 M KOH 1 M CH_3OH	-	0.114 A/mg	-	[43]
$(CoCrFeMnNi)_3O_4$	Nanoparticle	-	Soft chemical	1 M KOH 0.5 M CH_3OH	-	0.11 A/mg	91% after 4000 seconds	[41]

Additionally, $Pt_{52}Fe_{11}Co_{10}Ni_{11}Cu_{10}Ag_8$ exhibited the highest mass activity, which was higher than that of Pt, but lower than that of $Pt_{43}Ru_{57}$, indicating that increase in entropy could improve the catalytic activity.

9.3.1.1 Phase Structure

Typical HEA phase structures include face-centered cubic (*fcc*) structures, body-centered cubic (*bcc*) structures, and hexagonal close-packed (*hcp*) structures. *Fcc*-structured HEAs are the most widely studied for MOR (Table 9.2), which consist of *fcc* metals such as Ir, Pd, Pt, Rh, and first-row metals (Al, Co, Cr, Fe, Ni). It has been found that Os, Ru and Re are *hcp* in ambient conditions, which helps to improve the mechanical properties and chemical stability in alloys. Inspired by this, Yusenko et al. [27] synthesized a series of IrOsReRhRu HEA NPs with three different structures (*fcc*, *hcp*, and *fcc/hcp*) using high-temperature pyrolysis and investigated their MOR performance in an acidic electrolyte containing 1 M H_2SO_4 and 1 M CH_3OH. They found that the single-phase $Ir_{0.19}Os_{0.22}Re_{0.21}Rh_{0.2}Ru_{0.19}$ HEA with *hcp* structure exhibited the best MOR performance. The enhanced electrocatalytic activity could be attributed to the electronic effects through which the electronic configuration of Pt atoms was moderately adjusted, resulting in diminished adsorption of CO_{ads} and facile removal of CO_{ads} and thus, making CH_3OH in the solution interact with the catalysts surface more conveniently.

9.3.1.2 Synergistic Effect of Multiple Active Sites

The synergistic effect of multiple active sites contributed positively to the catalytic efficacy for MOR. Wang and coworkers prepared ultra-small and uniform $Pt_{18}Ni_{26}Fe_{15}Co_{14}Cu_{27}$ HEA NPs for alkaline MOR [28]. The $Pt_{18}Ni_{26}Fe_{15}Co_{14}Cu_{27}/C$ exhibited lower onset potential (412 mV at 0.1 A/mg_{Pt} mass activity), higher mass activity (15.04 A/mg_{Pt}) and specific activity (114.93 mA/cm^2) than Pt/C (545 mV, 1.45 A/ mg_{Pt}, 27.48 mA/cm^2), as shown in Figure 9.2a and b. Besides, the decay of mass activity in merely 6.4% after 1000 electrochemical cycles demonstrated the exceptional stability of the $Pt_{18}Ni_{26}Fe_{15}Co_{14}Cu_{27}/C$ (Figure 9.2c). The larger I_f/I_b of $Pt_{18}Ni_{26}Fe_{15}Co_{14}Cu_{27}/C$ (3.26) than that of Pt/C (2.31) indicated the enhanced catalyst activity of CO remove after CH_3OH dehydrogenation of $Pt_{18}Ni_{26}Fe_{15}Co_{14}Cu_{27}/C$. The excellent MOR performance was attributed to the synergistic effect of multi-active sites on the HEA surface. It is proposed that Pt sites are responsible for the electron transfer for the HEA surfaces due to its positive offset, Fe sites facilitate the intermediate adsorption stability and the strong resistance to CO poisoning with an independent electronic structure within its surrounding environment, Cu sites support the electroactivity of the HEA, Co sites enhance the electron transfer efficiency in MOR, and Ni sites preserve the electroactive electron center (Figure 9.2d and e).

9.3.1.3 Compressive Strain Engineering

Appropriate compressive strain is also beneficial for enhancing the catalytic activity of alloys, which alters the electronic band structure of metal atoms [29]. To verify the key role of compressive strain in HEMs, Wang et al. [30] proposed a surface strain engineering strategy to prepare PtFeCoNiCu HEAs by controlling the thermal reduction temperature. HEA-400 was synthesized through heating HEA precursor at 400°C, while HEA-700 was obtained via annealing HEA-400 at 700°C. HEA-700 exhibited a 0.94% compressive strain and had a better alloy structure, resulting in the diffraction peaks of HEA-700 shifted to higher angles in the XRD pattern, compared with HEA-400. The different surface compressive strain between HEA-700 and HEA-400 endows HEM-700 higher catalytic activity than HEA-400. Electrochemical results show that HEA-700 possesses higher mass activity (1.40 A/mg) and specific activity (3.29 mA/cm^2) than HEA-400 (1.30 A/mg for mass activity, 2.35 mA/cm^2 for specific activity) (Figure 9.3a). Additionally, HEA-700 has better CO anti-poisoning performance in MOR than HEA-400 and Pt/C, which was proved by the I_f/I_b of the samples. The radial distribution function (RDF) results show that HEA-700 had a shorter Pt–Pt bond length than HEA-400 and Pt/C, which results from the compressive strain and could improve the electrocatalytic activity by tuning d-band center [31,32]. The bond length of Pt–Pt

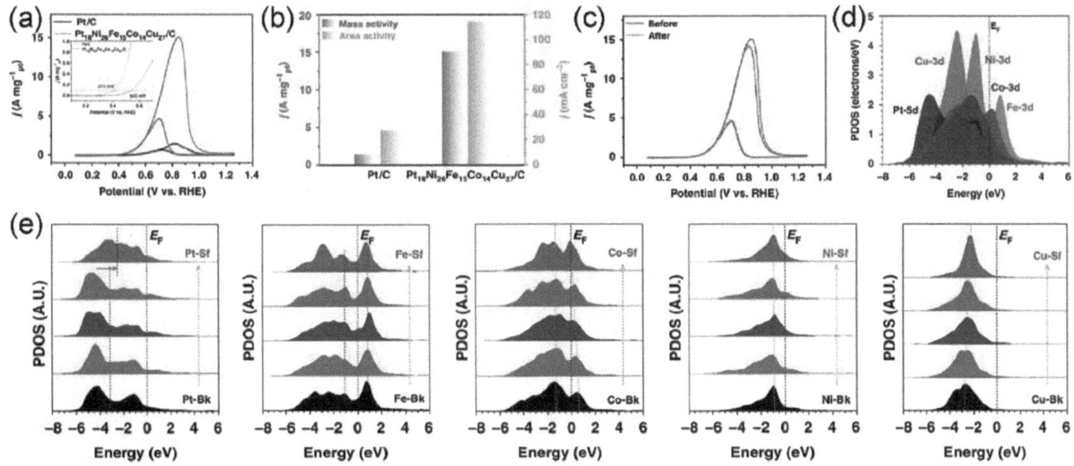

FIGURE 9.2 (a) CV curves and (b) mass activity and specific activity of $Pt_{18}Ni_{26}Fe_{15}Co_{14}Cu_{27}/C$ and Pt/C in an alkaline medium. (c) CV curves of $Pt_{18}Ni_{26}Fe_{15}Co_{14}Cu_{27}/C$ before and after 1000 cycles. (d) The partial projected density of states (PDOSs) and (e) the site-dependent PDOSs of different metals in HEA. Adapted from reference [28]. Copyright 2020, The Authors, some rights reserved; exclusive licensee [Springer Nature]. Distributed under a Creative Commons Attribution License 4.0 (CC BY).

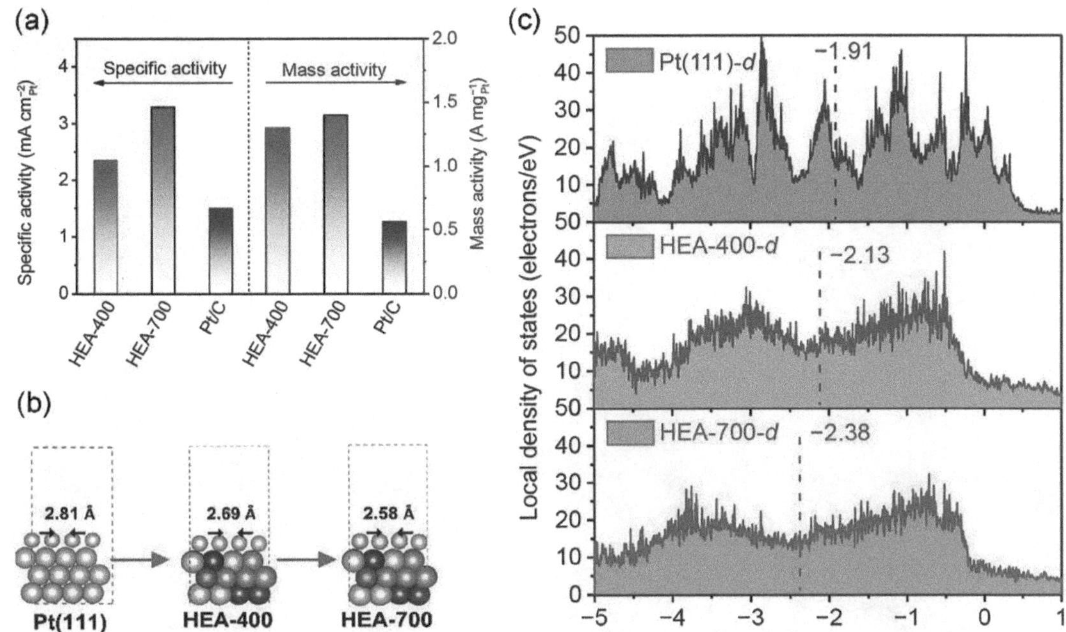

FIGURE 9.3 (a) Specific activity and mass activity of HEA-700, HEA-400, and Pt/C; (b) Geometrically optimized atomic structures of Pt(111), HEA-400 and HEA-700; (c) the position of the d-band center, as calculated by DFT. Adapted with permission [30]. Copyright 2021, Springer Nature.

was set according to the experimental values to perform DFT calculations (Figure 9.3b), and the results indicate that the reduced *d*-band center of Pt caused by the compression strain could result in a weaker CO adsorption on HEA-700 relative to HEA-400 and Pt(111) (Figure 9.3c). This work presents a novel approach and direction for designing excellent HEA catalysts.

9.3.1.4 Synthesis Methods

The challenges in HEAs synthesis hinder the exploiting in catalytic mechanisms and properties of HEA catalysts. Peng's group recently synthesized a senary HEA (PtCrTaVFeAl) NPs using a hydrogen cold plasma-enabled synthesis method, as illustrated in Figure 9.4 [33]. Typically, the well-mixed Pt and reactive metal precursor were introduced into a dielectric barrier discharge (DBD) reactor purged by Ar, followed by feeding hydrogen into the reactor to generate cold plasma at room temperature and atmospheric pressure (Figure 9.4b). The reactor temperature quickly rose to ~40°C within 10 seconds and then increased gradually until it stabilized at around 200°C

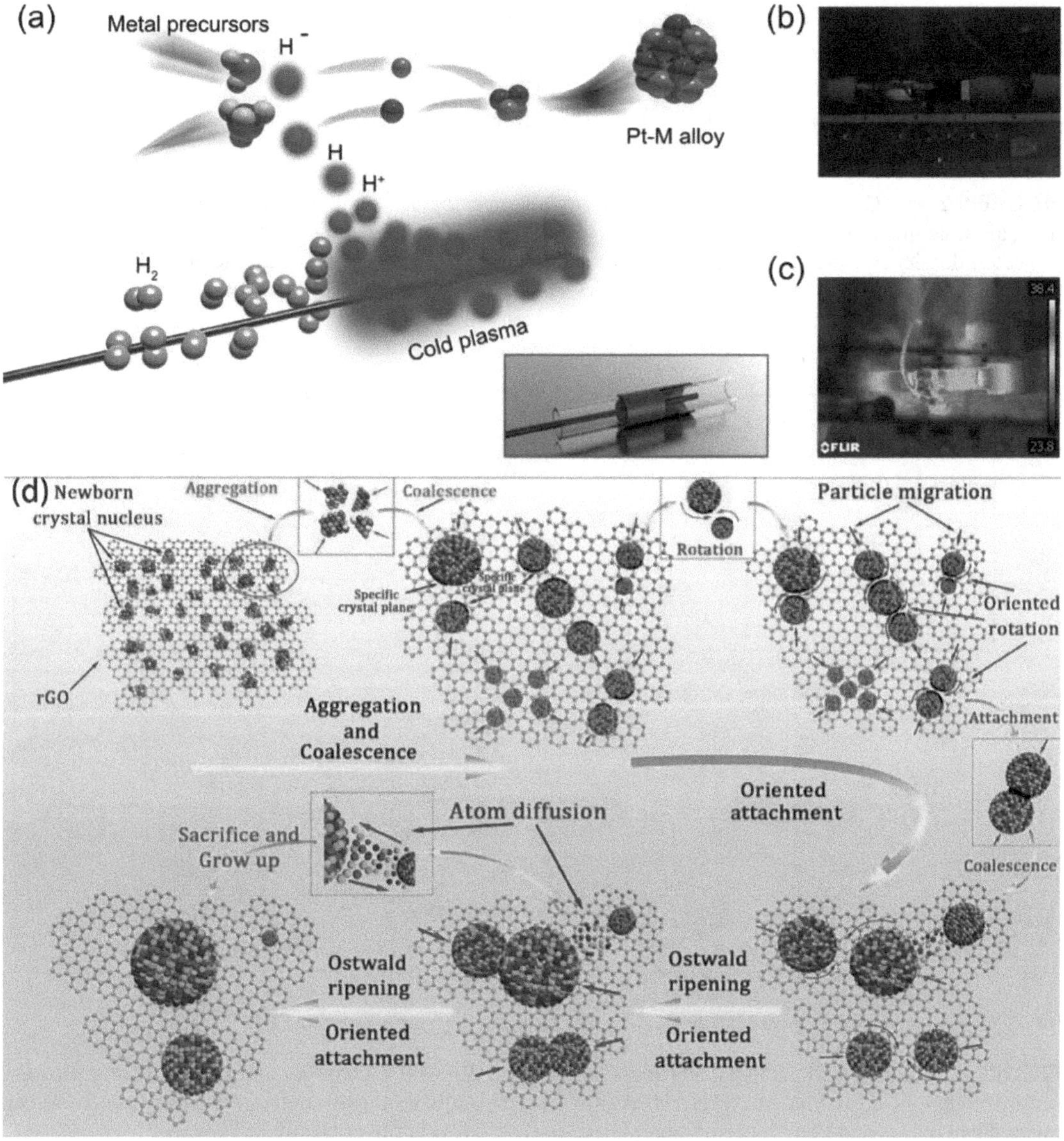

FIGURE 9.4 (a) Schematic illustration of hydrogen cold plasma synthesis process; (b) Plasma generator at work, emitting glowing H_2 cold plasma; (c) Infrared thermal image of the DBD reactor captured after 10 seconds of operation; Adapted with permission [33]. Copyright 2022, American Chemical Society. (d) Growth mechanism of Pt-based HEA-NPs during TDR process. Adapted with permission [25]. Copyright 2022, American Chemical Society.

(Figure 9.4c). The reduction process reached completion within 10 minutes. The effective formation of Pt-HEA was ascribed to the generation of hydrogen anions in the hydrogen plasma, which had an extremely high reduction capacity, with a standard reduction potential of –2.3 V to reduce both platinum and active metal precursors into nanoalloys. The composition ratio of Pt-HEA was finally determined as $Pt_{28}Cr_{20}Ta_{20}V_{11}Fe_{11}Al_{10}$. In acidic MOR, $Pt_{28}Cr_{20}Ta_{20}V_{11}Fe_{11}Al_{10}$/C exhibited current densities of 2.82 mA/cm^2_{Pt} at 0.85 V vs. RHE, more than three times higher than Pt/C (0.84 mA/cm^2_{Pt}), benefiting from the remarkable increase in the intrinsic activity. Dispersed Pt active sites and the reduced *d*-band center of the HEA catalyst would help to weaken the adsorption of CO_{abs} and alleviate the toxic effect of CO_{abs} [2].

A spray drying technique combined with the thermal decomposition reduction (SD-TDR) method was developed by Chen's group to prepare a series of Pt-based HEA NPs with up to ten dissimilar elements loaded on carbon supports [25]. The authors proposed that the rapid and homogeneous precipitation of the elements without segregation or localized elemental aggregation was facilitated by rapid heating and instantaneous drying. They used three stages to describe the HEAs growth in the TDR process: aggregation and coalescence, oriented attachment, and Ostwald ripening, as show in Figure 9.4d. In this experiment, the preparation of precursor NPs with a homogeneous mixture of elements was the most important step in the SD process, and the dispersion of polymetallic precursor NPs (MPNPs) was crucial for the formation of NPs. During MOR, the metal atoms M around Pt atom in HEA possess more negative potentials preferring to adsorb –OH to form M–OH_{ads}, which was beneficial for expediting the removal of CO_{ads} from Pt, thereby improving the electroactivity and CO toxicity resistance. Results showed that the specific current density and mass current density of the senary HEA were eight times and five times higher than those of commercial Pt/C in the acidic solution, indicating the high electrocatalytic performance of the PtCoCuRuNiFe HEA. As expected, the ratio of I_f and I_b of the senary HEA (1.3) was higher than that of the commercial Pt/C (0.88), suggesting good tolerance to the poisoning species.

9.3.2 Noble Metal-Based Nanoporous-HEA

Apart from regulating the composition, one of the key researches focuses of HEAs catalysts is achieving precise control of nanocrystal morphology. Nanoporous-PtRuCuOsIr HEA is prepared by Chen et al. [34] using the chemical dealloying method of mechanically alloyed AlCuPtRuOsIr precursor. As shown in Figure 9.5a, the structure of distinctive bi-continuous ligament channels could be discerned. The ligaments were comprised of multiple minuscule nanocrystals, each just nanometer in size. The spacing of certain lattice fringes was 0.221 nm. The high-resolution transmission electron microscope (HRTEM) picture and fast Fourier transform (FFT) pattern confirm the nano-crystalline essence of the interlinked ligaments. The nanoporous-PtRuCuOsIr HEA has excellent MOR electroactivity, with the specific activity of 3.0 mA/cm^2 and mass activity of 857.5 A/mg_{Pt}.

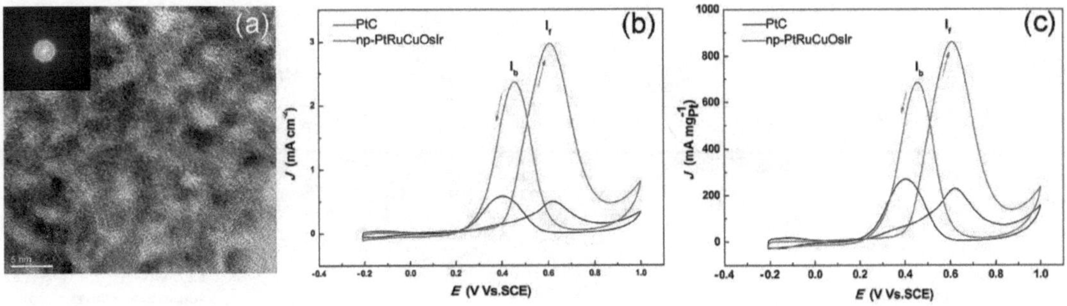

FIGURE 9.5 (a) HRTEM image of the PtRuCuOsIr HEA. (b and c) CV of the PtRuCuOsIr and Pt/C for MOR in an acidic solution. Adapted with permission [34]. Copyright 2014, Elsevier.

(Figure 9.5b and c). Pt was responsible for the dissociation and chemical adsorption of CH_3OH in the multi-component HEA catalyst. Additionally, the alloyed 3d metal component generates surface oxyhydroxides, significantly enhancing its ability to extract active oxygen from the acidic electrolyte water [35]. Subsequently, the active oxygen effectively oxidized the carbonaceous residue into CO_2 [36]. In addition, the channels and porous features in the material provide excellent transport paths for molecules and electrons, which greatly promotes reaction kinetics. Moreover, the material's channels and porous characteristics act as excellent pathways for the transportation of molecules and electrons, thereby significantly amplifying reaction kinetics.

9.3.3 Noble Metal-Based HEA NWs

One-dimensional NWs with structural anisotropy have a significant positive effect on electron/mass transfer and utilization efficiency, thus enhancing the electrocatalytic activity. Xu's group successfully prepared PdPtCuAgAu HEA NW networks via a mild one-pot synthesis using $C_{16}NCOOH(Br)$ as the soft templates, which guided the growth of NWs (Figure 9.6a) [37]. They pointed out that the carboxyl groups and Br^- in the surfactants were crucial for the formation of NWs. The former stabilized micelles intermolecular through hydrogen bonds, and bonded with metal precursors via coordination, while the latter formed $PdBr_4^{2-}$ with lower redox potentials, leading to a mild reduction, which are beneficial for the growth of the NWs. Compared with Pd NPs, the mass activity and stability of PdPtCuAgAu HEA NWs were improved (Figure 9.6b), which mainly resulted from the morphological advantage of NWs and the high-entropy advantage of HEA with sluggish diffusion effect and lattice distortion effect. Additionally, Li and coworkers prepared ultrathin PtRuRhCoNi HEA NWs using hexadecyl trimethyl ammonium bromide (CTAB) acting as structure directing agent and $Mo(CO)_6$ acting as a reducing agent [38]. PtRuRhCoNi NWs/C exhibited higher mass activity of 6.65 A/mg_{PtRuRh} (8.20 A/mg_{Pt}) and specific activity of 6.56 mA/cm^2_{PtRuRh} than PtRuRhCoNi NPs/C (6.07 A/mg_{PtRuRh}, 5.66 mA/cm^2_{PtRuRh}) in alkaline conditions. Moreover, it maintained higher residual mass activity (90.2%, 0.96 A/mg_{PtRuRh}) than PtRuRhCoNi NPs/C (89.1%, 0.59 A/mg_{PtRuRh}) after 2000 CV cycles and 1h electrolysis. Such excellent MOR performance of PtRuRhCoNi NWs/C should be attributed to the strong self-complementary effect caused by the strong coupling between bonding and anti-bonding orbitals from the multi-elements in PtRuRhCoNi-HEAs.

9.3.4 Noble Metal-Based HEI Nanoplates

Although HEAs have outstanding catalytic selectivity and efficiency over conventional alloys, the atoms in alloys maintain a disordered distribution in their lattice structure, which constrain the feasible atomic arrangements of metallic solutes, thereby hindering the realization of their advantageous properties [39]. Intermetallic compounds feature with a well-defined geometric atomic

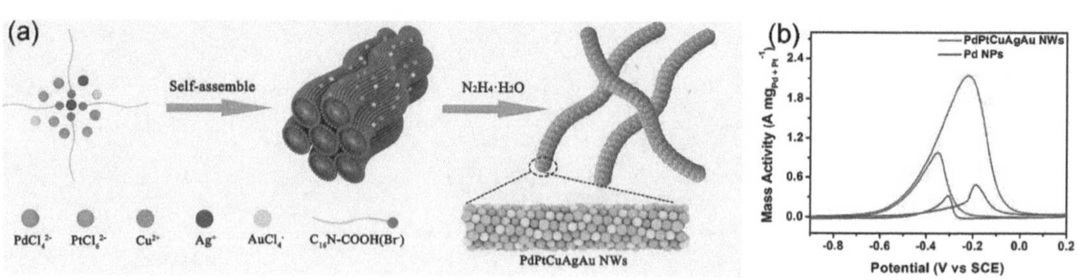

FIGURE 9.6 (a) Schematic illustration of PdPtCuAgAu HEA NWs formation process; (b) CV curves of PdPtCuAgAu NWs and Pd NPs in alkaline media. Adapted with permission [37]. Copyright 2022, Elsevier.

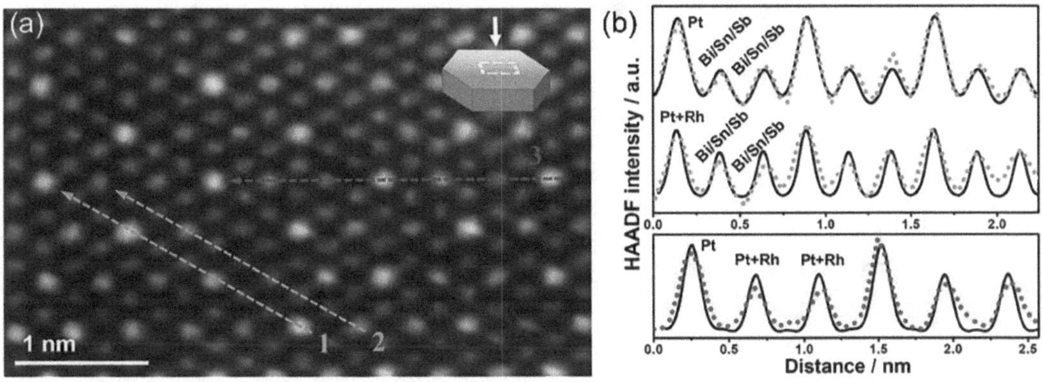

FIGURE 9.7 (a) The HAADF-STEM image of the center part of the PtRhBiSnSb HEI nanoplates used to simulate atomic arrangements; (b) HAADF image of Rh-substituted Pt columns. Adapted with permission [40]. Copyright 2022, John Wiley and Sons.

configuration and can be finely manipulated to clarify the relationship between structure and properties. Merging the strengths of HEAs and intermetallic compounds, HEIs emerge as a highly prospective option for electrocatalysts. Chen et al. [40] systhesized *hcp*-structured PtRhBiSnSb HEI nanoplates using a wet-chemistry method and successfully identified the atomic structure of the HEI nanoplates as (PtRh)(BiSnSb). The authors resolved the atomic arrangements of the PtRhBiSnSb HEI nanoplates by simulating the high-angle annular dark (HAADF) image of the central parts viewed along the [001] direction (Figure 9.7a). They identified three types of atomic configurations in the HEI through HAADF intensity line profiles, as shown in Figure 9.7b. Based on the information, they constructed the theoretical atomic model and the identified atomic structure was obtained when the periodic stacking pattern was consistent with the model. The isolated without contiguous Pt sites on the catalyst' surface and enlarged Pt–Pt distance was beneficial for improving the tolerances to CO poisoning through reducing or even eliminating CO_{ads} because at least three contiguous Pt sites were required to form CO_{ads}. Additionally, the incorporation of Rh optimized the electronic structures and enhanced the electron transfer efficiency in PtRhBiSnSb HEI nanoplates, thus also improving their oxidation capability. Besides, the synergistic effect of Bi, Sn, and Sb made the electronic structures of the active sites robust.

9.3.5 Non-Noble Metal-Based HEMs

As mentioned above, noble metals-based HEA have exhibited excellent electrocatalytic activity toward MOR, however, the high price of noble metals limits their application in industry. So, there is a need to develop the catalysts that are sufficient in nature, cheaper, and easier for scalable production. In recent years, non-noble metal-based catalysts, particularly Ni-based ones, have garnered significant interest and attention from researchers, due to their favorable surface oxidation properties, cost-effectiveness, and prevalence in the Earth's crust. However, the fabrication of non-noble metal-based HEAs remains a formidable challenge, mainly due to the difficulties of controllable synthesis, particularly when it comes to precisely controlling particle size and morphology during the fabrication process.

Talluri et al. [41] prepared a spherical Ni-based spinel high-entropy oxide (HEO) with an overall composition of $(CoCrFeMnNi)_3O_4$ using a soft chemical method and first reported its MOR performance in an alkaline condition. In comparison with Co_3O_4 and NiO nanostructures synthesized through the same method and $NiCo_2O_4$ nanostructure reported by Qian [42], $(CoCrFeMnNi)_3O_4$ exhibited the highest mass activity of about 110 mA/mg, and a moderate onset potential of 0.45 V, which was lower than Co_3O_4 and NiO, while higher than $NiCo_2O_4$. Besides, the $(CoCrFeMnNi)_3O_4$ HEO showed a good stability as indicated by a mere 9% reduction in specific current density after 4000 consecutive seconds. Mushiana et al. [43] employed a modified sol-gel method to synthesize $Ni_2Fe_{0.5}Co_{0.5}$–BP HEM, which involved metal chloride precursors and $B(OMe)_3$ serving as the borate precursor, while H_3PO_4 was utilized as the acid catalyst for both hydrolysis and gelation processes. The electrocatalytic activity of $Ni_2Fe_{0.5}Co_{0.5}$–BP was the highest compared to that of $Ni_2Co_{0.5}$–BP, Ni_2–BP, and $Ni_2Fe_{0.5}$–BP. The authors believed that the increase in the number of accessible active sites and the synergetic effect of HEMs were responsible for the improvement in MOR.

9.4 SUMMARY AND PROSPECTS

We have methodically outlined the mechanism of MOR and condensed the recent advancements of HEMs electrocatalysts for MOR in this chapter. By compiling published articles, we have put forth design tactics to enhance electrocatalysts for HEAs, such as the manipulation of phase structures, the utilization of multiple active sites in a synergistic manner, and the application of compressive strain. Although there has been considerable interest in using HEAs as MOR catalysts, several obstacles remain that must be surmounted before their widespread commercialization, including the large-scale and facile synthesis methods, HEAs stability under operating conditions, and the correlation between structure and MOR performance. Further research will be required to explore the potential of these materials and to address the challenges that need to be overcome.

1. Developing facile, low-cost, and large-scale methods for producing HEMs that can stimulate their use in industry;
2. Designing various non-noble metal-based HEMs with comparable activity to Pt catalysts, or lower noble metal loading HEM catalysts;
3. Modulating synergistic effect of the multi-elements in HEMs to improve the electrochemical activity, stability in MOR as well as CO anti-poisoning of nanoscale HEMs;
4. Exploiting precise simulation models and calculation methods to modify the composition and surface properties of HEM by calculating the adsorption energy of the catalytic site and considering the surrounding chemical environment of the active site.

REFERENCES

[1] L. Chen, X. Liang, D. Wang, Z. Yang, C. He, W. Zhao, J. Pei, Y. Xue, Platinum-ruthenium single atom alloy as a bifunctional electrocatalyst toward methanol and hydrogen oxidation reactions, *ACS Appl. Mater. Interfaces* 14 (2022) 27814–27822.
[2] H. Huang, X. Hu, J. Zhang, N. Su, J. Cheng, Facile fabrication of platinum-cobalt alloy nanoparticles with enhanced electrocatalytic activity for a methanol oxidation reaction, *Sci. Rep.* 7 (2017) 45555.
[3] J. Tiwari, R. Tiwari, G. Singh, K. Kim, Recent progress in the development of anode and cathode catalysts for direct methanol fuel cells, *Nano Energy* 2 (2013) 553–578.
[4] Y. Tong, X. Yan, J. Liang, S. Dou, Metal-based electrocatalysts for methanol electro-oxidation: Progress, opportunities, and challenges, *Small* 17 (2021) 1904126.
[5] S. Munjewar, S. Thombre, R. Mallick, A comprehensive review on recent material development of passive direct methanol fuel cell, *Ionics* 23 (2017) 1–18.
[6] J. Varcoe, R. Slade, Prospects for alkaline anion-exchange membranes in low temperature fuel cells, *Fuel Cells* 5 (2005) 187–200.
[7] H. Bahrami, A. Faghri, Review and advances of direct methanol fuel cells: Part II: Modeling and numerical simulation, *J. Power Sources* 230 (2013) 303–320.
[8] A. Hamnett, Mechanism and electrocatalysis in the direct methanol fuel cell, *Catal. Today* 38 (1997) 445–457.
[9] E. Yu, U. Krewer, K. Scott, Principles and materials aspects of direct alkaline alcohol fuel cells, *Energies* 3 (2010) 1499–1528.

[10] X. Fu, C. Wan, A. Zhang, Z. Zhao, H. Huyan, X. Pan, S. Du, X. Duan, Y. Huang, Pt3Ag alloy wavy nanowires as highly effective electrocatalysts for ethanol oxidation reaction, *Nano Res.* 13 (2020) 1472–1478.

[11] L. Ruan, E. Zhu, Y. Chen, Z. Lin, X. Huang, X. Duan, Y. Huang, Biomimetic synthesis of an ultrathin platinum nanowire network with a high twin density for enhanced electrocatalytic activity and durability, *Angew. Chem. Int. Ed.* 52 (2013) 12577–12581.

[12] C. Tan, Y. Sun, J. Zheng, D. Wang, Z. Li, H. Zeng, J. Guo, L. Jing, L. Jiang, A self-supporting bimetallic Au@Pt core-shell nanoparticle electrocatalyst for the synergistic enhancement of methanol oxidation, *Sci Rep.* 7 (2017) 6347.

[13] Q. Wu, X. Huang, T. Wan, D. Xiang, X. Li, K. Wang, X. Yuan, P. Li, M. Zhu, Enhancing electrocatalytic methanol oxidation on PtCuNi core-shell alloy structures in acid electrolytes *Inorg. Chem.* 61 (2022) 2612–2618.

[14] T. Jarvi, S. Sriramulu, E. Stuve, Reactivity and extent of poisoning during methanol electro-oxidation on platinum (100) and (111): A comparative study *Colloid. Surface A* 134 (1998) 145–153.

[15] K. Mikita, M. Nakamura, N. Hoshi, In situ infrared reflection absorption spectroscopy of carbon monoxide adsorbed on Pt(S)-[*n*(100)×(110)] electrodes, *Langmuir* 23 (2007) 9092–9097.

[16] H. Shi, F. Liao, W. Zhu, C. Shao, M. Shao, Effective PtAu nanowire network catalysts with ultralow Pt content for formic acid oxidation and methanol oxidation, *Int. J. Hydrogen. Energy* 45 (2020) 16071–16079.

[17] J. Zhang, X. Qu, Y. Han, L. Shen, S. Yin, G. Li, Y. Jiang, S. Sun, Engineering PtRu bimetallic nanoparticles with adjustable alloying degree for methanol electrooxidation: Enhanced catalytic performance, *Appl. Catal. B-Environ.* 263 (2020) 118345.

[18] M. Hanifah, J. Jaafar, M. Othman, A. Ismail, M. Rahman, N. Yusof, F. Aziz, N. Rahman, One-pot synthesis of efficient reduced graphene oxide supported binary Pt-Pd alloy nanoparticles as superior electro-catalyst and its electro-catalytic performance toward methanol electro-oxidation reaction in direct methanol fuel cell, *J. Alloy. Compd.* 793 (2019) 232–246.

[19] X. Teng, A. Shan, Y. Zhu, R. Wang, W. Lau, Promoting methanol-oxidation-reaction by loading PtNi nano-catalysts on natural graphitic-nano-carbon, *Electrochim. Acta* 353 (2020) 136542.

[20] M. Rana, P. Patil, M. Chhetri, K. Dileep, R. Datta, U. Gautam, Pd-Pt alloys nanowires as support-less electrocatalyst with high synergistic enhancement in efficiency for methanol oxidation in acidic medium, *J. Colloid Interf. Sci.* 463 (2016) 99–106.

[21] N. Dimitrova, M. Dhifallah, T. Mineva, T. Boiadjieva-Scherzer, H. Guesmi, J. Georgieva, High performance of PtCu@TiO2 nanocatalysts toward methanol oxidation reaction: From synthesis to molecular picture insight, *RSC Adv.* 9 (2019) 2073–2080.

[22] A. Ostovari Moghaddam, E. Trofimov, Toward expanding the realm of high entropy materials to platinum group metals: A review, *J. Alloy. Compd.* 851 (2021) 156838.

[23] J. Yeh, S. Chen, S. Lin, J. Gan, T. Chin, T. Shun, C. Tsau, S. Chang, Nanostructured high-entropy alloys with multiple principal elements: Novel alloy design concepts and outcomes, *Adv. Eng. Mater.* 6 (2004) 299–303.

[24] H. Li, J. Lai, Z. Li, L. Wang, Multi-sites electrocatalysis in high-entropy alloys, *Adv. Funct. Mater.* 31 (2021) 2106715.

[25] P. Zhao, Q. Cao, W. Yi, X. Hao, J. Li, B. Zhang, L. Huang, Y. Huang, Y. Jiang, B. Xu, Z. Shan, J. Chen, Facile and general method to synthesize Pt-based high-entropy-alloy nanoparticles, *ACS Nano* 16 (2022) 14017–14028.

[26] C. Tsai, K. Yeh, P. Wu, Y. Hsieh, P. Lin, Effect of platinum present in multi-element nanoparticles on methanol oxidation, *J. Alloy. Compd.* 478 (2009) 868–871.

[27] K. Yusenko, S. Riva, P. Carvalho, M. Yusenko, S. Arnaboldi, A. Sukhikh, M. Hanfland, S. Gromilov, First hexagonal close packed high-entropy alloy with outstanding stability under extreme conditions and electrocatalytic activity for methanol oxidation, *Scripta Mater.* 138 (2017) 22–27.

[28] H. Li, Y. Han, H. Zhao, W. Qi, D. Zhang, Y. Yu, W. Cai, S. Li, J. Lai, B. Huang, L. Wang, Fast site-to-site electron transfer of high-entropy alloy nanocatalyst driving redox electrocatalysis, *Nat. Commun.* 11 (2020) 5437.

[29] P. Strasser, S. Koh, T. Anniyev, J. Greeley, K. More, C. Yu, Z. Liu, S. Kaya, D. Nordlund, H. Ogasawara, M. Toney, A. Nilsson, Lattice-strain control of the activity in dealloyed core-shell fuel cell catalysts, *Nat. Chem.* 2 (2010) 454–460.

[30] D. Wang, Z. Chen, Y. Huang, W. Li, J. Wang, Z. Lu, K. Gu, T. Wang, Y. Wu, C. Chen, Y. Zhang, X. Huang, L. Tao, C. Dong, J. Chen, C. Singh, S. Wang, Tailoring lattice strain in ultra-fine high-entropy alloys for active and stable methanol oxidation, Sci. *China Mater.* 64 (2021) 2454–2466.

[31] Q. Wang, Q. Zhao, Y. Su, G. Zhang, G. Xu, Y. Li, B. Liu, D. Zheng, J. Zhang, Hierarchical carbon and nitrogen adsorbed PtNiCo nanocomposites with multiple active sites for oxygen reduction and methanol oxidation reactions, *J. Mater. Chem. A* 4 (2016) 12296–12307.

[32] M. Shao, Q. Chang, J. Dodelet, R. Chenitz, Recent advances in electrocatalysts for oxygen reduction reaction, *Chem. Rev.* 116 (2016) 3594–3657.

[33] D. Wu, L. Yao, M. Ricci, J. Li, R. Xie, Z. Peng, Ambient synthesis of Pt-reactive metal alloy and high-entropy alloy nanocatalysts utilizing hydrogen cold plasma, *Chem. Mater.* 34 (2022) 266–272.

[34] X. Chen, C. Si, Y. Gao, J. Frenzel, J. Sun, G. Eggeler, Z. Zhang, Multi-component nanoporous platinum-ruthenium-copper-osmium-iridium alloy with enhanced electrocatalytic activity towards methanol oxidation and oxygen reduction, *J. Power Sources* 273 (2015) 324–332.

[35] K. Ley, R. Liu, C. Pu, Q. Fan, N. Leyarovska, C. Segre, E. Smotkin, Methanol oxidation on single-phase Pt–Ru–Os ternary alloys, *J. Electrochem. Soc.* 144 (1997) 1543.

[36] H. Lei, S. Suh, B. Gurau, B. Workie, R. Liu, E. Smotkin, Deuterium isotope analysis of methanol oxidation on mixed metal anode catalysts, *Electrochim. Acta* 47 (2002) 2913–2919.

[37] D. Fan, K. Guo, Y. Zhang, Q. Hao, M. Han, D. Xu, Engineering high-entropy alloy nanowires network for alcohol electrooxidation, *J Colloid Interf. Sci.* 625 (2022) 1012–1021.

[38] H. Li, M. Sun, Y. Pan, J. Xiong, H. Du, Y. Yu, S. Feng, Z. Li, J. Lai, B. Huang, L. Wang, The self-complementary effect through strong orbital coupling in ultrathin high-entropy alloy nanowires boosting pH-universal multifunctional electrocatalysis, *Appl. Catal. B-Environ.* 312 (2022) 121431.

[39] Z. Jia, T. Yang, L. Sun, Y. Zhao, W. Li, J. Luan, F. Lyu, L. Zhang, J. Kruzic, J. Kai, J. Huang, J. Lu, C. Liu, A novel multinary intermetallic as an active electrocatalyst for hydrogen evolution, *Adv. Mater.* 32 (2020) 2000385.

[40] W. Chen, S. Luo, M. Sun, X. Wu, Y. Zhou, Y. Liao, M. Tang, X. Fan, B. Huang, Z. Quan, High-entropy intermetallic PtRhBiSnSb nanoplates for highly efficient alcohol oxidation electrocatalysis, *Adv. Mater.* 34 (2022) 2206276.

[41] T. Mushiana, M. Khan, M. Abdullah, N. Zhang, M. Ma, Facile sol-gel preparation of high-entropy multielemental electrocatalysts for efficient oxidation of methanol and urea, *Nano Res.* 15 (2022) 5014–5023.

[42] L. Qian, L. Gu, L. Yang, H. Yuan, D. Xiao, Direct growth of NiCo2O4 nanostructures on conductive substrates with enhanced electrocatalytic activity and stability for methanol oxidation, *Nanoscale* 5 (2013) 7388.

[43] B. Talluri, K. Yoo, J. Kim, High entropy spinel metal oxide (CoCrFeMnNi)3O4 nanoparticles as novel efficient electrocatalyst for methanol oxidation and oxygen evolution reactions, *J. Environ. Chem. Eng.* 10 (2022) 106932.

[44] D. Wu, L. Yao, M. Ricci, J. Li, R. Xie, Z. Peng, Ambient synthesis of Pt-reactive metal alloy and high-entropy alloy nanocatalysts utilizing hydrogen cold plasma, *Chem. Mater.* 34 (2022) 266–272.

[45] H. Qiu, G. Fang, Y. Wen, P. Liu, G. Xie, X. Liu, S. Sun, Nanoporous high-entropy alloys for highly stable and efficient catalysts, *J. Mater. Chem. A* 7 (2019) 6499–6506.

[46] N. Katiyar, S. Nellaiappan, R. Kumar, K. Malviya, K.G. Pradeep, A. Singh, S. Sharma, C. Tiwary, K. Biswas, Formic acid and methanol electro-oxidation and counter hydrogen production using nano high entropy catalyst, *Mater. Today Energy* 16 (2020) 100393.

[47] A. Wang, H. Wan, H. Xu, Y. Tong, G. Li, Quinary PdNiCoCuFe alloy nanotube arrays as efficient electrocatalysts for methanol oxidation, *Electrochim. Acta* 127 (2014) 448–453.

10 High Entropy Materials for Electrocatalytic Hydrogen Generation

Abhijit Ray

10.1 INTRODUCTION

Three major green energy technological revolutions of coming decades, namely the water electrolysis to produce hydrogen, fuel-cell-based transports, and metal-air batteries for high energy density storage should rely on conversion between water, oxygen, and hydrogen. In compliance, three reactions, namely the oxygen evolution reaction (OER), oxygen reduction reaction (ORR), and hydrogen evolution reaction (HER), are the basic operations to be performed. All these reactions require overcoming a thermodynamic energy barrier, where a catalyst must play a vital role in reducing the same. Precious Pt, Pd, some of their alloys, and oxides of some noble metals like RuO_2 and IrO_2 are the benchmark catalysts for the ORR, HER, and OER. They are qualified by their low overpotential, good selectivity, and long-term stability under punitive operating conditions [1–4]. However, their high cost and low natural abundance are the two main reasons why they cannot be considered for widespread commercial applications. Therefore, a continuous effort by global scientific communities has been centered on developing non-noble and non-precious metal-based catalysts with equitable performance in the above reactions [5–8]. Although most of the transition metal-based electrocatalysts developed for HER, OER, and ORR have shown reasonably good performance in terms of overpotential, turnover number, Tafel slope, electron kinetics, etc., many of them suffer long-term stability in electrolytic environments due to dissolution problem [9].

Most of the above-mentioned materials are either alloys of two/three metals or binary/ternary compounds. In contrast, some alloys formed as single-phase solid solutions of at least five elements mixed in an equi-elemental ratio, which are so-called 'High Entropy Alloys (HEA)' show extraordinary thermodynamic stability [10]. The key advantage of HEAs over bi- or trimetallic alloys is the maximization of configurational entropy, which bestows the ensuing materials with astonishing physicochemical and mechanical stability, including corrosion and wear resistance, thermal strength, etc. In this context, the HEAs could be next generation electrocatalysts for HER, OER, and ORR applications.

Principally HEAs applied in the above catalytic applications are either alloys of noble and precious metals (Pt, Pd, Rh, Ru, Ir, etc.) or non-noble transition metals (such as Fe, Co, Ni, Mn, Cr, Mo, W, etc). In some cases, the addition of Group-IIIA metals such as Al/Ga or In coupled with either of the above two classes. Doping the HEAs with elements, such as Al, has been shown to induce a broad structural phase transition in them which may have significantly different physical and/or chemical properties [11]. In contrast to bulk HEAs, developing nanoparticulate HEAs offer superior electrocatalytic response owing to their diverse morphologies, high surface areas, a larger number of active sites, etc. [12].

DOI: 10.1201/9781003391388-10

The following sections give a brief overview of the thermodynamic mainstay of the HEAs followed by modulation of binding energies in them, best possible synthetic strategies, a relation between their structure and properties, selectivity aspects, approach to their computational design, an account of their major applications in HER and an outlook at the end.

10.2 THEORETICAL BACKBONE OF HEAs

In the theory of thermodynamics, entropy is defined as a state variable to describe the magnitude of the disorder of the system. According to Boltzmann's statistics, entropy is denoted by the expression:

$$S = k_B \ln W \tag{10.1}$$

where is Boltzmann constant and W is the number of all possible microscopic configurations of the system (also known as microstates in the macrostate). In a system of materials, the entropy contains different origins. The entropy of a material system of mixing, ΔS_{tot} includes four contributions: configuration entropy, ΔS_{conf}, vibration entropy, ΔS_{vib}, electronic entropy, ΔS_{elec}, and magnetic entropy, ΔS_{mag}. Its balance can be expressed as:

$$\Delta S_{tot} = \Delta S_{conf} + \Delta S_{vib} + \Delta S_{elec} + \Delta S_{mag} \tag{10.2}$$

In a solid solution, like alloy, the configuration entropy predominates over other contributions, and the total entropy change may be simply equated to it to avoid complexity in the calculations. This major entropy contribution can be expressed further as,

$$\Delta S_{conf} = -R \sum_{i=1}^{N} c_i \ln(c_i) \tag{10.3}$$

Here R is the universal gas constant, c_i is the molar fraction of the ith component of the alloy, and N is the total number of elements present in the alloy system. This entropy obviously maximizes when $c_1 = c_2 = c_3 = \ldots = c_N$. The alloy then named, 'High Entropy Alloy,' is a single-phase solid solution with an equimolar ratio of the elements. It eventually provides,

$$\Delta S_{conf} = R \ln N \tag{10.4}$$

With this definition, one can visualize the total entropy of mixing as shown in Figure 10.1 for various possible numbers of constituents (N). It shows that the entropy of mixing can exceed $1.5R$ when N is taken as 5 or more to refer to them as HEAs.

It is noteworthy that to meet the above requirement of HEAs, one can also consider the inequal molar ratio of the major constituent, provided the atomic % of each element is greater than 5% and less than 35% [10]. There can also be minor elements added provided their atomic % is set below 5%. Such an addition of these minor elements can drastically bring several functionalities to the HEAs.

Moreover, the stability of a thermodynamic system may be expressed in terms of Gibb's free energy change of mixing,

$$\Delta G_{mix} = \Delta H_{mix} - T\Delta S_{mix} \tag{10.5}$$

where ΔH_{mix} the Helmholtz free energy refers to the total energy of the system (enthalpy of mixing) and T is the temperature. Obviously, for HEAs, a high ΔS_{mix} becomes beneficial in lowering ΔG_{mix} and eventually the system gains remarkable stability.

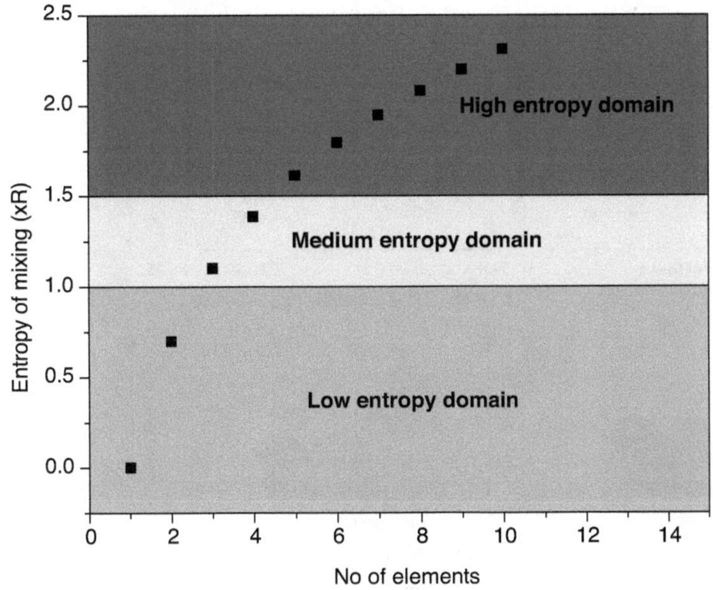

FIGURE 10.1 Regions of the definition of entropy of mixing in a solid solution for various possible numbers of constituent elements.

10.3 EFFECT OF THE BINDING ENERGY OF CONSTITUENTS IN HIGH ENTROPY ALLOYS

The binding energies and Gibb's free energy are used as descriptors for the overall catalytic activity of an electrocatalyst. One may refer to the HEAs and HEOs as complex solid solution (CSS) of elements where the binding energy distribution within the components plays a vital role in the catalytic activity and stability as well [13]. The compositional ratio of elements in the CSS would directly affect the polarization curve (the voltammogram) of the HEA applied as the electrocatalyst. When all five elements are placed in an equiatomic ratio (i.e., 20% of each), such as $A_{20}B_{20}C_{20}D_{20}E_{20}$ (see Figure 10.2c left column) for an alloy ABCDE the voltammogram uniformly distributes all the 'S'-shaped polarizations in different potential ranges (Figure 10.2a) corresponding to the binding energy distribution (Figure 10.2b). In this case, the determination of overpotential becomes broad. When a low binding energy element is introduced in a large fraction in the HEA, such as $A_{11}B_{56}C_{11}D_{11}E_{11}$, where B has lower binding energy than C, D, and E (the central column of Figure 10.2), the overpotential is determined at a large negative value. Therefore, such a composition is unfavorable. In a similar fashion, therefore if the large binding energy element is introduced at a larger fraction, such as $A_{11}B_{11}C_{11}D_{11}E_{56}$ (the right column of Figure 10.2), the overpotential is low enough, and such a composition is most favorable for the HEA to show high activity.

10.4 SYNTHESIS STRATEGIES FOR HEAS

The synthesis of HEAs can follow the alloy formation strategy provided by Hume-Rothery rules for binary systems. In general, a similarity in atomic size, electronegativity, and a preferred crystal structure are some of the most required criteria to enhance the entropy of the resultant CSS. Usually, solution-based processes of synthesis are challenging due to different rates of decomposition of the metal precursors. According to $\Delta G = \Delta H - T\Delta S$, a negative Gibb's free energy requires that

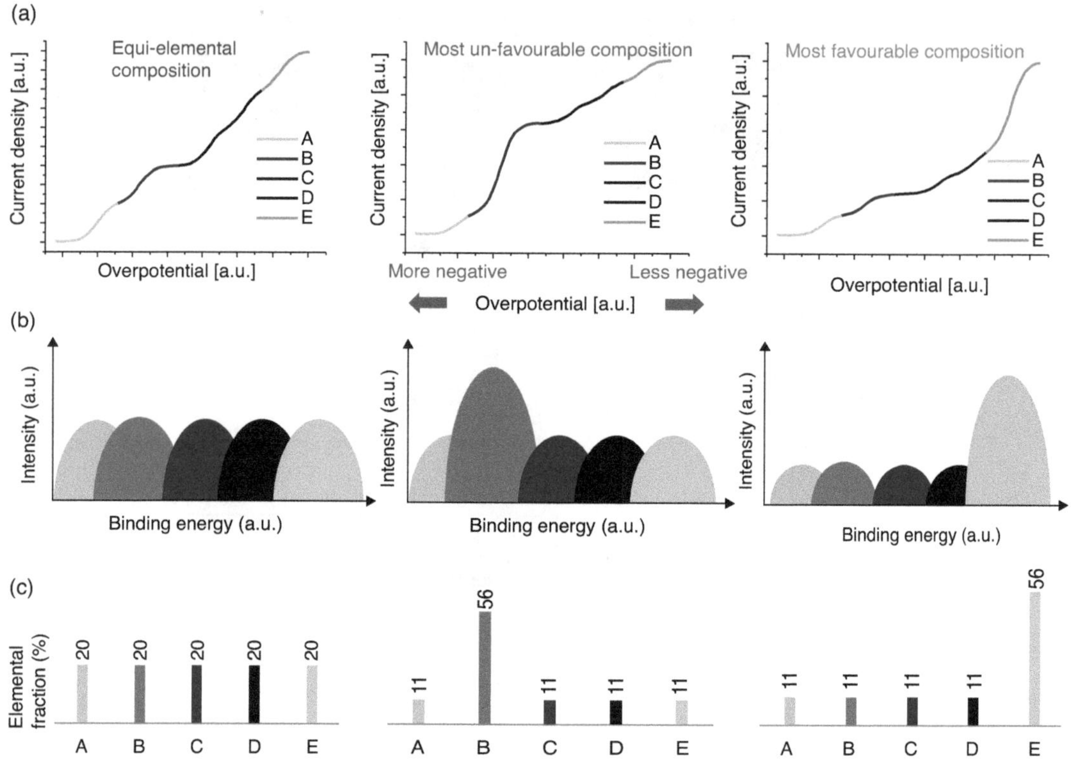

FIGURE 10.2 (a) Representative voltammogram or polarization curves originated from (b) various types of binding energy distributions of (c) possible elemental composition ratios in the HEAs.

the formation occurs at a high temperature or at a high energy of decoupling the metal ions from their precursors. Another important aspect of the phase formation while selecting transition metal components is their chance of oxidation. The formation of a single-phase CSS brings exceptional properties of HEAs, such as remarkable durability and catalytic activity. However, such complex, multicomponent alloys cannot be prepared by simply mixing various elements or metals with an equi-elemental ratio. With the Hume-Rothery rule in mind, the following parameters should be considered to synthesize stable HEAs:

i. Electronegativity of constituents, which will dictate the electrochemical properties of the alloy,
ii. Electron-to-atom ratio,
iii. Valence electron concentration (VEC), which controls the phase stability of the resultant alloy.

The electronegativity differences between constituents induce significant charge redistribution and create highly active sites over specific elements with optimized energy barriers for simultaneously stabilizing intermediates, such as OH* and H*, which greatly enhances the efficiency of water splitting [14]. In general, HEAs tend to form either face-centered cubic (fcc) or body-centered cubic (bcc) or a mix of both lattices. A VEC <6.87 causes the formation of bcc HEAs, whereas VEC $\geq$ 8.0 causes fcc phase formation [15,16]. There are a few other criteria theoretically established with consideration of mixing entropy of configuration and enthalpy of formation, which may also be useful while synthesizing the HEAs [17,18]. In the following sections, we account for standard synthetic processes adopted in HEAs in the form of nanoparticles and nanostructures, especially for electrocatalytic HER applications.

10.4.1 Approach-1: Top-Down

It is the simplest approach to making HEA nanoparticles from solid/bulk by breaking them down into smaller fragments or particles with the help of an external energy source. The source of energy can be mechanical, chemical, thermal, or optical (e.g., LASER ablation). Out of them mechanical and LASER-based techniques are popular as they provide high throughput and hence commercially viable.

10.4.2 Approach-2: Bottom-Up

In this approach, the alloy or the compound is formed from its precursor in atomic or ionic, or molecular species. The precursor is typically a liquid or gas in the processes such as hydrothermal, solvothermal, carbothermal shock, sputtering, auto-combustion, etc. It is ionized or dissociated or sublimed or evaporated and then condensed to form either an amorphous or a crystalline nanoparticle or a film. A wide variation of desired size, shape, and pore size distribution of the HEAs is possible in this approach by simply varying the initial parameters, such as molar concentration and feed rate of the reactant. An additional advantage of this approach is the resulting HEAs with less amount of surface defects, uniform chemical composition, and sharp size distribution (e.g., from 100 to 0.1 nm [19]). Table 10.1 lists several methods adopted in the above two approaches to synthesize HEAs applied as the HER/OER electrocatalysts.

TABLE 10.1
Popular Methods of Synthesizing HEAs for HER/OER Electrocatalysis Application

Method of Preparation	Principles & Advantages of the Process	HEA Example & Category	Reference
High energy ball milling	Inexpensive. Stimulates the mechanical activation of the powder mixture or induces chemical reaction by scaling down and transformation of the powder reactants.	CoCrFeNiMo/ Non-noble	[20]
Arc-melting	Electric discharge at low voltage, high current. Very high temperatures (4000°C) production in a short time (10–20 seconds).	IrMoNiCoAl/Noble AuPtPdCuNiCoFeAl/ Noble	[21] [22]
Electrochemical dealloying	Dissolution (leaching) and diffusion processes on a rigid lattice. Formation of porous/mesoporous microstructure for high surface area applications.	NiCoFeMoMn/ Non-noble	[23]
LASER ablation	Discharge of bulk of materials from the surface, usually initiated by the interaction of short-duration intense LASER pulses. High production rate and maintaining stoichiometry of composition.	CrCoFeNiMn and CrCoFeNiMnMo/ Non-noble	[24]
Hydrothermal/ Solvothermal	Chemical reaction at a high-temperature, high-pressure reaction environment in a closed chamber. Less energy intensive (mild operating condition, in most cases below 300°C)	AuPtPdRhRu/Noble PdCuMoNiCo/Noble	[25] [26]
Carbothermal shock	Two-step process of rapid joule heating followed by quenching of metal precursors on oxygenated carbon supports, with adjustable shock durations and ramp rates.	PtLaNiCoFe/Noble	[27]

(Continued)

TABLE 10.1 (*Continued*)
Popular Methods of Synthesizing HEAs for HER/OER Electrocatalysis Application

Method of Preparation	Principles & Advantages of the Process	HEA Example & Category	Reference
Sputtering	Bombarding together a selection of metal targets surface with energetic inert gas ions (Ar^+) causing physical ejection of surface atoms or small clusters, followed by deposition in a medium like ionic liquids (in case of thin film, a substrate). The cleanest fabrication method to obtain least defects in the resulting alloy or compound.	FeNiMoCrAl/ Non-noble	[28]
Auto-combustion	Chemical gelation followed by ignition (usually from nitrate salt of the metals)	CoCrCuNiAl/ Non-noble	[29]

TABLE 10.2
Few Examples of Noble-, Non-Noble and Mixed Noble HEAs and Their Physico-Chemical Characteristics

Category	Example	Method of Preparation	Specific Properties	References
Noble	AuPtAgPdCu nanoparticles (NP)	Arc-melting, followed by cryo-milling	Single-phase fcc crystal structure	[30]
	AuPtPdCuNiAl -NP	Powder metallurgy, fast cooling, and dealloying	-	[22]
	AuPtSnPdCuNiCoFe	Carbothermal Shock	Uniformly distributed NP of size ~ 5 nm	[31]
Non-noble	CuNiCoFeCr	Wet chemical synthesis	Spherical NP with an average size of 26.7 ± 3.3 nm	[32]
	NiCoFeMnCr	Co-sputtering	Very small size NP with average size 1.7 nm with high crystallinity	[33]

10.5 CATEGORIES OF HIGH ENTROPY ALLOYS: NOBLE METAL-BASED, NOBLE METAL-FREE AND THEIR COMBINATION

Noble metals by virtue are characterized by favorable adsorption and desorption of hydrogen atoms from their binding sites, making them very close to the top of volcano plots (Exchange current density as a function of Gibbs free energy). Their presence even in the HEAs would prefer such binding characteristics. Although they have promising physical characteristics, their low abundance is a big concern. On the other hand, noble metal electrocatalysts often suffer from operational instability under extreme working conditions. For example, Pt itself and its alloys are susceptible to dissolution, and agglomeration, and have poor resistance to electrode poisoning. Thus, it is highly desirable to replace the noble metal-based electrocatalysts with non-noble ones. It is noteworthy that there are several noble and non-noble metal combinations that may produce an equitable or even greater performance. Most interestingly, a combination of noble and non-noble metals have been extensively studied to ensure high catalytic activities originated from a synergistic effect and the *d*-band center modulation, which facilitate the electron transfer from their Fermi level to modified anti-bonding orbitals. Table 10.2 lists some of the important developments of HEAs in these two categories: their preparation condition and their specific properties.

10.6 COMPUTATIONAL DESIGN APPROACHES FOR HEAS

Due to the severe complexity of multi-elemental alloys and compounds, manual screening of the materials is often tricky. Therefore, computational studies (such as first principles studies) and machine learning-based design can help in identifying the optimal HEA compositions. A large composition space among active metals (Ru, Rh, Co, Ni, Ir, Pd, Cr, Fe, Cu, and Mo) can be created by using such a computation tool [34]. The screening is done in a three-step optimization:

Step 1: Phase selection by composition pre-screening

Not all compositions can produce the stable phase, rather leading to phase separation. The computation should consider the atomic size mismatch (less than about 6%) and mixing enthalpy ($-11.6 < \Delta H_{mix} < 3.2$ kJ/mol). In this way, millions of variations (*combinatorial explosion*) can be made with machine learning algorithm.

Step 2: Density functional theory approach for thermodynamic alloy formation

Gibb's free energy is calculated for several alloys with 4, 5, etc. number of elements, while binary alloy first satisfied the criteria in step 1. The mixing enthalpy, ΔH_{mix} is required to be lowered whereas the entropy of mixing ΔS_{mix} should increase to ensure the formation of such phases at lower temperature.

Step 3: Kinetic structural simulation and high-temperature synthesis

In the final step, the structural stability of the alloy is examined, and hence the effective combination is narrowed down. The Monte-Carlo (MC) and Molecular Dynamic (MD) combined approach is adopted to randomly mix all components optimized in step 2 and then MC provides an alloy at high T mixing, followed by MD at a low T that stabilizes the structure by high entropy adaptation.

10.7 APPLICATIONS OF HEA IN THE HYDROGEN EVOLUTION REACTION

Hydrogen generation or the 'HER' from water is a thermodynamically up-hill reaction, where every mol of clean water requires 237 kJ of energy to break the hydrogen-oxygen bond in water. It also requires additional energy to overcome the change in entropy of the reaction. Therefore, the process cannot proceed below at least 286 kJ/mol if no external energy source is used. External energy sources should be used to provide the required energy to surmount this barrier. Catalysts play a significant role in the reduction of extra energy requirements. Specifically, an electrocatalyst just needs an external bias as that is used in an electrolyzer. Theoretically, the minimum bias required to initiate the water splitting is 1.23 V (vs. RHE). However, practically the bias requirement is more than 1.23 V, and the additional bias is also known as 'overpotential'. For HER, Pt-based catalysts by virtue of their electronic structure can perform the water splitting at the lowest overpotential among any other elements in the periodic table. But the high cost of precious metals greatly limits the sustainable production of hydrogen by using Pt. A low overpotential is also characteristic of other noble metals such as Pd, Ru, etc. showing just above that of Pt. Alloying these noble metals with a variety of non-noble metals is an effective strategy to reduce the cost of H_2 production besides a nearly complementary low overpotential. Mechanistically, it not only largely increases the utilization rate of precious metals but also alters the coordination and electronic environments of precious metals. Thus, adjusts the adsorption energy of intermediate species, and the catalytic performance of the material.

The HER necessarily should occur at the cathode of the electrolysis cell where the reduction process (protons to gaseous H_2) is held. However, this reaction occurs in most of the cases in three steps. In the first step, a proton on an active site is reduced to form the adsorbed nascent hydrogen (H*) (also called the *Volmer step*). The next steps are either a second proton/electron transfer (the *Heyrovsky step*) and/or boding two such H* (the *Tafel step*) to produce molecular H_2. Among them, the Volmer step is usually considered to be the rate-limiting step for HER, and thermodynamic energy barriers are needed to overcome in this step. A highly effective electrocatalyst may perform this task to successfully conduct the forward reaction.

$$H^+ + e^- + (*) \rightarrow H^* \tag{10.6}$$

$$H^* + H^+ + e^- \rightarrow H_2 \tag{10.7}$$

$$H^* + H^* \rightarrow H_2 \tag{10.8}$$

(Here, '*' indicates an associated energy)

The HEAs are multicomponent alloys, and their hydrogen adsorption and desorption property can be adjustable through the elemental ratio as discussed before. In more detail, they can be selectively mixed with specific elements in a specific proportion to control the adsorption of intermediates at active sites through a cooperative effect of their electronic orbitals. In the following two sections, mechanisms specific to noble- and non-noble metal-based HEAs are discussed. Below, we classify HER through two different groups of HEAs: the noble metal-based HEAs and the purely non-noble metal-based HEAs.

10.7.1 Noble Metal-Based HEAs in HER

The noble metal-based category principally considers Pt- and Pd-based HEAs. The cost for NPs would be low due to the partial usage of the noble metal. In general, they are high entropy metallic glasses (HEMG) or multinary (quinary, senary, septenary, octonary, and denary) alloys with granular support (MAGS) or ultra-fine HEAs (UFHEAs), etc. in literature. Strategically they are synthesized in the following routes:

i. HEMGs: Nanodroplet-mediated electrodeposition [35].
ii. MAGSs: Fast Moving Bed Pyrolysis [36].
iii. UFHEAs: Simple one-pot oil phase strategy [37].

In HEAs containing both Pt and Pd, the presence of transition metals and Pd can reduce the anti-bonding state of Pt, and more electrons can fill the anti-bonding state, thereby promoting the desorption of hydrogen species (H*) and encourage hydrogen release [38,39]. A theoretical prediction has also shown that the presence of non-noble transition metals (NNTM) can sequentially play a role in the adsorption-desorption of water, binding of H*s and finally desorption of H_2 from all different NNTM sites. For example, as shown in Figure 10.3a, the DFT-based calculation made in $Pt_{18}Ni_{26}Fe_{15}Co_{14}Cu_{27}$ HEA nanoparticles by Lai and Wang groups [37] shows that s- and p- orbitals of H_2O eventually drop near all major d-orbital projection of the HEA's elements. This indicates that the transfer of electrons from HEAs to adsorbed water would have been well established. The Fe–d orbitals are highest lying, and therefore H_2O molecules are initially adsorbed on Fe sites, which activates the dissociation of water molecules and promotes the stability of OH1 in adjacent empty sites. The binding energy map (Figure 10.3b) suggests that adjacent Ni and Cu sites can stabilize H*, and subsequently formed H_2 can be destabilized from Pt site (a negative binding energy indicates destabilized bonds and vice versa). This is how the whole HEA nicely performs a cooperative tuning of charge transfer to complete Volmer, Heyrovsky, and Tafel steps. Similar activity is also observed in Pd-based noble metal as well as other non-noble metal HEAs (Table 10.3).

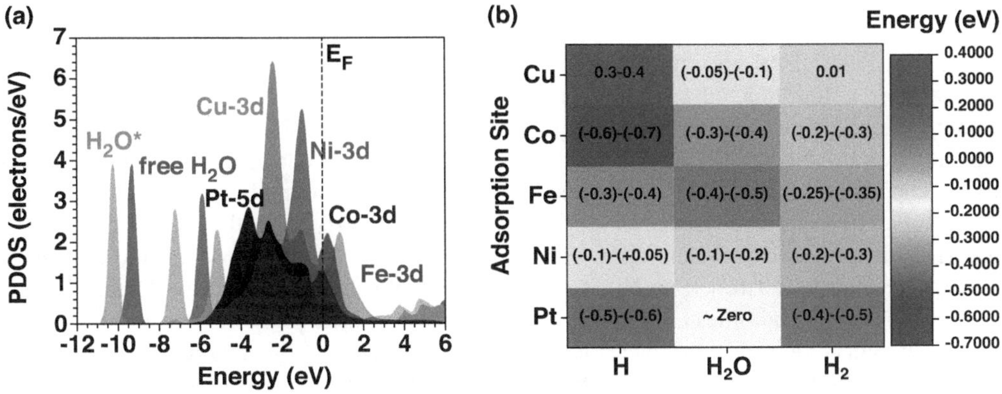

FIGURE 10.3 The DFT results for $Pt_{18}Ni_{26}Fe_{15}Co_{14}Cu_{27}$ nanoparticles with H_2O adsorbed. (a) The projected density of states (PDOSs) of various elements and species for water adsorption. (b) The binding energy mapping of HER. Adapted with permission [37]. Copyright The Authors, some rights reserved; exclusive licensee Springer Nature. Distributed under a Creative Commons Attribution License 4.0 (CC BY).

TABLE 10.3
The High Entropy Alloys Applied in Hydrogen Evolution Reaction under Noble as Well as Non-Noble Metal-Based Categories

Example (Nature)	Category	Performance	Reference
PtAuPdRhRu (nanoparticles)	Noble mixed	An onset potential of 25 mV (vs. RHE) and a Tafel slope of 62 mV/dec	[43]
CoFeLaNiPt (HEMGs)	Noble mixed	Pt working as the active site	[27]
CoCrFeNiAl (Self-supported nanostructure)	Non-noble	An overpotential of 73 mV to reach a current density of 10 mA/cm after HF treatment and in situ electrochemical activation	[44]
FeCoPdPtIr on Graphene oxide	Noble mixed	Excellent activity and stability	[45]
PtNiFeCoCu (Ultra-small ~3.4 nm nanoparticle)	Noble mixed	Exhibits the 10.96 A mgPt −1 high activity at −0.07 V versus RHE	[46]
PdFeCoNiCu (nanoparticles)	Noble mixed	18 mV overpotential in 1 m KOH solution	[40]
IrMoNiCoAl (nanoparticles)	Noble mixed	A low Tafel slope of 28mV/dec	[22]
CuAlNiMoFe (nanoporous electrode)	Non-noble	56 mV overpotential to reach a current density of 100 mA/cm^2 in 1 m KOH solution	[41]
NiFeMoCoCr (nanoparticles)	Non-noble	An overpotential of only 107 mV in acidic solutions and 172 mV in the basic solution at current density of 10 mA/cm	[47]
FeCoNiMnRu nanoparticles (embedded on carbon nanofibers)	Non-noble	An overpotentials of 71 mV at 100 mA/cm^2	[14]

10.7.2 Non-Noble Metal-Based HEAs in HER

Noble metal-based HEAs are important candidates as electrocatalysts for HER because of their unique abilities to fine-tune adsorption energy, long-term stability in corrosive environments, and abundant lattice distortion. The mechanism associated with non-noble-based HEAs is quite like that of noble metal-based HEAs as discussed above. However, there can be an additional flexibility of taking advantage of the d-orbitals of more transition metals in the HEA. For example, CuAlNiMoFe HEA electrode has been reported to show low overpotential of 56 mV even at a very high current density of 100 mA/cm^2 [40]. Here, Mo effectively dissociates water into H* and HO* which are successively adsorbed on Ni(Fe) and Mo, respectively. The former performs Heyrovsky and Tafel steps (i.e., it combines H* s to form H_2), whereas the latter promotes further water adsorption. The Cu matrix facilitates fast electron transfer by virtue of its high conductivity.

In summary, the thermodynamics of the HER are optimized by lowering the energy barrier of water splitting, and the reaction kinetics can be enhanced by improving the interaction with protons. At the time of hydrolysis, the adsorption of OH* competes with that of H* on a single active site. Therefore, it weakens the production of H_2 in successive steps. In addition to that, it is very difficult to balance OH*, H*, and other intermediates on a single active site [41]. This can be fixed by designing active sites, such as that in HEAs where each element will perform different functions in a cooperative manner to obtain control over the reaction intermediates. By providing independent oxygen affinity sites that can bind to OH* for hydrolysis, the adsorption of H* can be greatly promoted. Thus, improving the kinetics of the rate-limiting Volmer step [42].

10.8 OUTLOOK AND FUTURE DIRECTIVES

High entropy alloys have been identified as one of the most important classes of material with extraordinary thermodynamic stability along with excellent activity in highly demanding catalytic reactions in the field of energy conversion and storage, including the hydrogen evolution reactions among others like OER, CO_2 reduction reaction, ORR, nitrogen reduction reaction, redox reactions in batteries and supercapacitors, etc. This class of materials needs non-conventional synthesis techniques to ensure multi-site activity, size and composition regulation, strain and defect engineering, etc. Moreover, their characterizations are equally important to understand and forecast such catalytic activities. Besides, combinatorial materials design, where number explosion can be a tricky situation, Machine Learning can help in converging the choice of elements and their compositional ratio in the HEAs.

Amid the above three important considerations of synthesis, characterization, and materials design, the new HEAs should be developed with the following directives in the future. Although there may be many other significant points, one may note the following in the priority:

i. The present HEA preparation methods need a highly controlled, stringent, and sometimes harsh synthetic atmosphere. The nanostructure forms of HEAs would be demanding in the future, such as 2D nanosheets, 1D nano-rods & wires, high surface area, and hollow forms, etc., which may need hybrid solution-based routes. More non-noble HEAs should be explored as well to bring their commercial viability.
ii. Besides high-resolution microscopies (TEM, HRTEM, STEM, etc.) and energy-dispersive elemental analysis (e.g., EDX), chemical information is not very well understood in the HEAs. The determination of exact chemical composition and bonding in HEAs is complex. High energy X-Ray absorption studies (XANES and EXAFS) are required to be reported along with joint interpretation by XPS. Many new HEAs' powder X-ray diffraction data is not available, and such a new database would help materials science communities. Similarly, their vibrational (such as Raman spectroscopy) and other fine-structure data would also be highly demanding.

iii. The high entropy materials (HEMs) will not be limited to HEAs in the near future. Instead, high entropy-oxides (HEOs), -nitrides (HENs), -sulfides (HESs), -phosphides (HEPs), -carbides (HECs), -organic frameworks (HEOFs), etc. would be emerging very soon in catalytic applications. Their design and synthesis are the most promising aspects in the coming days.
iv. More detailed DFT-based calculations are required to explain many unrevealed activities in the HEAs. Also, the interface and van-der Waals heterostructures of various possible HEMs would emerge unmatched devices and electrocatalysts for energy in the future.

ACKNOWLEDGMENT

Mr. Atulkumar Mishra, Senior Research Fellow at Solar Research & Development Centre (SRDC), PDEU is acknowledged for his help in editing style and referencing in this chapter.

REFERENCES

[1] J. Greeley, T.F. Jaramillo, J. Bonde, I. Chorkendorff and J.K. Nørskov, Computational high-throughput screening of electrocatalytic materials for hydrogen evolution, *Nat Mater* 5 (2006) 909–913.
[2] Z.W. She, J. Kibsgaard, C. F. Dickens, I. Chorkendorff, J. K. Nørskov, T. F. Jaramillo, Combining theory and experiment in electrocatalysis: Insights into materials design, *Science* 355 (2017) eaad4998.
[3] C.C.L. McCrory, S. Jung, I.M. Ferrer, S.M. Chatman, J.C. Peters and T.F. Jaramillo, Benchmarking hydrogen evolving reaction and oxygen evolving reaction electrocatalysts for solar water splitting devices, *J Am Chem Soc* 137 (2015) 4347–4357.
[4] J. Yu, Q. He, G. Yang, W. Zhou, Z. Shao and M. Ni, Recent advances and prospective in ruthenium-based materials for electrochemical water splitting, *ACS Catal* 9 (2019) 9973–10011.
[5] F. Yu, L. Yu, I.K. Mishra, Y. Yu, Z.F. Ren, H.Q. Zhou, Recent developments in earth-abundant and non-noble electrocatalysts for water electrolysis, *Mater Today Phys* 7 (2018) 121–138.
[6] Z. Wu, X. F. Lu, S. Zang, X. W. (David) Lou, Non-noble-metal-based electrocatalysts toward the oxygen evolution reaction, *Adv Funct Mater* 30 (2020) 1910274.
[7] L. Zhang, J. Zhang, D.P. Wilkinson and H. Wang, Progress in preparation of non-noble electrocatalysts for PEM fuel cell reactions, *J Power Sources* 156 (2006) 171–182.
[8] K. Zeng, X. Zheng, C. Li, J. Yan, J. Tian, C. Jin, P. Strasser, R. Yang, Recent advances in non-noble bifunctional oxygen electrocatalysts toward large-scale production, *Adv Funct Mater* 30 (2020) 2000503.
[9] S. Cherevko, Stability and dissolution of electrocatalysts: Building the bridge between model and "real world" systems, *Curr Opin Electrochem* 8 (2018) 118–125.
[10] Y. Zhang, T.T. Zuo, Z. Tang, M.C. Gao, K. A. Dahmen, P. K. Liaw, Z.P. Lu, Microstructures and properties of high-entropy alloys, *Prog Mater Sci* 61 (2014) 1–93.
[11] C.J. Tong, Y.L. Chen, S.K. Chen, J.W. Yeh, T.T. Shun, C.H. Tsau, S.Y. Chang, Microstructure characterization of AlxCoCrCuFeNi high-entropy alloy system with multiprincipal elements, *Metall Mater Trans A Phys Metall Mater Sci* 36 (2005) 881–893.
[12] Y. Zhou, X. Shen, T. Qian, C. Yan, J. Lu, A review on the rational design and fabrication of nanosized high-entropy materials, *Nano Res* 16 (2023) 7874–7905.
[13] T. Löffler, A. Ludwig, J. Rossmeisl, W. Schuhmann, What makes high-entropy alloys exceptional electrocatalysts? *Angew Chem –Int. Ed* 60 (2021) 26894–26903.
[14] J. Hao, Z. Zhuang, K. Cao, G. Gao, C. Wang, F. Lai, S. Lu, P. Ma, W. Dong, T. Liu, M. Du, Unraveling the electronegativity-dominated intermediate adsorption on high-entropy alloy electrocatalysts, *Nat Commun* 13 (2022) 2662.
[15] S. Guo, C. Ng, J. Lu and C.T. Liu, Effect of valence electron concentration on stability of fcc or bcc phase in high entropy alloys, *J Appl Phys* 109 (2011) 103505.
[16] G.M. Tomboc, T. Kwon, J. Joo and K. Lee, High entropy alloy electrocatalysts: A critical assessment of fabrication and performance, *J Mater Chem A Mater* 8 (2020) 14844–14862.
[17] Y.F. Ye, Q. Wang, J. Lu, C.T. Liu and Y. Yang, The generalized thermodynamic rule for phase selection in multicomponent alloys, *Intermetallics (Barking)* 59 (2015) 75–80.
[18] Y.F. Ye, Q. Wang, J. Lu, C.T. Liu and Y. Yang, Design of high entropy alloys: A single-parameter thermodynamic rule, *Scr Mater* 104 (2015) 53–55.

[19] H.-D. Yu, M.D. Regulacio, E. Ye and M.-Y. Han, Chemical routes to top-down nanofabrication, *Chem Soc Rev* 42 (2013) 6006–6018.

[20] J. Tang, J.L. Xu, Z.G. Ye, X.B. Li and J.M. Luo, Microwave sintered porous CoCrFeNiMo high entropy alloy as an efficient electrocatalyst for alkaline oxygen evolution reaction, *J Mater Sci Technol* 79 (2021) 171–177.

[21] Z. Jin, J. Lv, H. Jia, W. Liu, H. Li, Z. Chen et al., Nanoporous Al-Ni-Co-Ir-Mo high-entropy alloy for record-high water splitting activity in acidic environments, *Small* 15 (2019) 1904180.

[22] H.J. Qiu, G. Fang, Y. Wen, P. Liu, G. Xie, X. Liu et al., Nanoporous high-entropy alloys for highly stable and efficient catalysts, *J Mater Chem A Mater* 7 (2019) 6499–6506.

[23] H. Liu, H. Qin, J. Kang, L. Ma, G. Chen, Q. Huang et al., A freestanding nanoporous NiCoFeMoMn high-entropy alloy as an efficient electrocatalyst for rapid water splitting, *Chemical Engineering Journal* 435 (2022) 134898.

[24] J. Johny, Y. Li, M. Kamp, O. Prymak, S.X. Liang, T. Krekeler et al., Laser-generated high entropy metallic glass nanoparticles as bifunctional electrocatalysts, *Nano Res* 15 (2022) 4807–4819.

[25] M. Liu, Z. Zhang, F. Okejiri, S. Yang, S. Zhou and S. Dai, Entropy-maximized synthesis of multimetallic nanoparticle catalysts via a ultrasonication-assisted wet chemistry method under ambient conditions, *Adv Mater Interfaces* 6 (2019) 1900015.

[26] X. Zuo, R. Yan, L. Zhao, Y. Long, L. Shi, Q. Cheng, D. Liu, C. Hu, A hollow PdCuMoNiCo high-entropy alloy as an efficient bi-functional electrocatalyst for oxygen reduction and formic acid oxidation, *J Mater Chem A Mater* 10 (2022) 14857–14865.

[27] M. W. Glasscott, A. D. Pendergast, S. Goines, A. R. Bishop, A. T. Hoang, C. Renault, J. E. Dick, Electrosynthesis of high-entropy metallic glass nanoparticles for designer, multi-functional electrocatalysis, *Nat Commun* 10 (2019) 2650.

[28] S.Y. Li, T.X. Nguyen, Y.H. Su, C.C. Lin, Y.J. Huang, Y.H. Shen, C. P. Liu, J. J. Ruan, K. S. Chang, J. M. Ting, Sputter-deposited high entropy alloy thin film electrocatalyst for enhanced oxygen evolution reaction performance, *Small* 18 (2022) 2106127.

[29] B. Niu, F. Zhang, H. Ping, N. Li, J. Zhou, L. Lei, J. Xie, J. Zhang, W. Wang, Z. Fu, Sol-gel autocombustion synthesis of nanocrystalline high-entropy alloys, *Sci Rep* 7 (2017) 3421.

[30] J. K. Pedersen, T. A. A. Batchelor, A. Bagger, J., High-entropy alloys as catalysts for the CO2 and CO reduction reactions, experimental realization, *ACS Catal* 10 (2020) 2169–2176.

[31] Y. Yao, Z. Huang, P. Xie, S.D. Lacey, R.J. Jacob, H. Xie, F. Chen, A. Nie, T. Pu, M. Rehwoldt, D. Yu, M. R. Zachariah, C. Wang, R. S. Yassar, J. Li, L. Hu, Carbothermal shock synthesis of high-entropy-alloy nanoparticles, *Science* 359 (2018) 1489–1494.

[32] M.P. Singh, C. Srivastava, Synthesis and electron microscopy of high entropy alloy nanoparticles, *Mater Lett* 160 (2015) 419–422.

[33] T. Löffler, H. Meyer, A. Savan, P. Wilde, A. Garzón Manjón, Y.T. Chen, E. Ventosa, C. Scheu, A. Ludwig, W. Schuhmann, Discovery of a multinary noble metal-free oxygen reduction catalyst, *Adv Energy Mater* 8 (2018) 1802269.

[34] Y. Yao, Z. Liu, P. Xie, Z. Huang, T. Li, D. Morris, Z. Finfrock, J. Zhou, M. Jiao, J. Gao, Y. Mao, J. Miao, P. Zhang, R. S. Yassar, C. Wang, G. Wang, L. Hu, Computationally aided, entropy-driven synthesis of highly efficient and durable multi-elemental alloy catalysts, *Sci Adv* 6 (2020) eaaz0510.

[35] M.W. Glasscott, A.D. Pendergast, S. Goines, A.R. Bishop, A.T. Hoang, C. Renault, J.E. Dick, Electrosynthesis of high-entropy metallic glass nanoparticles for designer, multi-functional electrocatalysis, *Nat Commun* 10 (2019) 2650.

[36] S. Gao, S. Hao, Z. Huang, Y. Yuan, S. Han, L. Lei, X. Zhang, R. S. Yassar, J. Lu, Synthesis of high-entropy alloy nanoparticles on supports by the fast moving bed pyrolysis, *Nat Commun* 11 (2020) 2016.

[37] H. Li, Y. Han, H. Zhao, W. Qi, D. Zhang, Y. Yu, W. Cai, S. Li, J. Lai, B. Huang, L. Wang, Fast site-to-site electron transfer of high-entropy alloy nanocatalyst driving redox electrocatalysis, *Nat Commun* 11 (2020) 5437.

[38] N. Du, C. Wang, X. Wang, Y. Lin, J. Jiang and Y. Xiong, Trimetallic tristar nanostructures: Tuning electronic and surface structures for enhanced electrocatalytic hydrogen evolution, *Adv Mater* 28 (2016) 2077–2084.

[39] Y. Pan, K. Sun, Y. Lin, X. Cao, Y. Cheng, S. Liu, L. Zeng, W.C. Cheong, D. Zhao, K. Wu, Z. Liu, Y. Liu, D. Wang, Q. Peng, C. Chen, Y. Li, Electronic structure and d-band center control engineering over M-doped CoP (M = Ni, Mn, Fe) hollow polyhedron frames for boosting hydrogen production, *Nano Energy* 56 (2019) 411–419.

[40] D. Zhang, Y. Shi, H. Zhao, W. Qi, X. Chen, T. Zhan, S. Li, B. Yang, M. Sun, J. Lai, B. Huang, L. Wang, The facile oil-phase synthesis of a multi-site synergistic high-entropy alloy to promote the alkaline hydrogen evolution reaction, *J Mater Chem A Mater* 9 (2021) 889–893.
[41] X. Wang, Y. Zheng, W. Sheng, Z. J. Xu, M. Jaroniec, S. Z. Qiao, Strategies for design of electrocatalysts for hydrogen evolution under alkaline conditions, *Mater Today* 36 (2020) 125–138
[42] H. Li, J. Lai, Z. Li, L. Wang, Multi-sites electrocatalysis in high-entropy alloys, *Adv Funct Mater* 31 (2021) 2106715.
[43] Y. Wang, W. Luo, S. Gong, L. Luo, Y. Li, Y. Zhao, Z. Lie, Synthesis of high-entropy alloy nanoparticles by step-alloying strategy as superior multifunctional electrocatalyst, *Adv Mater* (2023) 2302499.
[44] P. Ma, M. Zhao, L. Zhang, H. Wang, J. Gu, Y. Sun, W. Ji, Z. Fu, Self- supported high-entropy alloy electrocatalyst for highly efficient H2 evolution in acid condition, *J Materiomics* 6 (2020) 736–742.
[45] W. Sheng, M. Myint, J.G. Chen and Y. Yan, Correlating the hydrogen evolution reaction activity in alkaline electrolytes with the hydrogen binding energy on monometallic surfaces, *Energy Environ Sci* 6 (2013) 1509–1512.
[46] H. Li, Y. Han, H. Zhao, W. Qi, D. Zhang, Y. Yu, W. Cai, S. Li, J. Lai, B. Huang, L. Wang, Fast site-to-site electron transfer of high-entropy alloy nanocatalyst driving redox electrocatalysis, *Nat Commun* 11 (2020) 5437.
[47] G. Zhang, K. Ming, J. Kang, Q. Huang, Z. Zhang, X. Zheng, X. Bi, High entropy alloy as a highly active and stable electrocatalyst for hydrogen evolution reaction, *Electrochim Acta* 279 (2018) 19–23.

11 High Entropy Materials for Oxygen Evolution Reactions

Bhagyashri. B. Kamble, Arun Karmakar, and Subrata Kundu

11.1 INTRODUCTION

To create a better way of life, advance social change, and achieve sustainable development, humanity needs clean energy. Indeed, meeting ongoing globalization and its energy demands will require the use of a variety of carbon-based fuels [1,2]. Massive consumption of these carbon-based fuels to generate electricity releases large amounts of CO_2 gas into the environment, further deteriorating normal climatic conditions by increasing global temperature. In addition, since the supply of these carbon-based fuels is limited, the sources of these fuels must be non-renewable in nature, meeting these high energy demands with high efficiency. Therefore, it is urgent to develop efficient alternative resources that can meet these high energy demands with high efficiency. Oxygen evolution reaction (OER) is a key energy conversion step for both the electrochemical water splitting and recharging process of metal-oxygen batteries [3,4]. However, slow OER usually requires a high overvoltage [5]. Highly active catalysts are required to accelerate OER kinetics and reduce reaction overpotentials. Noble metal oxides such as RuO_2 and IrO_2 are known to exhibit high electrocatalytic activity for OER. However, high material costs and inadequacy of natural resources block their commercialization. Recently, intensive research has been carried out to harvest OER electrocatalysts employing earth-abundant elements. Among them, oxidized Ni and Co-based alloys and (oxy) hydroxides (such as NiFe, NiCo, and CoFe) are emerging as the most promising candidates [6]. Catalysts' compositions and surface electronic structures have strong relationships with their activities. Recently, it was reported that introducing a proper third metal species into NiFe- or CoFe-based (oxy) hydroxides plays a positive role in further enhancing the OER activity [7]. For instance, a conjunctional screen of over 3000 trimetallic oxides concluded that the most active OER catalysts under alkaline conditions were oxides containing nickel, iron, and other metal species [8,9]. A gelled CoFeW(oxy)hydroxide with upgraded OER performance compared to binary CoFe-based layered double hydroxides was synthesized. Based on theoretical calculations, it was assumed that the incorporation of tungsten into NiFe-based (oxy) hydroxides results in nearly optimized delta G of the intermediates during OER. Ternary NiCoFe (oxy)-hydroxides have also been proven to be very promising for OER [10]. Despite much progress achieved, further improving and tuning the catalytic activity is still urgently needed. Recently, a new class of materials, so-called high-entropy materials (HEMs), is receiving continuously increasing attention. HEMs give rise to attractive features, including the preference for single-phase solid solutions with simple crystal structures, having attributes exceeding their constituent elements, as well as the possibility of tailoring functional properties.

Many HEMs, including alloys, oxides, oxyfluorides, borides, carbides, nitrides, sulfides, and phosphides, have been reported in a broad range of utilizations [11–13]. Having multiple elements with different characters gives the HEMs four core effects of (i) high entropy, (ii) lattice distortion, (iii) sluggish diffusion, and (iv) cocktail effect (Figure 11.1). Therefore, a harmonious mixture of multi-elements results in HEM with excellent features as functional materials in energy storage and conversion systems. Dai et al. have explored the idea of utilizing HEMs (MnFeCoNi) for OER compared to highly active RuO_2 catalysts. The designed catalyst exhibited a low overpotential of 302 mV at 10 mA/cm^2 and the Tafel slope is only 83.7 mV/dec [14]. This chapter covers electrocatalysis to produce oxygen from the fundamentals to recent developments and future directions for commercial scale.

DOI: 10.1201/9781003391388-11

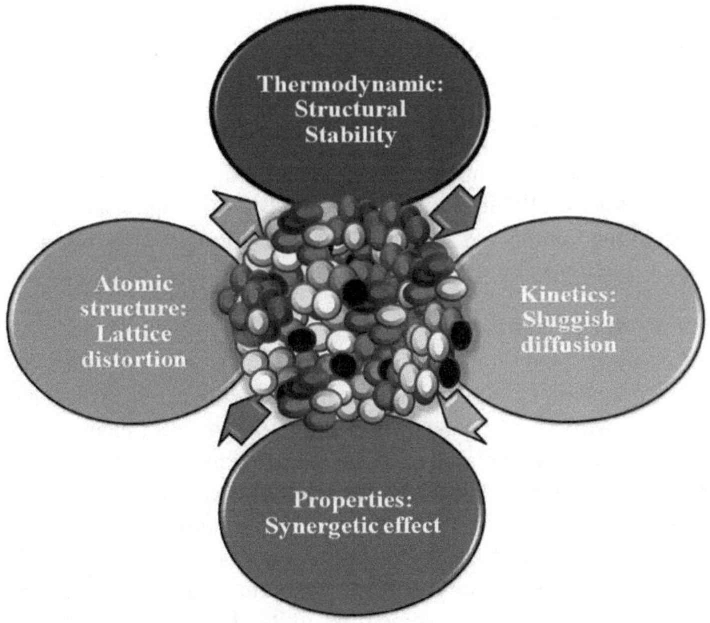

FIGURE 11.1 The schematic showcases the four core effects of HEMs. The flowchart shows the importance of HEMs in terms of various physical parameters.

11.2 HIGH ENTROPY MATERIAL

The classic alloying method has been used for many years to create novel composites with enhanced properties by adding trace amounts of other elements to the base element. Yeh and his associates were the first to use the broad high-entropy notion [15,16]. It is based on the observation that, as equimolar elements are introduced to the alloy system, the entropy of mixing (ΔS_{mix} in equation 11.1) increases.

$$\Delta S_{\text{mix}} = -R\sum_{i=1}^{n} C_i \ln C_i = -R\sum_{i=1} \frac{1}{n}\ln\frac{1}{n} = R\ln n \tag{11.1}$$

(In is the number of elements, c_i is the concentration of component i, and R is the gas constant)

Yet, high-entropy materials (HEMs) are not just determined by the quantity or make-up of their constituent elements and are not just restricted to inorganic materials. It is possible to see the composition of high-entropy organic materials as a viable strategy for creating intriguing new materials, such as carbon compounds and polymers. Entropy is defined as the degree of system disruption from this perspective. The idea of meso-entropy carbon compounds was first suggested by Feng and Zhuang [17]. Meso-entropy of materials can be engineered between high-entropy and low-entropy forms by considering the entropy value of a specific structure of carbon allotropes. The structure of the carbon material, such as the number of carbon ring members, can be changed to control the degree of disorder and entropy value. The lattice packing entropy of crystalline polymers can grow and be influenced by various types of polymers with various repeating unit ratios. Many fundamental HEM characteristics, structural impacts, and design tactics have been explored recently [18]. HEMs are great choices for a range of energy-related applications due to their appealing and advantageous characteristics such as synergy, high-temperature stability, and corrosion resistance [19,20].

This work's main objective is to review all published HEM's stated performance and novel applications in the literature on catalysis, and energy conversion. We next go over other potential uses for these materials in various energy-related systems, considering their established or conceivable potential functionality as HEMs. But first, we outline and categorize broad standards for the appropriate elemental choice in HEMs to produce advantageous single solid-solution phases.

11.2.1 HEM Synthesis Strategies

In recent decades, many advanced synthetic strategies have been proposed. Nanostructured HEMs are easier to study than macro-HEMs. However, the synthesis of nanostructured HEMs remains a challenging task due to the complexity of uniformly integrating different elements with different chemical and physical functions (Figure 11.2).

11.2.1.1 Polymer Mediated Synthesis

HEMs made up of various metal atoms are produced using a polymer-assisted synthesis method, which involves doping copolymers with metal ions or salts, followed by a two-step annealing procedure [21,22]. For instance, Mirkin et al. showed how to synthesize multi-element nanoparticle libraries made up of any combination of five metallic elements using a polymer nanoreactor-mediated synthesis approach (Au, Ag, Co, Cu, and Ni) [23]. It has been demonstrated that by switching up the synthesis domain, the elemental composition and nanoparticle size may be easily controlled. Systematic research has revealed a strong correlation between the high entropy alloy (HEA) structure and the miscibility of metallic elements. Template-mediated techniques can also be used to create many metal nanostructures. Dendrimers with phenylazomethine units (DPA) and organometallic frameworks (MOFs), which are produced by the coordination of metal ions and organic ligands, may be suitable for HEM in this technique [24]. To expose more active sites of transitional species, template-mediated approaches can first produce HEMs with clearly defined structures. Yet, it also encourages higher mass and charge transfer, which significantly improves electrocatalytic performance.

11.2.1.2 Conventional Alloying Methods

Physical vapor deposition (PVD), arc discharge or laser deposition, and mechanical alloying techniques are examples of traditional alloying procedures [25]. A top-down method, magnetron co-sputtering from an elemental source onto the substrate enables the synthesis of HEM films using PVD. Different deposition speeds can be provided by controlling the elemental sputtering target to fine-tune the HEM film's ultimate composition. Arc discharge and laser deposition techniques are also employed to create HEM nanoparticles in addition to PVD.

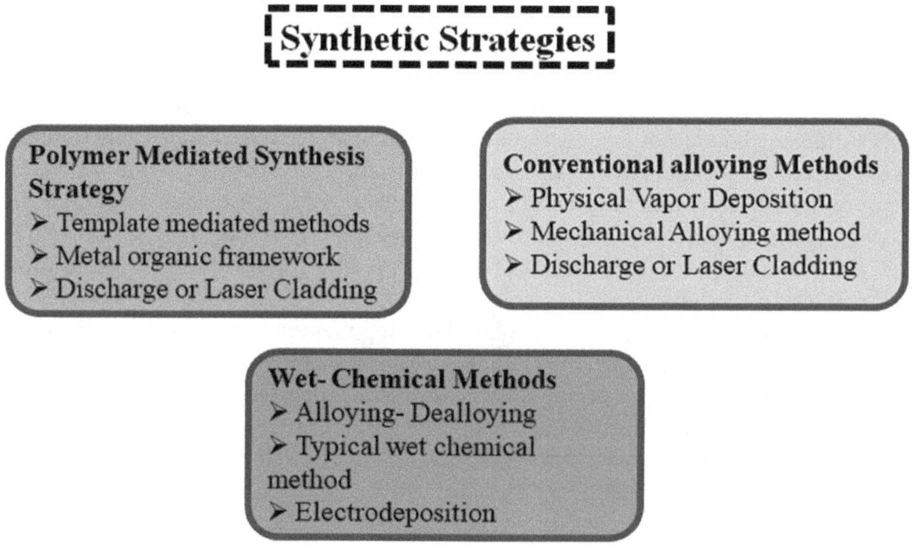

FIGURE 11.2 Schematically showing the synthesis methods for HEMs.

The target is a well-mixed polymetallic powder that can be vaporized at the extremely high temperatures produced by an arc discharge or laser deposition. These nanomaterials (HEMs) can also be created using mechanical alloying and physical mixing techniques involving ball milling of bulk targets. In addition to retaining the inherent features of pure metal powders, HEM made by the mechanical alloying approach also benefits from the plentiful active sites, potent synergistic effects, and entropy-stabilizing qualities of HEM.

11.2.1.3 Conventional Wet-Chemical Methods

Typical wet chemical processes, electro-coating, and alloying-dealloying techniques were all examples of conventional wet chemical processes [26–28]. By co-reducing metal salts and capping agents, which can regulate crystal formation and aggregation, nanoscale alloys are typically produced in solution using wet-chemical processes [29–30]. By merely altering the solvent, reducing agent, and capping agent, it is simple to modify the morphology, size, and crystal facets of HEM. The wet-chemical technique thus seems to be the most promising method for producing HEMs with perfect nanostructures.

The electrodeposition method is also frequently employed to manufacture high-quality HEM. Moreover, it was demonstrated that the electrodeposition approach may successfully produce HEM that is disordered and amorphous. To build HEMs with porous architectures, it is suggested to combine melting, quick cooling, chemical etching, and alloying-dealloying operations. Qiu et al. [31] constructed porous AlNiColrMo HEM nanostructures using this technique. To be more precise, they have combined and melted equal amounts of Ni, Co, Ir, and another element X (X: Mo, Nb, V, etc.), then remelted the mixture with a sizable amount of Al to create the precursor alloys ($Al_{96}Ni_1Co_1Ir_1X_1$).

11.2.2 Parameters Affecting HEM Formation

11.2.2.1 Enthalpy Mixing

The mixing of enthalpy (ΔH_{mix}) was first investigated by Zhang et al. [32] ΔH_{mix} can be estimated for each multicomponent system using the following equation Miedema's model from the following equation (11.2) [33].

$$\Delta H_{\text{mix}} = \sum_{i=1, i \neq j}^{n} 4 \Delta H_{\text{mix}C_i C_j}^{AB} \tag{11.2}$$

where n is the number of components, c_i and c_j are each component's concentrations, and HAB mix is the mixing enthalpy of the two components (i & j) in their AB-type binary mixture. Later, Guo and Liu Guo investigated the role of the enthalpy of mixing as one of the determining parameters on the production of various phases in multicomponent alloys using comparable statistical analyses on a bigger database [34].

According to their analysis, solid solutions (data mixed with intermetallic) only develop when the enthalpy of mixing is moderately positive or neutral (Figure 11.3). Hence, it is easier to produce

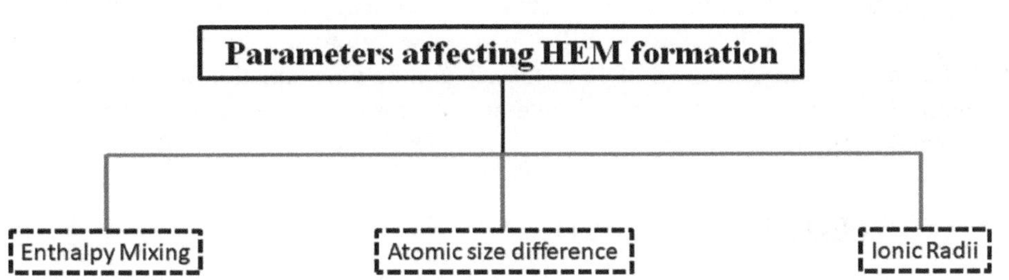

FIGURE 11.3 Significant factors influencing HEM formation such as enthalpic, atomic, and ionic radii contribution.

single-phase solid solutions with smaller magnitudes of H_{mix}. On the other hand, the binding energy between elements increases with increasing negative H_{mix}, and the miscibility gap between elements increases with increasing positive H_{mix}, making elements less likely to randomize in solution. Amorphous phases arise when the mixing enthalpy is more negative.

Enthalpy is produced during mixing and is ideally near zero. The huge positive H_{mix} of metals with various structures might not be balanced by substantial entropic effects or mixing and phase transitions might need to happen at considerably higher temperatures. On the other hand, the creation of single-phase mixed oxides under entropy dominance is not appropriate for the significant negative H_{mix} of mostly soluble metal oxides. Metals having various crystal forms (such as rock salt, wurtzite, etc.) or cation coordination are suitable options for this purpose.

Entropy-stabilized oxides are single-phase rock salt mixed formations that form in a way that favors entropy. For the entropic stability of single-phase HEM products, at least one of the prototype metal cation oxides should have a distinct crystal structure from the others.

11.2.2.2 Atomic Size Difference

The difference in atomic sizes among the components of the mixture turned out to be a crucial and important factor affecting the phase choice of multicomponent alloys. The disordered structure of solid solutions results from the random distribution of the constituent atoms, which have an equal chance of being found in one of the crystal lattice positions if the component atomic sizes are relatively identical. Atomic size mismatch effects can be seen in nonideal actual mixes of some components. Zhang et al. calculated the parameter for HEAs using the following equation (11.3). [35]

$$\delta = 100\sqrt{\sum_{i=1}^{n} c_i\left(1-\frac{r_i}{\bar{r}}\right)^2},\ \bar{r} = \sum_{i=1}^{n} c_i r_i \tag{11.3}$$

Here n signifies the number of components, c_i is the concentration of component I r_i is the atomic radius of component I and r are the average atomic size of the n components in alloy. They were the first to report the impact of this parameter on the phase stability of multicomponent alloys. Many researchers have assessed the use of another parameter to determine crucial values for phase selection. A key aspect determining the lattice strain of HEA is the difference in atomic sizes. Because of the dispersion of electrons and phonons, lattice distortion can lower thermal and electrical conductivity [36].

11.2.2.3 Ionic Radii

The ionic radii of metal cations have an impact on the lattice distortion and formation of single-phase HEO, much as the variation in atomic size radii of HEAs. The choice of metal cations with similar cation radii is taken into consideration in order to produce single-phase HEOs with cubic rock salt or fluorite structures. The differences in ionic radii for each of the multiple cation sites are calculated independently as rA and rB for ABO_3-type perovskite HEOs with a primitive cubic structure in which one or both A and B cation sites are made up of multiple oxides. The creation of single-phase high-entropy perovskite oxide, however, failed to occur within the range of atomic size differences of 6.5% but rather even in greater rB of 11%–13%. The following equation was used to apply additional structural variables of Goldschmidt's tolerance factor (*t*) for the stability of perovskite cubic structure to multicomponent systems.

11.3 OXYGEN EVOLUTION REACTION

The fundamental principle of water splitting includes the production of hydrogen and oxygen by applying electricity from external sources. The half-cell reaction of overall water splitting can be signified by equations (11.4) and (11.5):

$$\text{Anode}: H_2O \rightarrow \frac{1}{2}O_2 + 2H^+ + 2e^- \quad \left(E_{O_2/H_2O} = 1.23 \text{ V vs RHE}\right) \tag{11.4}$$

$$\text{Cathode}: H_2 \rightarrow 2H^+ + 2e^- \quad \left(E_{H_2/H_2} = 0 \text{ V vs RHE}\right) \tag{11.5}$$

The standard electrode potential required for producing molecule of O_2 and one molecule of H_2 is about 1.23 versus reversible hydrogen electrode (RHE). However, it is impracticable to govern the reaction at a potential due to the complex and sluggish kinetics of half-cell reaction. In addition, with kinetic hindrance of water splitting another important aspect to control the rate of the reaction is the electrolyte. The reaction mechanism differs broadly with nature of electrolytes such as acidic, basic, or neutral.

11.3.1 Mechanism

In general, the mechanism of OER takes place via four proton and four electron coupled process. It is noteworthy to mention that the OER mechanism is not effortless as HER due to complicated four electron/proton transfer. The important factor of commercialization of efficient electrolyzer is based on the efficiency of anodic probe. However, the electron transfer path possibly differs with the electrolyte for OER. The overall mechanistic steps are shown as equation (11.6–11.11) where M designates active site.

$$M + H_2O \xrightarrow{\text{Acid}} M - OH + H^+ + e^- \tag{11.6}$$

$$M - OH \xrightarrow{\text{Acid}} M - O + H^+ + e^- \tag{11.7}$$

$$2M - O \xrightarrow{\text{Acid}} 2M + O_2 \tag{11.8}$$

$$M + OH \xrightarrow{\text{Alkali}} M - OH + H^+ + e^- \tag{11.9}$$

$$M - OH + OH \xrightarrow{\text{Alkali}} M - O + H_2O + e^- \tag{11.10}$$

$$2M = O \xrightarrow{\text{Alkali}} 2M + O_2 \tag{11.11}$$

In alkaline environment, after the construction of M–OH bond, additional reactions occur either through the formation of metal oxide or through metal oxyhydroxide intermediate. Among the given mechanisms, the one with RDS decides the total overpotential of the electrocatalytic water oxidation reaction. Therefore, while designing a suitable electrocatalyst, we need to consider the effective S–O bond formation with optimum bond energy, i.e., bond energy that is neither too high nor too low.

11.4 DEVELOPMENT OF HEM FOR WATER SPLITTING

The de-alloying synthesis method for HEM has been disclosed by Qiu et al. ($Al_{98}Ni_{1.5}Fe_{0.5}$ and $Al_{97.5}Ni_{1.5}Co_{0.5}Fe_{0.5}$) [37]. The technique of synthesis entails melting pure metals in an induction melting furnace with Ar protection, followed by melt spinning to create the alloy ribbons. The uniform open-pore structure effectively exposes the active sites, faster diffusion of the electrolyte, and production of oxygen gas. Commercial RuO_2, ternary np-AlNiFe, and quaternary np-AlNiFeCo were examined for their OER performance (Figure 11.4a and b). Quinary np-AlNiFeCoMo, in contrast, had the lowest onset potential (around 1.44 V). The proposed electrocatalyst requires just 1.47 V (240 mV), which is much less than other catalysts (NiFeOx=350 mV, N-doped carbon catalysts 380 mV), according to the study. Figure 11.4b demonstrates that the quinary np-Tafel AlNiFeCoMo's slope is also considerably less than the catalyst (np-AlNiFe, np-AlNiCoFe, np-AlNiCoFeCu and np-AlNiCoFeCuMo).

Electrochemical impedance spectra (EIS) of the np-AlNiCoFeX (X=Mo, Nb, Cr) reveal substantially smaller arcs than those of other electrodes (Figure 11.4c), demonstrating low impedance resistance and excellent efficiency of charge transport between the catalysts and electrolyte for these electrodes. Thus, it is logical to conclude that the combination of these five elements (such as AlNiCoFeMo oxides) facilitates both quick interface reactions between the alkaline solution and the electrodes as well as fast electron transit (like a mixed valence conductor). When compared to other combinations, these effects clearly increase OER activity.

In addition, the author also studied the entropy effect by adding various metals such as Cr, Nb, V, Zr, and Mn due to their nontoxicity, high abundant in composite (Figure 11.4d and e). The results signify that catalytic enhancement of the quaternary np-AlNiCoFe toward the OER mechanism takes place in the order of Cr or Nb > V Zr Cu > Mn. This significant catalytic enhancement of Cr and Nb is attributed not only to the entropy effect but also to electronic influence. Figure 11.4f represents the excellent durability of as prepared np-AlNiCoFe.

Due to the need for a high-temperature synthetic approach and the incomplete knowledge of HEMs, the significant promise of high-entropy perovskite fluorides (HEPFs) in electrocatalysis has not yet been fulfilled. Because of this, Wang et al. [38] have described a facile, low-temperature mechanochemical production of K(MgMnFeCoNi)F_3 (HEPFs). Effective OER catalysis is demonstrated by the suggested K(MgMnFeCoNi)F_3. According to the OER performance investigation,

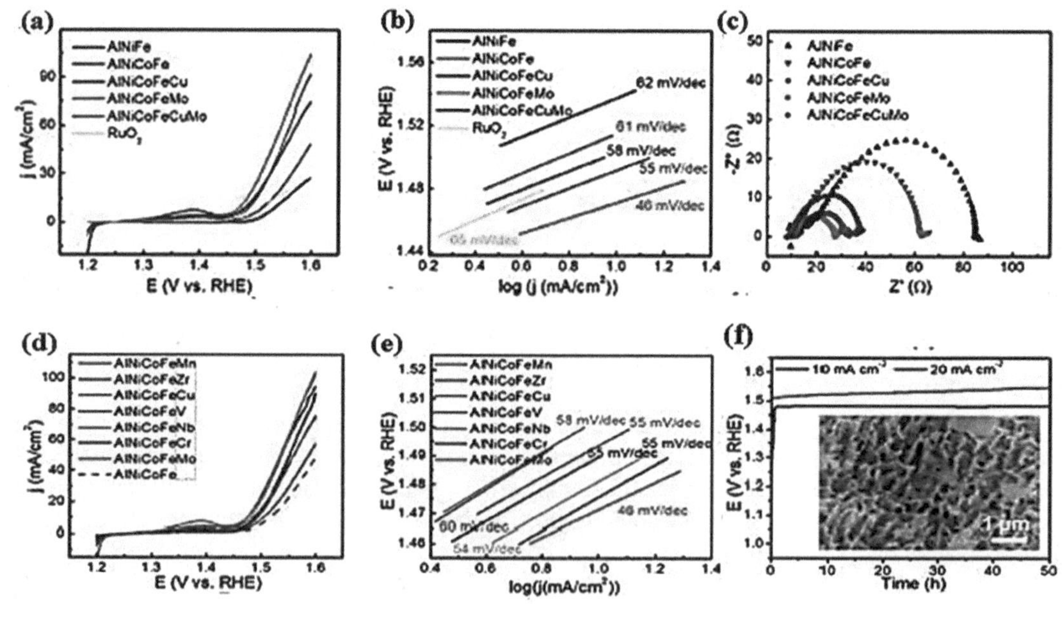

FIGURE 11.4 (a and d) Electrochemical measurements. LSV curves of np-HEAs, np-AlNiCoFe, np-AlNiFe, and RuO_2 electrodes; (b and e) The corresponding Tafel curves; (c) Electrochemical impedance spectra (f) At 10 mA/cm^2, the potential change with time and in the inset the SEM image after 50 hours testing. 1.0 M KOH solution was used for all these tests. Adapted with permission [37]. Copyright (2019), American Chemical Society.

$K(MgMnFeCoNi)F_3$ requires an overpotential of 369 mV, which is significantly less than KMF_3. $K(MgMnFeCoNi)F_3$ had a matching Tafel slope of 61 mV/dec, which was lower than those of KMF_3 (77 mV/dec). Unexpectedly, adding salt to the A sites of ABF3 increased their OER performance even further. At a current density of 10 mA/cm^2, $K_{0.8}Na_{0.2}(MgMnFeCoNi)F_3$ had the lowest overpotential, outperforming both commercial IrO_2 and the conventional perovskite oxide in terms of OER performance, $K_{0.8}Na_{0.2}(MgMnFeCoNi)F_3$ had a Tafel slope of 55 mV/dec, which was lower than those of $K(MgMnFeCoNi)F_3$ and $Na(MgMnFeCoNi)F_3$, showing an increase in OER kinetics as a result of significant sodium doping. During ten hours, a steady voltage was used to assess the HEPF catalysts' stability. This astounding OER performance is ascribed to the cubic shape with layering and a hollow structure which provides scattered active sites and mass transfer during OER catalysis.

The high surface area and active sites of metal-organic frameworks (MOFs), which are made up of metal ions and organic group linkers, make them desirable catalysts and electrocatalysts for reactions like OER. MOF template approach for the synthesis of HEA particles on carbon fabric with sub-5 nm has been described by Huang et al. [39]. To create HEA NPs with thin graphite linkers on the porous structure, pentametallic MOF/CC precursors were pyrolyzed at 450°C (Figure 11.5a). In an alkaline electrolyte (1M KOH), HEA NPs with FCC crystalline architecture adorned on the porous structure of the carbon linker were investigated as electrocatalysts for OER and compared to ternary and quaternary NP catalysts made using the same procedure. The results demonstrated that HEA NPs were more effective than other ternary and quaternary NP catalysts, with a Tafel slope of 43 mV/dec (Figure 11.5b). The embellished OER activity and low overpotential of the HEA NP catalyst are accredited to the atomic defects in the structure and high surface tension. In extension to higher catalytic activity, HEA NPs showed long-term robustness and good activity withholding over 24 hours (Figure 11.5c).

A high-entropy MOF (MnFeCoNiCu-MOF) was created by Zhao et al. [40] using an equimolar mixture of five metal ions and 1,4-benzenedicarboxylic acid (1,4-BDC) at ambient temperature

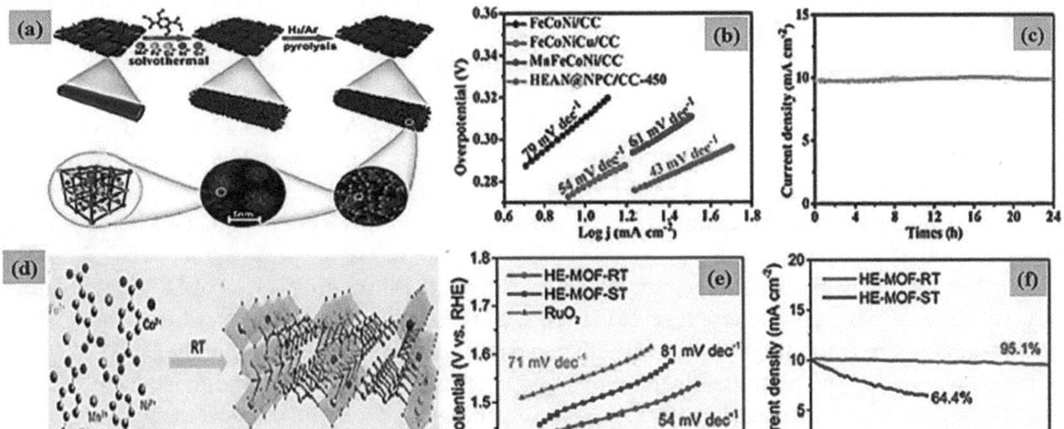

FIGURE 11.5 (a) Schematic of MOF-template synthesis of HEA NPs on carbon cloth; (b) Tafel plots of FeCoNi/CC, FeCoNiCu/CC, MnFeCoNi/CC, and HEAN@NPC/CC-450 electrocatalysts for OER test in 1 M KOH electrolyte; (c) current density retention of (HEAN@NPC/CC-450) over time. Adapted with permission [39]. Copyright (2020), Royal Society of Chemistry. (d) Schematic of high entropy MOF synthesized by room-temperature solution method; (e) Tafel plots of HE-MOF-RT, HE-MOF-RT and RuO_2 electrocatalysts for OER test in O_2-saturated 1 M KOH electrolyte; and (f) chronoamperometric curves of HE-MOF-RT and HE-MOF-ST at a current density of 10 mA/cm^2. Adapted with permission [40]. Copyright (2020), Royal Society of Chemistry.

(HE-MOF-RT). (Figure 11.5d). Another polymetallic MOF material was created utilizing a solvothermal solution heating technique (HE-MOF-ST). Characterization of the produced materials shows that the room-temperature approach successfully forms HEMs by delivering uniform and random distributions of metal ions with a close-to-equimolar structure. In alkaline solution (1 M KOH), the OER activity of HE-MOFs was examined, and its results were compared to those of polymetallic MOFs from solvothermal processes and traditional RuO2 catalysts. According to the results, HE-MOF exhibited a low overpotential of 245 mV at a current density of 10 mA/cm^2 and a small Tafel slope of 54 mV/dec, which were both significantly lower than those of HE-MOF ST. (293 mV and 81 mV/dec) and commercial RuO2 (346 mV and 71 mV/dec) (alkaline system) (Figure 11.5e). Also, compared to those made using the solvothermal approach and cumulative elemental distribution, HE-MOFs synthesized at ambient temperature with uniform elemental distribution demonstrated greater activity retention (Figure 11.5f).

Due to their excellent catalytic activity, transition metal sulfides with multi-element characteristics represent a promising class of catalysts for OERs. According to the Sabatier principle, high entropy metal sulfides (HEMS) with various metallic components can provide superior compositional control for the best adsorption of reaction intermediates, which will increase OER activity [41,42]. Because of the thermodynamic incompatibility of the constituent polymetallic elements in the sulfide structure, their synthesis is still challenging. Thus, Liangbing Hu et al. [43] described a solvothermal combination of metal salts followed by pulsed thermal annealing at 1650 K for 0.055 seconds for the synthesis of (CrMnFeCoNi)S*x*.

According to computational study, the overpotentials of the proposed electrocatalyst (unitary, binary, ternary, and quaternary M_xS_y) exhibit an inverse relationship with the number of metallic components. This suggests that the metallic elements work in harmony with one another. The pronounced OER activity and excellent durability of our HEMS indicate their excellent potential as an electrocatalyst for efficient water splitting. The author used density functional theory to comprehend the important role of multielemental mixing in activating HEMS with a synergistic effect (DFT). Figure 11.6a, where the Co atoms are chosen as OER active sites and are surrounded by five S atoms on the surface of (100), is consistent with the consensus in the literature. We estimated the free energy evolution process to examine the thermodynamics of OER on multielemental transition sulfides, where OER started with H_2O adsorption and progressed through several deprotonation processes and O–O coupling to produce *OH, *O, *OOH, and O2 successively (Figure 11.6b). When compared to the unary sulfides, such as Cr9S8 (2.74 V), Mn9S8 (2.27 V), and Fe9S8, we discovered that the limiting potential of OER on the HEMS is 1.55 V. Notably, the second or third proton-electron transfer step serves as the rate-determining phase. Figure 11.6b displays the same free energy as the limiting potential of 1.55 V indicating that it has a good chance of being used as an OER catalyst.

The Sabatier principle's relevance to volcanoes is revealed through a series of DFT calculations as a function of adsorption energy, or Eo (Figure 11.6c). The calculation shows that the adsorption energy, or Eo, is too large for Ni while being weak for Mn, Fe, and especially Cr. The HEMS $(Cr_2MnFe_2Co_2Ni_2)S_8$ with multielemental synergistic effect, on the other hand, is situated close to the volcano's summit with nearly neutral adsorption energy and may be the best catalyst option for OER. The movement in charge state among several metallic atoms can be used to further explain the multi-metallic synergy impact on adsorption energy Eo. In accordance with Bader charge analysis, the charge density difference map between $(Cr_2MnFe_2Co_2Ni_2)S_8$ and Co_9S_8 near the Co atom shows that more electrons were accumulated (yellow iso-surface Figure 11.6d). In $(Cr_2MnFe_2Co_2Ni_2)S_8$, the apparent interfacial electron transfer from Fe and Mn atoms to Co atoms results in a larger positive charge on the Fe (0.19e) and Mn (0.34e) sites and a more negative charge on the Co (0.31 e) site. This is the basis for Bader analysis (Figure 11.6e). Moreover, density of state (DOS) analysis was performed to determine the metal active site's d-band center, and a greater value indicated a stronger catalyst-adsorbate interaction [38]. The d-band center of $(Cr_2MnFe_2Co_2Ni_2)S_8$ (Figure 11.6f) exhibits its balanced OER intermediate adsorption energy, which is neither too strong nor too weak,

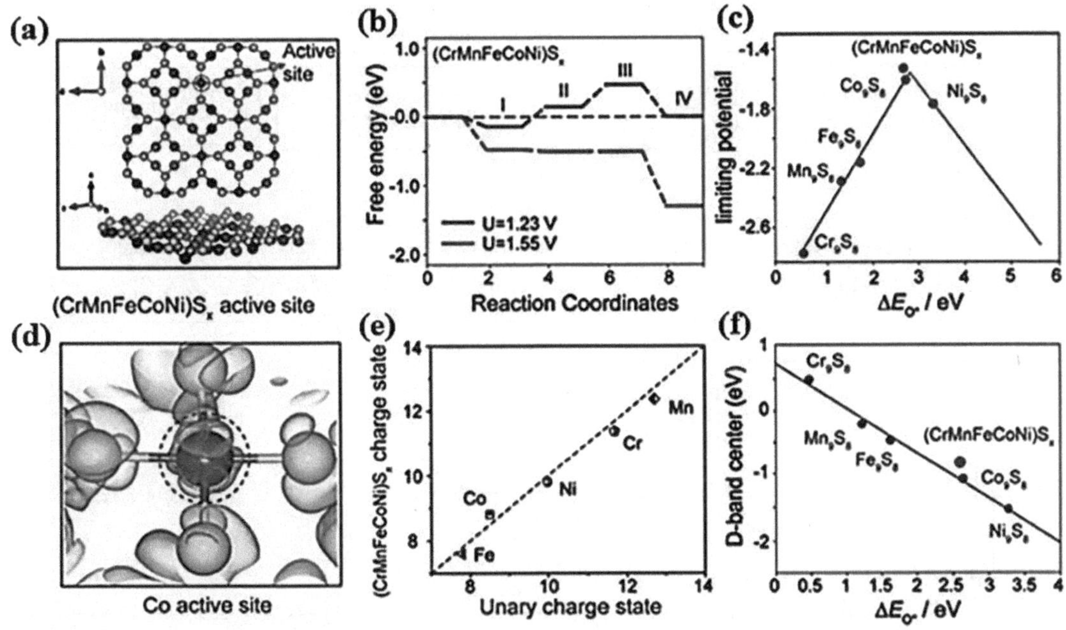

FIGURE 11.6 (a) Schematic of the constructed ($Cr_2MnFe_2Co_2Ni_2)S_8$ model by treating cationic Co centers as active sites. (b) Free energy evolution in a hypothesized OER mechanism. (c) Calculated catalytic activity volcano plot of unary M_xS_y and HEMS ($Cr_2MnFe_2Co_2Ni_2)S_8$. (d) Charge density difference between ($Cr_2MnFe_2Co_2Ni_2)S_8$ and Co_9S_8 near the cobalt center. Iso-surface level is 0.001 eV $Å^{-3}$ Yellow and cyan represent electrons accumulation and depletion, respectively. (e) Calculated relative charge states of Cr, Mn, Fe, Co, and Ni in ($Cr_2MnFe_2Co_2Ni_2)S_8$. (f) Comparison of the energies of d-band center energy of ($Cr_2MnFe_2Co_2Ni_2)S_8$ and unary M_xS_y catalysts as a function of adsorption energy (ΔEo*). Adapted with permission [43]. Copyright (2021), Wiley-VCH.

in good accord with the catalytic volcano plot. The DFT analysis provided a fundamental understanding of our HEMS as a powerful OER catalyst as a result. The computational prediction for the synergistic effect for good catalytic activity and extended durability is validated by HEMS (CrMnFeCoNi)S*x* nano-catalyst. The topic of water-splitting reactions of oxygen evolution (OER) is covered in detail to help focus the conversation. Although there are many different HEM electrocatalyst evaluation methodologies, over the past ten years, many helpful trends have evolved as best practices. This section's goal is to condense these patterns into a comprehensive approach for assessing HEMs, which need special consideration because of their complicated surface structure [44]. Figure 11.7a illustrates some of the HEM's observed characteristics for the benchmarking technique, including activity and stability of the electrochemical active surface area (ECSA). Evaluation of the aforementioned factors may satisfy the following requirements: (i) to reduce the amount of time and tests required to assess a catalyst's activity and near-term stability; (ii) to employ common electrochemical techniques and tools that a normal researcher working in the field of electrocatalysis can easily access.

To measure and evaluate electrocatalytic activity, it is crucial to accurately determine the ECSA of a specific material. The double-layer capacitance of the catalyst is typically measured using EIS or cyclic voltammetry (CV) in a non-faradaic zone. CV can be used to measure the system's open circuit potential (OCP) and the ECSA. Seven scan rates of CV were used (Figure 11.7b). A linear function was shown when the cathodic and anodic currents were plotted against the scan rate, with the slope indicating the double-layer capacitance (CDL) The CoFeLaNiPt HEMG-NPs' ECSA was calculated using CV to be 0.0013 cm^2. The so-called S number, a new metric for electrocatalyst

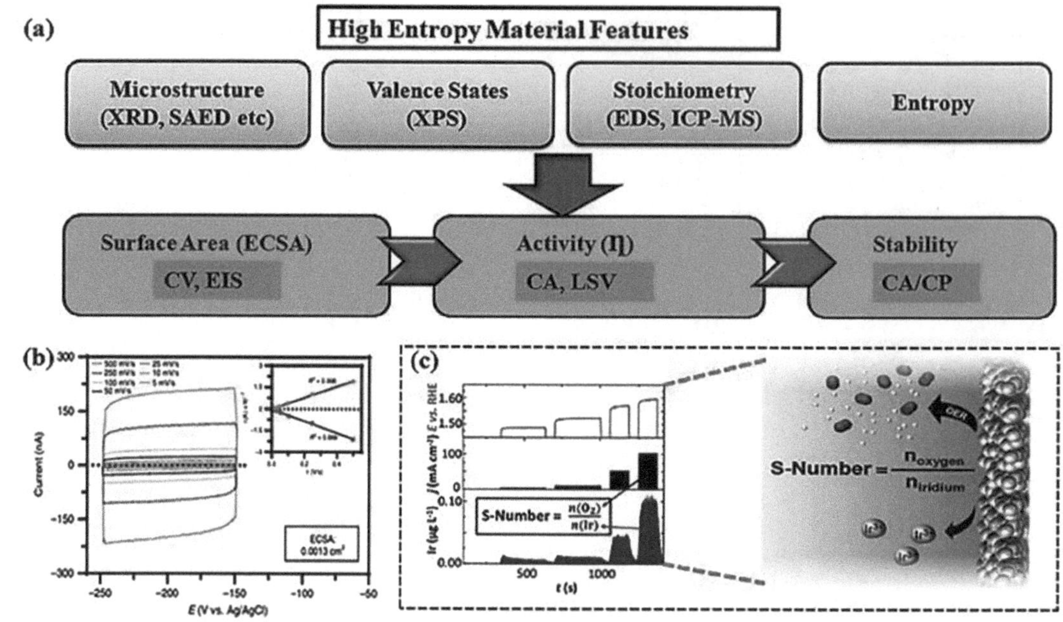

FIGURE 11.7 (a) Basic protocol for the classification and benchmarking of HEM electrocatalysis. (b) Cyclic voltammogram over multiple scan rates used to extract the ECSA for a CoFeLaNiPt high-entropy metallic glass electrocatalyst. Adapted with permission [44]. Copyright (2019), Springer-Nature; Distributed under a Creative Commons Attribution License 4.0 (CC BY) https://creativecommons.org/licenses/by/4.0/; (c) Correlated electrochemical and ICP-MS data used to obtain the S-Number for an IrOx film. Adapted with permission [45]. Copyright (2018), Springer-Nature.

stability benchmarking, allows for direct lifespan evaluation, demonstrative comparison of stability attributes, and insights into degradation mechanisms.

As the complete lifetime of a material can be easily determined using this parameter, the Stability-number represents a paradigm-shifting development in the evaluation of electrocatalysts. The capacity to assess material dissolution and related mechanisms (and potentially uncover parasitic reactions and their products) makes this method invaluable, even though the instrumental infrastructure may be more important [45]. The ratio of the amount of developed oxygen (derived from Qtotal) to the amount of dissolved metal is known as the "Stability-number" (S-number) (could be extracted from ICP-MS data). The higher the S number, more is the stability of the active center of the electrocatalyst (Figure 11.7c). Finally, it is important to note that although we have only examined a small number of specific reports for the study of water splitting thus far in this chapter, there are many other publications that are accessible that highlight the significance of HEM. Owing to the chapter's space constraints, we were only able to provide this brief overview; however, readers who are truly interested in learning more about the significance of HEM in depth can consult Table 11.1 and the accompanying sources.

11.5 CONCLUSION AND PERSPECTIVE

To achieve optimal performance, HEMs offer a comprehensive platform for adjusting the antisite disordering of atoms within a crystalline structure. However, the relationship between performance and surface structure is still unclear. Achieving this structure-property link and carefully controlling their structure and local electrical structure are crucial steps in the development of HEMs. Below, we list a few difficulties in brief. These problems also offer a great opportunity to investigate HEMs with the aid of sophisticated computations, syntheses, and characterization methods.

TABLE 11.1
Electrocatalytic Activity of Some Typical High-Entropy Materials for the OER

Sr. No	Composition	Overpotential	Tafel Slope	Ref. No
1	AlCrCuFeNi	270	77.5	[46]
2	FeCoNiCrVB	237	24.2	[47]
4	$(FeCoNiCrMn)_3O_4$	288	60	[48]
5	CoCuFeMoOOH	199	48.8	[49]
6	$(CoCuFeMnNi)_3O_4$	350	59.5	[50]
7	$La(CrMnFeCo_2Ni)O_3$	325	51.2	[51]
8	$K(MgMnFeCoNi)F_3$, $K(MgMnCoNiZn)F_3$	314	55	[38]
9	Ag@CoCuFeAgMoOOH	218	35	[52]
10	FeCoNiPB amorphous oxide	235	53	[53]

i. HEM thermal stability. The idea of employing configurational entropy to stabilize a single-phase solid solution is well acknowledged, and the term "high-entropy" is frequently used in the literature. Nevertheless, it is difficult to establish the precise contributions of entropy and enthalpy to the total Gibbs free energy. This includes configurational, vibrational, electronic, and magnetic entropy. The surface strain requires additional attention.
ii. For electrocatalysis, HEMs' surface shape and reaction-time stability are crucial. Although the surface structure of HEMs can be more intricate than the catalysts now in use, this issue is crucial for catalysis research.
iii. In particular, from the standpoint of materials science, it is important to recognize that determining the high-entropy nature of the created materials should be regarded as the core of the research and backed by detailed and reliable experimental or theoretical evidence.
iv. A significant difficulty that needs to be addressed is finding the actual active locations of a high-entropy water-splitting catalyst.
v. For the goal of water electrolysis, novel materials with increased compositional space should be investigated in order to fully use the high-entropy concept. It is strongly advised to do in situ and operando research to better comprehend the catalytic sites of HEMs.
vi. It is crucial that higher processing tools be used to help in the design and screening of possible high-entropy catalysts as well as the comprehension of their catalytic properties.

A better knowledge of the structure-property correlations of HEMs for water electrolysis should be the goal of all these initiatives. We anticipate additional intriguing developments in the creation of HEMs for applications such as energy storage and energy conversion for water electrolysis.

REFERENCES

[1] Li X., Hao X., Abudula A., Guan G. Nanostructured catalysts for electrochemical water splitting: Current state and prospects. *J. Mater. Chem. A.* 4 (2016) 11973–12000.
[2] Dincer I. Green methods for hydrogen production. *Int. J. Hydrogen Energy* 37 (2012) 1954–1971.
[3] Seitz L.C., Dickens C.F., Nishio K., Hikita Y., Montoya J. A highly active and stable IrOx/SrIrO3 catalyst for the oxygen evolution reaction. *Science* 353 (2016) 1011–1014.
[4] Zhu Y., Yue K., Xia C., Zaman S., Yang H., Wang X., Yan Y., Yu Xia B. Recent advances on MOF derivatives for non-noble metal oxygen electrocatalysts in zinc-air batteries. *Nano-Micro Lett.* 137 (2021) 8472.
[5] Lyu F., Wang Q., Choi S. M., Yin Y. Noble-metal-free electrocatalysts for oxygen evolution. *Small* 15(2019) 1804201.
[6] Jamesh Xiaoming Sun M.-I. Recent progress on earth abundant electrocatalysts for oxygen evolution reaction (OER) in alkaline medium to achieve efficient water splitting - A review. *J. Power Sources* 400 (2018) 31–68.

[7] He Z., Zhang J., Gong Z., Lei H., Zhou D., Zhang N., Mai W., Zhao S., Chen Y. Activating lattice oxygen in NiFe-based (oxy)hydroxide for water electrolysis. *Nat. Commun.* 13 (2022) 2191.

[8] Gerken J. B., Shaner S. E., Masse R. C., Porubsky N. J., Stahl S. S. A survey of diverse earth abundant oxygen evolution electrocatalysts showing enhanced activity from Ni-Fe oxides containing a third metal. *Energy Environ. Sci.* 7 (2014) 2376–2382.

[9] Wang A. L., Xu H., Li G. R. NiCoFe layered triple hydroxides with porous structures as high-performance electrocatalysts for overall water splitting. *ACS Energy Lett.* 1 (2016) 445–453.

[10] Zhang B., Zheng X., Voznyy O., Comin R., Bajdich M., Garcia-Melchor M., HanL., Xu J., Liu M., Zheng L., Garcia de Arquer F. P., Dinh C. T., Fan F., Yuan M., Yassitepe E., Chen N., Regier T., Liu P., Li Y., De Luna P., Janmohamed A., Xin H. L., Yang H., Vojvodic A., Sargent E. H. Homogeneously dispersed multimetal oxygen-evolving catalysts. *Science* 352 (2016) 333–337.

[11] Gao M., Miracle D., Maurice D., Yan X., Zhang Y., Hawk J. High-entropy functional materials. *J. Mater. Res.* 33 (2018) 3138–3155.

[12] Yan X., Constantin L., Lu Y., Silvain J., Nastasi M., Cui B. (Hf0.2Zr0.2Ta0.2Nb0.2Ti0.2)C high-entropy ceramics with low thermal conductivity, *J. Am. Ceram. Soc.* 101 (2018) 4486–4491.

[13] Sarker P., Harrington T., Toher C., Oses C., Samiee M., Maria J.-P., Brenner D. W., Vecchio K. S., Curtarolo S. High-entropy high-hardness metal carbides discovered by entropy descriptors, *Nat. Commun.* 9 (2018) 4980.

[14] Dai W., Lu T., Pan Y. Novel and promising electrocatalyst for oxygen evolution reaction based on MnFeCoNi high entropy alloy. *J. Power Sources* 430 (2019) 104–111.

[15] Yeh J.-W., Chen S.-K., Lin S.-J., Gan J.-Y., Chin T.-S., Shun T.-T., Tsau C.-H., Chang S.-Y. Nanostructured high-entropy alloys with multiple principal elements: Novel alloy design concepts and outcomes, *Adv. Eng. Mater.* 6 (2004) 299–303.

[16] Cantor B., Chang I. T. H., Knight P., Vincent A. J. B. Microstructural development in equiatomic multicomponent alloys. *Mater. Sci. Eng., A* 375-377 (2004) 213–218.

[17] Feng B., Zhuang X. The philosophy of carbon: Meso- entropy materials. *Faraday Discuss.* 227 (2019) 80–90.

[18] Yeh J., Lin S. Breakthrough Applications of High-Entropy Materials. *J. of Mater. Res.* 33 (2018) 3129–3137.

[19] Gao M. C., Liaw P. K., Yeh J. W., Zhang Y. *High-Entropy Alloys: Fundamentals and Applications*, 2016, Switzerland, Springer Cham.

[20] Sarkar A., Wang Q., Schiele A., Chellali M. R., Bhattacharya S. S., Wang D., Brezesinski T., Hahn H., Velasco L., Breitung B. High-entropy oxides: Fundamental aspects and electrochemical properties. *Adv. Mater.* 31 (2019) 1806236.

[21] Li M, Li Z, Fu G, Tang Y. Recent advances in amino-based molecules assisted control of noble-metal electrocatalysts. *Small* 17 (2021) e2007179.

[22] Jiang B, Guo Y, Kim J, Whitten AE, Wood K, Kani K, Rowan A.E., Henzie J, Yamauchi Y. Mesoporous metallic iridium nanosheets. *J. Am. Chem. Soc.* 140 (2018) 12434–12441.

[23] Chen P-C, Liu X, Hedrick JL, Xie Z, Wang S, Lin Q-Y, Hersam MC, Dravid VP, Mirkin CA. Polyelemental nanoparticle libraries. *Science* 352 (2016) 1565–1569.

[24] Wang HF, Chen L, Pang H, Kaskel S, Xu Q. MOF-derived electrocatalysts for oxygen reduction, oxygen evolution and hydrogen evolution reactions. *Chem. Soc. Rev.* 49 (2020) 1414–1448.

[25] Joyner J, Oliveira EF, Yamaguchi H, Kato K, Vinod S, Galvao DS, Salpekar D, Roy S, Martinez U, Tiwary CS, Ozden S, Ajayan PM. Graphene supported MoS2 structures with high defect density for an efficient her electrocatalysts. *ACS Appl Mater Interfaces* 12 (2020) 12629–12638.

[26] Ganguli S, Ghosh S, Tudu G, Koppisetti H, Mahalingam V. Design principle of monoclinic NiCo2Se4 and Co3Se4 nanoparticles with opposing intrinsic and geometric electrocatalytic activity toward the OER. *Inorg Chem.* 60 (2021) 9542–9551.

[27] Xu H, Zhao Y, Wang Q, He G, Chen H. Supports promote single-atom catalysts toward advanced electrocatalysis. *Coord. Chem. Rev.* 451 (2022) 214261.

[28] Wang C, Shang H, Wang Y, Xu H, Li J, Du Y. Interfacial electronic structure modulation enables CoMoOx/CoOx/RuOx to boost advanced oxygen evolution electrocatalysis. *J Mater Chem A.* 9 (2021) 14601–14606.

[29] Luo X, Liu C, Wang X, Shao Q, Pi Y, Zhu T, Li Y, Huang X. Spin regulation on 2D Pd-Fe-Pt nanomeshes promotes fuel electrooxidations. *Nano Lett.* 20 (2020) 1967–1973.

[30] Feng Y, Yang H, Zhang Y, Huang X, Li L, Cheng T, Shao Q. Te-doped Pd nanocrystal for electrochemical urea production by efficiently coupling carbon dioxide reduction with nitrite reduction. *Nano Lett.* 20 (2020) 8282–8289.

[31] Jin Z, Lv J, Jia H, Liu W, Li H, Chen Z, Lin X, Xie G, Liu X, Sun S, Qiu HJ. Nanoporous Al-Ni-Co-Ir-Mo high-entropy alloy for record-high water splitting activity in acidic environments. *Small* 15 (2019) e1904180.

[32] Zhang B. Y., Zhou Y. J., Lin J. P., Chen G. L., Liaw P. K. Solid - solution phase formation rules for multi-component alloys. *Adv. Eng. Mater.* 10 (2008) 299–303.

[33] Miedema A. R., de Châtel P. F., de Boer F. R. Cohesion in alloys - fundamentals of a semi-empirical model. *Phys. B+C*, 100 (1980) 1–28.

[34] Guo S., Liu C. T. Phase stability in high entropy alloys: Formation of solid-solution phase or amorphous phase. *Prog. Nat. Sci. Mater. Int.* 21 (2011) 433–446.

[35] Rost C. M., Sachet E., Borman T., Moballegh A., Dickey E. C., Hou D., Jones J. L., Curtarolo S., Maria J. Entropy - stabilized oxides. *Nat. Commun.* 6 (2015) 1–8.

[36] Karmakar A., Sankar S. S., Kumaravel S., Madhu R., Mahmoud K. H., El-Bahy Z. M., Kundu S. Ruthenium-doping-induced amorphization of vs4 nanostructures with a rich sulfur vacancy for enhanced hydrogen evolution reaction in a neutral electrolyte medium. *Inorg. Chem.* 61 (2022) 1685–1696.

[37] Qiu H. J., Fang G., Gao J., Wen Y., Lv J., Li H., Xie G., Liu X., Sun S. Noble metal-free nanoporous high-entropy alloys as highly efficient electrocatalysts for oxygen evolution reaction. *ACS Materials Lett.* 1 (2019) 526–533.

[38] Wang T., Chen H., Yang Z., Liang J., Dai S. High-entropy perovskite fluorides: A new platform for oxygen evolution catalysis. *J. Am. Chem. Soc.* 142 (2020) 4550–4554.

[39] Huang K., Zhang B., Wu J., Zhang T., Peng D., Cao X., Zhang Z., Li Z., Huang Y. Exploring the impact of atomic lattice deformation on oxygen evolution reactions based on a sub-5 Nm pure face-centred cubic high-entropy alloy electrocatalyst. *J. Mater. Chem. A* 8 (2020) 11938–11947.

[40] Zhao X., Xue Z., Chen W., Bai X., Shi R., Mu T. Ambient fast, large-scale synthesis of entropy-stabilized metal-organic framework nanosheets for electrocatalytic oxygen evolution. *J. Mater. Chem. A* 7(2019) 26238–26242.

[41] Greeley J., Mavrikakis M. Alloy catalysts designed from first principles. *Nat. Mater.* 3 (2004) 810–815.

[42] Du X., Huang J., Zhang J., Yan Y., Wu C., Hu Y., Yan C., Lei T., Chen W., Fan C., Xiong J. Modulating electronic structures of inorganic nanomaterials for efficient electrocatalytic water splitting. *Angew. Chem., Int. Ed.* 58 (2019) 4484.

[43] Mingjin C., Chunpeng Y., Boyang L., Qi D., Meiling W., Sooyeon H., Hua X., Xizheng W., Guofeng W., Liangbing H. High-entropy metal sulfide nanoparticles promise high-performance oxygen evolution reaction. *Adv. Energy Mater.* 21 (2021) 2002887.

[44] Glasscott M. W., Pendergast A. D., Goines S., Bishop A.R., Hoang A.T., Renault C., Dick J.E. Electrosynthesis of high-entropy metallic Glass nanoparticles for designer, multi-functional electrocatalysis. *Nat. Commun.* 10 (2019) 2650.

[45] Geiger S., Kasian O., Ledendecker M., Pizzutilo E., Mingers A. M., Fu W.T., Diaz-Morales O., Li Z., Oellers T., Fruchter L. The stability number as a metric for electrocatalyst stability benchmarking. *Nat. Catalysis* 1 (2018) 508–515.

[46] Zhong X., Zhu Y. A., Dai W. J., Yu J., Lu T., Pan Y. Electrochemically reconstructed high-entropy amorphous FeCoNiCrVB as a highly active oxygen evolution catalyst. *New J. Chem.* 46 (2022) 8398.

[47] Sharma L., Katiyar N. K., Parui A., Das R., Kumar R., Tiwary C. S., Singh A. K., Halder A., Biswas K. Low-cost high entropy alloy (HEA) for high-efficiency oxygen evolution reaction (OER). *Nano Res.* 15 (2021) 4799.

[48] Duan C. Q., Li X. L., Wang D., Wang Z. Y., Sun H. Y., Zheng R. G., Liu Y. G. Nanosized high entropy spinel oxide (FeCoNiCrMn)3O4 as a highly active and ultra-stable electrocatalyst for the oxygen evolution reaction. *Sustainable Energy Fuels* 6 (2022) 1479.

[49] Zhang L. J., Cai W. W., Bao N. Z., Top-level design strategy to construct an advanced high-entropy Co-Cu-Fe-Mo (Oxy)hydroxide electrocatalyst for the oxygen evolution reaction. *Adv. Mater.* 33 (2021) 2100745.

[50] Wang D., Liu Z. J., Du S. Q., Zhang Y. Q., Li H., Xiao Z. H., Chen W., Chen R., Wang Y. Y., Zou Y. Q., Wang S. Y. Low-temperature synthesis of small-sized high-entropy oxides for water oxidation. *J. Mater. Chem. A* 7 (2019) 24211.

[51] Nguyen T. X., Liao Y. C., Lin C. C., Su Y. H., Ting J. M. Advanced high entropy perovskite oxide electrocatalyst for oxygen evolution reaction. *Adv. Funct. Mater.* 31 (2021) 2101632.

[52] Zhang L., Cai W., Bao N., Yang H. Implanting an electron donor to enlarge the d-p hybridization of high-entropy (Oxy) hydroxide: A novel design to boost oxygen evolution. *Adv. Mater.* 34 (2022) 2110511.

[53] Wang Q. Q., Li J. Q., Li Y. J., Shao G. M., Jia Z., Shen B. L. Non-noble metal-based amorphous high-entropy oxides as efficient and reliable electrocatalysts for oxygen evolution reaction. *Nano Res.* 15 (2022) 8751–8759.

12 High Entropy Materials for Oxygen Reduction Reactions

Ji-Chang Ren, Guolin Cao, and Wei Liu

12.1 INTRODUCTION

Chemical reactions occurring on solid surfaces are fundamental to heterogeneous catalysis, which plays a central role in various industrial processes such as energy conversion, chemical manufacturing, and environmental protection. The rational design of heterogeneous catalysts relies heavily on the fine-tuning of surface characteristics, which is generally achieved through the optimization of adsorption energy and microkinetic modeling. In recent years, significant progress has been made in the field of computational modeling, particularly in quantum chemistry methods like density functional theory (DFT), which has enabled better understanding of active sites, interfacial chemical bonding, and elementary reaction steps on solid surfaces. This chapter provides a review of recent advances in computational modeling of surface adsorption and chemical reactions on solid surfaces of high-entropy alloys (HEAs) for heterogeneous catalysis.

HEAs are composed of multiple elements (typically five or more) that are disorderly arranged on specific crystal lattices and have recently shown great promise as catalysts for chemical reactions, particularly the oxygen reduction reaction (ORR) [1–4]. For instance, Yao et al. [5] achieved high-throughput synthesis of multi-metallic nanocluster with a homogeneous alloy structure using a thermal shock heating technique and obtained two promising catalysts for accelerating the ORR process. Löffler et al. [1] investigated a noble-metal-free HEA CrMnFeCoNi using a sputtering synthetic method to compete with Pt-based catalysts in ORR activity. Thomas et al. [6] proposed a framework to optimize HEA catalysts by combining high-throughput experiments and modelings.

While experimental studies of HEA catalysts have rapidly grown, theoretical modeling lags behind, leaving the superior catalytic performances of HEAs still unclear. Predicting HEA catalytic performance is challenging due to the extremely large chemical space of HEAs and diverse local reaction environments, even for a single HEA material. Commonly applied *ab initio* methods, like DFT calculations, have limited prediction ability for large and complex alloy systems, as well as complicated reaction processes. Therefore, it is essential to combine some other techniques, such as high-throughput calculations, machine learning models, and statistical analysis, to study the surface adsorption and chemical reaction processes on HEA surfaces.

In this chapter, we will provide a summary of recent developments and applications of data-driven machine learning and reaction descriptor model methodologies for ORR activity. The focus will be on structure-activity correlations and the interplay of atomic-scale factors. To generate data, we will introduce high-throughput computation frameworks and their applications for molecule adsorptions on HEA surfaces. HEA surfaces offer a natural platform for optimizing the binding strengths of adsorbates, with tunable adsorption energies and a wide energy distribution even for the same active site. Machine learning models will be used to efficiently screen adsorption energies over thousands of active sites and perform statistical analysis of catalytic behaviors. These models have proven to be powerful tools for predicting adsorption energies and uncovering atomic factors that affect reaction activity. Finally, we discuss the anomalous behavior of HEA surfaces and current challenges and strategies for modeling chemical reactions on HEA surfaces.

 DOI: 10.1201/9781003391388-12

12.2 ADSORPTION ON THE SURFACES OF HEAS

The ORR process involves four proton-electron transfers to reduce oxygen to water or two proton-electron transfers to produce hydrogen peroxide in acidic solutions. On the catalyst surface, the key steps of ORR involve the adsorption/desorption of three main intermediates, namely COOH*, *OH and *O adsorption systems. In this section, we will explore the adsorption of these species on HEA surfaces and highlight the unique adsorption behaviors that distinguish HEA surfaces from conventional alloy substrates. In theoretical studies of molecule adsorption on solid surfaces, with an accurate knowledge of the adsorption geometry, or the specific adsorption site (*i.e.*, top, bridge, and fcc/hcp-hollow sites), is a prerequisite for additional analysis of the adsorbate properties. However, this prerequisite is not applicable to the study of HEA surfaces due to their highly diverse local atomic environment. This means that the adsorption on HEA surfaces is no longer a deterministic process, but rather a statistical behavior that depends on the actual probability distributions of the active sites.

Due to the significant differences in adsorption behavior, conventional scaling rules [7–18] used in metallic catalysts need to be re-examined for the study of HEA catalysts. For example, on transition-metal surfaces, the energies of the transition state are scaled with the adsorption energies of the key intermediates, known as the Brønsted-Evans-Polanyi (BEP) relation [19]. This scaling relation allows for the representation of the kinetic chemical process with the thermodynamic properties of the intermediates, typically the qualities of the adsorption energies. Therefore, adsorption energy has long been applied as a reactivity descriptor for heterogenous catalysis in the past few decades. However, given the distinct adsorption behaviors of HEA surfaces, it is necessary to re-evaluate the use of conventional descriptors in the study of HEA catalysts.

Furthermore, Nørskov's group [7] demonstrated the existence of scaling properties between the adsorption energies of hydrogen-containing molecules, such as CH_x (x=0, 1, 2, 3) and OH_x (x=0, 1), on transition-metal surfaces. This results in stronger restrictions on surface reactivity, as it indicates limitations when optimizing one reaction step without compromising others. For instance, on transition-metal surfaces, O* prefers the hollow site, while *OH prefers to coordinate to the top site, as shown in Figure 12.1. A linear relation exists between the adsorption energies of O* and

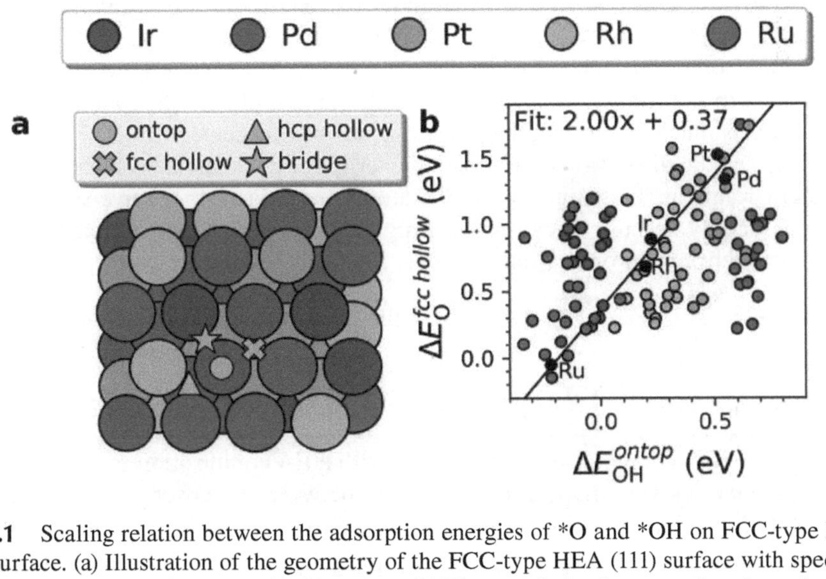

FIGURE 12.1 Scaling relation between the adsorption energies of *O and *OH on FCC-type IrPdPtRhRu (111) HEA surface. (a) Illustration of the geometry of the FCC-type HEA (111) surface with specific adsorption sites, including the top, hollow, and bridge sites; (b) The correlation between the adsorption energies of *O on hollow sites and the adsorption energies of *OH on top site. The results reveal that the scaling relation on the HEA surface is significantly weaker than that observed on pure metals. Adapted with permission [20]. Copyright (2022), Elsevier.

OH, expressed as $\Delta E_{O*} = \gamma \Delta E_{*OH} + \zeta$, where γ is proportional to interfacial orbital hybridization between adsorbate and the active site of the substrate, and ζ is independent of the adsorbed substrate selected. To test if this scaling relation holds for HEA catalysts, Jack and co-workers [20] performed high-throughput DFT calculations for the ORR process on the IrPdPtRhRu (111) surface. As illustrated in Figure 12.1a, the atoms that form the hollow site might be different on this HEA surface. For the HEA containing five elements, there are $C_5^3 = 35$ different threefold hollow sites and $C_5^1 = 5$ different top sites. It is easy to understand that a scaling relation between the adsorption energies of O adsorbed on the hollow site and *OH adsorbed on the top site does not exsit, as there are different kind of sites between hollow (35) and top (5) positions. Indeed, based on DFT adsorption energy calculations, it can be observed in Figure 12.1b that the correlation between the two adsorption conditions is weak: some hollow sites tend to bind with *O stronger while others tend to bind with *O weaker than the results of scaling relation. Interestingly, when *O is fixed on top site, the scaling relation is recovered, with the same slope as that observed in transition-metal surface. Moreover, the scaling relation between the adsorption energies of *OH and *OOH is preserved on the HEA surface, as both adsorbates prefer to adsorb on top site.

The findings presented above provide an initial insight into HEA catalysis, indicating that the scaling relation can still hold on HEA surfaces if the intermediates bind to the surface with the same coordination. Nevertheless, it is crucial to carefully examine the adsorption coordination on HEA surfaces. Unlike pure metal surfaces, where the adsorption coordination is uniform, on HEA surfaces, the local chemical environment surrounding the adsorption site, as well as the atoms binding with the adsorbate, play a vital role in determining the adsorbate–surface interaction. Therefore, the geometric factor must be considered when investigating adsorption on HEA surfaces.

In metallic systems, the strong screening effect implies that the nearsightedness principle [21] should be satisfied in HEAs. This principle suggests that the binding strength for the same adsorption site is only slightly perturbed by the variation of the nearby atomic environment, resulting in relatively narrow distributions of adsorption energy values. This principle holds true for CoMoFeNiCu HEAs, as studied by Saidi [22]. Figure 12.2a shows that the distributions of adsorption energies of carbon atom on the entire HEA surfaces are relatively narrow for all systems with different composition concentrations. However, the small range of energy distribution is not general for all HEA systems but is only associated with those that have a strong adsorption center. For systems with relatively weak adsorption centers, like AgAuCuPdPt HEAs in Figure 12.2b, the peak positions of the adsorption energies are much higher than those of CoMoFeNiCu HEAs, indicating weaker adsorption centers in the former. Furthermore, the dispersion of the adsorption energies on weak adsorption centers is much larger than that on strong adsorption centers.

In HEA surfaces, the dispersion of adsorption energy for the same active site is induced by the local atomic environment. This implies that the atoms comprising the active site have partly lost their individual identity because of the surrounding atomic environment. To examine the effect of the local atomic environment on scaling relations, Saidi introduced an average adsorption energy $\Delta\hat{E}$, defined as $\Delta\hat{E} = \sum_{l}^{n_s} f_l \Delta E_l \Big/ \sum_{l}^{n_s} f_l$, where n_s is the number of random slab models with different atomic arrangements, and $f = \prod_{k}^{5} (c_k)^{N_k}$ counts the number of possibilities of generating a surface microstructure with N_k metal atoms consistent with HEA composition c_k. Unlike conventional scaling relations that describe the linear relationship between the adsorption energies on a local active site, $\Delta\hat{E}$ is introduced to estimate the average binding strength of the entire HEA surface. Thus, the relationships of $\Delta\hat{E}$ represent a global scaling relation between different adsorbates.

The CoMoFeNiCu HEA system displays a linear relationship between the averaged adsorption energies of CH_x and OH_x species, as illustrated in Figure 12.2c and e, which is similar to pure transition-metal systems [7]. However, caution must be taken before assuming that HEA and pure metals

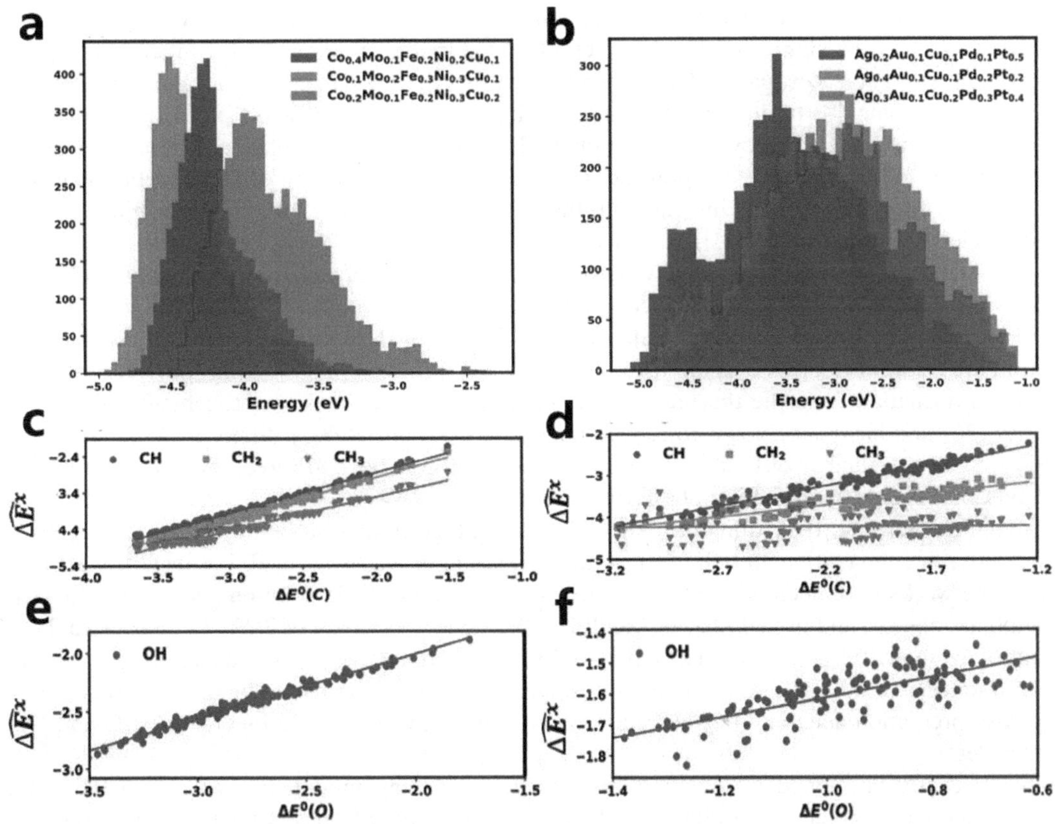

FIGURE 12.2 Adsorption energy distributions of carbon atom on CoMoFeNiCu HEAs (a) and AgAuCuPdPt HEAs (b) with varying composition concentrations. The results showed that the scaling relation was maintained in CoMoFeNiCu adsorption systems, as seen in (c) and (e), due to the narrow distributions of the adsorption energy. However, for all tested adsorbates in AgAuCuPdPt adsorption systems, as seen in (d) and (f), the linear relations were found to break down. Adapted with permission [22]. Copyright (2022), Springer Nature.

behave similarly in catalytic processes. This assumption relies on complete monolayer adsorption, where bond formation and bond breaking occur on the same active site, and intermediates do not diffuse during the reaction. However, due to the broad range of adsorption energy distributions, weakly binding sites to intermediates may exist, for which the desorption barrier is small. On the other hand, Figure 12.2d and f indicate that scaling relations on AgAuCuPdPt surfaces are not easily established because of the relatively weak adsorption center, especially for the active site formed with Ag or Au elements.

12.3 MACHINE LEARNING PREDICTIONS OF PHASE STRUCTURES IN HEA CATALYSIS

The catalytic performance of HEAs is strongly influenced by their microstructure and surface properties. Ideally, a single-phase solid solution with FCC or BCC structures maximizes configuration entropy. However, the phase structures of HEAs can vary depending on the synthetic method, and the presence of multiple principal elements can make the design of HEAs with desirable single-phase solid solution challenging due to the complex phase diagrams [23,24].

Despite remarkable advancements in quantum chemistry methods and increasing computing power, there is still a significant gap between idealized model systems and the inherent complexities of real HEA surfaces, particularly when considering the highly intricate local atomic environment. To overcome this limitation, there is a pressing need for more efficient predictive models that can achieve the same accuracy as conventional DFT calculations. In this regard, machine learning algorithms offer an attractive solution for capturing the complex nature of HEA surface catalysis. These algorithms can provide insights into the nature of active sites, predict adsorption energies, and accelerate the discovery of HEA catalysts.

The stability of an HEA is largely due to the high degree of disorderly distributed elements within a simple crystal structure, which maximizes the mixing entropy and minimizes the Gibbs free energy of the system. However, determining whether an HEA can form a stable solid solution within the framework of DFT is almost impossible. The parameter method [25] provides estimates of phase formation from the thermo-physical parameters of the constituent elements of potential HEAs [26] and has been shown to reduce the phase overlap between different phase structures. However, as the number of thermo-physical factors increases beyond three, the empirical method becomes unreliable in predicting phase structures.

Machine learning algorithms have an innate advantage in dealing with the complex relationship between composition and phase structure formation in HEAs [27–31]. For instance, artificial neural networks have been utilized to predict the formation of solid solution phases, intermetallic compound phases, and mixed phases, yielding a prediction accuracy of 75% [32]. Remarkably, by combining an artificial HEA dataset (created through synthetic minority oversampling technique) with a dataset of 322 as-cast samples, a support vector machine-based model [28] was able to increase prediction accuracy beyond 90% for distinguishing solid solutions with BCC and FCC structures.

We recently developed a random forest machine learning model, termed as AS-RF model [33], which can predict single-phase solid solution HEAs based on a database of high-throughput as-sputtered HEA samples. One of the distinctive features of the AS-RF model is its ability to estimate physical feature importance contributions, which allows the model to be interpreted by determining the order and effect size of the features associated with the outcome. Using the AS-RF model, we predicted 224 quinary phase diagrams, encompassing approximately 32,000 single-phase solid solution HEAs with BCC and FCC structures in the Cr–Co–Fe–Ni–Mn–Cu–Al composition space.

The flowchart outlining the main process is presented in Figure 12.3. The random forest model is an ensemble learning method that utilizes decision trees, where the number and depth of the trees control the complexity of the model. To ensure reliable decision-making, we employed the Gini-index method [34] for node splitting, where the expression is $\text{Gini} = 1 - \sum_{m=1}^{M} (p_m)^2$. Here, p_m denotes the probability of the training sample belonging to class m, and M is the number of phase structures. The decision-making in each node is achieved by minimizing Gini. To evaluate the model's quality, we applied 10-fold cross-validation to balance the bias and variance of the predictions. In the random forest model, each tree is treated as an independent estimator and votes for one phase structure. The model then collects the votes from all the trees and makes the final phase structure prediction using the formula $\hat{y} = \arg\max_{j} \sum_{i=1}^{M} \omega_i p_{ij}^{\text{tr}}$, where p_{ij}^{tr} is the probability of the predicted class label j from the i-th tree, and $\hat{y}$ is the final predicted phase structure.

As presented in Figure 12.3, our model's predictions exhibit strong agreement with experimental observations. Subsequently, by plotting the model-predicted 67,000 alloys individually onto the network based on their phase structures, various correlations between elements can be established, as illustrated in Figure 12.4. Specifically, for HEAs with single-phase solid solutions, the interaction

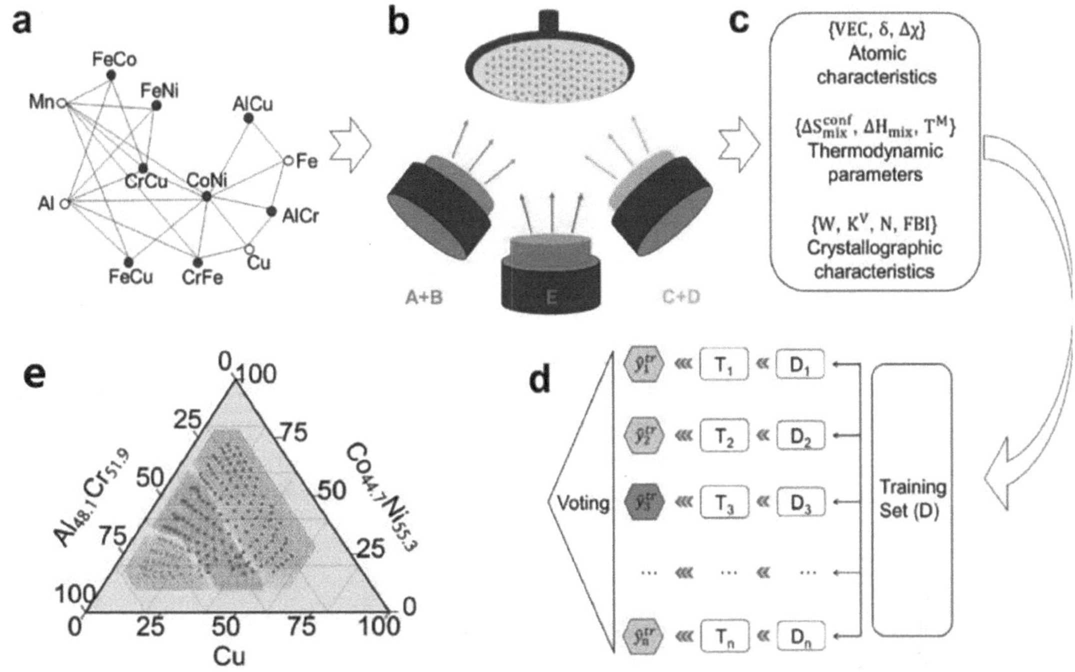

FIGURE 12.3 A schematic diagram of the random forest model for the phase diagram predictions. (a) Elemental network for the formation of five-principal HEAs. (b) A schematic graph represents the fabrication processes of as-sputtered HEAs. (c) A list of thermos-physical parameters for the descriptions of the HEA samples. (d) The architecture of the AS-RF machine learning model. D_n is the subset of the training data. T_n is a single decision tree. $\hat{y}_n^{tr}$ is decision made by T_n. The filled colors in the figure denote different phases of HEAs. During the training process, to reduce the similarity and correlations between decision trees, 60% of the training data have been randomly selected for each tree. To make the final predictions in the RF model, each tree first collects the results from the leaf nodes and votes for the one with largest probabilities, and then the RF collects the votes from each tree and provides the final prediction. (e) Illustration of one phase diagram of the predicted results (background). The experimental results are denoted with the corresponding dot points. Adapted with permission [33]. Copyright (2022), Elsevier.

between Al, Cr, and Fe elements is found to have the most significant impact on the formation of BCC structures, while the correlation between Cu, Ni, and Co predominates the formation of FCC structures.

12.4 MACHINE LEARNING PREDICTIONS OF ADSORPTION ENERGY ON HEA SURFACES

With the confirmation of solid solution stability in HEAs, Thomas et al. [35] have taken a pioneering step by employing machine learning model to predict the adsorption energy for key ORR intermediates on HEA surfaces. Specially, they investigated the adsorption of *O and *OH on the (111) surface of IrPdPtRhRu HEA with an FCC structure. To begin with, they conducted DFT calculations on fifty 4-layered 3×4 slab models, where the atoms of the top two layers were permuted. The energy difference between the 50 slabs was found to be within 0.03 eV/atom. Subsequently, *O and *OH adsorbates were positioned on the top and FCC hollow sites, respectively, and the adsorption energies were determined through high-throughput DFT calculations.

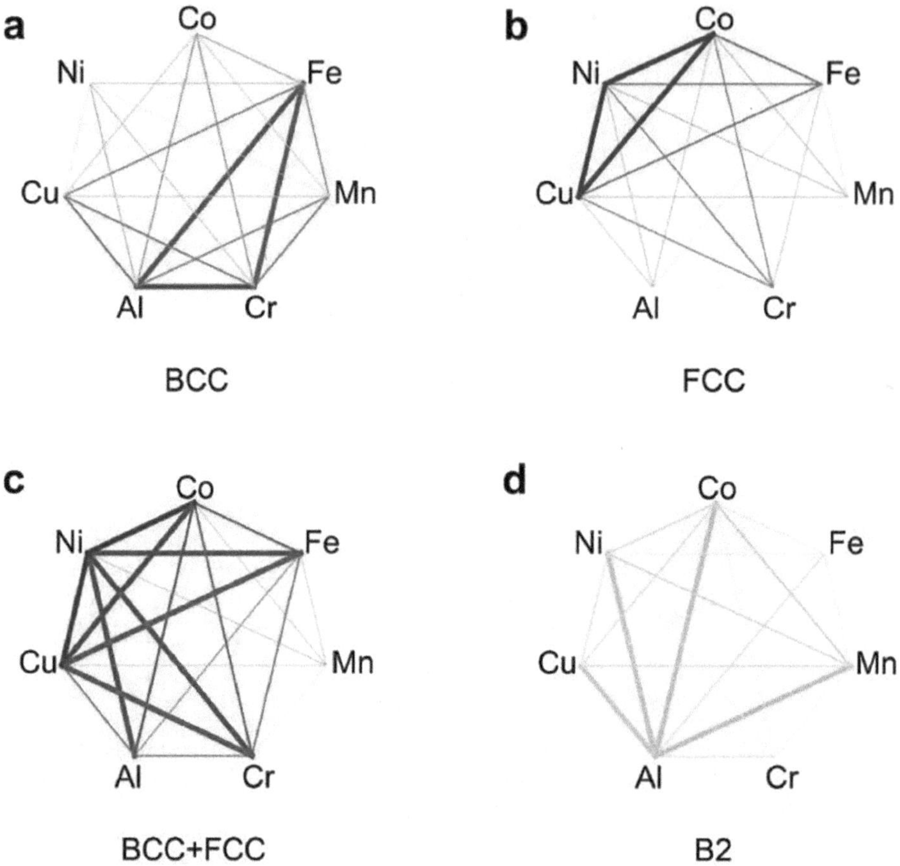

FIGURE 12.4 The relationships between element correlations and phase formations of different types of solid solution. Panel (a) shows BCC-type single-phase solid solutions, panel (b) shows FCC-type single-phase solid solutions, panel (c) shows BCC+FCC-type mixing phase solid solutions, and panel (d) shows the B2 intermediate metallic phase. In each panel, the weights of the lines in the networks represent the different correlation strengths between elements. A larger weight indicates a stronger correlation between elements. Adapted with permission [33]. Copyright (2022), Elsevier.

They developed a linear machine learning model to predict top adsorption. The formula for this prediction is expressed as follows:

$$\Delta E^{i}_{\text{pred}} = \sum_{k}^{\text{metals}} C_{1,\,k} N^{i}_{1,\,k} + \sum_{k}^{\text{metals}} C_{2,\,k} N^{i}_{2,\,k} + \sum_{k}^{\text{metals}} C_{3,\,k} N^{i}_{3,\,k} \tag{12.1}$$

In this equation, $C_{1,\,k}$, $C_{1,\,k}$, *and* $C_{1,\,k}$ represent the weighting parameters determined through model optimization, while $N^{i}_{1,\,k}, N^{i}_{1,\,k}, N^{i}_{1,\,k}$ represent the number of each element k found in each zone numbered 1, 2, 3 in Figure 12.5a. For the hollow site, the model expression is slightly different. The first term runs over 35 possibilities, and two additional zones, denoted as 4 and 5, are included.

The accuracy of the predicted results was evaluated by comparing them with the results obtained from DFT calculations, as shown in Figure 12.5b. The predicted results for the adsorption energies of *O and *OH on IrPdPtRhRu (111) surface showed very high accuracy, with a root-mean-square deviation (RMSE) of 0.063 and 0.076 eV for *OH and *O, respectively. The high prediction ability of the simple linear model is understandable, as the specific atom locations of nearby atoms at the

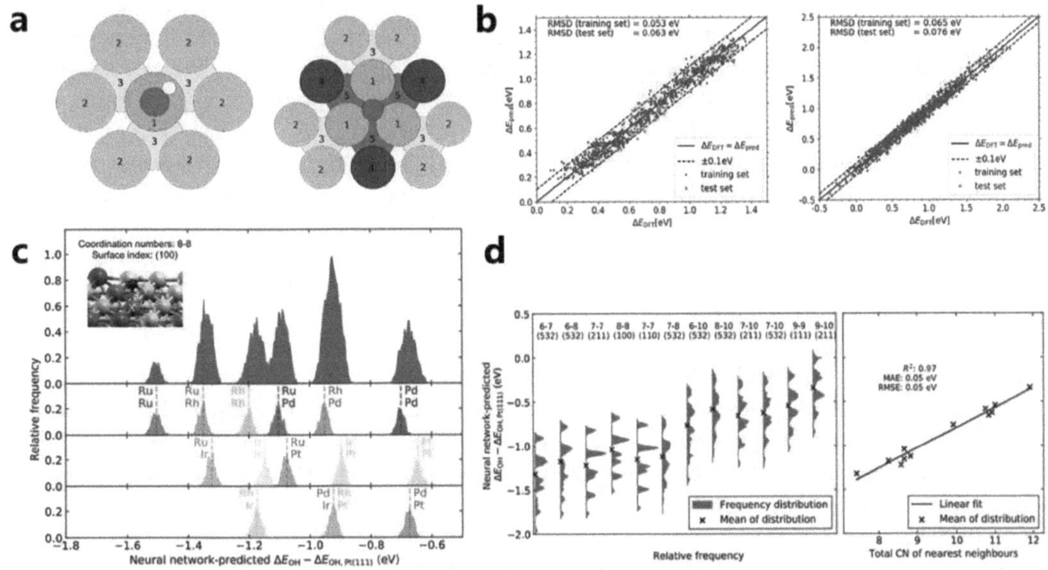

FIGURE 12.5 The machine learning predictions of adsorption energies as well as statistical analysis of the adsorption behaviors on IrPdPtRhRu (111) surface. (a) Parameterization of the surface configurations using nearest neighbors for *OH on the top position and *O on the hollow position. Adapted with permission [35]. Copyright (2019), Elsevier. (b) Plotting of the machine learning predicted adsorption energies against DFT-calculated resluts. Adapted with permission [35]. Copyright (2019), Elsevier. (c) Frequency distribution of *OH adsorption energy resulting from the ligand effect on IrPdPtRhRu (100) surface. The top row shows the overall energy distribution, which is further broken down in terms of the identity of the two active-site atoms. The dotted lines indicate the average values of the corresponding peaks. Adapted with permission [36]. Copyright (2020), Elsevier. (d) Frequency distribution of *OH adsorption energy for each coordination environmnet, in which the mean values correlate linearly with the total CN of nearest neighbors. Adapted with permission [36]. Copyright (2020), Elsevier.

binding site are of less importance. Starting from the second term in the model, each term represents the averaged perturbation for the leading term, indicating that the diverse local environment impacts the active center in a manner of mean-field perturbations in the case of IrPdPtRhRu HEAs.

However, while this simple linear model has demonstrated success, it is limited in its ability to provide physical insights into the adsorption mechanisms on HEA surfaces, which in turn restricts its application for exploring and designing HEA catalysts. Typically, two factors contribute to surface reaction activity: geometric and electronic. For pure metal surfaces, catalysis theories have been well developed, such as the d-band model for the electronic effect and the coordination number model for the geometric effect. However, the diverse local active environment of HEA surfaces makes it challenging to understand the adsorption mechanisms.

Understanding the adsorptions on HEA surfaces presents a key challenge due to the subtle interplay between ligand and coordination effects. The ligand effect arises from the random spatial distribution of different metal elements, while the coordination effect refers to different crystal facets. To disentangle these two effects, Lu et al. [36] developed a high-quality neural network model to predict the adsorption energies of *OH adsorbed on the bridge sites of IrPdPtRhRu with different surfaces. Their model achieves a high accuracy, with a mean absolute error (MAE) and RMSE of 0.094 and 0.117 eV, respectively.

Using this model, the researchers [36] conducted separate investigations into the effects of ligand and coordination by independently varying each factor. They identified changes in adsorption energies, depicted in Figure 12.5c, and reported that the width and shape of the adsorption distributions were similar. For active sites containing identical elements, denoted as X–X, the average adsorption energy followed the order of Ru–Ru<Rh–Rh<Ir–Ir<Pd–Pd<Pt–Pt, which is in accordance with the trend found in their respective pure metal surfaces (Ru<Rh<Ir<Pd<Pt). For active sites consisting of different elements, denoted as X–Y, the resulting adsorption energy was found to fall between that of X–X and Y–Y, indicating that the adsorption energy is contributed to by the individual effects of the binding atoms. These findings shed light on the complex interplay of ligand and coordination effects in HEA surfaces and highlight the utility of machine learning models in predicting adsorption behaviors.

To explore the coordination effect, they first defined the total coordination number (CN) of the nearest neighbor using the equation $\overline{CN}_{neighbours} = \sum_{i}^{nearest\ neighbours} CN_i / CN_{max}$, where CN_{max} is equal to 12, which is the maximum possible CN for the FCC structure. As shown in Figure 12.5d, on the (111) surface the adsorption energies are notably lower compared to those on the pure Pt(111) surface, indicating the ligand effect of the HEA. By tuning the values of $\overline{CN}_{neighbours}$, lower coordinated environments exhibited stronger binding ability with *OH, which is consistent with the trend observed on pure metal surfaces [37]. Therefore, the nearest-neighbor CNs play a significant role in the modification of the adsorption energy.

In the last part of this subsection, we will discuss practical techniques for analyzing the large datasets generated by machine learning models and gaining new insights into the complex correlations and statistical behaviors of adsorption processes on HEA surfaces.

In a recent work by Singh's group [38], they utilized a similar neural network model to predict the adsorption energies of COOH*, CO*, and CHO* adsorbed on the FeCoNiCuMo (111) surface. As shown in Figure 12.6a–c, high accuracy was achieved for all three adsorptions. Based on the model's predictions, the researchers conducted statistical analysis of the large adsorption dataset. As depicted in Figure 12.6d, the top adsorption site contributed the most to the adsorption energy due to the strong metal-carbon bond. Moreover, the COOH* and CHO* intermediates were found to be more favorable to adsorb on the nearest-neighbor sites (2–7) compared to the CO* intermediate. This suggests the possibility of breaking the scaling relation on the FeCoNiCuMo (111) surface during the CO_2RR process.

To gain a deeper understanding of the influence of neighboring atoms on the reaction activity of the central site, Pearson correlation analysis was conducted to estimate the correlation between the neighboring sites and the adsorption energy. As shown in Figure 12.6e, Cu and Mo exhibit the largest positive and negative correlations, respectively, for COOH* and CHO* adsorption. This suggests that the neighbor Cu atom weakens the binding strength of COOH*/CHO* intermediates on the active center, while the neighboring Mo atom enhances the binding between COOH*/CHO* intermediates and the active center. These statistical methods enable us to gain a clearer understanding of adsorption on HEA surfaces, which can help in developing more reasonable and typical structural models for expensive DFT calculations, like transition energy barrier calculations.

12.5 APPLICATIONS OF DESCRIPTOR METHODS IN CATALYSIS OF HEAS

In this section, we will discuss the use of descriptor methods in the study of ORR activity on HEA surfaces. Traditionally, adsorption energies of key intermediates in ORR have been used as reactivity descriptors in heterogenous catalysis. From an electronic interaction perspective, the simple yet efficient d-band theory of chemisorption, pioneered by Hammer and Nørskov, has been successfully applied to understand the reactivity of d-block metals. Subsequently, parameters such as d-band center [39], d-band upper edge [16], and the number of valence d-electrons [40] have been proposed as descriptors for chemisorbed systems. Compared to black-box machine learning-based adsorption predictions, physics-inspired descriptors can provide efficient adsorption energy predictions and more physical insights into bonding formation/breaking at the active

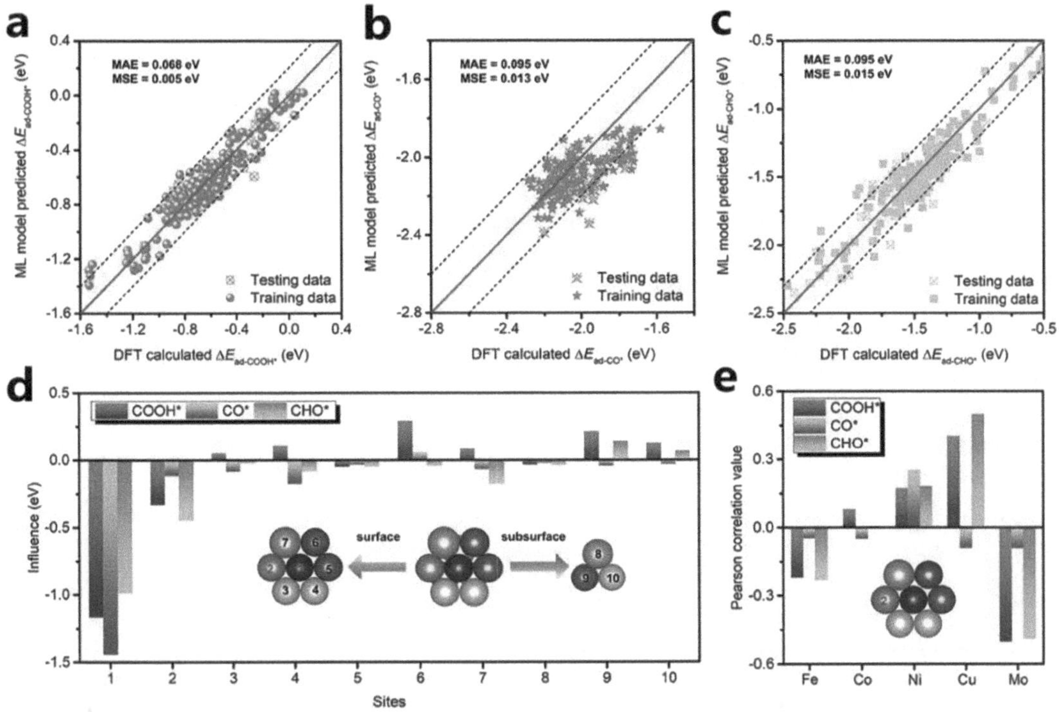

FIGURE 12.6 The adsorption energies of COOH* (a), CO* (b), and CHO* (c) on FeCoNiCuMo (111) surface predicted by neural network modes and plotted against the corresponding DFT-calculated adsorption energy. The dotted lines indicate a deviation of ±0.2 eV. (d) The average influence of each site around the active center, denoted with 1–10 in the insert plot. (e) Pearson correlation analysis for the atoms on active site 2 with regard to the adsorption energy of COOH*, CO*, and CHO*. Adapted with permission [38]. Copyright (2022), American Chemical Society.

site. However, applying the d-band theory directly to HEA catalysis is not feasible due to the distinct adsorbate chemical environment compared to pure metals. Unfortunately, to the best of our knowledge, there are currently no studies on the electronic descriptors for ORR activity on HEA surfaces. In this section, we will introduce recently developed descriptor methods for HEA catalysis, which will be beneficial for a deeper understanding of the electronic descriptors for ORR activity on HEA surfaces.

Gao and co-workers [41,42] generalized Nørskov's d-band theory for bimetallic alloys and HEAs by incorporating three key factors that influence adsorption energies: the valence and electronegativity of surface atoms, the coordination of active sites (i.e., different crystal facets), and the valence of adsorbates. Similar to d-band theory, they found that the contribution of surface d-orbitals to the adsorption energy ΔE_d is proportional to the coupling matrix V_{ad} between adsorbate and surface. They also found that the spatial extension of the surface d-orbital r_d is correlated with ΔE_d according to the relation $\Delta E_d \propto V_{ad}^2 \propto r_d^3/L^7$, where L is the adsorption distance. To estimate r_d they approximated it as the number of valance electrons, including both d- and s-electrons, and they empirically estimated L in terms of Mulliken electronegativity. Finally, they introduced an electronic descriptor in the form of $\psi = S_v^2/\chi^\beta$, where $S_v$is the number of valence electrons and β is a parameter with a value of 1/2 for Ag and Au and 1 for other transition metals, based on the fact that the d-orbital's contribution to the binding strength is much less in Ag and Au than in transition metals.

To predict the adsorption energy, they proposed a linear model $\Delta E = k\psi + b$, and tested it on a wide range of adsorption systems, including CH_x (x=0, 1, 2, 3), CO, COH, CHO, CHOH, COOH,

NH_x (x=0, 1, 2), NNH_2, OH_x (x=0, 1), OOH, OCH_3, P, F, and Cl on various substrates. They also tested the generalizability of their model by introducing extended surfaces, such as low-index surfaces (111), (100), and (110), cavity (111), stepped surfaces with (211), (532), (553), and (711) steps, kinks, surfaces with metal adatoms, and surface alloys with Pt/Pd as host and up to 27 doping elements in the first layer of surface.

As shown in Figure 12.7a, the adsorption energies of *C, $*CH_x$, *CO on various metal surfaces strongly correlate with the proposed electronic descriptor ψ. It is noteworthy that the values of slop k and intercept b are dependent on both the substrate and the adsorbate. To gain insight into the physical significance of these parameters, the authors generalized Nørskov's scaling relations between intermediates [7] by incorporating both electronic and geometric effects. The electronic effect is controlled by the slop k, which is expression as $k = 0.1(X_m - X)/(X_m + 1)$ (where X_m is the maximum number of valence electrons), while the geometric effect is determined by the intercept b with a formula of $b = \lambda\overline{CN} + \theta$ ($\overline{CN}$ is the generalized CN [37]). The pre-factor λ is described by the saturated-bond number of the central atom with $\lambda = 0.2(X + 1)/(X_m + 1)$. The final physics-inspired descriptor is

$$\Delta E = 0.1 \times \frac{X_m - X}{X_m + 1} \times \psi + 0.2 \times \frac{X + 1}{X_m + 1} \times \overline{CN} + \theta \quad (12.2)$$

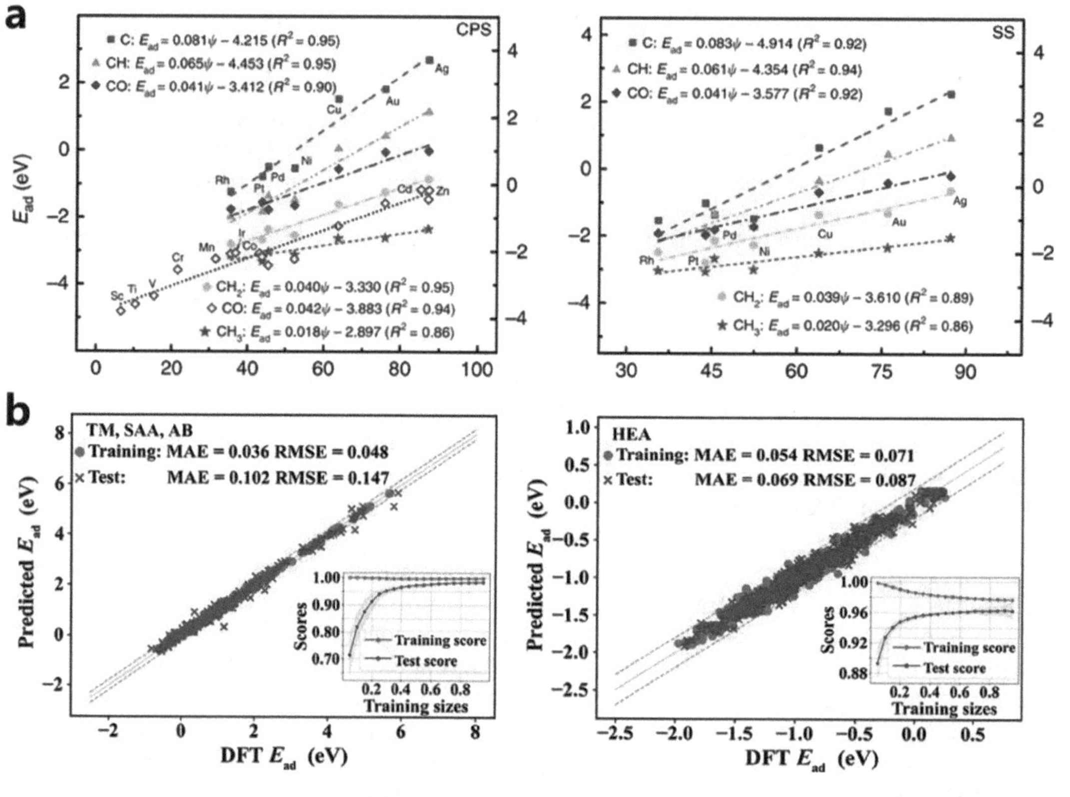

FIGURE 12.7 (a) Adsorption energies of *C, CO*, *CH, *CH2 and *CH3 on transition metals as a function of the electronic descriptor ψ. In each panel, the adsorption energies of each adsorbate are linearly correlated with the electronic descriptor ψ. Adapted with permission [42]. Copyright (2020), Springer Nature. (b) The prediction abilities for the adsorption energies CO* on transition-metal surfaces (left) and on IrPdPtRhRu HEA surface using the XGBoost regression machine learning model combined with the electronic descriptor ψ. Adapted with permission [41]. Copyright (2022), Royal Society of Chemistry.

where θ is the only parameter that requires to be determined and all the rest parameters have clear physical meaning and can be readily accessible.

They then generalized this model to the adsorptions on HEA surfaces [41] by modifying ψ with the expression of $\psi = \left(\prod_{i=1}^{N} S_{v_i}\right)^{2/N} / \left(\prod_{i=1}^{N} \chi_i\right)^{1/N}$, where N is the number of the atoms at active centers and S_{v_i} and χ_i are the i-th atom's valance number and electronegativity, respectively. They took the proposed descriptors as input features for the machine learning model, as shown in Figure 12.7b, extremely high accuracy has been achieved (MAE = 0.102 eV). With this optimized model, they further transfer it to predict the adsorption energies on HEA surfaces. The RMSE of the predicted adsorption energies on HEA surfaces is 0.149 eV, indicating a high transferability of the model.

12.6 PERSPECTIVES ON MODELING ORR ON HEA SURFACES

HEAs have emerged as promising candidates for heterogeneous catalysis due to their ability to create tailorable active sites by exploring the vast multidimensional composition space. While experimental work in this field is growing rapidly and outstanding catalytic performance has been observed, theoretical modeling of chemical reactions on HEA surfaces lags far behind. Given the enormous multidimensional composition space, only a tiny fraction of HEAs have been studied so far. To accelerate the exploration of HEA catalysts, the rational design of HEA catalysts requires the development of interpretable machine learning models and physics-inspired reaction descriptors. Furthermore, the subtle interplay of atomic-scale factors at the interface between adsorbates and HEA surfaces, ranging from HEA surface local morphology to the presence of high adsorbate coverage, requires greater attention. To construct clearer structure-activity relationships, more advanced nonlinear machine learning models, such as deep learning and graph neural networks, should be utilized to account for atomistic configurations comprising diverse adsorbates, adsorption locations, coordination environments, and HEA surface morphologies.

To gain a better understanding of the mechanism behind ORR activity on HEA surfaces, it is important to focus on the chemical nature of the interaction between the key adsorbates (including *O, *OH, and COOH*) and the HEA surface, whether from an electronic interaction or reaction kinetic theory perspective. To achieve this, there are central questions that need to be addressed: How does the Sabatier principle apply to HEA surfaces, and can HEAs overcome the limitations of the Sabatier principle? Answering these questions requires obtaining more information on the scaling relations of HEAs, which can be achieved through a combination of high-throughput DFT calculations, interpretable machine learning models, and physics-inspired descriptors [43].

REFERENCES

1. T. Löffler, H. Meyer, A. Savan, P. Wilde, A. Garzón Manjón, Y.-T. Chen, E. Ventosa, C. Scheu, A. Ludwig, W. Schuhmann, Discovery of a multinary noble metal-free oxygen reduction catalyst. *Advanced Energy Materials.* 8 (34) (**2018**) 1802269.
2. A. G. Manjón, T. Löffler, M. Meischein, H. Meyer, J. Lim, V. Strotkötter, W. Schuhmann, A. Ludwig, C. Scheu, Sputter deposition of highly active complex solid solution electrocatalysts into an ionic liquid library: effect of structure and composition on oxygen reduction activity. *Nanoscale. 12* (46) (**2020**) 23570–23577.
3. T. Löffler, F. Waag, B. Gökce, A. Ludwig, S. Barcikowski, W. Schuhmann, comparing the activity of complex solid solution electrocatalysts using inflection points of voltammetric activity curves as activity descriptors. *ACS Catalysis. 11* (3) (**2021**) 1014–1023.
4. T. Löffler, A. Savan, H. Meyer, M. Meischein, V. Strotkötter, A. Ludwig,W. Schuhmann, Design of complex solid-solution electrocatalysts by correlating configuration, adsorption energy distribution patterns, and activity curves. *Angewandte Chemie International Edition. 59* (14) (**2020**) 5844–5850.

5. Y. Yao, Z. Huang, T. Li, H. Wang, Y. Liu, H. S. Stein, Y. Mao, J. Gao, M. Jiao, Q. Dong, J. Dai, P. Xie, H. Xie, S. D. Lacey, I. Takeuchi, J. M. Gregoire, R. Jiang, C. Wang, A. D. Taylor, R. Shahbazian-Yassar, L. Hu, High-throughput combinatorial synthesis of multimetallic nanoclusters. *Proceedings of the National Academy of Sciences. 117* (12) (**2020**) 6316–6322.
6. T. A. A. Batchelor, T. Löffler, B. Xiao, O. A. Krysiak, V. Strotkötter, J. K. Pedersen, C. M. Clausen, A. Savan, Y. Li, W. Schuhmann, J. Rossmeisl, A. Ludwig, Complex-solid-solution electrocatalyst discovery by computational prediction and high-throughput experimentation. *Angewandte Chemie International Edition. 60* (13) (**2021**) 6932–6937.
7. F. Abild-Pedersen, J. Greeley, F. Studt, J. Rossmeisl, T. R. Munter, P. G. Moses, E. Skúlason, T. Bligaard, J. K. Nørskov, Scaling properties of adsorption energies for hydrogen-containing molecules on transition-metal surfaces. *Physical Review Letters. 99* (1) (**2007**) 016105.
8. J. K. Nørskov, T. Bligaard, J. Rossmeisl, C. H. Christensen, Towards the computational design of solid catalysts. *Nature Chemistry. 1* (1) (**2009**) 37–46.
9. J. Greeley, I. E. L. Stephens, A. S. Bondarenko, T. P. Johansson, H. A. Hansen, T. F. Jaramillo, J. Rossmeisl, I. Chorkendorff, J. K. Nørskov, Alloys of platinum and early transition metals as oxygen reduction electrocatalysts. *Nature Chemistry. 1* (7) (**2009**) 552–556.
10. J. K. Nørskov, F. Abild-Pedersen, F. Studt, T. Bligaard, Density functional theory in surface chemistry and catalysis. *Proceedings of the National Academy of Sciences. 108* (3) (**2011**) 937–943.
11. A. Kulkarni, S. Siahrostami, A. Patel,J. K. Nørskov, Understanding catalytic activity trends in the oxygen reduction reaction. *Chemical Reviews. 118* (5) (**2018**) 2302–2312.
12. I. E. L. Stephens, A. S. Bondarenko, U. Grønbjerg, J. Rossmeisl, I. Chorkendorff, Understanding the electrocatalysis of oxygen reduction on platinum and its alloys. *Energy & Environmental Science. 5* (5) (**2012**) 6744–6762.
13. A. J. Medford, A. Vojvodic, J. S. Hummelshøj, J. Voss, F. Abild-Pedersen, F. Studt, T. Bligaard, A. Nilsson, J. K. Nørskov, From the Sabatier principle to a predictive theory of transition-metal heterogeneous catalysis. *Journal of Catalysis. 328* (**2015**) 36–42.
14. F. Calle-Vallejo, D. Loffreda, M. T. M. Koper, P. Sautet, Introducing structural sensitivity into adsorption-energy scaling relations by means of coordination numbers. *Nature Chemistry. 7* (5) (**2015**) 403–410.
15. V. Viswanathan, H. A. Hansen, J. Rossmeisl, J. K. Nørskov, Universality in Oxygen Reduction Electrocatalysis on Metal Surfaces. *ACS Catalysis. 2* (8) (**2012**) 1654–1660.
16. H. Xin, A. Vojvodic, J. Voss, J. K. Nørskov, F. Abild-Pedersen, Effects of *d*-band shape on the surface reactivity of transition-metal alloys. *Physical Review B. 89* (11) (**2014**) 115114.
17. Z.-J. Zhao, S. Liu, S. Zha, D. Cheng, F. Studt, G. Henkelman, J. Gong, Theory-guided design of catalytic materials using scaling relationships and reactivity descriptors. *Nature Reviews Materials. 4* (12) (**2019**) 792–804.
18. N. B. Halck, V. Petrykin, P. Krtil,J. Rossmeisl, Beyond the volcano limitations in electrocatalysis - oxygen evolution reaction. *Physical Chemistry Chemical Physics. 16* (27) (**2014**) 13682–13688.
19. T. Bligaard, J. K. Nørskov, S. Dahl, J. Matthiesen, C. H. Christensen, J. Sehested, The Brønsted-Evans-Polanyi relation and the volcano curve in heterogeneous catalysis. *Journal of Catalysis. 224* (1) (**2004**) 206–217.
20. J. K. Pedersen, T. A. A. Batchelor, D. Yan, L. E. J. Skjegstad,J. Rossmeisl, Surface electrocatalysis on high-entropy alloys. *Current Opinion in Electrochemistry. 26* (**2021**) 100651.
21. E. Prodan, W. Kohn, Nearsightedness of electronic matter. *Proceedings of the National Academy of Sciences. 102* (33) (**2005**) 11635–11638.
22. W. A. Saidi, Emergence of local scaling relations in adsorption energies on high-entropy alloys. *npj Computational Materials. 8* (1) (**2022**) 86.
23. S. Guo, Q. Hu, C. Ng,C. T. Liu, More than entropy in high-entropy alloys: Forming solid solutions or amorphous phase. *Intermetallics. 41* (**2013**) 96–103.
24. Y. F. Ye, Q. Wang, J. Lu, C. T. Liu,Y. Yang, High-entropy alloy: Challenges and prospects. *Materials Today. 19* (6) (**2016**) 349–362.
25. S. Guo, Phase selection rules for cast high entropy alloys: an overview. *Materials Science and Technology. 31* (10) (**2015**) 1223–1230.
26. S. Guo, C. T. Liu, Phase stability in high entropy alloys: Formation of solid-solution phase or amorphous phase. *Progress in Natural Science: Materials International. 21* (6) (**2011**) 433–446.
27. C. Wen, Y. Zhang, C. Wang, D. Xue, Y. Bai, S. Antonov, L. Dai, T. Lookman,Y. Su, Machine learning assisted design of high entropy alloys with desired property. *Acta Materialia. 170* (**2019**) 109–117.

28. Y. Li, W. Guo, Machine-learning model for predicting phase formations of high-entropy alloys. *Physical Review Materials. 3* (9) (**2019**) 095005.
29. Z. Zhou, Y. Zhou, Q. He, Z. Ding, F. Li,Y. Yang, Machine learning guided appraisal and exploration of phase design for high entropy alloys. *npj Computational Materials. 5* (1) (**2019**) 128.
30. S. Feng, H. Fu, H. Zhou, Y. Wu, Z. Lu,H. Dong, A general and transferable deep learning framework for predicting phase formation in materials. *npj Computational Materials. 7* (1) (**2021**) 10.
31. Y. Sun, Z. Lu, X. Liu, Q. Du, H. Xie, J. Lv, R. Song, Y. Wu, H. Wang, S. Jiang, Z. Lu, Prediction of Ti-Zr-Nb-Ta high-entropy alloys with desirable hardness by combining machine learning and experimental data. *Applied Physics Letters. 119* (20) (**2021**) 201905.
32. W. Huang, P. Martin,H. L. Zhuang, Machine-learning phase prediction of high-entropy alloys. *Acta Materialia. 169* (**2019**) 225–236.
33. J.-C. Ren, J. Zhou, C. J. Butch, Z. Ding, S. Li, Y. Zhao,W. Liu, Predicting single-phase solid solutions in as-sputtered high entropy alloys: High-throughput screening with machine-learning model. *Journal of Materials Science & Technology. 138* (**2023**) 70–79.
34. L. Breiman, J. H. Friedman, R. A. Olshen, C. J. Stone In *Classification and Regression Trees*, **1984**. https://doi.org/10.1201/9781315139470
35. T. A. A. Batchelor, J. K. Pedersen, S. H. Winther, I. E. Castelli, K. W. Jacobsen, J. Rossmeisl, High-entropy alloys as a discovery platform for electrocatalysis. *Joule. 3* (3) (**2019**) 834–845.
36. Z. Lu, Z. W. Chen,C. V. Singh, Neural network-assisted development of high-entropy alloy catalysts: Decoupling ligand and coordination effects. *Matter. 3* (4) (**2020**) 1318–1333.
37. S. Cheon, T.-H. Kim, S.-H. Lee, H. W. Yeom, Chiral solitons in a coupled double Peierls chain. *Science. 350* (6257) (**2015**) 182–185.
38. Z. W. Chen, Z. Gariepy, L. Chen, X. Yao, A. Anand, S.-J. Liu, C. G. Tetsassi Feugmo, I. Tamblyn, C. V. Singh, Machine-learning-driven high-entropy alloy catalyst discovery to circumvent the scaling relation for CO2 reduction reaction. *ACS Catalysis. 12* (24) (**2022**) 14864–14871.
39. B. Hammer, J. K. Nørskov, Electronic factors determining the reactivity of metal surfaces. *Surface Science. 343* (3) (**1995**) 211–220.
40. F. Calle-Vallejo, N. Inoglu, H.-Y. Su, I. Man, M. Koper, J. Kitchin, J. Rossmeisl, Number of outer electrons as descriptor for adsorption processes on transition metals and their oxides. *Chemical science. 4* (**2013**) 1245.
41. X. Li, B. Li, Z. Yang, Z. Chen, W. Gao, Q. Jiang, A transferable machine-learning scheme from pure metals to alloys for predicting adsorption energies. *Journal of Materials Chemistry A. 10* (2) (**2022**) 872–880.
42. W. Gao, Y. Chen, B. Li, S.-P. Liu, X. Liu,Q. Jiang, Determining the adsorption energies of small molecules with the intrinsic properties of adsorbates and substrates. *Nature Communications. 11* (1) (**2020**) 1196.
43. T. Chen, C. Guestrin, XGBoost: A Scalable Tree Boosting System. In *Proceedings of the 22nd ACM SIGKDD International Conference on Knowledge Discovery and Data Mining*, Association for Computing Machinery: San Francisco, California, USA, **2016**; pp 785–794.

13 High Entropy Materials for Oxygen Reduction Reaction

Likai Wang, Xuzi Cong, Yuanwei Ma, and Zhongfang Li

13.1 INTRODUCTION

To meet the growing energy demand while reducing reliance on fossil fuels, more energy storage and efficient energy conversion devices are needed to develop renewable energy.[1,2] Fuel cells are considered one of the most promising technologies and have attracted significant attention due to their favorable properties, such as high power density, high efficiency, and friendly environment.[3–5] To improve the performance and overall efficiency of fuel cells, high-performance catalysts are essential to be explored and designed to reduce oxygen with four-electron transfer pathways and improve the efficiency and durability of PEMFC.[6–9] Researchers are exploring large amounts of new functional materials with tailored properties to promote the four-electron transfer pathway and optimize the performance fuel cell systems.[10–16] The research of bimetallic alloys has attracted wide attention due to the high activity and low loading of Pt, such as PtCo, PtNi, and PtCu.[7,8,17–19] However, transition metals in binary alloys are prone to corrosion in harsh catalytic environments.[20–23] As a new class of functional materials, HEMs which are composed of multi-components have attracted considerable attention due to their unexpected properties.[24–28] Hereby, we will introduce the basic information about HEMs, such as concept, structure, and the application of high entropy materials with different structures for ORR.

13.1.1 Introduction of HEMs

HEMs-based catalysts with a wide range of compositional modulations and complex surfaces could rationally tune the electronic and geometric structures to a larger degree, which facilitates the optimization of catalytic activity.[27,29–33] Alloying is one of the most effective methods to fabricate multicomponent systems with tailored properties, in which multi-element atoms are randomly distributed in an ordered pattern for intermetallics.

In 2004, Cantor and coworkers explored the single-phase face-centered cubic (FCC) solid solution of $Fe_{20}Cr_{20}Mn_{20}Ni_{20}Co_{20}$ alloy which consists of five components, and they proposed that the combination of multi-elements in equimolar proportions could form multicomponent alloys with improved mechanical performance compared to conventional binary alloys.[31] Meanwhile, Yeh et al. also reported similar research and first introduced the new concept of high entropy alloys (HEAs) which refer to an alloy containing five or more metal elements in an equal molar ratio, and the concentration of each element ranging from 5 to 35 at%.[24] With the development of nanoscience and nanotechnology, the controlled synthesis of nanoscale HEMs could be achieved via various nontraditional procedures and the size, shape, and configuration of HEMs have been precisely modulated.[32,34–39] HEMs-based materials exhibited highly efficient catalytic activity for electrocatalysis reactions. According to Boltzmann's formula, the entropy of the system (ΔS_{mix}) during the formation of the alloy system can be calculated as follows:[40,41]

$$\Delta S_{\text{mix}} = -k_B \ln = -R\sum_{i=1}^{n} C_i \ln c_i \tag{13.1}$$

DOI: 10.1201/9781003391388-13

where k_B is the Boltzmann constant and ω is the probability of a given state, R is the molar gas constant, c_i is the molar fraction of the i_{th} component, n is the number of element types. ΔS_{mix} of alloy systems can be maximized by mixing multiple elements with an equal molar ratio, $c_{i=}$ $1/n$.

$$\Delta S_{\text{mix}} = -R\left(\frac{1}{n}\ln\frac{1}{n}+\frac{1}{n}\ln\frac{1}{n}+\ldots+\frac{1}{n}\ln\frac{1}{n}\right) = R\ln n \tag{13.2}$$

Based on the formula (equation 13.2), ΔS_{mix} can be expressed as the function of the number of alloy components. When the number of alloy components reaches 5 or more, the ΔS_{mix} reaches a large value ($\Delta S_{\text{mix}} \geq 1.61R$). For the entire multicomponent alloy system, the mixed Gibbs free energy (ΔG_{mix}) can be expressed as:

$$\Delta G_{\text{mix}} = \Delta H_{\text{mix}} - T\Delta S_{\text{mix}} \tag{13.3}$$

According to the above fomula, if the mixed enthalpy (ΔH_{mix}) keeps unchanged, enlarging the ΔS_{mix} of alloy system can decrease their mixed Gibbs free energy and contribute to enhance higher stability of the alloy system.[28,42]

Single-phase solid solution HEAs could form various crystal structures depending on their composition, such as FCC, body-centered cubic (BCC), and hexagonal close-packed (HCP) structures.[41,42] Different atoms could randomly occupy lattice positions and form a single-phase solid solution. The most typical structures of HEA nanocatalysts reported are FCC, demonstrating good mechanical stability and catalytic activity.[25,35,41–43] While BCC and HCP structures are rarely reported. Based on the structure of HEAs, researchers have identified four core effects, namely the high entropy effect, lattice distortion effect, sluggish diffusion effect, and "cocktail" effect. The high entropy effect is derived from an earlier concept of high entropy alloys (HEAs), indicating the contribution of mixed entropy to the formation of alloy phase.[35,40,44] Based on the Gibbs phase rule, the number of phases (P) in a given alloy system at constant pressure can be expressed as:

$$P = C + 1 - F \tag{13.4}$$

Where C is the number of elements in the alloy system, F is the degree of freedom for the alloy system. A quinary alloy system could form six equilibrium phases by the prediction of classical Gibbs phase rule (equation 13.4). However, from the free energy expression (equation 13.3), if ΔS_{mix} of quinary alloy system is large enough to compensate for the effect of ΔH_{mix}, the quinary alloy system tend to form a uniformly mixed single-phase solid solution, especially at sufficiently high temperatures. As a result, high mixed entropy promotes to form a single phase, and the number of phases in high entropy alloys can be significantly reduced.

Lattice distortion effect is caused by atoms occupying deviation due to the size difference between different atoms in high entropy materials, where the lattice points shift from the ideal position and result in severe lattice distortion.[34] The lattice distortion could lead to the uneven level of the same atomic plane, which causes the X-ray on the uneven crystal plane to produce obvious Bragg scattering, weakening and broadening XRD diffraction peak of HEAs.[45] The lattice distortion effect not only greatly increases the hardness of HEMs but also dramatically changes their physical and chemical properties. Sluggish diffusion effect of HEMs is closely related to the lattice distortion, which leads to the increase in the diffusion activation energy barrier of atoms in the lattice, decreasing the effective diffusion rate of atoms and promoting the dynamics stability of HEMs.[28,46,47] Yao and coworkers simulated the self-diffusion coefficients of Ru atom in binary $Ru_{0.25}Ni_{0.75}$ alloy and quinary $Ru_{0.25}Rh_{0.25}Co_{0.2}Ni_{0.2}Ir_{0.1}$ alloy by Monte Carlo-Molecular Dynamics and found that the diffusion coefficients of Ru in quinary alloys are lower than that of in the binary alloys, indicating that sluggish diffusion effect improves the stability of HEMs. Cocktail effect of HEAs represents the synergistic effect due to the interaction of various elemental components. The combination of mixed elements may bring about some unexpected properties by tuning their electronic/geometric structures to enhance their performance. The cocktail effect broadly explains the synergetic effect of each element in HEMs, but its specific physical significance is not clear.[40,46] It is acknowledged that lattice distortion optimizes the electronic structure of the active site, and high entropy effect and sluggish diffusion effect improve the thermodynamic and kinetic stability of HEMs catalyst.

13.1.2 Significance of ORR and Reaction Mechanism

It is very significant to understand the ORR reaction mechanism and confirm the catalytic active sites. Pt-based catalyst system is the mainstream for ORR in an acidic medium, while many non-noble metal catalysts can be stable enough to exert their high catalytic activity under alkaline conditions. Compared with the hydrogen oxidation reaction (HOR) at the anode of PEMFC, O_2 reduction is a multi-electron process that involves multiple electron transfers and the generation of intermediate products. Electrocatalysts are needed to overcome its slow kinetics and higher overpotential.[6,48] The differences between ORR mechanisms are based on the number of proton-coupled electron transfer steps that precede the O-O bond break step. Typically, the direct four-electron pathway in acidic media was from O_2 to H_2O: $O_2+H^++4e^-=2H_2O$, while the two-electron pathway was from O_2 to H_2O_2, followed by $2e^-$ reduction to change H_2O_2 to H_2O: $O_2+2H^++2e^-=H_2O_2$; $H_2O_2+2H^++2e^-=2H_2O$. It is known that the Tafel plot is about 60 mV/dec and the reaction order is 1.5 with respect to H* activity at high potential. Three different adsorbed intermediate species are considered to be *OOH, *O, and *OH, where *OOH is formed first, and then the O-O bond is being broken.[6,48] By comparison, the Tafel slope is about 120 mV/dec and the reaction order is 1 at low potential, while the ORR mechanism is a dissociation mechanism where the O–O bond is broken before *OH is formed.

The combination of experiment and theoretical calculation provides important clues for the structure engineering of HEMs-based catalysts. It is well acknowledged that the adsorption energy of key intermediates in ORR should not be too strong or too weak.[49,50] In particular, HEAs and high entropy alloy oxides (HEOs) have been widely applied to ORR. The surface, interface, and defects of HEMs are investigated by various advanced characterization techniques at the atomic level, indicating the link between regulation of the geometric and electronic structure and modification of the ORR performance.

13.2 THE APPLICATION OF HEAS IN ORR

IIIORR at the cathode is a critical process due to its complex reaction pathways and slow electron transfer kinetics. High-performance cathode catalysts are needed to accelerate the reaction rate. Over the past few decades, extensive research efforts have been devoted to exploring HEM electrocatalysts. Nanostructured HEAs are the most prevalent and valid materials for ORR as shown in Figure 13.1.[29,30,38,51–53] We review the reported HEAs catalysts for ORR.

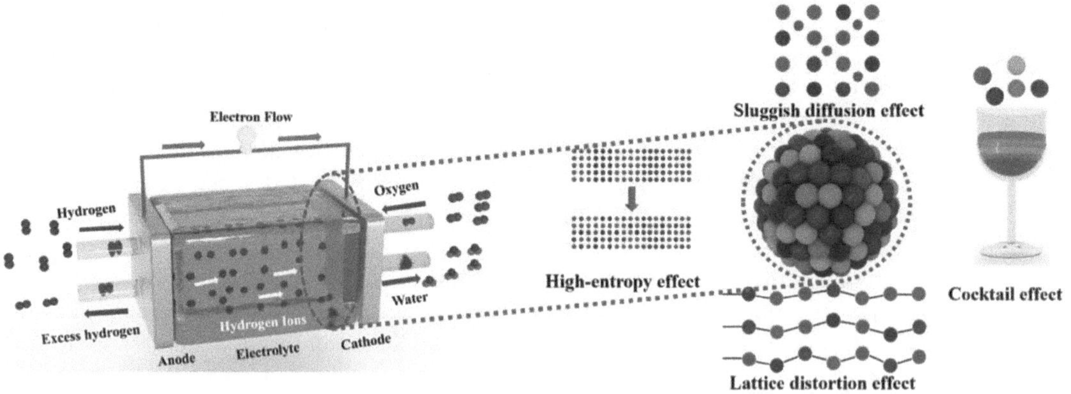

FIGURE 13.1 High entropy materials for ORR.

13.2.1 HEAs Nanoparticles for ORR

In 2017, Schuhmann group suggested quaternary PdAuAgTi thin film system could be applied as a candidate for ORR electrocatalysts by using SDC.[54] They first reported the multinary Cr–Mn–Fe–Co–Ni alloy nanoparticles could be used to catalyze the ORR, demonstrating high-performance activity toward ORR close to that of Pt/C. Related quaternary alloys exhibited obviously decreased ORR activity compared with that of Cr–Mn–Fe–Co–Ni alloy nanoparticles, implying that this might be attributed to a synergistic effect of all five elements.[29] Rossmeisl et al. proposed theoretical research on HEAs as ORR electrocatalysts using density functional theory (DFT) calculations. The adsorption energies of *OH and *O on a random subset of available binding sites were modeled on IrPdPtRhRu HEAs surface. They established a dataset of adsorption energy and a function model, which could be used to predict the catalytic activity and further optimize the HEA composition. Compared with the overpotential of pure Pt (111), the results confirmed that $Ir_{0.2}Pd_{0.32}Pt_{0.093}Rh_{0.196}Ru_{0.289}$ exhibited a reduction in overpotential of ~40 mV.[55]

Yao et al. reported the high-throughput synthesis of a series of ultrafine and alloyed multimetallic nanoclusters supported on carbon supports by thermal shock heating. The series of HEAs displayed similar size and dispersed nanoparticles, and each element was well distributed throughout the HEA nanoparticles. Meanwhile, they investigated electrochemical analysis by a conventional rotating disk electrode (RDE), PtPdRhNi and PtPdFeCoNi exhibited highly efficient activity for ORR, the overall ORR electron number was calculated to be 3.9~4.0. Linear sweep voltammograms (LSV) curves of HEAs catalysts demonstrate higher limiting current densities, slightly more positive onset potentials, and half-wave potentials than those of commercial Pt/C. Tafel slopes of PtPdRhNi, PtPdFeCoNi, and commercial Pt/C are about 32 mV/decade, 31 mV/decade, and 37 mV/decade, respectively. It is further confirmed that PtPdRhNi, PtPdFeCoNi, and commercial Pt/C have similar reaction mechanisms in ORR. Additionally, the stability was also tested by chronoamperometry at 0.6 V (vs. RHE). The current density of PtPdRhNi and PtPdFeCoNi decreased by 36% and 29% after operation for 15 h, respectively. However, the current density of Pt decreased by 39.1% upon 15 h operation, implying that HEAs catalysts have better duration than pure single metal.[53]

Skrabalak et al. also developed a novel synthetic route to overcome the challenge that limited the phase separation in the colloidal route due to the different rates of metal precursor.[56,57] As shown in Figure 13.2, a series of monodisperse HEAs, including quinary, senary, and septenary phases, were achieved by the conversion of core@shell nanoparticles. Their composition can be precisely controlled by modulating the ratio between the metal precursors and the seeds. The ORR activities of quinary HEAs system were tested in O_2-saturated 0.1 M KOH solution, suggesting that the PdCuPtNiFe and PdCuPtNiCo NPs exhibited highly specific and mass activities compared with other samples and commercial Pt/C.

To develop an efficient, stable, and earth-abundant catalyst toward ORR, Xia group reported a facile and feasible procedure to fabricate uniform and nanoscaled PtFeCoNiCu HEA nanoparticles as electrocatalysts for the ORR by using an impregnation reduction method.[58] As shown in Figure 13.3, PtFeCoNiCu HEA demonstrated high ORR activity with the Pt mass activity of 1.738 A mg_{Pt}^{-1} at the potential of 0.9 V, which is much higher than that of Pt/C (0.116 A mg_{Pt}^{-1}). It also had excellent stability with only negative-shifted (3 mV) after 100,000 cycles of accelerated durability test. It is confirmed that transition metals can be well alloyed with Pt without obvious phase separation as shown by TEM and X-ray-based techniques.

Structural engineering of HEA nanoparticles could also increase the intrinsic activity of each active site on the catalyst surface, and it has been shown to be an effective method to improve ORR activity and durability. Luo group also explorated a simple route to achieve the construction of ordered HEA NPs supported by a novel 2D nitrogen-doped mesoporous carbon (mNC) nanosheets. The OHEA-mNC electrocatalyst demonstrated high-performance ORR activity and durability compared with those of disordered HEAs and commercial Pt/C, which was attributed to the fine

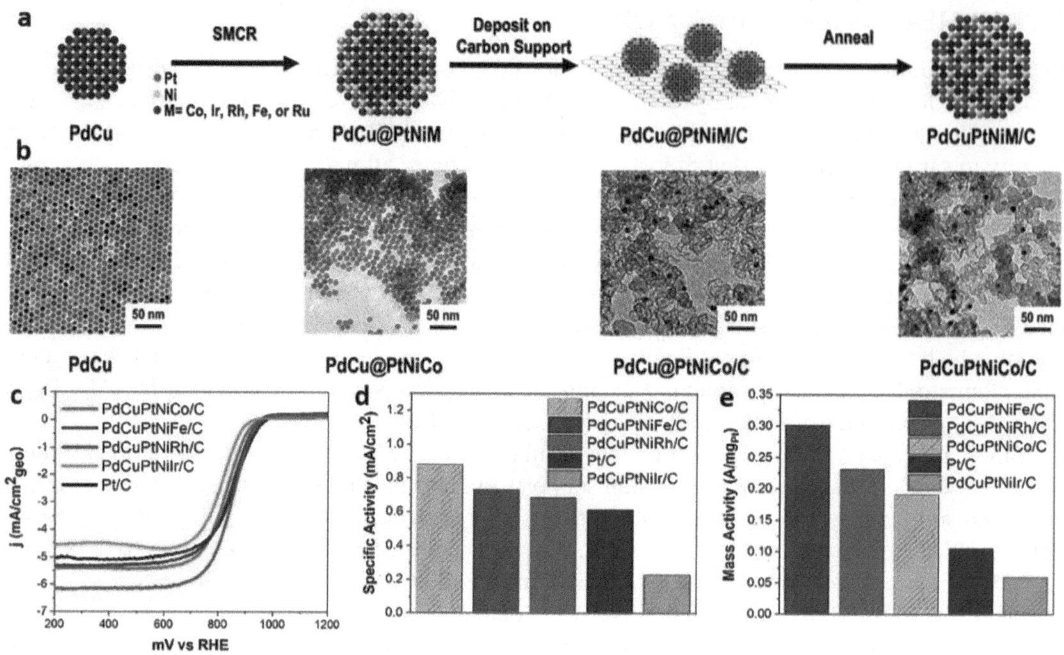

FIGURE 13.2 (a) Scheme illustration and (b) TEM images of HEA NPs, (c) the ORR activity of HEA NPs. Adapted with permission [56]. Copyright (2022) American Chemistry Society.

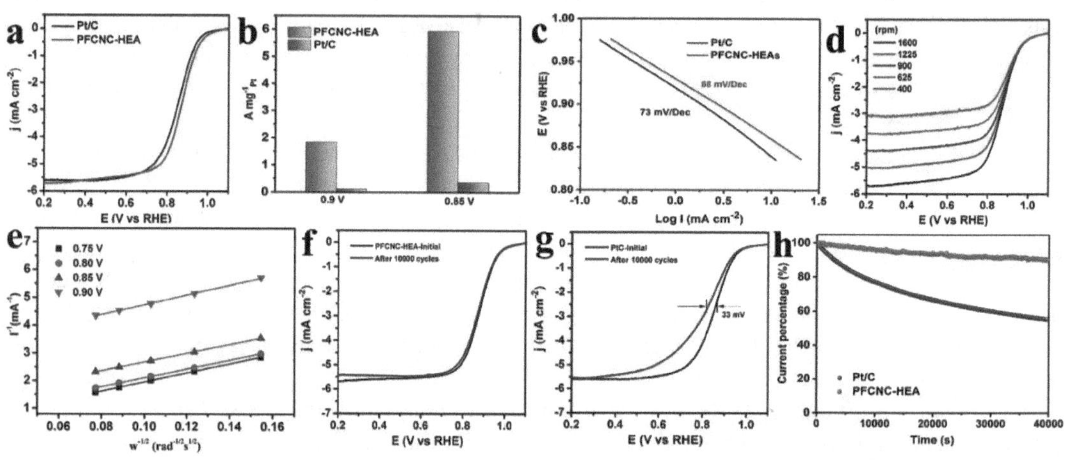

FIGURE 13.3 (a–e) The ORR activity of PFCNC-HEA and Pt/C. (f-h) the corresponding durability of PFCNC-HEA and Pt/C. Adapted with permission [58]. Copyright (2023) Elsevier.

spatial location and regulated order structure. According to the theoretical calculations, the ordered HEA NPs displayed the clear task partition for each metal site and the high structural stability to enhance their electocatalytical activity, further confirming that ordered HEA NPs have a promising potential for ORR catalysts.[59]

Due to the limit of high-temperature procedure, Yu et al. also synthesized a series of PtPdFeCoNi HEA nanoparticles by using high-temperature injection method, where the reductant (L-AA) and metal salts were injected into the oleylamine (OAm) solution at 280°C, PtPdFeCoNi HEA nanoparticles with a particle size of about 12±4 nm were obtained. Each element was also well distributed through the entire PtPdFeCoNi nanoparticles, no phase segregation was observed due to the ultrafast co-reduced procedure. Meanwhile, the ECSA area of PtPdFeCoNi nanoparticles was calculated to be about 68.6 $m^2\ g_{Pt}^{-1}$ higher than that of commercial Pt/C (53.9 $m^2\ g_{Pt}^{-1}$), indicating that it has more active sites for catalysis. PtPdFeCoNi nanoparticles perform a half-wave potential ($E_{1/2}$) of 0.920 V vs. RHE more positive than that of Pt/C (0.870 V vs. RHE), and the mass activity of PtPdFeCoNi nanoparticles was about 1.23 A mg_{Pt}^{-1}, which also outperforms the U.S. Department of Energy target in 2025 (0.44 A mg_{Pt}^{-1}). Furthermore, the high stability of PtPdFeCoNi nanoparticles was also identified by only 6 mV negative-shifted of $E_{1/2}$ after 50,000 cycles of accelerated durability test.[52]

13.2.2 Nanoporous HEAs for ORR

With the exploration of the preparation method, more nanostructured HEAs can be achieved to improve their performance. Among them, nanoporous metals have also shown a promising potential for catalysis and energy storage due to their large surface area and abundant active sites. Qiu and coworkers reported a facile and scalable route to obtain a series of nanoporous high entropy alloys (np-HEAs) by top-down dealloying strategy, where AlNiCuPtPdAu, AlNiCuPtPdAuCoFe, and all-non-noble metal AlNiCuMoCoFe np-HEAs with precisely controlled composition were synthesized without any surfactant.[60] The dealloyed np-HEAs showed a homogeneous nanoporous structure with alloy ligaments and dark nanopores, which could increase the surface area and facilitate the mass transfer for electrochemical catalysis reaction. From the high angle annular dark-field (HAADF)-STEM EDS mapping of np-HEA, each element is well distributed, and there is no apparent phase separation or elemental segregation. For the ORR performance, the LSV curves of np-HEAs displayed a positively shifted $E_{1/2}$ (~0.9 V vs. RHE) compared with that of Pt/C (~0.82 V) in oxygen-saturated 0.1 M $HClO_4$ aqueous solution. The Pt mass activity of np-HEAs was about 2.24 A mg_{Pt}^{-1}, which was 10 times that of Pt/C (~0.22 A mg_{Pt}^{-1}). Additionally, the activity of np-HEA remained ~92.5% after 100 000 cycles of accelerated durability test, confirming that np-HEA demonstrated long-term durability compared with Pt/C.

To further decrease the Pt content in np-HEAs, Qiu group employed the top-down chemical dealloying method to fabricate a series of nanoporous Al–Cu–Ni–Pt–X (X: Mn, Co, Mo, Pd, Au, etc.) with different compositions, the composition effect of np-HEA was investigated in detail.[51] As shown in Figure 13.4, the LSV curves of all as-prepared samples and Pt/C were investigated in oxygen-saturated 0.1 M $HClO_4$ solution, and the np-HEAs (Al–Cu–Ni–Pt–Mn) system demonstrated highly efficient ORR activity. Compared with these samples, the np-AlCuNiPtMn catalyst displayed the best ORR activity with an $E_{1/2}$ of ~0.945 V. The Pt mass activity of the np-AlCuNiPtMn ECSA is calculated to be about 3.5 A mg_{Pt}^{-1} which is much higher than that of Pt/C (0.22 A mg_{Pt}^{-1}), confirming the synergistic effect by merging Al, Cu, Ni, Mn, and Pt into one single phase. Moreover, as depicted in Figure 13.4g and h, the ORR activities of these np-HEAs catalysts were also tested in 0.1 M KOH aqueous solution. The np-AlCuNiPtMn sample also displayed the best ORR activity with an $E_{1/2}$ of 0.954 V. Meanwhile, the half-wave potential of the np-AlCuNiPtMn sample hardly had negative-shift after 30,000 cycles of accelerated durability test, indicating the long-term stability and durability. According to first principles simulations, the co-existence of Cu and Ni in the subsurface layer of np-HEAs could contribute to weaken the binding of chemical species at the surface layer. The addition of Mn was used as the fifth element in np-HEAs, which decreases the d-band center of Pt and tunes the electronic properties of catalysts surfaces that optimize the ORR activity.

To further decrease the cost of electrocatalysts, Qiu group also fabricated a series of noble-metal-free nanoporous HEAs with oxidized surfaces by the top-down dealloying method.[30] As displayed in Figure 13.5, AlFeCoNiCr exhibits the uniformly 3D open-pore structure with the pore size range from 200 to 400 nm, where there are many ultrafine nanoscale pores in the solid walls of 3D open-pore structure. The np-AlFeCoNiCr sample demonstrated an obvious reduction peak in the CV curve, and the electron transfer number (n) was calculated to be ~3.7 based on K-L

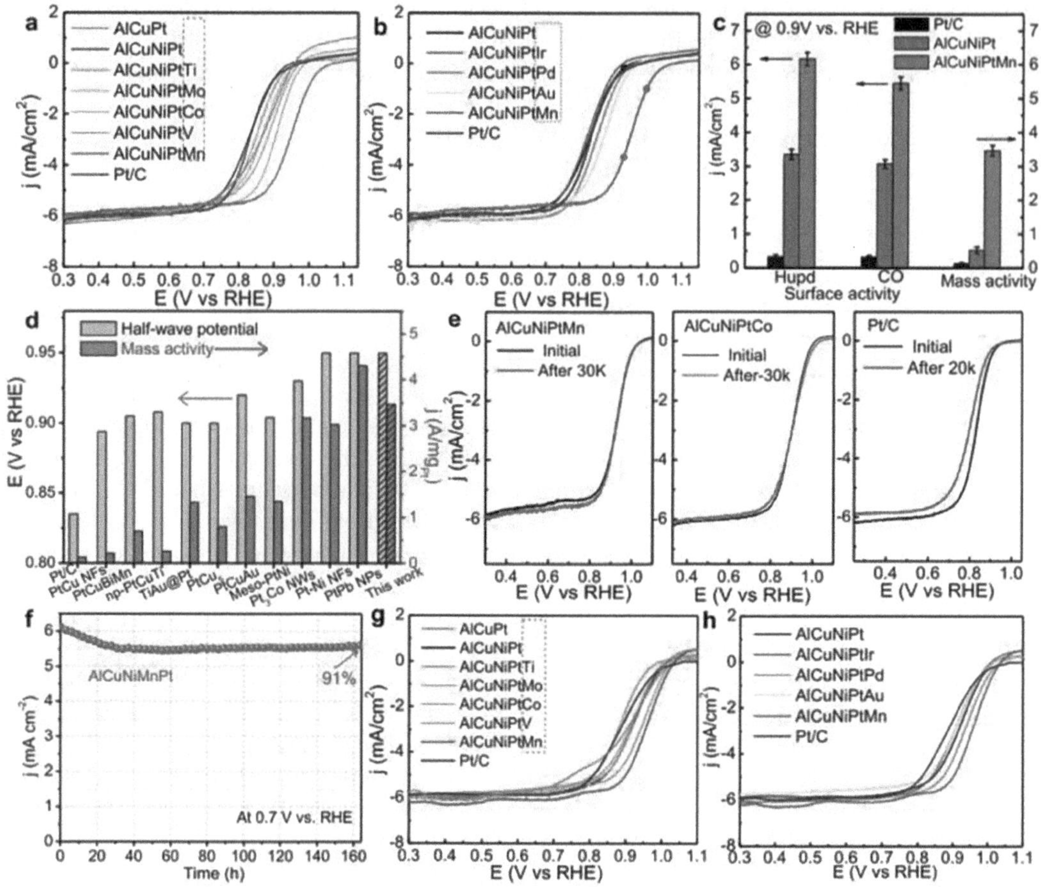

FIGURE 13.4 (a–f) The ORR activity of all samples and the corresponding durability in acid solution. (g-h) The ORR activity of all samples in alkaline solution. Adapted with permission [51]. Copyright (2020) Elsevier.

plots, suggesting that ORR at the cathode is $4e^-$ pathway on the np-AlFeCoNiCr sample surface. These works confirmed that the nanoporous HEAs are a promising platform for the exploration of efficient catalysts.

13.2.3 High Entropy Intermetallics for ORR

Generally, HEAs typically form a disordered FCC phase due to the random distribution of multi-elements in near-equimolar compositions. On the contrary, ordered intermetallic alloys have definite crystal structures with a long-range ordered arrangement of atoms, which could accurately control structural, geometric, and electronic effects. High entropy intermetallics (HEIs) are a new class of materials that combine the characteristics of both HEAs and ordered intermetallic alloys. These materials inherit the merits and properties of both alloy types, which achieve high stability, better corrosion resistance, and unique electronic properties. Therefore, HEIs exhibit highly efficient activity and excellent stability for electrocatalysis.

However, ordered intermetallic nanoparticles are usually prepared by high-temperature pyrolysis, which easily results in the agglomeration of nanoparticles. Benefiting from the strong interaction between Pt and S atom doped in the porous carbon, Liang group first successfully fabricated Pt-based intermetallics nanoparticles with a mean size of < 5 nm supported by S-doped porous carbon. This synthesis strategy can be widely applied to various metal elements consisting of Pt

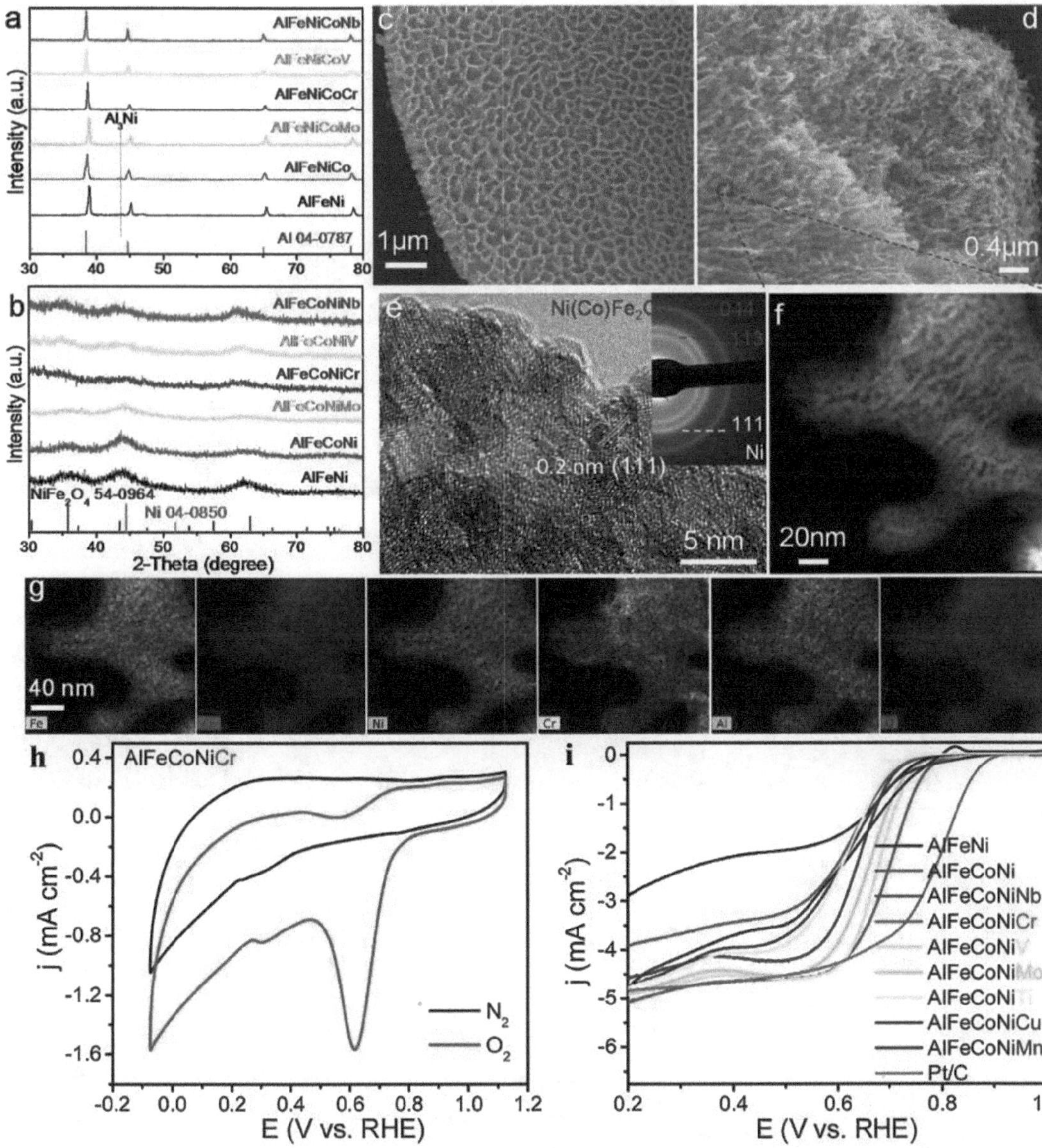

FIGURE 13.5 (a–g) the structure and morphology characterization of HEAs. (h-i) the ORR activity of all samples in alkaline solution. Adapted with permission [30]. Copyright (2020) Elsevier.

with 16 other metal elements. The intermetallic libraries demonstrated highly efficient activity and excellent stability for electrocatalysis in PEMFC.[61] Similarly, Huang group explored a series of intermetallic $Pt_4FeCoCuNi$ nanoparticles with tunable ordering degrees to study the effects of the crystal structure of HEAs on catalytic activity and durability. The ordering transformation of HEI $Pt_4FeCoCuNi$ nanoparticles could be directly tuned by modulating the sintering temperature and duration, which could be investigated by aberration-corrected STEM at the atomic level. The valence band spectrum confirmed the change in the electronic structure of intermetallic $Pt_4FeCoCuNi$ with the variation of their crystal structure. Highly ordered $Pt_4FeCoCuNi$ displayed a downshift of the d-band center compared with pure Pt, which weakened the adsorption of intermediates for ORR. Highly ordered $Pt_4FeCoCuNi$ demonstrated a mass activity (3.78 A mg_{Pt}^{-1} at 0.9 V vs. RHE) higher

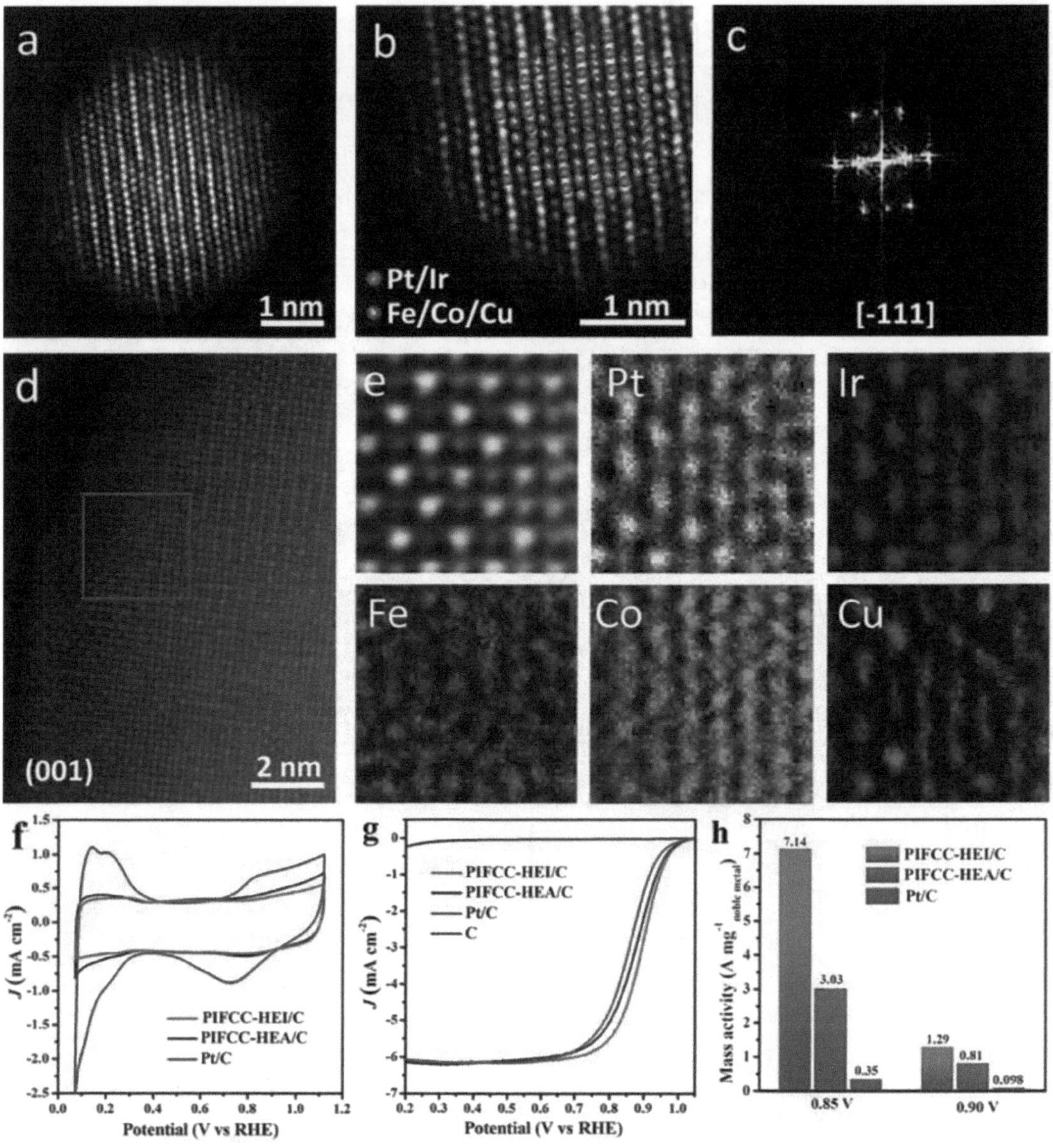

FIGURE 13.6 (a–e) The microstructure of PIFCC-HEI NP, (f-h) the ORR activity of PIFCC-HEI NP. Adapted with permission [63]. Copyright (2023) American Chemical Society.

than that of partially ordered, disordered $Pt_4FeCoCuNi$, and 20 wt% Pt/C (1.84, 0.86, 0.20 A mg_{Pt}^{-1}, respectively). Such an enhanced performance is ascribed to the optimized electronic structure and surface structure of highly ordered $Pt_4FeCoCuNi$ structure.[62]

Even though these HEIs have promising potential for ORR catalysts, the study of HEIs still faces some limitations and challenges. Xia group constructed high active facets to further improve the activity of ordered PtIrFeCoCu HEI (PIFCC-HEI) NPs, which were well dispersed on the carbon support with a particle size of 6 nm.[63] As shown in Figure 13.6, the PIFCC-HEI/ C exhibited high mass activity of 7.14 A $mg_{\text{noble metals}}^{-1}$ at the potential of 0.85 V, which is much higher than those of Pt/C and PIFCC-HEA/ C (0.35 A mg_{Pt}^{-1} and 3.03 A $mg_{\text{noble metals}}^{-1}$, respectively). Additionally, they also demonstrated excellent stability over 60,000 cycles of accelerated durability tests. The superior performance of PIFCC-HEI/C should be ascribed to its ultrahigh active facets as evidenced by experiment and DFT calculations results, the (001) facet could provide the lowest activation barriers, downshift of d-band centers, and the optimal modulation of electron structures for oxygen reduction reaction.

13.3 THE APPLICATION OF HEOs IN ORR

High entropy oxides (HEOs) are inherited from HEAs. It is also well acknowledged that HEOs have demonstrated high activity for ORR due to their unusual structural diversities.[64] Compared with HEAs which usually have single-site occupancies in the FCC structure, HEOs demonstrate separately cation and anion sublattices, which promote enhanced structural characteristics. HEOs also include at least five metallic elements with equal or near-equal ratio. Obviously, the high entropy property also results in the four core effects of HEOs: the high entropy effect, lattice distortion effect, sluggish diffusion effect and cocktail effect, so high entropy oxides demonstrate high activity for a variety of catalysis reactions compared with the traditional materials. In 2015, Maria and coworkers first fabricated the entropy-induced stabilization of single-phase multicomponent oxides, where Zn, Cu, Mg, Ni, and Co were randomly packed in the oxide lattice.[28,64] Hence, more studies concentrated on the study and application of HEOs.[65,66]

HEO structures are particularly beneficial for enhancing the catalytic activity of ORR.[66–69] Hu group developed a series of HEO NPs consisting of ten metal elements (Hf, Zr, La, V, Ce, Ti, Nd, Gd, Y, and Pd), which were fabricated through the rapid high-temperature heating process. The high-temperature process and high entropy effect could contribute to form the single phase of the uniform mixing of multi-elements without any phase segregation. HEO NPs supported by commercial carbon black demonstrated highly stable and efficient catalytical activity for ORR. The 10-HEO/C electrocatalysts exhibited a half-wave potential (0.85 V vs. RHE), which was slightly higher than those of Pd/C (0.82 V) and PdO/C (0.83 V), and the Pd mass activity of 10-HEO/C is 0.49 A mg_{Pd}^{-1} at 0.85 V which is much higher than those of Pd/C and PdO/C. The durability of the 10-HEO/C was also evaluated by chronoamperometry measurement, demonstrating the current density of the 10-HEO/C remained 86% after the operation of 100 h. The 10-HEO NPs were well dispersed on the carbon black support without obviously agglomeration, and the elemental distribution of 10-HEO NPs was also homogenous, indicating excellent interfacial and structural stability.[68] Dutta et al. also crafted a quinary FeCrCoMnNi-based HEO derived from multi-metallic metal-organic frameworks precursor. And the variants of HEOs could also be tuned by changing the initial precursor cocktail via this synthetic strategy. As shown in Figure 13.7, the quinary HEO exhibited an excellent activity and stability for both ORR and OER in alkaline conditions.[69] In conclusion, HEOs are considered to be potential materials with promising properties for ORR.

13.4 CONCLUSIONS

As a new class of functional materials, HEMs with a wide range of compositional modulations and inherent complex surfaces, have attracted considerable attention for their application in electrocatalytic reactions. In this review, the applications of HEMs for ORR have been discussed in detail. Ongoing studies mainly explore different HEM compositions, nanostructuring techniques, and surface modifications to enhance their ORR activity. Through theoretical calculations and predictions, the active sites in HEMs can be identified and guide the design. Advanced experimental and in-situ characterization techniques could accurately detect the surface structure and identify the active site, which are employed to study the structure-property relationships and modify HEM catalysts. HEMs provide a promising avenue for the design and development of efficient and durable catalysts for the oxygen reduction reaction.

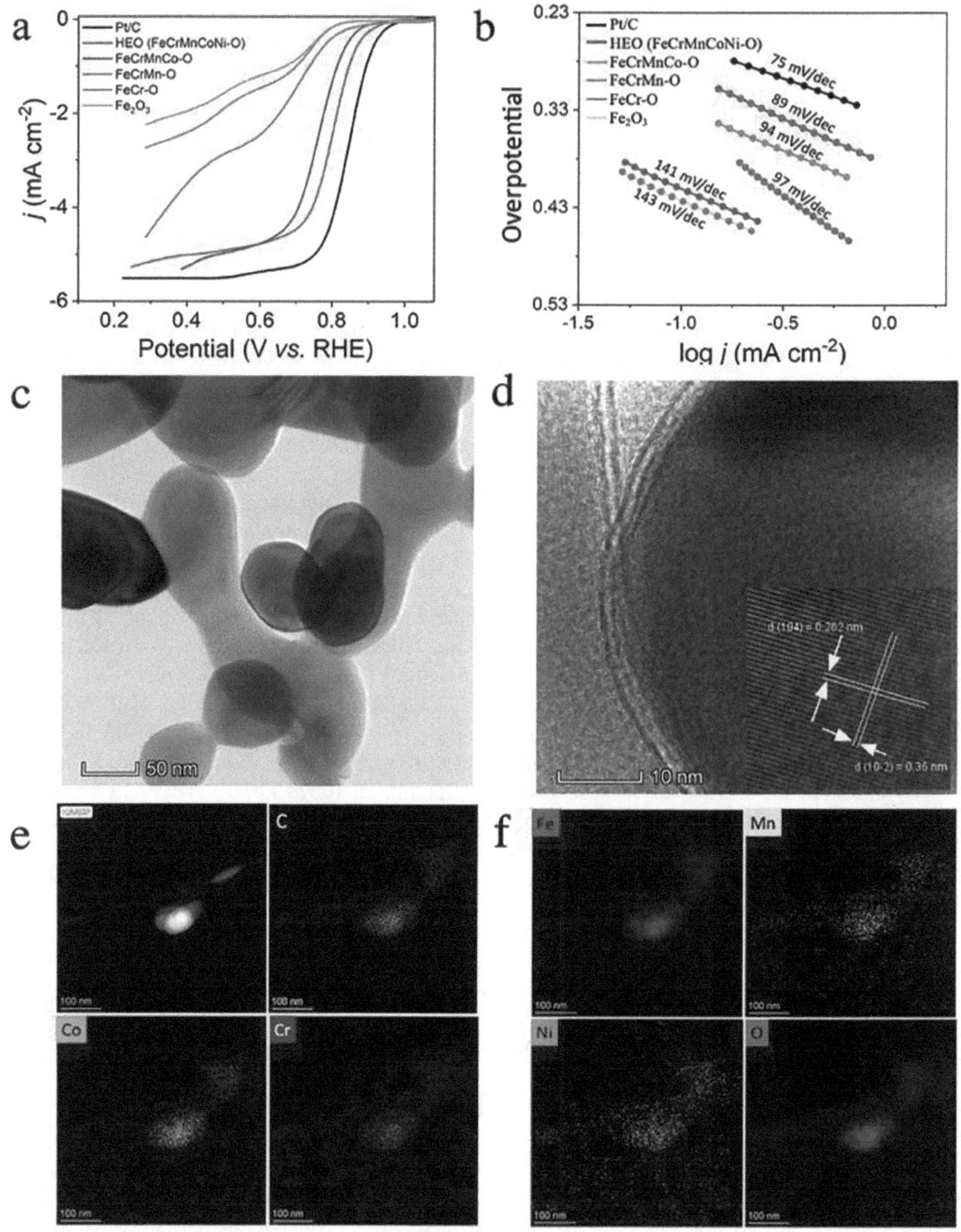

FIGURE 13.7 (a-b) ORR activities for all samples, (c-f) the microstructure and elemental distribution of HEO-750. Adapted with permission [69]. Copyright (2023) Elsevier.

ACKNOWLEDGMENTS

This work is supported by National Natural Science Foundation of China (No. 21805170, 22172093, and 21776167), Qingdao Postdoctoral Innovation Project (No. QDBSH20220202031).

REFERENCES

1 S. Chu and A. Majumdar, Opportunities and Challenges for a Sustainable Energy Future, *Nature* 488 (2012) 294–303.

2 K. Jiao, J. Xuan, Q. Du, Z. Bao, B. Xie, B. Wang, Y. Zhao, L. Fan, H. Wang, Z. Hou, S. Huo, N. P. Brandon, Y. Yin and M. D. Guiver, Designing the Next Generation of Proton-Exchange Membrane Fuel Cells, *Nature* 595 (2021) 361–369.

3 J. R. Varcoe and R. C. Slade, Prospects for Alkaline Anion-Exchange Membranes in Low Temperature Fuel Cells, *Fuel cells* 5 (2005) 187–200.

4 M. K. Debe, Electrocatalyst Approaches and Challenges for Automotive Fuel Cells, *Nature* 486 (2012) 43–51.

5 F. Xiao, Y. Wang, G.-L. Xu, F. Yang, S. Zhu, C.-J. Sun, Y. Cui, Z. Xu, Q. Zhao, J. Jang, X. Qiu, E. Liu, W. S. Drisdell, Z. Wei, M. Gu, K. Amine and M. Shao, Fe-N-C Boosts the Stability of Supported Platinum Nanoparticles for Fuel Cells, *J. Am. Chem. Soc.* 144 (2022) 20372–20384.

6 M. Shao, Q. Chang, J.-P. Dodelet and R. Chenitz, Recent Advances in Electrocatalysts for Oxygen Reduction Reaction, *Chem. Rev.* 116 (2016) 3594–3657.

7 L. Huang, S. Zaman, X. Tian, Z. Wang, W. Fang and B. Y. Xia, Advanced Platinum-Based Oxygen Reduction Electrocatalysts for Fuel Cells, Acc. *Chem. Res.* 54 (2021) 311–322.

8 X. Zhao and K. Sasaki, Advanced Pt-Based Core-Shell Electrocatalysts for Fuel Cell Cathodes, Acc. *Chem. Res.* 55 (2022) 1226–1236.

9 L. Yan, P. Li, Q. Zhu, A. Kumar, K. Sun, S. Tian and X. Sun, Atomically Precise Electrocatalysts for Oxygen Reduction Reaction, *Chem* 9 (2023) 280–342.

10 L. Wang, Y. Sun, S. Zhang, H. Di, Q. Liu, X. Du, Z. Li, K. Yang and S. Chen, Co/Co2p Nanoparticles Encapsulated within Hierarchically Porous Nitrogen, Phosphorus, Sulfur Co-Doped Carbon as Bifunctional Electrocatalysts for Rechargeable Zinc-Air Batteries, *ChemElectroChem* 8 (2021) 4286–4295.

11 J. Lu, W. Zhou, L. Wang, J. Jia, Y. Ke, L. Yang, K. Zhou, X. Liu, Z. Tang, L. Li and S. Chen, Core-Shell Nanocomposites Based on Gold Nanoparticle@Zinc-Iron-Embedded Porous Carbons Derived from Metal-Organic Frameworks as Efficient Dual Catalysts for Oxygen Reduction and Hydrogen Evolution Reactions, *ACS Catal.* 6 (2016) 1045–1053.

12 T. Zhang, Z. Li, L. Wang, P. Sun, Z. Zhang and S. Wang, Spinel Mnco2o4 Nanoparticles Supported on Three-Dimensional Graphene with Enhanced Mass Transfer as an Efficient Electrocatalyst for the Oxygen Reduction Reaction, *ChemSusChem* 11 (2018) 2730–2736.

13 L. Wang, Z. Tang, W. Yan, Q. Wang, H. Yang and S. Chen, Co@Pt Core@Shell Nanoparticles Encapsulated in Porous Carbon Derived from Zeolitic Imidazolate Framework 67 for Oxygen Electroreduction in Alkaline Media, *J. Power Sources* 343 (2017) 458–466.

14 L. Wang, Z. Tang, W. Yan, H. Yang, Q. Wang and S. Chen, Porous Carbon-Supported Gold Nanoparticles for Oxygen Reduction Reaction: Effects of Nanoparticle Size, *ACS Appl. Mater. Inter.* 8 (2016) 20635–20641.

15 C. Wang, Y. Liu, Z. Li, L. Wang, X. Niu and P. Sun, Novel Space-Confinement Synthesis of Two-Dimensional Fe, N-Codoped Graphene Bifunctional Oxygen Electrocatalyst for Rechargeable Air-Cathode, *Chem. Eng. J.* 411 (2021) 128492.

16 S. Zhang, L. Wang, L. Fang, Y. Tian, Y. Tang, X. Niu, Y. Hao and Z. Li, A Facile Method to Prepare Ultrafine Pd Nanoparticles Embedded into N-Doped Porous Carbon Nanosheets as Highly Efficient Electrocatalysts for Oxygen Reduction Reaction, *J. Electrochem. Soc.* 167 (2020) 054508–054514.

17 M. Zysler, E. Carbo-Argibay, P. J. Ferreira and D. Zitoun, Dealloyed Octahedral Ptcu Nanoparticles as High-Efficiency Electrocatalysts for the Oxygen Reduction Reaction, *ACS Appl. Nano Mater.* 5 (2022) 11484–11493.

18 T. Xia, K. Zhao, Y. Zhu, X. Bai, H. Gao, Z. Wang, Y. Gong, M. Feng, S. Li, Q. Zheng, S. Wang, R. Wang and H. Guo, Mixed-Dimensional Pt-Ni Alloy Polyhedral Nanochains as Bifunctional Electrocatalysts for Direct Methanol Fuel Cells, Adv. *Mater.* 35 (2023) e2206508.

19 Z. Zhao, Z. Liu, A. Zhang, X. Yan, W. Xue, B. Peng, H. L. Xin, X. Pan, X. Duan and Y. Huang, Graphene-Nanopocket-Encaged Ptco Nanocatalysts for Highly Durable Fuel Cell Operation under Demanding Ultralow-Pt-Loading Conditions, *Nat. Nanotechnol.* 17 (2022) 968–975.

20 J. Wang, G. Zhang, H. Liu, Z. Li, L. Wang, J. Tressel and S. Chen, High-Performance Electrocatalytic Reduction of Co2 to Co by Ultrathin Pdcu Alloy Nanosheets, *Sep. Purif. Technol.* 320 (2023) 124186.

21 W. Yan, Z. Tang, L. Wang, Q. Wang, H. Yang and S. Chen, Pdau Alloyed Clusters Supported by Carbon Nanosheets As efficient Electrocatalysts For oxygen reduction, *Int. J. Hydro. Energy* 42 (2017) 218–227.

22 C. Huang, H. Liu, Y. Tang, Q. Lu, S. Chu, X. Liu, B. Shan and R. Chen, Constructing Uniform Sub-3 nm Ptzn Intermetallic Nanocrystals Via Atomic Layer Deposition for Fuel Cell Oxygen Reduction, *Appl. Catal. B-Environ.* 320 (2023) 121986.

23 Z. Qiao, C. Wang, C. Li, Y. Zeng, S. Hwang, B. Li, S. Karakalos, J. Park, A. J. Kropf, E. C. Wegener, Q. Gong, H. Xu, G. Wang, D. J. Myers, J. Xie, J. S. Spendelow and G. Wu, Atomically Dispersed Single Iron Sites for Promoting Pt and Pt3co Fuel Cell Catalysts: Performance and Durability Improvements, *Energy Environ. Sci.* 14 (2021) 4948–4960.

24 J.-W. Yeh, S.-K. Chen, S.-J. Lin, J.-Y. Gan, T.-S. Chin, T.-T. Shun, C.-H. Tsau and S.-Y. Chang, Nanostructured High-Entropy Alloys with Multiple Principal Elements: Novel Alloy Design Concepts and Outcomes, *Adv. Eng. Mater.* 6 (2004) 299–303.

25 D. Wu, K. Kusada, T. Yamamoto, T. Toriyama, S. Matsumura, S. Kawaguchi, Y. Kubota and H. Kitagawa, Platinum-Group-Metal High-Entropy-Alloy Nanoparticles, *J. Am. Chem. Soc.* 142 (2020) 13833–13838.

26 T. Löffler, A. Ludwig, J. Rossmeisl and W. Schuhmann, What Makes High-Entropy Alloys Exceptional Electrocatalysts? *Angew. Chem. Int. Ed.* 60 (2021) 26894–26903.

27 E. P. George, D. Raabe and R. O. Ritchie, High-Entropy Alloys, *Nat. Rev. Mater.* 4 (2019) 515–534.

28 Y. Sun and S. Dai, High-Entropy Materials for Catalysis: A New Frontier, *Sci. Adv.* 7 (2021) eabg1600.

29 T. Löffler, H. Meyer, A. Savan, P. Wilde, A. Garzón Manjón, Y.-T. Chen, E. Ventosa, C. Scheu, A. Ludwig and W. Schuhmann, Discovery of a Multinary Noble Metal-Free Oxygen Reduction Catalyst, *Adv. Energy Mater.* 8 (2018) 1802269.

30 G. Fang, J. Gao, J. Lv, H. Jia, H. Li, W. Liu, G. Xie, Z. Chen, Y. Huang, Q. Yuan, X. Liu, X. Lin, S. Sun and H.-J. Qiu, Multi-Component Nanoporous Alloy/(Oxy)Hydroxide for Bifunctional Oxygen Electrocatalysis and Rechargeable Zn-Air Batteries, *Appl. Catal. B-Environ.* 268 (2020) 118431.

31 B. Cantor, I. T. H. Chang, P. Knight and A. J. B. Vincent, Microstructural Development in Equiatomic Multicomponent Alloys, *Mater. Sci. Eng. A* 375-377 (2004) 213–218.

32 H. Li, H. Huang, Y. Chen, F. Lai, H. Fu, L. Zhang, N. Zhang, S. Bai and T. Liu, High-Entropy Alloy Aerogels: A New Platform for Carbon Dioxide Reduction, *Adv. Mater.* 35 (2023) 2209242.

33 X. Fu, J. Zhang, S. Zhan, F. Xia, C. Wang, D. Ma, Q. Yue, J. Wu and Y. Kang, High-Entropy Alloy Nanosheets for Fine-Tuning Hydrogen Evolution, *ACS Catal.* 12 (2022) 11955–11959.

34 N. Kumar Katiyar, K. Biswas, J.-W. Yeh, S. Sharma and C. Sekhar Tiwary, A Perspective on the Catalysis Using the High Entropy Alloys, *Nano Energy* 88 (2021) 106261.

35 Y.-C. Qin, F.-Q. Wang, X.-M. Wang, M.-W. Wang, W.-L. Zhang, W.-K. An, X.-P. Wang, Y.-L. Ren, X. Zheng, D.-C. Lv and A. Ahmad, Noble Metal-Based High-Entropy Alloys as Advanced Electrocatalysts for Energy Conversion, *Rare Metals* 40 (2021) 2354–2368.

36 N. L. N. Broge, M. Bondesgaard, F. Søndergaard-Pedersen, M. Roelsgaard and B. B. Iversen, Autocatalytic Formation of High-Entropy Alloy Nanoparticles, *Angew. Chem. Int. Ed.* 59 (2020) 21920–21924.

37 P. Zhao, Q. Cao, W. Yi, X. Hao, J. Li, B. Zhang, L. Huang, Y. Huang, Y. Jiang, B. Xu, Z. Shan and J. Chen, Facile and General Method to Synthesize Pt-Based High-Entropy-Alloy Nanoparticles, *ACS Nano* 16 (2022) 14017–14028.

38 L. Tao, M. Sun, Y. Zhou, M. Luo, F. Lv, M. Li, Q. Zhang, L. Gu, B. Huang and S. Guo, A General Synthetic Method for High-Entropy Alloy Subnanometer Ribbons, *J. Am. Chem. Soc.* 144 (2022) 10582–10590.

39 W. Shi, H. Liu, Z. Li, C. Li, J. Zhou, Y. Yuan, F. Jiang, K. Fu and Y. Yao, High-Entropy Alloy Stabilized and Activated Pt Clusters for Highly Efficient Electrocatalysis, *SusMat* 2 (2022) 186–196.

40 X. Huang, G. Yang, S. Li, H. Wang, Y. Cao, F. Peng and H. Yu, Noble-Metal-Based High-Entropy-Alloy Nanoparticles for Electrocatalysis, *J. Energy Chem.* 68 (2022) 721–751.

41 K. Zhao, X. Li and D. Su, High-Entropy Alloy Nanocatalysts for Electrocatalysis, *Acta Phys. Chim. Sin.* 37 (2020) 2009077.

42 Y. Ma, Y. Ma, Q. Wang, S. Schweidler, M. Botros, T. Fu, H. Hahn, T. Brezesinski and B. Breitung, High-Entropy Energy Materials: Challenges and New Opportunities, *Energy Environ. Sci.* 14 (2021) 2883–2905.

43 H. Li, J. Lai, Z. Li and L. Wang, Multi-Sites Electrocatalysis in High-Entropy Alloys, *Adv. Funct. Mater.* 31 (2021) 2106715.

44 X. Wang, W. Guo and Y. Fu, High-Entropy Alloys: Emerging Materials for Advanced Functional Applications, *J. Mater. Chem. A* 9 (2021) 663–701.

45 B. Wang, Y. Yao, X. Yu, C. Wang, C. Wu and Z. Zou, Understanding the Enhanced Catalytic Activity of High Entropy Alloys: From Theory to Experiment, *J. Mater. Chem. A* 9 (2021) 19410–19438.

46 K. Kusada, M. Mukoyoshi, D. Wu and H. Kitagawa, Chemical Synthesis, Characterization, and Properties of Multi-Element Nanoparticles, *Angew. Chem. Int. Ed.* 61 (2022) e202209616.

47 Y. Yao, Z. Liu, P. Xie, Z. Huang, T. Li, D. Morris, Z. Finfrock, J. Zhou, M. Jiao, J. Gao, Y. Mao, J. Miao, P. Zhang, R. Shahbazian-Yassar, C. Wang, G. Wang and L. Hu, Computationally Aided, Entropy-Driven Synthesis of Highly Efficient and Durable Multi-Elemental Alloy Catalysts, Sci. *Adv.* 6 (2020) eaaz0510.

48 A. Kulkarni, S. Siahrostami, A. Patel and J. K. Nørskov, Understanding Catalytic Activity Trends in the Oxygen Reduction Reaction, *Chem. Rev.* 118 (2018) 2302–2312.

49 J. Greeley, I. E. L. Stephens, A. S. Bondarenko, T. P. Johansson, H. A. Hansen, T. F. Jaramillo, J. Rossmeisl, I. Chorkendorff and J. K. Nørskov, Alloys of Platinum and Early Transition Metals as Oxygen Reduction Electrocatalysts, *Nat. Chem.* 1 (2009) 552–556.

50 J. K. Nørskov, J. Rossmeisl, A. Logadottir, L. Lindqvist, J. R. Kitchin, T. Bligaard and H. Jónsson, Origin of the Overpotential for Oxygen Reduction at a Fuel-Cell Cathode, *J. Phy. Chem. B* 108 (2004) 17886–17892.

51 S. Li, X. Tang, H. Jia, H. Li, G. Xie, X. Liu, X. Lin and H.-J. Qiu, Nanoporous High-Entropy Alloys with Low Pt Loadings for High-Performance Electrochemical Oxygen Reduction, *J. Catal.* 383 (2020) 164–171.

52 Y. Yu, F. Xia, C. Wang, J. Wu, X. Fu, D. Ma, B. Lin, J. Wang, Q. Yue and Y. Kang, High-Entropy Alloy Nanoparticles as a Promising Electrocatalyst to Enhance Activity and Durability for Oxygen Reduction, *Nano Res.* 15 (2022) 7868–7876.

53 Y. Yao, Z. Huang, T. Li, H. Wang, Y. Liu, H. S. Stein, Y. Mao, J. Gao, M. Jiao, Q. Dong, J. Dai, P. Xie, H. Xie, S. D. Lacey, I. Takeuchi, J. M. Gregoire, R. Jiang, C. Wang, A. D. Taylor, R. Shahbazian-Yassar and L. Hu, High-Throughput, Combinatorial Synthesis of Multimetallic Nanoclusters, *Proc. Natl. Acad. Sci. U.S.A.* 117 (2020) 6316–6322.

54 J. Li, H. S. Stein, K. Sliozberg, J. Liu, Y. Liu, G. Sertic, E. Scanley, A. Ludwig, J. Schroers, W. Schuhmann and A. D. Taylor, Combinatorial Screening of Pd-Based Quaternary Electrocatalysts for Oxygen Reduction Reaction in Alkaline Media, *J. Mater. Chem. A* 5 (2017) 67–72.

55 T. A. A. Batchelor, J. K. Pedersen, S. H. Winther, I. E. Castelli, K. W. Jacobsen and J. Rossmeisl, High-Entropy Alloys as a Discovery Platform for Electrocatalysis, *Joule* 3 (2019) 834–845.

56 S. L. A. Bueno, A. Leonardi, N. Kar, K. Chatterjee, X. Zhan, C. Chen, Z. Wang, M. Engel, V. Fung and S. E. Skrabalak, Quinary, Senary, and Septenary High Entropy Alloy Nanoparticle Catalysts from Core@Shell Nanoparticles and the Significance of Intraparticle Heterogeneity, *ACS Nano* 16 (2022) 18873–18885.

57 Y. Chen, X. Zhan, S. L. A. Bueno, I. H. Shafei, H. M. Ashberry, K. Chatterjee, L. Xu, Y. Tang and S. E. Skrabalak, Synthesis of Monodisperse High Entropy Alloy Nanocatalysts from Core@Shell Nanoparticles, *Nanoscale Horiz.* 6 (2021) 231–237.

58 T. Chen, F. Ning, J. Qi, G. Feng, Y. Wang, J. Song, T. Yang, X. Liu, L. Chen and D. Xia, Ptfeconicu High-Entropy Solid Solution Alloy as Highly Efficient Electrocatalyst for the Oxygen Reduction Reaction, *iScience* 26 (2023) 105890.

59 G. Zhu, Y. Jiang, H. Yang, H. Wang, Y. Fang, L. Wang, M. Xie, P. Qiu and W. Luo, Constructing Structurally Ordered High-Entropy Alloy Nanoparticles on Nitrogen-Rich Mesoporous Carbon Nanosheets for High-Performance Oxygen Reduction, *Adv. Mater.* 34 (2022) 2110128.

60 H.-J. Qiu, G. Fang, Y. Wen, P. Liu, G. Xie, X. Liu and S. Sun, Nanoporous High-Entropy Alloys for Highly Stable and Efficient Catalysts, *J. Mater. Chem. A* 7 (2019) 6499–6506.

61 L.-N. W. Cheng-Long Yang, Peng Yin, Jieyuan Liu, Ming-Xi Chen, Qiang-Qiang Yan, Zheng-Shu Wang, Shi-Long Xu, Sheng-Qi Chu, Chunhua Cui, Huanxin Ju, Junfa Zhu, Yue Lin, Jianglan Shui, Hai-Wei Liang, Sulfur-Anchoring Synthesis of Platinum Intermetallic Nanoparticle Catalysts for Fuel Cells, *Science* 374 (2021) 459–464

62 Y. Wang, N. Gong, H. Liu, W. Ma, K. Hippalgaonkar, Z. Liu and Y. Huang, Ordering-Dependent Hydrogen Evolution and Oxygen Reduction Electrocatalysis of High-Entropy Intermetallic Pt4fecocuni, *Adv. Mater* 35 (2023) 2302067.

63 G. Feng, F. Ning, Y. Pan, T. Chen, J. song, Y. Wang, R. Zou, D. Su and D. Xia, Engineering Structurally Ordered High-Entropy Intermetallic Nanoparticles with High-Activity Facets for Oxygen Reduction in Practical Fuel Cells, *J. Am. Chem. Soc.* 145 (2023) 11140–11150.

64 Z.-Y. Liu, Y. Liu, Y. Xu, H. Zhang, Z. Shao, Z. Wang and H. Chen, Novel High-Entropy Oxides for Energy Storage And Conversion: From fundamentals to Practical Applications, *Green Energy Environ.* (2023) 1341–1357.

65 T. X. Nguyen, Y.-C. Liao, C.-C. Lin, Y.-H. Su and J.-M. Ting, Advanced High Entropy Perovskite Oxide Electrocatalyst for Oxygen Evolution Reaction, *Adv. Funct. Mater.* 31 (2021) 2101632.

66 Z. Chen, J. Wu, Z. Chen, H. Yang, K. Zou, X. Zhao, R. Liang, X. Dong, P. W. Menezes and Z. Kang, Entropy Enhanced Perovskite Oxide Ceramic for Efficient Electrochemical Reduction of Oxygen to Hydrogen Peroxide, *Angew. Chem. Int. Ed.* 61 (2022) e202200086.
67 Y. Zhang, J. Lyu, Y.-L. Zhao, K. Hu, Z. Chen, X. Lin, G. Xie, X. Liu and H.-J. Qiu, In Situ Coupling of Ag Nanoparticles with High-Entropy Oxides as Highly Stable Bifunctional Catalysts for Wearable Zn-Ag/Zn-Air Hybrid Batteries, *Nanoscale* 13 (2021) 16164–16171.
68 T. Li, Y. Yao, B. H. Ko, Z. Huang, Q. Dong, J. Gao, W. Chen, J. Li, S. Li, X. Wang, R. Shahbazian-Yassar, F. Jiao and L. Hu, Carbon-Supported High-Entropy Oxide Nanoparticles as Stable Electrocatalysts for Oxygen Reduction Reactions, *Adv. Funct. Mater.* 31 (2021) 2010561.
69 K. V. R. Siddhartha Sairam, S. K. T. Aziz, I. Karajagi, A. Saini, M. Pal, P. C. Ghosh and A. Dutta, A Quinary High Entropy Metal Oxide Exhibiting Robust and Efficient Bidirectional O2 Reduction and Water Oxidation, *Int. J. Hydro. Energy* 48 (2023) 10521–10531.

14 High Entropy Materials for CO_2 Conversion

Telem Simsek, Seval H. Guler, Omer Guler, and Tuncay Simsek

14.1 INTRODUCTION

Human activities constantly produce greenhouse gases, which cause temperature increases that have a significant impact on the global climate system. The main reason for this is fossil fuels, and the stoichiometry of combustion shows that burning 1 ton of carbon in fossil fuels results in over 3.5 tons of CO_2 [1]. Data from the International Energy Agency (IEA) show that the concentration of CO_2 in the atmosphere was 394 ppmv in 2012, representing a 40% increase compared to the mid-18th century and a significant increase in the past century [2]. Currently, this value is 412 ppmv. In other words, it is known that the accumulation in the atmosphere is more than 1 teraton at present [1]. According to a model developed by the IEA, in order to limit the temperature increase to 2°C by 2050, CO_2 levels should not exceed 15 gigatons annually. In other words, global greenhouse gas emissions need to be reduced by at least 50% by 2050 when compared to 1990 levels [3]. If CO_2 emissions continue at this rate, it is certain that the temperature increase will exceed the target of 2°C by 2050. While the Paris Climate Agreement signed in 2015 is an effective step toward reducing CO_2 emissions, it is not sufficient to prevent climate change.

It is estimated that the amount of CO_2 released into the atmosphere in 2022 has reached a record high of 37.5 billion tons. The annual release of CO_2 into the atmosphere continues to increase every year, except for the COVID-19 pandemic in 2020, which resulted in a 5% decrease in CO_2 emissions compared to the previous year. Fossil fuel combustion accounts for the largest share of CO_2 emissions, at approximately 60%. Currently, over 80% of the world's energy needs are met by fossil fuels (33% oil, 24% natural gas, and 23% coal), with coal being the most polluting of all. Coal emits 820 metric tons of greenhouse gas emissions for every gigawatt of electricity generated.

To limit the temperature increase to 2°C by 2050, CO_2 levels should not exceed 15 gigatons per year. Currently, three methods can be employed to reduce CO_2: controlling CO_2 emissions, carbon capture and storage (CCS), and chemical conversion and utilization of CO_2. Options for reducing human-generated CO_2 emissions include increasing energy efficiency, replacing fossil fuels with low-carbon energy, and eliminating CO_2 from fossil fuel combustion through permanent underground storage, known as Carbon Capture and Storage. These measures are aimed at reducing emissions. However, it is more important to reduce the amount of CO_2 in the atmosphere. Converting CO_2 into C1-type chemicals (contain just a single carbon atom, such as carbon monoxide (CO), carbon dioxide (CO_2), methane (CH_4), methanol (CH_3OH), formic acid (HCOOH)) by using CO_2 as a raw material and reducing CO_2 emissions is an effective way to reduce atmospheric CO_2 levels [4]. Currently, over 100 million tons of CO_2 are used annually as a raw material in the production of chemicals such as urea, salicylic acid, and methanol [5]. CO_2 can be converted into these chemicals by electrocatalysis, photocatalysis, and thermal catalysis. Among these, thermal catalysis is the most prominent method because of its fast kinetics. In addition, CO_2 is a highly stable molecule, and additional activation energy is required to convert it. Thermal catalysis is more

DOI: 10.1201/9781003391388-14

effective than electrocatalysis and photocatalysis, which have low energy efficiency. In addition to the general conversion methods mentioned above, it is also necessary to further specify the conservation methods of CO_2. These methods can be listed as CO_2 hydrogenation, esterification, oxidation and reduction, and methanation.

14.2 HIGH ENTROPY ALLOYS

High entropy alloys (HEAs) are now emerging as state-of-the-art alloys. Although HEAs were first discovered in 1995, intensive research on these alloys has been carried out for the last 10 years. The unique microstructures and properties of these alloys have shown their potential for use in many applications. To understand HEAs, traditional alloys need to be understood first. Until the discovery of HEAs, traditional alloys were known as new materials resulting from mixing two or more metals in certain proportions to enhance mechanical properties or other features. One of the metals in an alloy is the main metal, and the other metals added to the alloy are soluble metals within the main metal. In traditional alloys, the metal with the highest concentration in the alloy is the main metal, and the other metals are present in lower proportions and are, therefore, soluble metals. The main metal can dissolve other metals up to a certain extent. Metals that are insoluble within the main metal create a different phase, intermetallic compounds, or precipitate phases in the microstructure of the alloy. As a result, some properties of the alloy are positively affected, whereas others are negatively affected. In particular, the formation of hard and brittle intermetallics within the structure can increase the strength of the alloy but decrease its plasticity. In HEAs, the alloy consists of five or more metals, which are added to the alloy in equimolar or near-equimolar proportions. As a result, the mixing entropy of HEAs (1.6R) is higher than that of traditional alloys. Due to their high mixing entropy, the microstructure of HEAs consists of a single phase, such as fcc, bcc, hcp, or a mixture of these phases (known as the high entropy effect). These phases are disordered solid solutions. Intermetallics typically form ordered phases with lower mixing entropy. The mixing entropy for stoichiometric intermetallic compounds is zero. Therefore, in HEAs, single-phase solid solution phases are more stable than the formation of such phases [6]. In Table 14.1, the variation of the mixture entropy (ΔS_{mix}) of the alloy according to the number of components (n) forming the alloy is given.

One of the features of HEAs is that the number of atoms in the lattice is higher and more diverse than in traditional alloys, leading to slower diffusion. Yeh et al. [6] compared diffusion rates in three different types of alloys and found that HEAs have slower diffusion rates than stainless steels and pure metals. This slow diffusion in HEAs results in many different properties, such as slowing down grain growth, exceptional thermal stability, and the formation of nanostructures (Figure 14.1a). Another effect of HEAs is lattice distortion (Figure 14.1b). The lattice in HEAs is composed of many types of elements, each with different sizes. These size differences inevitably lead to lattice distortion, with larger atoms pushing neighboring atoms apart and leaving extra space around smaller atoms. The strain energy associated with lattice distortion increases the overall free energy of the HEA lattice and affects its properties. For instance, lattice distortion can impede dislocation motion and cause significant solid solution strengthening. Additionally, it leads to increased electron and phonon scattering, resulting in lower electrical and thermal conductivity.

TABLE 14.1
Relationship between n and ΔS_{mix} in HEAs [7]

n	1	2	3	4	5	6	7	8	9	10	11	12	13
ΔS_{mix}	0	0.69	1.1	1.39	1.61	1.79	1.95	2.08	2.2	2.3	2.4	2.49	2.57

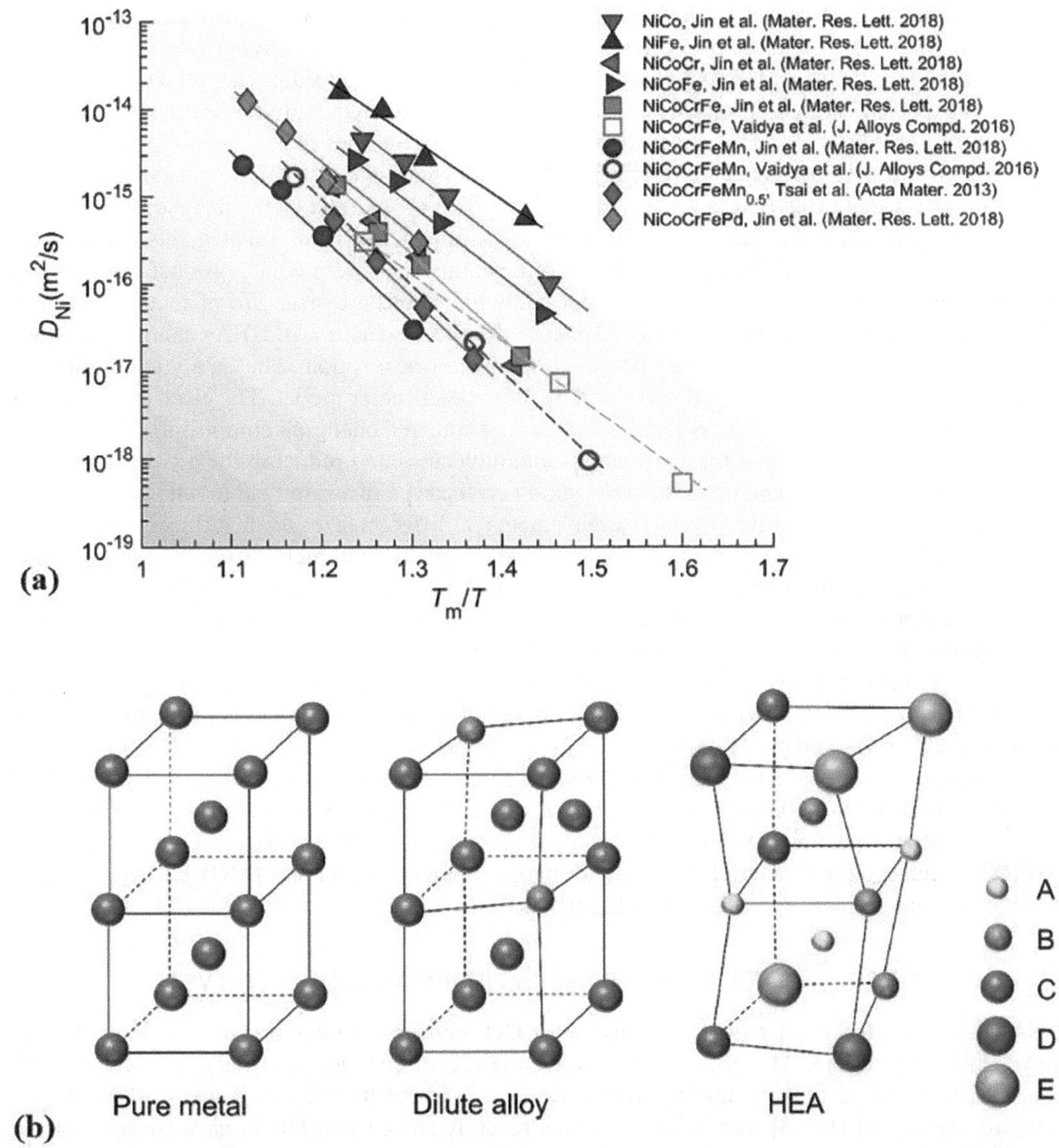

FIGURE 14.1 (a) Diffusion coefficients of Ni, D_{Ni}, in a series of fcc alloys, including high entropy alloys against the inverse homologous temperature, T_m/T, including NiCo, NiFe, NiCoFe, NiCoCr, NiCoFeCr, NiCoFeCrMn, NiCoFeCrPd, CoCrFeMnNi, CoCrFeNi, and $FeCoCrNiMn_{0.5}$ denotes melting temperature, (b) Schematics of lattice distortion in body-centered cubic pure metals, conventional dilute alloys, and high entropy alloys. (A–E) represent different element species in general. Adapted with permission [8]. Copyright (2021) Elsevier.

The presence of five or more different metals in HEAs, their differing properties and crystal structures from each other, and their random distribution in the alloy result in previously unpredictable properties of HEAs. This effect is called the cocktail effect [6]. The mentioned effects have enabled HEAs to have high strength, high hardness, good ductility, good wear resistance, superior fatigue resistance, excellent creep resistance, and exceptional corrosion resistance properties. In addition to mechanical properties, there are studies showing the high potential of HEAs in magnetic applications, radiation applications, and biomedical applications [9]. Recently, there are also studies suggesting that HEAs may have superior catalytic properties compared to traditional materials. The reasons for exhibiting superior properties in catalytic applications and having great potential can be explained as follows: (i) HEAs have multi-principal elements, and thus, the cocktail effect can lead to unexpected catalytic properties. They can show unexpected catalytic properties due to the superpositions of these elements rather than the individual properties of each element that make up the HEA. (ii) Most HEAs consist of combinations of transition metals that are considered as main elements for various catalytic reactions. (iii) The presence of multiple elements results in multiple atomic arrangements or surface microstructures containing active catalytic elements, which leads to the emergence of various adsorption modes for reactants and intermediate products. (iv) Mixing transition metals in different atomic ratios modifies the electronic structure of metals and gives HEAs a unique catalytic property. (v) The "distorted lattice" structure in HEAs can increase their potential energy and provide a lower potential barrier for adsorption. The catalytic activity and selectivity in HEAs are primarily determined by the d-band center model. The electronic structure of HEAs can be changed due to cage distortion, cage strain, and changing composition. The d-band shifts upward, creating a bond between metals and molecules and reducing the adsorption energy for reactants and intermediate products. Depending on the type of strain field (compressive vs. tensile), cage strain can be adjusted by selecting elements of different sizes to shift the existing d-band center upward, enabling stronger interaction between adsorbates and metallic components. In addition, the presence of different metallic elements on the surface (depending on surface composition and microstructure) can create a redistribution of surface charges with an alternative arrangement of accumulated and depleted areas. This adjustability can allow for improved selectivity. (vi) HEAs in the form of nanoparticles or nanosized particles are necessary for enhanced activity because the high surface area per volume provides efficient material usage and multiple active sites due to the shape effect of nanosized particles [9].

In the studies conducted, there are mainly seven catalytic reactions for HEAs, which are used as catalysts. These are ammonia decomposition reaction (ADR), ammonia oxidation reaction (AOR), oxygen reduction reaction (ORR), CO_2 reduction reaction (CO_2RR), hydrogen evolution reaction (HER), oxygen evolution reaction (OER), and methane oxidation reaction (MOR). These catalytic reaction mechanisms are summarized in Figure 14.2 [10].

14.3 CONVERSION OF CO_2 BY HIGH ENTROPY ALLOYS CATALYST

HEAs have attracted great potential catalysts for CO_2 conversion owing to their unique physical and chemical properties. The potential for the CO_2 reduction reaction (CO_2RR) to produce a variety of carbon-neutral fuels, including methane, methanol, and formic acid among others, makes it highly significant [11–13]. For instance, it was recently shown that HEAs have strong catalytic activity for electrochemical CO_2 reduction, and certain HEAs even outperform the catalytic activity of traditional precious metal catalysts like copper (Cu) [14]. Moreover, by altering their micro/nanostructure and chemical makeup, HEAs may be tailored to specific applications [8]. They have a large surface area, lots of active sites, and variable electronic and surface characteristics, making them suitable catalysts for the dissociation of CO_2. Current research has the potential to boost green technology for climate change mitigation.

For the design of CO_2 conversion catalysts with a desired activity, selectivity, and stability, the capacity to contain numerous primary elements in a single lattice is required. First, one or two primary constituents largely control the properties of conventional alloys. Low concentrations of dopants would only be used to fine-tune the predetermined features. The discovery of HEAs paves the way for so far unimaginable compositional diversity and structural complexity [15]. In addition, the increased configurational entropy can be used to make up for the multielement incompatibility that resulted from the significant enthalpic variances. By doing this, the concentrations of the primary elements can be increased, perhaps exceeding the miscibility barriers in the bulk phase [16]. This makes a wider variety of geometric and electrical properties

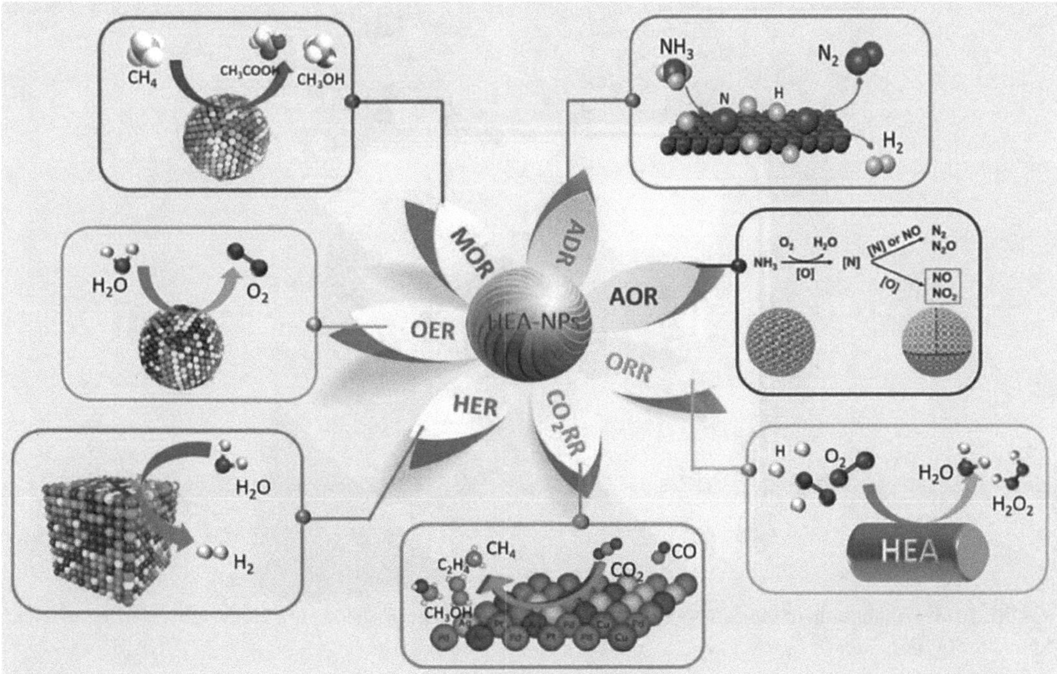

FIGURE 14.2 Seven catalytic reaction mechanisms of HEA-NPs. The seven catalytic applications include Ammonia Decomposition Reaction (ADR), Ammonia Oxidation Reaction (AOR), Oxygen Reduction Reaction (ORR), CO_2 Reduction Reaction (CO_2RR), Hydrogen Evolution Reaction (HER), Oxygen Evolution Reaction (OER), Methane oxidation reaction (MOR). Adapted with permission [10]. Copyright (2022) Elsevier.

available. Second, the large lattice distortion brought on by the difference in atomic size results in a thermodynamically unstable condition [17]. This might lower the energy barrier for molecule adsorption, activation, and conversion. The energy levels of bound intermediates are significantly altered by distortion-derived local strain, which is unavoidable in a diverse and complex system like HEAs [18]. Third, the slower diffusion kinetics caused by the greater diffusion activation energies of HEAs have been identified as the primary driver of the improved chemical, thermal, and mechanical stabilities. HEAs are appealing possibilities for better catalysts due to these unusual but desirable characteristics. Moreover, they are also useful platforms for investigating the enthalpy-entropy correlations on the catalytic side.

The effectiveness of HEAs as CO_2 reduction catalysts can be further enhanced by modifying their surface properties, such as by depositing thin films or nanoparticles of HEAs. The reduction of CO_2 on HEAs mainly involves adsorption of CO_2 on the surface, dissociation of CO_2 into CO and oxygen, reduction of CO into other products, and desorption of the products from the surface of HEA (Figure 14.3) [14]. Although the precise process by which CO_2 is reduced in HEAs is unknown, it is thought to involve a combination of electrochemical and catalytic activity.

The following is a list of some of the most typical CO_2 conversion reactions that HEAs can catalyze;

CO_2 reduction to carbon monoxide: For CO generation, COOH is first produced when a carbon atom binds to the catalyst surface, then CO is formed when COOH is dehydrogenated, and finally, CO is released when the catalyst is desorbed [19]. The initial critical intermediate for further reduction is CO and CO's binding energy controls its selectivity. The CO will be desorbed from the catalyst surface to produce CO if the binding energy is insufficient. The catalytically active sites are poisoned, and the HER takes over if the catalyst binds CO extremely strongly.

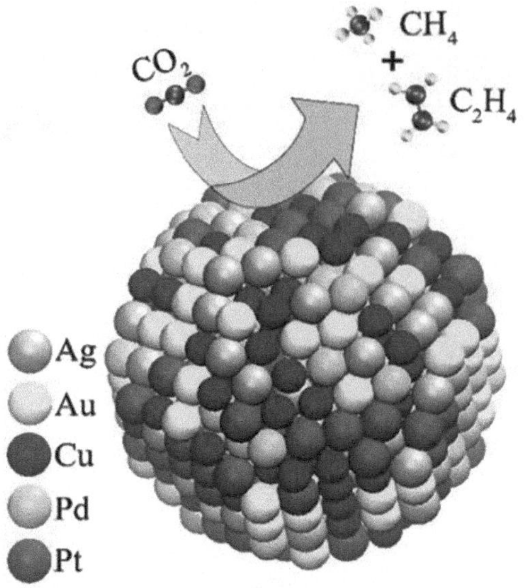

FIGURE 14.3 Schematic of the catalysis reaction. Adapted with permission [14]. Copyright (2020) American Chemical Society.

CO_2 methanation: HEAs can catalyze CO_2 methanation using H_2 as a reducing agent. The CO_2 methanation mechanism involves the adsorption of CO_2 molecules on the HEA surface, followed by the dissociation of H_2 to produce atomic hydrogen. The atomic hydrogen reacts with the adsorbed CO_2 molecules to form methane. The Sabatier reaction, commonly known as the exothermic CO_2 methanation reaction (equation 14.1):

$$CO_2 + 4H_2 \rightarrow CH_4 + 2H_2O, \quad H = 165.0 \ \text{kJ/mol} \tag{14.1}$$

has gained new attention as a result of the recent advancement of the power-to-gas concept [20].

CO_2 hydrogenation: HEAs can catalyze the reduction of CO_2 into hydrogen using H_2. The CO_2 hydrogenation process starts with the adsorption of CO_2 molecules on the HEA surface, then H_2 is split apart to form atomic hydrogen. Methanol (CH_3OH) and other useful compounds are created when the atomic hydrogen interacts with the adsorbed CO_2 molecules. For instance, one of the most promising ways to produce CO as a crucial feedstock for Fischer–Tropsch processes and as an intermediary step for the further synthesis of fuel and chemicals is the endothermic reverse water-gas shift reaction (equation 14.2) [21,22]:

$$CO_2 + H_2 \rightarrow CO + H_2O, \quad H = 41.0 \ \text{kJ/mol} \tag{14.2}$$

The following high entropy materials have been used in the conversion of carbon dioxide.

CoCuGaNiZn and AgAuCuPdPt alloys: Pedersen and colleagues investigated electrochemical CO_2 and CO reduction on two quinary fcc HEAs, CoCuGaNiZn and AgAuCuPdPt, using density functional theory (DFT) and supervised machine learning [23]. Nanocrystalline AuAgPtPdCu HEAs were employed by Nellaiappan et al. for the electrocatalytic reduction of CO_2 [4]. Nearly all of the CO_2 was converted to gaseous products by the HEA catalyst at low overpotentials (0.3 V vs. reversible hydrogen electrode, or RHE) (38.0% CH_4, 29.5%

C_2H_4, 4.0% CO, and 26.4% H_2). A Cu species with the Cu^{2+}/Cu^0 redox pair was shown to be the primary active site for CO_2 reduction. It is yet unclear how CO_2 reduction intermediates behave on the HEA surface, and cascade reactions may be able to produce higher-value chemicals.

CoNiCuRuPd alloys: Mori et al. synthesized supported HEA nanoparticles at low temperatures by utilizing the hydrogen spillover that occurs on TiO_2 via a linked proton-electron transfer mechanism [24]. There is evidence from both in-situ observations and theoretical simulations that Pd^{2+} ions are first reduced by H_2 to produce nuclei, and then hydrogen molecules are split apart to create active hydrogen atoms that allow the simultaneous reduction of nearby precursors. When hydrogenating CO_2, a CoNiCuRuPd/TiO_2 catalyst made in this way shows different selectivity and greatly increased stability compared to Pd/TiO_2. Theoretical studies also stressed that the combination of various metals causes the slow diffusion in these CoNiCuRuPd HEA nanoparticles and that lattice distortion plays a critical role in the superior robustness of this material. They found that the high entropy alloy nanoparticles were highly stable and efficient at converting CO_2 to methane.

(NiMgCuZnCo)O alloys: The mechanochemical production of the HE metal oxide (NiMgCuZnCo)O under ambient conditions is reported (Figure 14.4a) [25]. The use of single atoms/nanoclusters of up to 5 wt% of noble metal in (NiMgCuZnCo)O, displayed

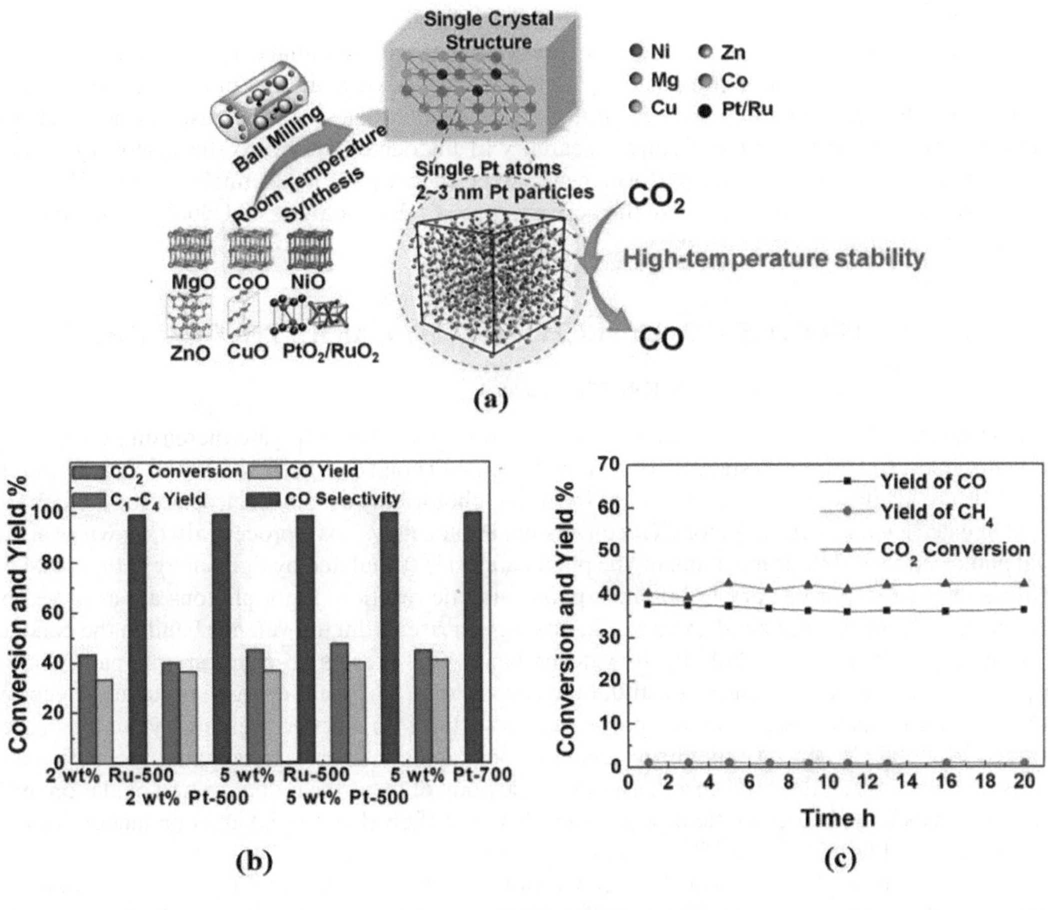

FIGURE 14.4 (a) Mechanochemical synthesis of Pt/Ru-(NiMgCuZnCo)O entropy-stabilized metal oxide solid solution. (b) CO_2 hydrogenation activity of 2 wt% Pt, 5 wt% Pt, 2 wt% Ru, and 5 wt% Ru under 500°C reaction temperature, (c) CO_2 hydrogenation stability at 500°C over 5 wt% Pt. Adapted with permission [25]. Copyright (2019), American Chemical Society.

good stability at high temperatures and created high catalytic activity in hydrogenation of atmospheric CO_2 to CO. Figure 14.4b illustrates the results of catalytic testing using 2 wt% Ru and 2 wt% Pt catalysts, which produced CO_2 conversions of 40.1% and 43.4% and CO yields of 33.9% and 36.6%, respectively. The yields of CO rose to 45.7% and 46.1% for 5 wt% Ru and 5 wt% Pt, respectively, with CO_2 conversions of 45.4% and 47.8%. There were not many additional gaseous products found, and all four of these catalysts had CO selectivities above 95%.

ZnCoCdNiCu-Zeolitic imidazolate frameworks: The creation of high entropy inorganic-organic hybrids is thought to be difficult synthetically due to the restricted high-temperature stabilities of molecular linkers. Using mechanochemistry, Dai and colleagues created high entropy ZnCoCdNiCu-zeolitic imidazolate frameworks (HE-ZIFs) in a natural environment [26]. The local short-range heating that occurs during ball milling speeds up molecular diffusion, causing Zn^{2+}, Co^{2+}, Cd^{2+}, Ni^{2+}, and Cu^{2+} to disperse randomly inside the 2-methylimidazole linker-coordinated ZIF lattice. The HE-ZIF retained a large surface area (1147 m^2/g), as an advantage for the adsorption of molecular species. To create carbonates, which are often utilized in the electrochemical and pharmaceutical sectors, HE-ZIFs were used in the cycloaddition of CO_2 with epoxides. In contrast to the single-cation ZIF and their physical mixtures, the five highly scattered metal sites in the HE-ZIF act as effective Lewis acidic sites in epoxide activation. This results in a higher yield for the conversion of CO_2.

The effectiveness of HEAs as CO_2 reduction catalysts can be enhanced by several strategies, including the optimization of composition, tuning of morphology, and the modification of surface properties. The deposition of HEAs as thin films or nanoparticles can increase surface area and improve catalytic performance. Future research will undoubtedly lead to the discovery of new HEAs with even better catalytic performance, paving the way for a sustainable future. However, further studies are needed to optimize the performance of HEA catalysts and elucidate the underlying reaction mechanisms.

14.4 CONVERSION OF CO_2 BY HIGH ENTROPY OXIDES PHOTOCATALYST

14.4.1 Photocatalytic Conversion Mechanism

Photocatalytic CO_2 reduction is considered a promising solution to mitigate increasing greenhouse gas emissions. CO_2 has high stability due to the high C=O bond energy (750 kJ/mol) [27]. Different methods are available for CO_2 conversion, including photocatalysis and electrocatalysis, which use light or electricity, respectively, for CO_2 conversion. Photocatalysis is a process also known as artificial photosynthesis. In this mechanism, the photocatalyst is stimulated by light energy (Figure 14.5). This is the first step necessary to initiate a photocatalytic reaction. Light photons are absorbed by the electron in the valence band. As a result, an electron rises from the valence band to the conduction band and then leaves a hole in the valence band. This creates an electron-hole pair. Charge separation and surface diffusion constitute the second step. The light energy-stimulated electrons move toward the active region on the photocatalyst surface. This active region is typically an area formed by metal elements. In the third step, electrons on the interface participate in reduction and holes in oxidation reactions. From a thermodynamic point of view, the reduction and oxidation reaction potentials of a photocatalytic process occur between the valence band and conduction band of a photocatalyst [28–32].

The first step in photocatalytic CO_2 conversion is the adsorption of CO_2 molecules onto the catalyst surface, which occurs in three modes. These are oxygen, carbon dioxide, and mixed coordination. All the resulting CO_2 adsorbates undergo bending that results in the loss of linear symmetry exhibited by free CO_2 molecules. This condition provides more activation with decreasing barriers [33]. These modes are explained in more detail in Figure 14.6. In the first case, the oxygen

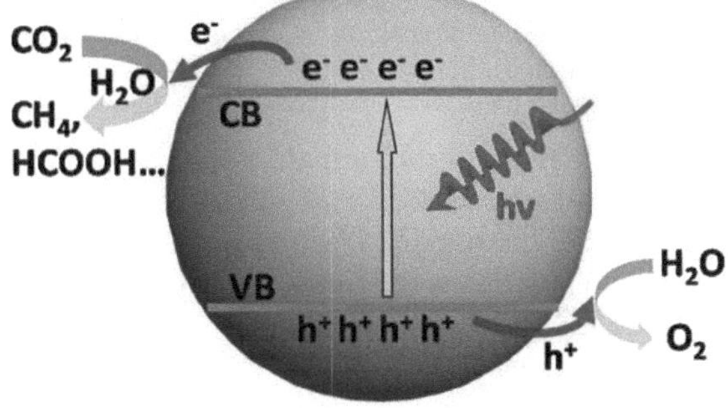

FIGURE 14.5 Schematic illustration of photoinduced generation of electron-hole pairs in semiconductor for CO_2 reduction with water. Adapted with permission [30]. Copyright (2015) Elsevier.

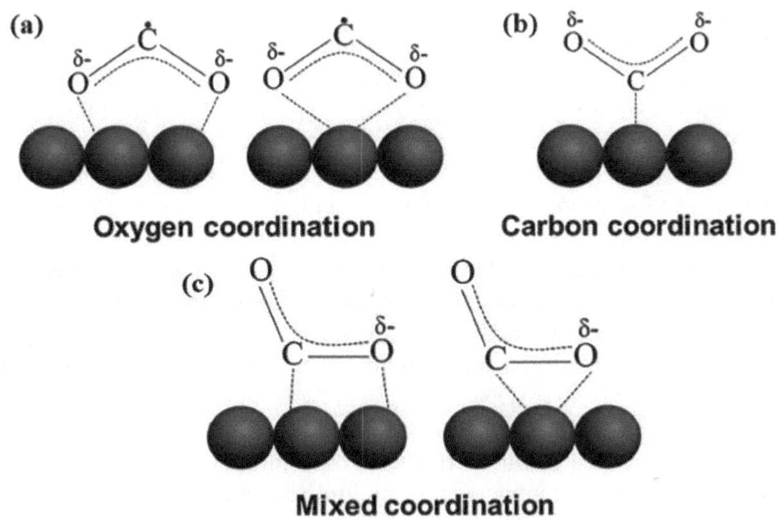

FIGURE 14.6 The schematic diagram reflects three different coordination modes of CO_2 molecules. Adapted with permission [32]. Copyright (2020) Royal Society of Chemistry.

atom in the CO_2 molecule is an electron donor and provides a pair of electrons to the surface Lewis acid center, resulting in two different bidentate binding structures known as oxygen coordination (Figure 14.6a). In the second case, the carbon atom is an electron acceptor, and when the Lewis acid center takes electrons from the Lewis base, a carbon-containing compound is formed, known as carbon coordination (Figure 14.6b). In the third case, both the oxygen and carbon atoms in the CO_2 molecule simultaneously donate and accept electrons, resulting in two different bidentate binding structures known as mixed coordination (Figure 14.6c). The coordination modes affect the formation of intermediate products in the reduction stage and lead to different reduction pathways [32].

According to reports, there are three primary routes for CO_2 photoreduction: (i) carbene pathway, (ii) formaldehyde pathway, and (iii) glycolaldehyde pathway [33]. The formaldehyde pathway includes the merging of CO_2 radicals with two protons leading to the creation of formic acid. Alternatively, formic acid can generate methanol and methane through the action of two protons. On the other hand, the carbene pathway involves the reaction of CO_2 radicals with a proton to break the C=O bond and produce CO. If the photocatalyst has moderate adsorption strength, CO is continuously reduced to a carbon radical by accepting electrons and protons. Methyl radicals can form methanol or methane by recombining with protons or hydroxyl groups, depending on the specific circumstances. The glycosyl pathway uses more C_2 compounds in the CO_2 reduction process.

Along this pathway, $CO_2^{\cdot -}$ radical combines with a proton to form a bidentate and then reunites with another proton to form monodentate formic acid. Free formyl radicals (HC˙O) are produced by electron transfer and oxygen transfer. These radicals can lead to other C_2 products, such as aldehydes, that can be converted to glyoxal [33]. However, in their study, Wang et al. reported that the intermediate profiles did not match those of the formaldehyde pathway with the studies they referenced and described the formyl pathway. In the formyl pathway, if the photocatalyst has moderate adsorption strength, the C-bound CO intermediate receives an electron and bonds to a proton on the C atom. The other proton-electron pair forms the formyl intermediate (CHO) that can be converted to CH_2O [34].

14.4.2 High Entropy Oxide Photocatalysts

Active photocatalysts are designed based on certain strategies, including increasing light absorbance, aligning band structures, narrowing band gaps, accelerating charge carrier migration, preventing electron-hole recombination, and providing active sites to enhance photocatalytic activity. It is known that high entropy oxides are used as photocatalysts for this purpose. Various HEOs can be created in the crystal structure through different combinations of metallic elements. Examples of such combinations include (Mg, Co, Ni, Cu, Zn), (Mg, Ti, Zn, Cu, Fe), and (Zr, Ce, Hf, Ti, La, Y, Gd, Ca, Mg, Mn), which form rock salt, fluorite, and spinel phase HEOs, respectively [35]. High entropy oxides (HEOs) were first defined by Rost et al. in 2015, using the definition of a single-phase oxide system containing five or more cations with a configuration entropy ($S_{config.}$) greater than 1.5 R (R, gas constant) [31,36]. The presence of different metal atoms in the same crystal structure in HEOs leads to atomic disorder, resulting in high entropy. Compared to doped transition metal oxides and low or medium-entropy oxides, the irregular chemical composition and atomic distribution of HEOs give them superior thermodynamic, crystal structure, and kinetic properties, resulting in properties such as high mechanical and high-temperature resistance [37]. High entropy ceramics, including HEOs and oxynitrides (HEONs), have recently been introduced as stable catalysts with high light absorption in the visible region. The catalytic advantage of HEOs is attributed to the confinement of different cation species with different radii within the same cage, leading to cage distortion and oxygen vacancy. The confinement of multiple different cation species in the same cage increases the configuration entropy. The cage structure in HEOs reduces the energy of the entire system. The increase in the number of oxygen vacancies in the catalytically active center (where the metal elements are located) supports the activation and transportation of materials. In addition, the use of catalytically active rare earth elements in addition to transition metals expands the diversity of the structures and catalytically active sites of HEOs [35].

HEOs fully meet the unique chemical and physical properties of high entropy alloys. HEOs can be used as catalysts due to their ability to maintain thermal stability at high temperatures and exhibit different reactivities due to their multi-component structures. Moreover, it is known that d0 and d10 electronic configuration oxides and oxynitrides show good photocatalytic activity, so selecting HEOs with five d0 cations is advantageous for designing photocatalysts [5]. In photocatalytic CO_2 reduction, active sites on the HEO surface adsorb CO_2 and are converted to CO via photocatalytic reduction. CO can then be used for the synthesis of hydrocarbons or other chemical products. The CO_2 conversion performance of HEAs depends on the alloy components and reaction conditions [37]. As mentioned before, these reactions occur via the electrons and holes formed on the surface of HEOs. In addition, CO_2 conversion performance is also dependent on other factors such as composition, temperature, and atmospheric conditions. Optimizing these factors can increase the CO_2 conversion performance of photocatalysts [29]. Currently, there are few studies on HEOs due to them being a new topic. These studies are summarized below.

Edalati has designed and synthesized a high entropy oxide system with multiple heterojunctions, Ti–Zr–Nb–Ta–WO_{12}, which is a first-of-its-kind for visible-light-driven photocatalytic oxygen production. They have designed this system as a photocatalyst that can produce oxygen under visible light without the need for a co-catalyst, and it has 10 types of hetero-bonds. The first selected combination of Ti, Zr, Hf, Nb, and Ta is defined by high light absorbance and a low bandgap [37]. Akrami investigated the photocatalytic CO_2 conversion of $TiZrHfNbTaO_{11}$. They found that this HEO structure exhibits activity in catalyzing the simultaneous conversion of CO_2 to CO and H_2O to H_2. They attributed this property to the suitability of the electron band structure, the long carrier lifetime, and the two-phase structure with defects and strains [31]. Edalati et al. developed $TiZrHfNbTaO_6N_3$ (HEON) to enhance the activity of the catalyst in the visible light region. They found that the produced high entropy oxynitride had a smaller band gap than $TiZrHfNbTaO_{11}$ with a band gap of 2.9 eV in their previous study. HEON with a band gap of 1.6 eV was defined as a semiconductor for photocatalysts. They also found that HEON could absorb a larger amount of light in the visible light region compared to binary oxides $TiZrHfNbTaO_6N_3$. The low electron-hole recombination rate of HEON makes it successful in producing photocurrent. It was stated that HEON is promising for photocatalytic CO_2 production due to its high stability and low band gap [31]. Jiang synthesized Fd-3m spinel-structured HEO $(Ga_{0.2}Cr_{0.2}Mn_{0.2}Ni_{0.2}Zn_{0.2})_3O_4$ via a sol-gel method. They noted that the calcination temperature of the catalyst affected the photocatalytic activity of CO_2. They reported that the calcination temperature altered the morphology, crystal phase, element distribution, and surface defects of the HEO structure. They stated that the improved photocatalytic performance of HEO was dependent on its porous structure, good light absorption capacity, abundant defect regions, and especially its appropriate energy band structure [38]. Dai synthesized a series of HEOs using solid-state synthesis, sol-gel, co-precipitation, and other methods. They found that HEOs ($MgNiCuZnCoO_x$) used as catalysts or supports exhibited higher temperature stability and water resistance activity in CO catalytic oxidation compared to traditional catalysts such as CuO and CeO, among others [36].

14.4.3 Photocatalyst Selection and Its Effect on Photocatalyst Efficiency

The photocatalyst is the central component in photocatalytic conversion, and its selection significantly impacts the efficiency of CO_2 conversion. To improve catalyst performance, strategies aim to increase the number and activity of active sites. The effectiveness of photocatalytic CO_2 conversion is largely determined by light harvesting, charge separation, and charge energy levels, all of which are influenced by charge carrier dynamics. Therefore, researchers have explored various approaches to tailor the electronic structures of photocatalysts and promote charge separation in materials. These methods, which have shown promising results in producing highly carbonaceous products, bear similarities to other photocatalytic reactions [33]. The variables that affect the efficiency of the photocatalyst can be summarized as follows:

- The catalytic activities, selectivities, surface areas, surface acid-base properties, and surface reactivity of high entropy oxides can have an effect on CO_2 conversion efficiency. Two dominant approaches to describe surface acidity and basicity are the Lewis and Brönsted-Lowry theories. A catalyst with a high surface area will have a more catalytically active area and thus provide higher conversion efficiency. Surface acid-base properties control the interaction of a catalyst with the reaction environment. For example, if basic groups exist on the surface of high entropy oxides, they can provide the acid catalysis required for the protonation of CO_2 and the formation of carbonic acid. Also, if acidic groups exist on the surface, they can provide the protonation required for CO_2 to react with reducing molecules [39]. The surface of high entropy oxides has a balance between stability and reactivity. This balance means that an increase in surface reactivity can lead to an increase in catalytic activity, but at the same time, an increase in surface reactivity can lead to a decrease in the stability of the catalyst. Additionally, surface reactivity indicates that enough active area is present for CO_2's catalytic conversion [31,32].

- Most photocatalysts require a photon energy that is primarily dependent on the catalyst's bandgap for electron excitation. The bandgap of many photocatalysts is wider than 3.2 eV, indicating the need for a light source with a wavelength of less than 400 nm for electron excitation. This restricts the use of solar radiation as a radiation source. Additionally, the rapid recombination of electrons and holes reduces the efficiency of CO_2 conversion. As a result, efforts in this field are mostly focused on the development of heterogeneous photocatalysts with a narrow bandgap and the facilitation of charge separation/transport [28].
- The oxygen vacancies are the reaction centers for the adsorption and dissociation of CO, H_2O, and CO_2. The most important factor that distinguishes high-configuration entropy oxides from low-entropy oxides is that they form a single-phase solid solution that ensures the stability of the catalyst. The formation of a solid solution with different atomic sizes and the resulting lattice distortion leads to the formation of many oxygen vacancies in the material, which ultimately increases the catalytic activity. In addition, the different diffusion rates of components with different atomic sizes and the resulting changes in chemical bonds contribute to the superior thermal and chemical stability of the catalyst. Furthermore, the cocktail effect directly increases the catalytic efficiency by enhancing the catalytic activity [36].
- The design of photoreactors has become an effective means to improve CO_2 conversion efficiency, as the reactor's performance is influenced by key parameters such as heat exchange, material selection, light source, and operating conditions. Photoreactors utilized in the photocatalytic reduction of CO_2 to valuable products are generally categorized into slurry, fixed-bed, and membrane types [28].

14.5 CONCLUSION AND OUTLOOK

The development of HEAs as catalysts for CO_2 conversion has shown significant promise in recent years. The experiments have definitely shown the significance of managing configurational entropy on the catalyst side, but further studies are required to completely comprehend the interaction between structural and compositional engineering and the optimization of catalytic behaviors. Innovative synthetic, computational, and characterization approaches will produce a larger and more precisely defined library of HEAs, with expected tailoring of the physical properties such as lattice, occupancy, surface, interface, active sites, and defect. The interactions between reactants, intermediates, and products, as well as the catalytic platform with adjustable entropic contributions will be beneficial in supplying details on effective and selective catalytic processes, making HEAs highly effective in catalyzing the conversion of CO_2 to value-added products such as CO, CH_4, and other hydrocarbons. However, there is still much to be explored in this field, including the optimization of HEA composition and structure and the development of novel synthesis methods.

Future research directions for HEAs in CO_2 conversion could include the investigation of new HEA compositions and structures and the development of advanced synthesis techniques such as high-throughput experimentation and machine learning-assisted synthesis. In addition, using advanced characterization techniques such as in situ spectroscopy and microscopy can provide a deeper understanding of the mechanisms underlying CO_2 dissociation over HEAs. These efforts can lead to discovery of new HEA catalysts with improved activity and selectivity for CO_2 conversion. Moreover, the application of HEAs in CO_2 conversion can also benefit from collaborations between materials scientists, chemists, and engineers. The development of scalable and cost-effective processes for the production and utilization of HEA catalysts can be achieved through interdisciplinary efforts. Additionally, the integration of HEA catalysts with other renewable energy sources, such as solar or wind power, can provide a sustainable approach to CO_2 conversion and utilization.

One of the major challenges in the use of HEAs for CO_2 conversion is their stability and durability under high-temperature and high-pressure conditions. HEAs are susceptible to oxidation and degradation, which can lead to a decrease in catalytic activity over time. Therefore, future research should focus on developing HEAs with improved stability and durability to ensure long-term catalytic performance.

Another area for future research is the utilization of HEAs in tandem with other catalysts and/or reactors to improve the efficiency of CO_2 conversion. For example, HEAs can be combined with zeolites or other porous materials to enhance their catalytic activity and selectivity. Additionally, HEAs can be used in combination with membrane reactors to separate and purify the reaction products, leading to higher yields of desired products.

Finally, the application of HEAs in CO_2 conversion can have a significant impact on the reduction of greenhouse gas emissions and the transition toward a sustainable energy economy. By utilizing CO_2 as a feedstock for the production of valuable chemicals and fuels, HEAs can contribute to the circular economy and reduce the dependence on fossil fuels. However, further research and development are needed to fully realize the potential of HEAs in CO_2 conversion and to address the technical and economic challenges associated with their commercialization. Overall, the development of HEAs as catalysts for CO_2 conversion represents a promising avenue for sustainable energy production and climate change mitigation. Further research and development in this field will be critical in achieving a more sustainable and low-carbon future.

REFERENCES

[1] E. Alper, O.Y. Orhan, CO2 utilization: Developments in conversion processes, *Pet.* 3 (2017) 109–126.

[2] J. Yin, M. Zheng, J. Chen, The effects of environmental regulation and technical progress on CO2 Kuznets curve: An evidence from China, *Energy Policy.* 77 (2015) 97–108.

[3] W. Li, H. Wang, X. Jiang, J. Zhu, Z. Liu, X. Guo, C. Song, A short review of recent advances in CO2 hydrogenation to hydrocarbons over heterogeneous catalysts, *RSC Adv.* 8 (2018) 7651–7669.

[4] W.G. Cui, G.Y. Zhang, T.L. Hu, X.H. Bu, Metal-organic framework-based heterogeneous catalysts for the conversion of C1 chemistry: CO, CO2, and CH4, *Coord. Chem. Rev.* 387 (2019) 79–120.

[5] M. Aresta, *Carbon dioxide as a chemical feedstock*, Weinheim, Wiley-VCH (2010).

[6] Y. Zhang, T.T. Zuo, Z. Tang, M.C. Gao, K.A. Dahmen, P.K. Liaw, Z.P. Lu, Microstructures and properties of high-entropy alloys, *Prog. Mater. Sci.* 61 (2014) 1–93.

[7] B.S. Murty, J.W. Yeh, S. Ranganathan, P. Bhattacharjee P, *High-entropy alloys*, Chapter 1, Elsevier (2019).

[8] W. Li, D. Xie, D. L, Y. Zhang, Y. Gao, Y.K. Liaw, Mechanical behavior of high-entropy alloys, *Prog. Mater. Sci.* 118 (2021) 100777.

[9] N.K. Katiyar, K. Biswas, J.W. Yeh, S. Sharma, C.S. Tiwary, A perspective on the catalysis using the high entropy alloys, *Nano Energy.* 88 (2021) 106261.

[10] Y. Wang, Y.H. Wang, High-entropy alloys in catalyzes and supercapacitors: Progress, prospects, *Nano Energy.* 104 (2022) 107958.

[11] A. Das, S.C. Mandal, B. Pathak, Unraveling the catalytically preferential pathway between the direct and indirect hydrogenation of CO2 to CH3OH using N-heterocyclic carbene-based Mn(I) catalysts: A theoretical approach, *Catal. Sci. Technol.* 11 (2021) 1375–1385.

[12] M. Umeda, Y. Niitsuma, T. Horikawa, S. Matsuda, M. Osawa, Electrochemical reduction of CO2 to methane on platinum catalysts without overpotentials: Strategies for improving conversion efficiency, *ACS Appl. Energy Mater.* 3 (2020) 1119–1127.

[13] L. Fan, C. Xia, P. Zhu, Y. Lu, H. Wang, Electrochemical CO2 reduction to high-concentration pure formic acid solutions in an all-solid-state reactor, *Nat. Commun.* 11 (2020) 3633.

[14] S. Nellaiappan, N.K. Katiyar, R. Kumar, A. Parui, K.D. Malviya, K.G. Pradeep, A.K. Singh, S. Sharma, C.S. Tiwary, B. Krishanu, High-entropy alloys as catalysts for the CO2 and CO reduction reactions: Experimental realization, *ACS Catal.* 10(6) (2020) 3658–3663.

[15] Y. Sun, S. Dai, High-entropy materials for catalysis: A new frontier, *Sci. Adv.* 7(20) (2021) 1–23.

[16] J.W. Yeh, Alloy design strategies and future trends in high-entropy alloys, *JOM.* 65 (2013) 1759–1771.

[17] Y. Xin, S. Li, Y. Qian, W. Zhu, H. Yuan, P. Jiang, R. Guo, L. Wang, High-entropy alloys as a platform for catalysis: Progress, challenges, and opportunities, *ACS Catal.* 10 (2020) 11280–11306.

[18] A. Khorshidi, J. Violet, J. Hashemi, A. A. Peterson, How strain can break the scaling relations of catalysis, *Nat. Catal.* 1 (2018) 263–268.

[19] B. Wu, J. Chen, L. Quian, Recent advances in heterogeneous electroreduction of CO2 on copper-based catalysts, *Catal.* 12(8) (2022) 860.
[20] I. Sreedhar, Y. Varun, S.A. Singh, A. Venugopal, B. M. Reddy, Developmental trends in CO2 methanation using various catalysts, *Catal. Sci. Technol.* 9 (2019) 4478–4504.
[21] S. S. Kim, H. H. Lee, S. C. Hong, A study on the effect of support's reducibility on the reverse water-gas shift reaction over Pt catalysts, *Appl. Catal. A-Gen.* 423 (2012) 100–107.
[22] G. Centi, E. A. Quadrelli, S. Perathoner, Catalysis for CO2 conversion: A key technology for rapid introduction of renewable energy in the value chain of chemical industries, *Energy Environ. Sci.* 6 (2013) 1711–1731.
[23] J. K. Pedersen, T. A. A. Batchelor, A. Bagger, J. Rossmeisl, High-entropy alloys as catalysts for the CO2 and CO reduction reactions, *ACS Catal.* 10 (2020) 2169–2176.
[24] K. Mori, N. Hashimoto, N. Kamiuchi, H. Yoshida, H. Kobayashi, H. Yamashita, Hydrogen spillover-driven synthesis of high-entropy alloy nanoparticles as a robust catalyst for CO2 hydrogenation, *Nat. Commun.* 12 (2021) 3884.
[25] H. Chen, W. Lin, H. Zhang, K. Jie, D.R. Mullins, X. Sang, S.Z. Yang, C.J. Jafta, C.A. Bridges, X. Hu, R.R. Unocic, J. Fu, P. Zhang, S. Dai, mechanochemical synthesis of high entropy oxide materials under ambient conditions: Dispersion of catalysts via entropy maximization, *ACS Mater. Lett.* 1, 1 (2019) 83–88.
[26] W. Xu, H. Chen, K. Jie, Z. Yan, T. Li, S. Dai, Entropy-driven mechanochemical synthesis of polymetallic zeolitic imidazolate frameworks for CO2 fixation, *Angewandte Chem. Int. Ed.* 58 (2019) 5018–5022.
[27] Z. Kovacic, B. Likozar, M. Hus, Photocatalytic CO2 reduction: A review of ab initio mechanism, kinetics, and multiscale modeling simulations, *ACS Catal.* 10 (2020) 14984–15007.
[28] F.O. Ochedi, D. Liu, J. Yu, A. Hussain, Y. Liu, Photocatalytic, electrocatalytic and photoelectrocatalytic conversion of carbon dioxide: A review, *Environ. Chem. Lett.* 19 (2021) 941–967.
[29] A. Kumar, G. Pandey, A review on the factors affecting the photocatalytic degradation of hazardous materials, *Mater. Sci. Eng. Int. J.* 1 (2017) 1–10.
[30] L. Yuan, Y. J. Xu, Photocatalytic conversion of CO2 into value-added and renewable fuels, *Appl. Surf. Sci.* 342 (2015) 154–167.
[31] S. Akrami, P. Edalati, M. Fuji, K. Edalati, High-pressure torsion for highly-strained and high-entropy photocatalysts, *KONA Powder Part. J.* (2023) 2024003.
[32] D. Qin, Y. Zhou, W. Wang, C. Zhang, G. Zeng, D. Huang, L. Wan, H. Wang, Y. Yang, L. Lei, S. Chen, D. He, Recent advances in two-dimensional nanomaterials for photocatalytic reduction of CO2: Insights into performance, theories, and perspective, *J. Mater. Chem. A.* 8 (2020) 19156–19195.
[33] T. Kong, Y. Jiang, Y. Xiong, Photocatalytic CO2 conversion: What can we learn from conventional COx hydrogenation?, *Chem. Soc. Rev.* 49 (2020) 6579–6591.
[34] Y. Wang, E. Chen, J. Tang, Insight on reaction pathways of photocatalytic CO2 conversion, *ACS Catal.* 1 (2022) 7300–7316.
[35] Y. Gao, Y. Liu, H. Yu, D. Zou, High-entropy oxides for catalysis: Status and perspectives, *Appl. Catal. A- Gen.* (2022) 118478.
[36] Y. Pan, J.X. Liu, T.Z. Tu, W. Wang, G.J. Zhang, High-entropy oxides for catalysis: A diamond in the rough, *J. Chem. Eng.* 451 (2023) 138659.
[37] P. Edalati, Y. Itago, H. Ishihara, T. Ishihara, H. Emami, M. Arita, M. Fuji, K. Edalati, Visible-light photocatalytic oxygen production on a high-entropy oxide by multiple-heterojunction introduction, *J. Photochem. Photobiol. A.* 433 (2022) 114167.
[38] Z. Jiang, R. Zhang, H. Zhao, J. Wang, L. Jia, Y. Hu, K. Wang, X. Wang, Preparation of (Ga0.2Cr0.2Mn0.2Ni0.2Zn0.2)3O4 high-entropy oxide with narrow bandgap for photocatalytic CO2 reduction with water vapor, *Appl. Surf. Sci.* 612 (2023) 155809.
[39] S.H. Albedwawi, A. AlJaberi, G.N. Haidemenopoulos, K. Polychronopoulou, High entropy oxides-exploring a paradigm of promising catalysts: A review, *Mater. Des.* 202 (2021) 109534.

15 High Entropy Materials as Electrodes for Supercapacitors

Amit K. Gupta, Priyanka Kumari, Aryan Singh, and Rohit R. Shahi

15.1 INTRODUCTION

The continuous consumption of non-renewable resources and their consequences (like global warming and contribution to greenhouse gases causing significant environmental changes) motivates scientists and researchers toward renewable and sustainable energy sources. The severe issue with renewable energy sources is that these sources cannot produce power 24 hours. The developed energy storage devices must be eco-friendly and will deliver high energy on demand. Batteries are efficient ways to store energy, and it is one of the dominant sectors in energy storage. However, a battery has certain limitations, such as low cycle life, low energy density, very high charging time, disruption in power supply, and use of non-eco-friendly electrode materials [1]. As compared to the commercially available Li batteries (~4 V), which are relatively expensive flammable, and have toxic materials, aqueous supercapacitors (1.5–2 V) offer higher power density, cyclability, and safety with lower cost; thus, the electrochemical capacitors /supercapacitors are a more sustainable option [1].

The supercapacitor is one of the efficient ways to store energy. It has many triumphs over batteries in terms of cyclic stability (SC > 300,000 hours vs. Battery 500 hours), power density, and fast storage capability (low charge/discharge time; SC: 1–10 seconds vs. battery 10–60 minutes) [2,3].

Applications of supercapacitors are not limited to storing renewable energy, they can also be used as an energy source in electronic devices, in regenerative braking, system backup, emergency doors, and as a power source for electric vehicles [4,5]. However, their operating performance could be better commercially for heavy industrial applications. The primary limitations of the supercapacitors are low energy density (10 Wh/kg) and the high cost of electrode materials, which prevent them from replacing batteries [6]. Different researchers have devoted many efforts in last decay to harness the supercapacitor technology and solve the technical problems associated with low energy density by nanostructured electrode materials that enhance specific capacity and increase operating voltage [6]. The literature review confirmed that considerable enhancement was observed, and the gap between supercapacitor and battery can be reduced by designing and developing electrodes with novel materials and structural engineering like decreasing particle size, modifying morphologies, and developing new hydride materials, and using new electrolytes [7,8]. Researchers have a growing interest in pseudo-capacitors mainly because of their higher energy densities than traditional electrochemical double layer capacitance (EDLC) supercapacitors [8]. Nano ferrite-based aqueous supercapacitors are very promising because oxygen and iron are the first and fourth abundant elements of the earth's crust, respectively. Moreover, ferrites have exceptionally high electrical conductivity due to Fe^{+2} and Fe^{+3} ions' coexistence in their spinel structure [8]. Among the various investigated ferrites for supercapacitive behavior, only nanoparticles Fe_3O_4 and $MnFe_2O_4$ exhibited pseudocapacitive behavior, and further improvement is required in developing nano ferrite-based supercapacitors [8]. In the same series, recently discovered high entropy materials (HEMs) (alloys and compounds of multi-component cation) are also identified as electrode materials for future supercapacitors. Here, we discussed the different aspects of HEMs which are vital for the electrochemical performance of HEM electrodes.

DOI: 10.1201/9781003391388-15

The main advantage of HEM is that we can obtain novel properties by combining different transition elements. These compounds showed impressive, extraordinary high-temperature Li-ion conductivity for solid-state electrodes in transition metal High Entropy Oxide (HEO) (TM–HEO). The first report for Li-ion conductivity demonstrated that several diffusion paths are created due to the insertion of Li-ions, and TMHEO has high Li-ion conductivity ($\geq 10^{-3}$ S/cm) [9]. The possible insertion of Li-ions into a rocksalt structure opens several diffusion pathways for Li-ions through the crystal lattice, giving rise to increased conductivity. The unique electrochemical properties of high entropy oxides (HEOs) open possibilities to use these materials as an electrode for supercapacitive properties.

15.2 HIGH ENTROPY MATERIALS DEFINITION AND PROPERTIES

Yeh et al. and Cantor et al. proposed the idea of HEAs in 2004 [10,11]. There are two ways to describe HEA systems first, based on composition, and second, based on configuration entropy [12]. Alloys with five or more principal elements with atomic percentages between 5% and 35% [10] and configuration entropy ≥1.5 R are known as HEAs. Furthermore, Murty et al. classified alloys as medium entropy material when the S_{config} value was $1.5\ R > S_{config} \geq 1\ R$ and low entropy materials for $S_{config} <$ 1 R. The configuration entropy of such a system can be evaluated using equation (15.1) [13].

$$S_{config} = -R\left[m\left(\sum_{i=1}^{m} \ln X_{i*}X_i\right)\right] \tag{15.1}$$

Here, X_i represents the mole fraction of the ith element, and R is the Universal Gas constant.

The configuration entropy value is found to be maximum when the constituent elements are in an equimolar ratio or the atomic fraction is equal for all the constituent elements. Then the formula for measuring the S_{config} is given by equation (15.2) [14]:

$$S_{config} = R \ln n \tag{15.2}$$

The configuration entropy plays a significant role in the structural stability of the system. From Gibb's equation (15.3), we can write,

$$\Delta G^{mix} = \Delta H^{mix} - T\Delta S^{mix} \tag{15.3}$$

From the above equation, it is clear that the Gibbs free energy is negative when the term $T\Delta S_{mix}$ is leading over the enthalpy, and it promotes the entropy stabilization factor. HEAs possess four core effects that make these alloys promising for potential applications. These effects are named as high entropy effect, cocktail effect, lattice distortion effect, and sluggish diffusion effect. The lattice distortion effect in HEA arose due to the difference in atomic size between the constituent elements, and elements have the same tendency to occupy the lattice site position. This effect leads to distortion in the structure of HEA. And helpful in tuning materials' optical, thermal, chemical, and electric nature [15]. For the sluggish diffusion effect, the rate of diffusion among the elements is not the same. This leads to the formation of nanograin, which makes it unique for oxygen evolution, electrocatalysis, reduction, and methanol oxidation [16]. Moreover, the cocktail effect is defined as the synergistic effect of alloying elements and hence significant impact on the properties [17].

The concept of entropy stabilization was first applied to multi-component oxides in 2015 by Rost et al. [18]. For such oxide, the effect of entropy has been contributed by both cation and anion elements in multi-component ceramic materials [19]. Generally, the entropy contribution of the anion site for O^{2-} is zero. However, oxygen vacancy or other anions (F^-) may contribute to configuration entropy value. The formula for calculating S_{config} is given by equation (15.4) for ceramic material [19].

$$S_{\text{config}} = -R\left[m\left(\sum_{i=1}^{n} \ln Y_{i*}Y_i\right)_{\text{cation site}} + \left(\sum_{j=1}^{m} \ln Y_{j*}Y_j\right)_{\text{anion site}}\right] \tag{15.4}$$

where Y_i and Y_j are the mole fraction of the element in the cation and anion site, respectively, the first reported $(Cu_{0.2}Mg_{0.2}Ni_{0.2}Zn_{0.2}Co_{0.2})O$ HEO has a rock salt structure [18]. They studied the effect of temperature on phase reversibility and explained how removing a single cation leads to the destabilization of the solid solution phase. After this report, there are several publications that showed immense potential in energy storage [20]. Based on structure, HEO can be classified into eight groups: rocksalt, fluorite, bixbyite, pyrochlore, magneto plumbite, O3-type layered, and perovskite spinel [20].

15.3 SYNTHESIS OF HEMS

There are several methods through which HEMs can be synthesized such as coprecipitation, solid-state method, nebulized spray pyrolysis, and flame spray pyrolysis [20,21]. These synthesis approaches have both benefits and limitations. In this section, we briefly discussed important synthesis approaches for synthesizing nanocrystalline HEMs.

15.3.1 Synthesis through Mechanical Alloying

In the mechanical alloying (MA) technique, the constituent metal powder undergoes continuous cold welding, fracturing, and rewelding to form homogeneous nanocrystalline HEA [22]. The nanosized particles are formed during this process, and further diffusion occurs, which exhibits the formation of solid solution phases [21]. Different parameters influence the particle size and alloying process such as process-control agent, ball-to-powder weight ratio, duration of ball-milling, size of balls, and number of revolutions per minute [21]. Mishra et al. synthesized AlCrFeMnNiTi HEA through a MA and reported the formation of the mixture of bcc and fcc phases [23]. The solid-state synthesis method is simple and frequently utilized in the synthesis of HEO. Liang et al. synthesized $(FeCoCrMnZn)_3O_4$ HEO which exhibited a spinel structure having particle size 0.65 μm [24].

15.3.2 Synthesis through Chemical Routes

Hydrothermal is a commonly used synthesis method for nanoparticle preparation, which provides freedom to tune the morphology of the synthesized nanoparticles. The starting material or precursor is dissolved in an organic solvent or deionized water and kept in the autoclave. If the reactant is dissolved in an aqueous medium is called hydrothermal, and when the reaction media is changed to a non-aqueous solvent is known as the solvothermal method [25]. Nguyen et al. synthesized non-equimolar HEO comprised of Cr, Co, Ni, Mn, and Fe through hydrothermal processes [26]. They found that the nanocrystalline HEO has particle size in the range of 100–200 nm and the electrode exhibited a specific capacity of 1235 mAh/g during charging and discharging [26].

The sol-gel method is a chemical synthesis route for the preparation of metal oxide and various composites developed with the metal oxide. The morphology, particle size, and texture can be controlled with this method with proper optimization [27]. In this process, three stages are involved, i.e., hydrolysis, condensation, and drying. It has been categorized into two forms: non-aqueous and aqueous sol-gel method. Mao et al. synthesized $(Co_{0.2}Cu_{0.2}Mg_{0.2}Ni_{0.2}Zn_{0.2})O$ HEO through the solution combustion method [28]. Rocksalt structure was formed for $(Co_{0.2}Cu_{0.2}Mg_{0.2}Ni_{0.2}Zn_{0.2})O$ HEO. Moreover, HEA can also be synthesized by reducing HEO synthesized through the sol-gel method. Niu et al. synthesized (CoCrCuNiAl) HEA through the sol-gel auto-combustion method [29]. A mixture of fcc and bcc phases was found with an average grain size of 14 nm.

15.3.3 Synthesis through Microwave-Assisted

The main drawbacks of conventional synthesis methods are long processing time, high energy consumption, required expensive instruments, and synthesized particles are large in size [30]. Microwave-assisted methods are novel synthesis approaches over conventional heating. The limitation of the conventional heating process is the thermal gradient, which leads to a slow and discontinuous reaction ambience [31]. Microwave irradiation is an effective and easy synthesis method for preparing HEM due to its different processing phenomena [32]. In this process, two different mechanisms occur when microwave interacts with the precursor, such as ionic conduction and dipolar interaction. The microwave irradiation method reduces the synthesis duration from days to seconds [32]. The benefit of this synthesis method is that it requires low temperature, high purity, and energy efficiency [32]. Kheradmandfard et al. synthesized (MgCuNiCoZn)O HEO through the microwave-assisted method at 2.45 GHz frequency [32]. The synthesized HEO was used as anode material for Lithium-Ion batteries and had good cyclability for 10000 cycles at 1 A/g current density. Veronesi et al. developed a series of FeCoNiCuAl, $FeCoCrNiAl_{2.5}$, and FeCrNiTiAlHEAs by direct heating of the constituent metal through microwave having a frequency of 2450 or 5800 MHz [33]. In this process, the cooling rate can be managed to optimize the ignition of the reaction.

15.4 HEMS AS AN ELECTRODE FOR SUPERCAPACITOR

15.4.1 HEAs as Electrode for Supercapacitors

Kong et al. synthesized equiatomic AlCoCrFeNiHEA through arc melting [34]. They prepared two ribbon samples of different thicknesses of 120 and 30 μm through rapid solidification technique at the wheel's speed of 7 and 33 m/s, respectively. The selective phase dissolution of AlCoCrFeNi HEA precursor was performed in 1 M H_2SO_4 solution at 313 K at different immersion time. (Al–Ni) ordered and (Cr–Fe) disordered bcc phase with lattice parameters 0.2846 and 0.2852 nm was formed for ribbon samples synthesized at the wheel's speed of 7 and 33 m/s, respectively. Formation of the nanoporous structure occurred with a regular pore channel and ligament of width 30 nm, formed by phase dissolution of a melt ribbon sample at speed 33 m/s in 1 M H_2SO_4 solution for 1 hour duration at 313 K. The specific surface area of the sample increased from 27.8 to 53.2 m^2/g when the wheel's speed was changed from 7 to 33 m/s. The supercapacitor property was studied in a 2 M KOH aqueous electrolyte for a binder-free nanoporous HEA electrode. The cyclic voltammetry (as shown in Figure 15.1a and b) and Galvanostatic charge and discharge (as shown in Figure 15.1c and d) were measured in the potential window –0.2 to 0.5 V. The nature of the CV curve was found to be pseudocapacitive type. The capacitance value for the sample synthesized at a wheel speed of 33 m/s is higher as compared to the sample synthesized at a wheel speed of 7 m/s due to the high BET-specific surface area of the metal ligament. However, the presence of an active metal oxide surface and the continuous chain of the conductive network of metal ligament results in the enhancement of the capacitance value. Moreover, they found the volumetric capacitance value of ~700 F/cm^3 at current density 1 mA/cm^2, which is large as compared to the hierarchical Ni nanoporous electrode [35].

Yuan et al. synthesized $FeCrCoMnNiAl_{0.75}$ HEA through induction melting exhibited bcc and fcc phases [36]. To fabricate the metallic frame of the bcc types, selective phase dissolution of $FeCrCoMnNiAl_{0.75}$ HEA precursor was performed in 0.5 M H_2SO_4 solution for 60 hours [36]. The three-dimensional metallic frame of the bcc phase has interconnected cubic void and walls of average size 121–150 and 31–45 nm respectively. The cyclic pulse electrodeposition was performed to decorate the MnO_2 thin layer on the metallic frame. The surface morphology of the MnO_2 and metallic frame composite, as shown in Figure 15.1e after the electrodeposition, showed a continuous frame worked structure. The benefit of the MnO_2 thin layer on the bcc metallic frame is, it works as the pseudo-capacitor. The super capacitive property was studied in three electrodes set up for

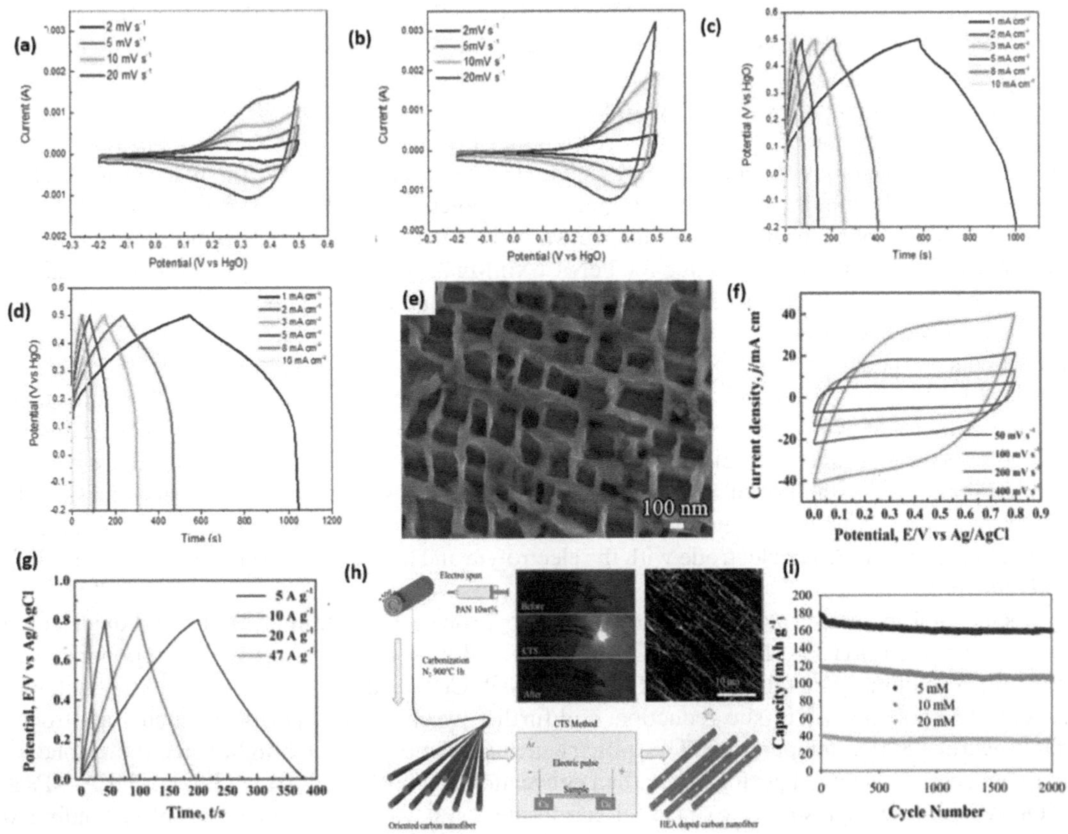

FIGURE 15.1 (a and b) represent the CV curve, and (c and d) represent the GCD curve for the AlCoCrFeNi HEA ribbon synthesized at wheel speeds of 7 & 33 m/s. Adapted with permission [34]. Copyright 2019, Elsevier. (e–g) represent the morphology of the surface after the MnO_2 electrodeposition on the HEA metallic frame, CV, and GCD profile for MnO_2 decorated on the HEA metallic frame, respectively. Adapted with permission [36]. Copyright 2021, Elsevier. (h) Pictorial view of the systematic synthesis process for HEA-NP/ACNF and (i) the capacitance retention at different 5 mM, 10 nM and 20 mM concentrations under FeNiCoMnMg chloride precursor. Adapted with permission [37]. Copyright 2020, Elsevier.

MnO_2/MF composite electrode in 0.5 M Na_2SO_4 aqueous electrolyte in the potential window 0–0.8 V. The cyclic voltammetry was measured at different scan rates, such as 50, 100, 200, and 400 mV/s. At higher scan rates, the nature of CV was of an ideal capacitive nature and regulates its quai-rectangular shape, as shown in Figure 15.1f. It was found that the capacitive contribution enhanced from 53 to 89.9% with the increment in scan rate. Moreover, the retention capacity of the electrode was also investigated and found to be 89.9% at the scan rate of 400 mV/s, dominant over the diffusion process for the composite MnO_2/MF electrode. Also, they reported the decrement in the specific capacitance value from 894 to 459 F/g as the scan rate increased from 50 to 400 mV/s for MnO_2/MF composite electrode. The results are higher than those reported MnO_2-based composite materials [36]. The Galvanic charge and discharge were measured at 5, 10, 20, and 47 A/g current density. As the current density changed from 5 to 47 A/g, the electrode's specific capacity value decreased from 961 to 675 F/g, as shown in Figure 15.1g. The higher value of specific capacitance at low current density was found because the ions have higher active sites for the redox reaction during diffusion toward the MnO_2 [36]. The higher value of specific capacitance for the MnO_2/MF composite electrode was found due to higher electronic conductivity and specific surface area. Moreover, properly utilizing MnO_2 decorated on metallic frame structure provides excellent stability and electronic conductivity to open framework structure.

Zeng et al. studied the super capacitive performance of FeNiCoMnMg HEA nanoparticle (HEA-NP) assembled on aligned carbon nanofibers (ACNF) through carbothermal shock (CTS)as shown in Figure 15.1h [37]. Carbon nanofiber (CNF) was used as a substrate, as it provides a higher specific surface area and excellent electrical and thermal conductivity [38]. The common method used for the preparation of ACNF is electrospinning (ES) and Chemical Vapor Deposition (CVD) [37]. The electrochemical performance was studied in the three-electrode setup in a 6 M KOH electrolyte. The cyclic voltammetry was measured in the potential window of 0.0 V–0.8 V at different scan rates. The super capacitive performance was examined in 5-, 10-, and 20-mM concentrations of FeNiCoMnMg chloride precursor for FeNiCoMnMg/ACNF electrode. They reported that the specific capacity value decreased with the increased concentration of FeNiCoMnMg chloride precursor solution. The specific capacity values were found to be 54, 127, and 203 F/g for 20-, 10-, and 5-mM concentrations of FeNiCoMnMg precursor solution, respectively, as shown in Figure 15.1i. FeNiCoMnMg HEA–NP/ACNF electrode exhibited higher specific capacitance (96 F/g) than the sample without ACNF (7 F/g) at 5 mM concentration. The developed electrode showed higher charge transfer resistance (18 Ω) for 20 mM concentrations with respect to 10 mM. The electrode showed good cyclic stability of 85% after 2000 cycles and a specific energy density of 21.7 Wh/Kg for five mM precursor concentration. The loss in cyclic stability arose due to the higher surface interaction of the HEA-NP electrode with the electrolyte and hence underwent irreversible solvation and reduction.

Shen et al. synthesized a composite of HEA nanoparticles with a thin layer of metal oxide at the outermost part of it (HEA–NPs@ MOL) and decorated it on the hyper cross-linked polymer carbon [39]. The quinary metallic cation (Fe^{2+}, Co^{2+}, Ni^{2+}, Cu^{2+}, and Sn^{2+}) gets adsorbed on the hyper cross-linked polymer via in-situ reduction, and further pyrolyzed was performed under a nitrogen atmosphere at 800°C temperature. The quinary metallic cations showed higher electronic conductivity and electrochemical performance than conventional single metal oxide [39]. For HEA-NPs@ MOL/ACNF hexagonal structure of Co_3Sn_2 has formed with two diffraction peaks corresponding to the graphite carbon plane. The nature of CV and GCD curves was found to be quasi-rectangular and isosceles triangles, respectively, which corresponds to pseudocapacitive behavior measured in the potential window of –0.1 to 0 V in 1 M KOH electrolyte. They calculated the specific capacitance value of 495.5 F/g at 0.5 A/g current density, and the capacitance value was dropped to 291.8 F/g with a current density of 10 A/g. They also studied the kinetics of the charge storage mechanism at different scan rates. They reported that the capacitance contribution for EDLC and fast redox reaction on the MOT layer correspond to a fast kinetic mechanism. Moreover, for slow kinetics, the capacitance mainly arose due to diffusion control reaction [39]. The contribution of the slow kinetic process declined with the increment of the scan rate. The HEA-NPs@ MOL/ACNF showed excellent cyclability for 15,000 cycles at 10 A/g with 94.9% of capacitance retention and 96.8% coulombic efficiency.

15.4.2 HEO as Electrode for Supercapacitors

Liang et al. synthesized $(FeCoCrMnZn)_3O_4$ HEO through the solid-state synthesis method [24]. They ball-milled the sample for 12 hours and calcined it at different temperatures in the range of 800–1000°C for 2 hours. They studied the supercapacitive properties of $(FeCoCrMnZn)_3O_4$/Ni Foam HESO electrode material in 1 M KOH electrolyte. Spinel structure was found with lattice constant 8.399 Å. The CV and GCD curves were performed for $(FeCoCrMnZn)_3O_4$/Ni Foam composite in the potential window 0.15–0.5 V and the nature of CV was found pseudocapacitive [24]. The specific capacitance value was found to be 352.9 F/g at a 5 mV/s scan rate. Moreover, at different current densities, 10, 5, 2, and 1 A/g, the specific capacitance was found to be 281.8, 301.1, 287.9, and 340.3 F/g, respectively. GCD's non-linear nature revealed the electrode's pseudo-capacitance behavior [24]. The developed electrode showed a good % cyclability performance of 69% after 1000 cycles at current density of 5 A/g.

Talluri et al. studied the electrochemical performance of nanocrystalline $(CoCrFeMnNi)_3O_4$ HESO synthesized through the reverse coprecipitation method [40]. The supercapacitive performance of the prepared electrode was studied in three electrodes set up using a 2 M KOH electrolyte. The CV measurement was performed at the potential window 0–0.5 V with different scan rates in the range of 5–100 mV/s as shown in Figure 15.2a. Higher capacitance value was found at low scan rates because, at this condition, ions get sufficient time to interact with the entire active material in the working electrode [40]. Also, the electrode showed pseudo-capacitance behavior. The found value of specific capacity value at the current density of 0.5, 1, 3, 5, 10, 15, 20 and 25 A/g was 239, 228, 210, 190, 152, 126, 108, and 90 F/g, respectively, as shown in Figure 15.2b. The electrode showed excellent cyclability of 76% for 1000 cycles at 1 A/g current density, as shown in Figure 15.2c. They also reported that the electrode exhibited specific energy of 24.1 Wh/kg at the current density of 0.5 A/g.

Lal et al. synthesized the nanocomposite of HEO with a carbon nanotube (CNT). (Ni, Fe, Co, Cr, and Al) HEO nanoparticles were used as a catalyst for the growth of CNT [41]. Firstly, they synthesized HEO (Ni-, Fe-, Co-, Cr- and Al-) through the sol-gel method and synthesized CNTs by using HEO. A pictorial representation of the synthesis route of HEO, HEO–CNT nanocomposite, and reduced HEO-CNTas is shown in Figure 15.2d. For all three HEO, HEO–CNT, and rHEO–CNT samples, the values of BET surface area were found to be 47.6, 151, and 158 m^2/g respectively. The electrochemical performance was studied in half cells in 1 M H_2SO_4 aqueous electrolyte for all three electrodes. The cyclic voltammetry and galvanostatic charge-discharge at a scan rate of 100 mV/s and current density of 1 A/g confirmed that the electrode nanocomposite has a pseudocapacitive nature. In contrast, HEO–CNT and rHEO–CNT showed both EDLC and pseudocapacitive behavior, as shown in Figure 15.2e and f. The three electrodes exhibited a specific

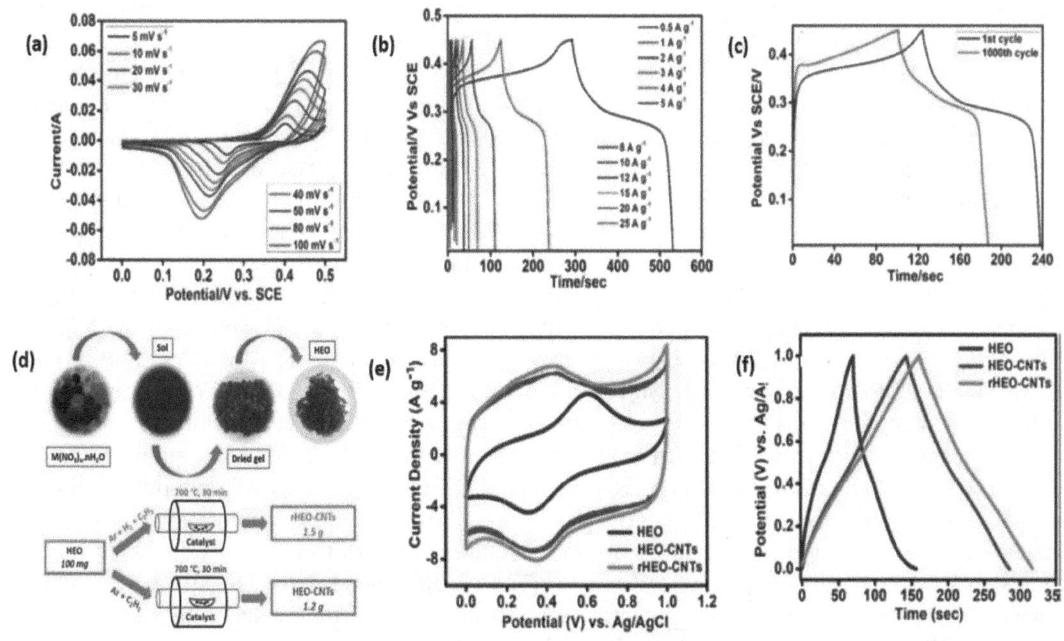

FIGURE 15.2 (a) and (b) represent the CV and GCD curve, (c) cyclability performance at the current density of 1 A/g for 1000 cycles for $(CoCrFeMnNi)_3O_4$ HEO. Adapted with permission [40]. Copyright (2021), Elsevier. (d) Descriptive diagram for the synthesis of HEO nanoparticles, HEO–CNT and rHEO–CNT nanocomposites, (e) and (f) represents the CV and GCD profile in 1 M H_2SO_4 electrolyte for HEO nanoparticles, HEO–CNT and rHEO–CNT nanocomposites. Adapted with permission [41]. Copyright (2019), American Chemical Society.

capacity of 91.5, 156, and 170.1 F/g for HEO, HEO–CNT, and rHEO–CNT, respectively. Among all the three electrode materials, rHEO–CNT exhibited the highest capacitance value due to a higher specific surface area than HEO–CNT. The electrochemical performance was also investigated for full cells in PVA/H_2SO_4 hydrogel electrolyte. The CV and GCD for HEO–CNT and rHEO–CNT were measured at a scan rate of 100 mV/s and a current density of 1 A/g. They concluded that there is no significant change in the shape of CV and GCD for both HEO–CNT and rHEO–CNT electrodes. The slight variation between the two electrodes about due to changes in specific surface area and the quantity of oxide present. The HEO–CNT electrode showed 67.6% capacitance retention in the current density range of 1 A/g to 5 A/g confirming the good rate capability in the full cell. Moreover, the HEO–CNT electrode exhibited excellent cyclability for 15,000 cycles and showed 100% coulombic efficiency at the current density of 5 A/g in the PVA/H_2SO_4 hydrogel electrolyte. However, the HEO–CNT electrode exhibited a specific capacity of 271 F/g in the 1-butyl-3-methylimidazolium [BMIM] [TFSI] electrolyte at the current density of 1 A/g.

15.4.3 High Entropy (Nitrides/Carbides) as Electrode for Supercapacitors

With the evolution of time, the idea of entropy stabilization has approached high entropy boride, nitride, and carbide [42]. Jin et al. synthesized High Entropy Metal Nitride (HEMN) through the mechanochemical synthesis approach via urea aid [42]. Five transition metal chloride (ACl_x, A = Mo, Cr, V, Nb, Zr, X = 3–5) was ball-milled for 30 minutes with urea and further pyrolyzed with nitrogen. Cubic structure was formed with lattice parameter 4.2607 Å. The BET-specific surface area and pore size were 278 m^2/g and 2 nm, respectively. The electrochemical performance was measured in a three-electrode setup in 1 M KOH electrolyte. The cyclic voltammetry was measured at different scan rates in the potential window −0.1–0 V. They found the value of specific capacity 230, 175, 113, 78, and 54 F/g at scan rates of 10, 20, 50, 100, and 200 mV/s, respectively. They reported that the HEMN electrode exhibited a higher capacitance value of 78 F/g at a scan rate of 100 mV/s as compared to Vn @ nanowire electrode (46.9 F/g) [43].

Sure et al. synthesized (TiNbTaZrHf)C high entropy carbide (HEC) through electrochemical processes [44]. They prepared the mixture of oxide and graphite in a cylindrical dish and heated at 523 K for 3 hours. The dish and graphite were taken as cathode and the anode in 1173 K $CaCl_2$ electrolyte for 6–8 hours at the applied voltage of 2.8–3.1 V. Single fcc phase with lattice parameter of 4.468 Å was formed for (TiNbTaZrHf)C HEC. The value of specific capacity was found to be 27.4, 42.2, 58.1, 71.0, and 95.2 F/g at scan rates of 100, 50, 30, 20 and 10 mV/s, respectively. The electrode showed capacity retention of 87% after 2000 cycles at the current density of 1 A/g.

Jiang et al. synthesized a series of High Entropy Prussian Blue Analogs (HEPBAs) through mechanochemistry followed by wet chemistry at room temperature, as shown in Figure 15.3a [45]. They developed a sequence of K(MgMnFeCoCu)Fe$(CN)_6$, K(MnFeCoCuNi)Fe$(CN)_6$, K(MgMnFeNiCu)Fe$(CN)_6$, and K(MgMnCoNiCu)Fe$(CN)_6$HEPBA. $K_3Fe(CN)_6$ was initially ball-milled with the transition metal chloride of Ist row metal to get a homogenous mixture. A homogeneous mixture was washed in water to remove K^+ and Cl^- ions to obtain the crystalline HEPBA. The chemical formula of HEPBAs is $K_3MFe(CN)_6$. Here, "M" symbolizes the quinary diverse metallic cation. HEPBA has a cubic phase structure. Some impure phases have been raised with the two metal-based cations at the 'M' site in PBA. In three-electrode configurations, the electrochemical performance was studied in 1 M Na_2SO_4 electrolyte. The cyclic voltammetry measurement was performed in the potential window 0.2–1 V at different scanning rates from 5 to 200 mV/s. The CV measurements showed pseudocapacitive behavior for the electrode. In CV, two redox peaks appeared for all the HEPBA sequence electrodes at a scan rate of 5 mV/s as shown in Figure 15.3b. At 5 mV/s, the HEPBA electrode containing Mg exhibited a higher specific capacitance value of 175 F/g as compared to single-element PBA. They also found that at higher scan rates at 200 mV/s, the K(MgMnFeCoNi)Fe$(CN)_6$, K(MgMnCoNiCu)Fe$(CN)_6$, K(MgMnFeCoCu)Fe$(CN)_6$ and K(MgMnFeNiCu)Fe$(CN)_6$ electrodes had value of capacitance in the range of 50–60 F/g which

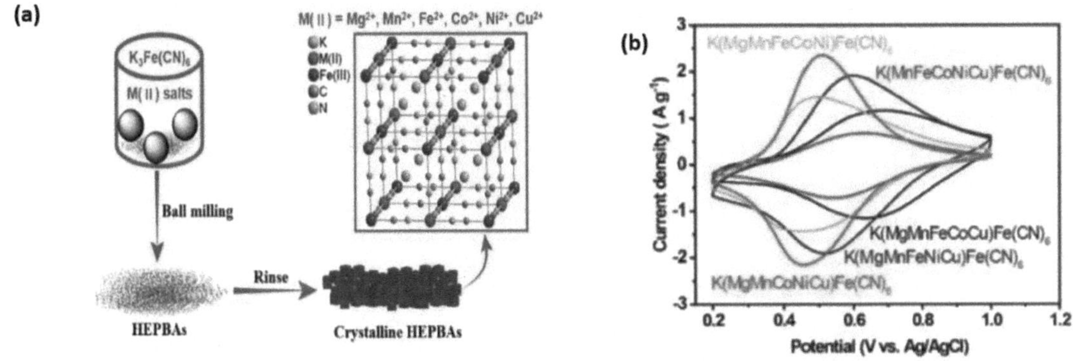

FIGURE 15.3 (a) Descriptive diagram of the synthesis of HEPBA sample, (b) CV curve of $KMFe(CN)_6$ series at 5 mV/s scan rate. Adapted with permission [45]. Copyright 2020, Elsevier.

is far better than $K(MnFeCoNiCu)Fe(CN)_6$ electrode due to the presence of Mg in the composition. They also compared $K(MgMnFeNiCu)Fe(CN)_6$HEPBA electrode in a wet and dry state. The electrode in the wet state exhibited higher specific capacitance (175 F/g) as compared to its dried form (82 F/g) at 5 mV/s scan rate. The enhancement in the capacitance value was due to the hydrogen bonding in the framework, which promoted the Grotthus Proton conduction during the oxidation and reduction process.

15.4.4 High Entropy MXene as the Electrode for Electrochemical Energy Storage

Etman et al. synthesized two-dimensional $Ti_{1.1}V_{0.7}Cr_xNb_{1.0}Ta_{0.6}C_3Tz$ (Tz=–F, –O, OH), High Entropy MXene and studied electrochemical performance for Zn_Ion Hybrid Supercapacitors (ZHSC) [46]. They initially prepared 3D HE MAX phase $Ti_{1.0}V_{0.7}Cr_{0.05}Nb_{1.0}Ta_{1.0}AlC_3$ through a solid-state reaction at 1500°C for 600 minutes in the presence of argon, and further 2D $Ti_{1.1}V_{0.7}Cr_xNb_{1.0}Ta_{0.6}C_3Tz$ (Tz= –F, –O, OH) HE MXene freestanding film was synthesized. The morphology of HE MXene films was comprised of extremely thin flakes of thickness 2.5 μm. The electrochemical performance was studied for HE MXene film in Zn-ion hybrid supercapacitors with 3 M zinc trifluoromethane sulfonate (Zn $(CF_3SO_3)_2$ and 15 M $ZnCl_2$ electrolytic solution. The CV (Figure 15.4a and b) and GCD (Figure 15.4c and d) measurements were performed in both 3(Zn $(CF_3SO_3)_2$ and 15 M $ZnCl_2$ electrolyte at the potential window (0.01–1.2 V). The nature of the CV curve in both the electrolyte solutions exhibited a pseudocapacitive behavior, and it retained a similar characteristic at a higher scan rate. Moreover, the nature of CV provided the information for the redox reactions of desertion/insertion of Zn^{2+} and H+ at the electrode surface in HE MXene to maintain electrical neutrality [47]. The specific capacity value were found to be 45, 54, 59, 68, and 73 and 43, 59, 61, 71, and 77 mAh/g at the current density of 10, 5, 3, 1, and 0.5 A/g in 3 M(Zn $(CF_3SO_3)_2$ and 15 M $ZnCl_2$ solution, respectively. The synthesized HE MXene exhibited excellent cyclability for 10,000 cycles at a current density of 10 A/g with 87 and 82% of capacitance retention for 3 M (Zn $(CF_3SO_3)_2$, and 15 M $ZnCl_2$ solution as shown in Figure 15.4e and f. They found that $Ti_{1.1}V_{0.7}Cr_xNb_{1.0}Ta_{0.6}C_3Tz$ HE MXene has high gravimetric capacitance than $Ti_3C_2T_z$ and V_2CT_z MXene [48].

15.5 SUMMARY

Based on the detailed discussion in the available reports, we can say that high entropy materials (HEMs) may behave as promising electrode materials for supercapacitors. The unique characteristics of HEM make them different and more beneficial from conventional electrode materials, such as structural stability, tuning the functional property, and variation in the composition. Here, we

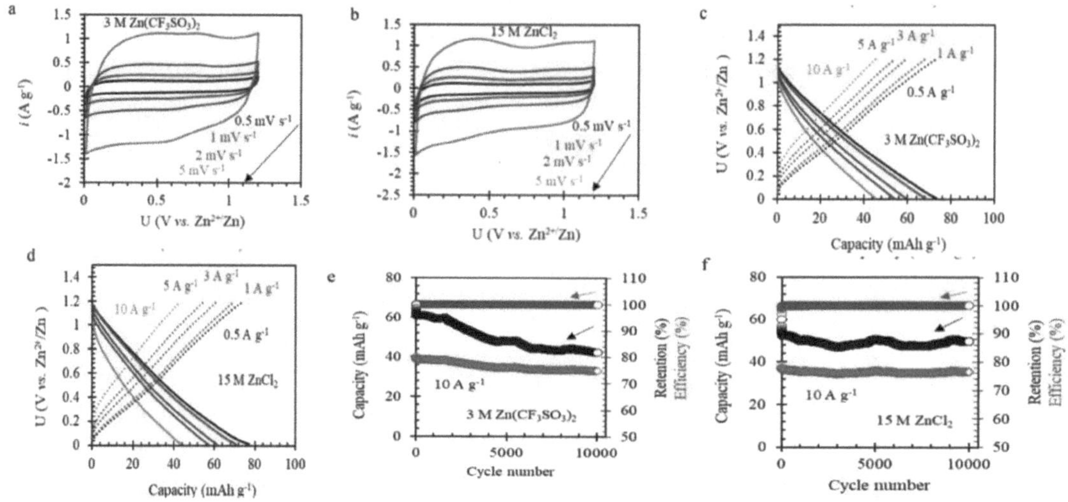

FIGURE 15.4 CV profile of (a) 3 M (Zn $(CF_3SO_3)_2$, (b) 15 M $ZnCl_2$ electrolyte, respectively, GCD curve of (c) 3 M (Zn $(CF_3SO_3)_2$ and (d) 15 M $ZnCl_2$ solution, respectively, cyclability and coulombic efficiency at 10 A/g current densities of (e) 3 M (Zn $(CF_3SO_3)_2$ and (f) 15 M $ZnCl_2$ solution, respectively. Adapted with permission [46]. Copyright 2022, Elsevier.

briefly discussed different HEA and HEO synthesis routes and summarized the critical finding in the supercapacitive properties of HEMs. Based on the results of the different studies, the following conclusions can be drawn and provide insight into the future development of these materials.

- The selective phase dissolution process helps to produce the nanoporous binder-free HEA. The nanoporous HEA may exhibit higher volumetric capacitance and cyclability due to the metal ligaments' higher surface area and active oxide surface.
- The assembling of HEA nanoparticles on aligned carbon nanofibers may also be an effective way to enhance the electrochemical performance of a particular HEA. The presence of CNF will provide a higher surface area and excellent electrical and thermal conductivity for the electrode material.
- HEC and HENmay also show promising electrochemical performance as compared to traditional carbide and nitride materials.
- The quinary cations have the potential to undergo oxidation and reduction reactions and promote the faradaic reaction and open a new area for designing and developing HEO electrodes for the high-performance supercapacitor application.
- The HEO nanoparticle has also been utilized as a catalyst for the growth of carbon nanotubes and as an energy storage material for application.
- The high entropy MXene is a two-dimensional material due to its unique characteristics such as structural stability, shortened electronic path, higher specific surface area, and presence of large active sites, which make it a promising material in the energy storage field.

ACKNOWLEDGMENTS

The authors would like to acknowledge the received financial assistance from the UGC-DAE CSR Project (CRS/2021-22/01/381)

REFERENCES

1. S.A. Delbari, L.S. Ghadimi, R. Hadi, S. Farhoudian, M. Nedaei, A. Babapoor, A. S. Namini, Q.V. Le, M. Shokouhimehr, M. S. Asl, M. Mohammad, Transition metal oxide-based electrode materials for flexible supercapacitors: A review, *J. Alloys Compounds* 857 (2021) 44.
2. H. Yu, X. Ge, C. Bulin, R. Xing, R. Li, G. Xin, B. Zhang, Facile fabrication and energy storage analysis of graphene/PANI paper electrodes for supercapacitor application, *Electrochim Acta* 253 (2017) 239.
3. J. Jan, Q. Wang, T. Wei, Z. Fan, Recent advances in design and fabrication of electrochemical supercapacitors with high energy densities, *Adv. Energy Mater.* 4(2014) 1300816.
4. S. A. H. Taleb, D. Brown, J. Dillet, P. Guillemet, J. Mainka, O. Crosnier, C. Douard, L. Athouël, T. Brousse, O. Lottin, Direct hybridization of polymer exchange membrane surface fuel cell with small aqueous supercapacitors, *Fuel Cells* 18 (2018) 299–305,
5. P. Simon, T. Favier, *Supercapacitors Based on Carbon or Pseudocapacitive Materials*, (2017), Hoboken, NJ, Wiley,
6. W. Raza, F. Ali, N. Raza, Y. Luo, KiHyun Kim, J. Yang, S. Kumar, A. Mehmood, E. E. Kwon, Recent advancements in supercapacitor technology, *Nano Energy* 52(2018) 441–473.
7. M.A.A.A. Abdah, N.H.N. Azman, S. Kulandaivalu, Y. Sulaiman, Review of the use of transition-metal-oxide and conducting polymer-based fibers for high-performance supercapacitor, Mater. Design 186 (2020) 108199.
8. K. Malaie, R. Ganjali, Spinel nano-ferrites for aqueous supercapacitors; linking abundant resources and low-cost processes for sustainable energy storage, *J. Energy Storage* 33 (2021) 102097.
9. D. Beradan, S. Franger, S. Meena, A.K. Dragoe, room temperature Lithium superionic conductivity in high entropy oxides, *J Mater. Chem. A* 4 (2016) 9536.
10. J-W. Yeh, S.-K. Chen, S.-J. Lin, J.-Y. Gan, T.-S. Chin, T.-T. Shun, C.-H. Tsau and S.-Y. Chang, Nanostructured high-entropy alloys with multiple principal elements: Novel alloy design concepts and outcomes, *Adv. Eng. Mater.* 6 (2004) 299–303.
11. B. Cantor, I. T. H. Chang, P. Knight and A. J. B. Vincent, Mater, Microstructural development in equiatomic multi-component alloys, *Sci. Eng. A* 375–377 (2004) 213–218.
12. J.-W. Yeh, Ann. Chim, Recent progress in high-entropy alloys, *Eur. J. Control* 31 (2006) 633–648.
13. B. S. Murty, J.-W. Yeh, R. Srikanth and P. P. Bhattacharjee, *High-Entropy Alloys*, (2019). Amsterdam, Elsevier.
14. M. C. Gao, J. Yeh, P. K. Liaw and Y. Zhang, *High-Entropy Alloys*, (2016). NY, Springer International Publishing.
15. Y. Zhang, T. T. Zuo, Z. Tang, M. C. Gao, K. A. Dahmen,P. K. Liaw and Z. P. Lu, Prog.Microstructures and properties of high-entropy alloys, *Mater. Sci.* 61 (2014) 1–93.
16. Y. Ma, Y. Ma, Q. Wang,S. Schweidler, M. Botros, T. Fu, H. Hahn, T. Brezesinski and B. Breitung, High-entropy energy materials: challenges and new opportunities, *Energy Environ. Sci.* 14 (2021) 2883–2905.
17. Y. F. Ye, Q. Wang, J. Lu, C. T. Liu and Y. Yang, High-entropy alloy: challenges and prospects, *Mater. Today* 19 (2016) 349–362.
18. C.M. Rost, E. Sachet, T. Borman, A. Moballegh, E. C. Dickey, D. Hou, J. L. Jones, S. Curtarolo and J.-P. Maria, Entropy-stabilized oxides, *Nat. Commun.* 6 (2015) 8485.
19. A. Sarkar, L. Velasco, D. Wang, Q. Wang, G. Talasila, L. Debiasi, C. Ku¨bel, T. Brzezinski, S. S. Bhattacharya, H. Hahn and B. Breitung, High entropy oxides for reversible energy storage, *Nat. Commun.* 9 (2018) 3400.
20. A. Sarkar, B. Breitung and H. Hahn, High entropy oxides: The role of entropy, enthalpy and synergy, *Scr. Mater.* 187 (2020) 43–48.
21. X. Wanga, W. Guo, and Y. Fu, High-entropy alloys: Emerging materials for advanced functional applications, *J. Mater. Chem. A*, 9 (2021) 663–701.
22. M. Vaidya, G. M. Muralikrishna, and B. S. Murty, High-entropy alloys by mechanical alloying: A review, *J. Mater. Res.* 34 (2019) 664–686.
23. R.K. Mishra, P.P. Sahay and RR Shahi, Alloying, magnetic and corrosion behavior of AlCrFeMnNiTi high entropy *alloy, J. Mater. Sci.* 54(3) (2018) 4433–4443.
24. B. Liang, Y. Ai, Y. Wang, C. Liu, S. Ouyang, and M. Liu, Spinel-type (FeCoCrMnZn)3O4 high-entropy oxide: Facile preparation and supercapacitor performance, *Materials* 13(24) (2020) 5798.
25. B. G. Rao, D. Mukherjee, B. M. Reddy, *Nanostructures for Novel Therapy, Novel Approaches for the Preparation of Nanoparticles*, (2017) Elsevier.

26. T.X Nguyen, J. Patra, J.K. Chang, and J.M. Ting, High entropy spinel oxide nanoparticles for superior lithiation-delithiationperformance, *J. Mater. Chem. A.* 8 (2020) 18963–18973.
27. R.J.P. Corriu, D. Leclercq, Recent developments of molecular chemistry for sol-gel processes. *Angew. Chem. Int. Ed.* 35 (1996) 1420–1436.
28. A. Mao, H-Z. Xiang, Z-G Zhang, K. Kuramoto, H. Yua, S. Ran, Solution combustion synthesis and magnetic property of rocksalt (Co0.2Cu0.2Mg0.2Ni0.2Zn0.2)O high-entropy oxide nanocrystalline powder, *J Magnetism Magnetic Mater.* 484 (2019) 245–252.
29. B. Niu, F. Zhang, H. Ping, N. Li, J. Zhou, L. Lei, J. Xie, J. Zhang, W. Wang, and Z. Fu, Sol-gel autocombustion synthesis of nanocrystalline high-entropy alloys, *Sci. Rep.* 7 (2017) 3421.
30. A. Sarkar, R. Djenadic, N.J. Usharani, K.P. Sanghvi, V.S.K. Chakravadhanula, A. S. Gandhi, H. Hahn, S.S. Bhattacharya, Nanocrystalline multi-component entropy stabilized transition metal oxides, *J. Eur. Ceram. Soc.* 37 (2017) 747–754.
31. X.L. Hu, J.C. Yu, Continuous aspect-ratio tuning and fine shape control of monodisperse alpha-Fe2O3 nanocrystals by a programmed microwave-hydrothermal method, *Adv. Funct. Mater.* 18 (2008) 880–887.
32. M. Kheradmandfard, H. Minouei, N. Tsvetkov, A. K. Vayghan,S.F. K- Bozorg, G. Kim, S. Ig Hong,D-E. Kim, Ultrafast green microwave-assisted synthesis of high-entropy oxide nanoparticles for Li-ion battery applications, *Mater. Chem. Phys.* 262 (2021) 124265.
33. P. Veronesi, R. Rosa, E. Colombini and C. Leonelli, Microwave-assisted preparation of high entropy alloys, *Technologies* 3 (2015) 182–197.
34. K. Kong, J. Hyun, Y. Kim, W. Kim, D. Kim, Nano-porous structure synthesized by selective phase dissolution of AlCoCrFeNi high entropy alloy and its electrochemical properties as supercapacitor electrode, *J. Power Sources* **437** (2019) 226927–226936.
35. H.-J. Qiu, Y. Ito, M.W. Chen, Hierarchical nanoporous nickel alloy as three-dimensional electrodes for high-efficiency energy storage, *Scr. Mater.* 89 (2014) 69–72.
36. Y. Yuan, Z. Xu, P. Han, Z. Dan, F. Qin, H. Chang, MnO2-decorated metallic framework supercapacitors fabricated from duplex-phase FeCrCoMnNiAl0.75 Cantor high entropy alloy precursors through selective phase dissolution, *J Alloys Compounds* 870 (2021) 159523–159532.
37. X. Xu, Y. Du, C. Wang, Y. Guo, J. Zou, K. Zhou,Z. Zeng, Y. Liu L. Li, High-entropy alloy nanoparticles on aligned electrospun carbon nanofibers for supercapacitors, *J. Alloys Compounds* 822 (2020) 153642–153650.
38. J. Ding, H.-p. Zhang, X. Li, Y. Tang, G. Yang, Cross-linked carbon nanofiber films with hierarchical pores as flexible electrodes for high-performance supercapacitors, *Mater. Des.* 141 (2018) 17e25.
39. E. Shen, X. Song, Q. Chen,M. Zheng, J. Bian, and H. Liu, Spontaneously forming oxide layer of high entropy alloy nanoparticles deposited on porous carbons for supercapacitors *ChemElectroChem* 8(1) (2021)260-269
40. B. Talluri, M.L. Aparna, N. Sreenivasulu, S.S. Bhattacharya, T. Thomas, High entropy spinel metal oxide (CoCrFeMnNi)3O4 nanoparticles as a high-performance supercapacitor electrode material, *Journal of Energy Storage* 42 (2021) 103004.
41. M. S. Lal and R. Sundara, High entropy oxidesa cost-effective catalyst for the growth of high yield carbon nanotubes and their energy applications, *ACS Appl. Mater. Interfaces* 11 (2019) 30846–30857.
42. T. Jin, X. Sang, R. R. Unocic, R. T. Kinch, X. Liu, J. Hu, H. Liu, and S. Da, Mechanochemical-assisted synthesis of high-entropy metal nitride via a soft urea strategy, *Adv. Mater.* 2018, 1707512.
43. X. Lu, T. Liu, T. Zhai, G. Wang, M. Yu, S. Xie, Y. Ling, C. Liang, Y. Tong, Y. Li, Improving the cycling stability of metal-nitride supercapacitor electrodes with a thin carbon shell, *Adv. Energy Mater.* 2014, *4*, 1300994.
44. J. Sure, D. S. M. Vishnu, H-K Kim and C. Schwandt, Facile electrochemical synthesis of nanoscale (TiNbTaZrHf)C high-entropy carbide powder, *Angewandte Chemie International Edition* 59 (2020) 11830–11835.
45. W. Jiang, T. Wang, H. Chen, X. Suo, J. Liang, W. Zhu, H. Li and S. Dai, Room temperature synthesis of high-entropy prussian blue analogues, *Nano Energy* 79 (2021) 105464.
46. A. S. Etman, J. Zhou, J. Rosen, Ti1.1V0.7CrxNb1.0Ta0.6C3Tz high-entropy MXene freestanding films for charge storage applications, *Electrochem. Commun.* 137 (2022) 107264.
47. M.J. Park, H. Yaghoobnejad Asl, A. Manthiram, Multivalent-ion versus proton insertion into battery electrodes, *ACS Energy Lett.* 5 (7) (2020) 2367–2375.
48. Q.i. Yang, Z. Huang, X. Li, Z. Liu, H. Li, G. Liang, D. Wang, Q. Huang, S. Zhang, S. Chen, C. Zhi, A wholly degradable, rechargeable Zn-Ti3C2 MXene capacitor with superior anti-self-discharge function, *ACS Nano* 13 (7) (2019) 8275–8283.

16 High Entropy Materials as Anode in Li-Ion Battery

Ababay Ketema Worku, Delele Worku Ayele, Minbale Admas Teshager, Molla Asmare Alemu, Biniyam Zemene Taye, and Xueqing Xu

16.1 INTRODUCTION

The increment in the alarming rate of its cost and rapid depletion of fossil fuels results in high demand for secondary batteries as power sources for portable electronics, and electric and hybrid electric vehicles [1]. Nowadays, research and development (R&D) has mainly focused on searching for materials with improved performance of secondary batteries for a certain application [2]. Among secondary batteries, lithium-ion batteries (LIBs), have got great attention from many researchers; due to lithium has the lightest weight, with a density of 0.534 g/cm^3, the highest voltage (3.6 V) and greatest energy density (240 Wh/kg) of all metals. Moreover, there is a high demand for lighter, thinner, and smaller products [3]. On the basis of these advantages, lithium-ion batteries (LIBs) have been widely used as power sources in a broad range of applications [4,5]. LIBs which are also known as "rocking chair" batteries due to lithium ions move back and forth between the two electrodes (cathode and anode) through the electrolyte during electrochemical reactions (charge and discharge) [6]. The charging and discharging processes in a typical LIB is shown in Figure 16.1. Initially, under the applied potential the lithium ions migrate from cathode to anode (charging); in the reverse process, the lithium ions move back from anode to cathode (discharging).

Efforts on improvement of the electrode materials (cathodes and anodes) and electrolytes can bring a change in energy storage technology and put forward higher requirements on next-generation LIBs [7]. In this regard, high-potential cathode materials and electrolytes have been reported. Recently, the high-entropy concept has been reported [8]. Along with this concept, a new class of materials called as high entropy materials (HEMs); specifically high entropy oxide (HEO) is gradually emerging into sight and bringing energy success [9]. The ideal composition adjustability and attractive synergistic effect make HEO promising to break through the integrated performance bottleneck of conventional anodes and provide new motivation for the design and development of electrochemical energy storage materials [10]. The driving forces behind the attraction of HEO are phase stability, the role of individual cations, potential mechanisms for controlling properties, as well as state-of-the-art synthetic strategies and modification approaches [11]. HEOs showed different development stages for the LIB anode (Figure 16.2). The structure tenability of HEOs provides them a great possibility for energy storage applications [12]. For instance, the large number of atomic sites on the surface of HEOs provides a huge opportunity to tune the surface electronic structures and electrochemical activities of the materials.

16.2 SYNTHESIS AND STRUCTURES OF HEOS

16.2.1 Synthesis of HEOs

Different physical and chemical synthesis methods have been developed to fabricate HEOs. Binary metal compounds are used as precursors for the synthesis of HEOs. Some of the synthesis methods are solid-state sintering, sol-gel, co-precipitation, sputtering deposition, and ball-milled.

DOI: 10.1201/9781003391388-16

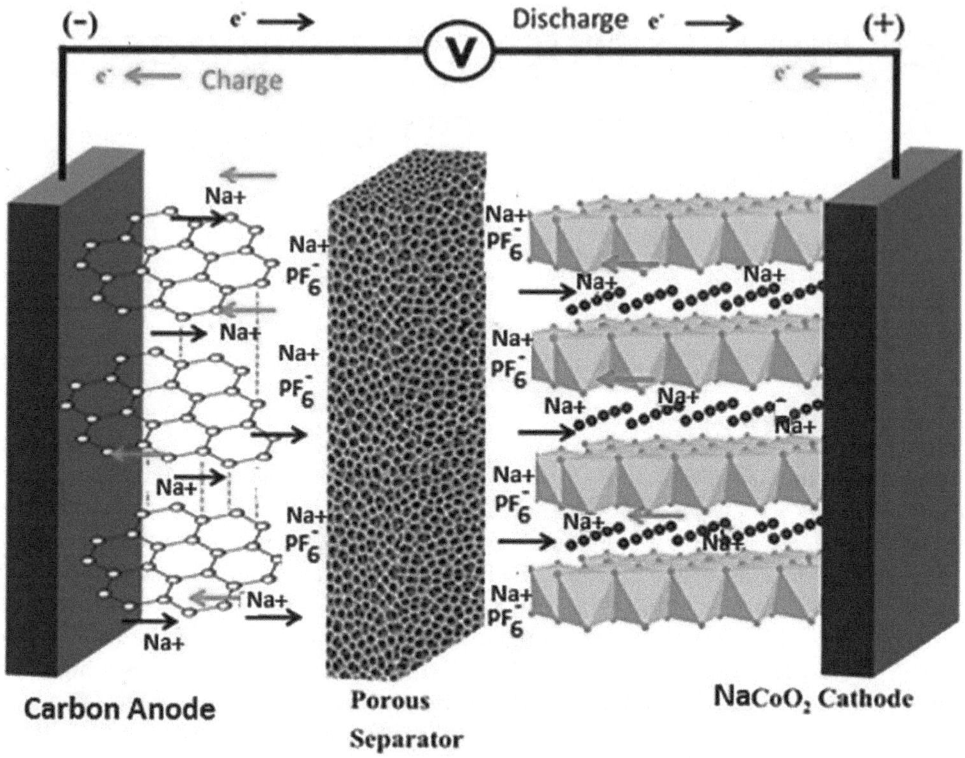

FIGURE 16.1 Schematic representation of a typical lithium-ion battery. Adapted from reference [7]. Copyright The Authors, some rights reserved; exclusive licensee [ACS]. Distributed under a Creative Commons Attribution License 4.0 (CC BY) https://creativecommons.org/licenses/by/4.0/.

For example, for solid-state methods, metal oxides are ball-milled, pelletized, and sintered for many hours; for this purpose, sintering temperature and soaking time are the key factors [14].

16.2.2 Structures of HEOs

For battery applications, the intrinsic properties of materials are strongly dependent on their structure, which in turn dictates the electrochemical behaviors. Moreover, the exploration of the structure-activity relationship is crucial for material design and performance optimization [15]. HEOs have different structures such as rock salt, fluorite, layered, spinel and perovskites are the most common structures (Figure 16.3).

Entropy-driven structural stabilization effects have been evidenced for HEOs. HEOs with rock salt structure are the most studied oxides, due to easy of synthesis and formation [16]. They have excellent chemical stability, providing them with great potential in electrochemical applications. HEOs with spinel and perovskite structure also attracted extensive attention in energy storage and conversion. Layered oxides have excellent thermal stability and are well known in lithium-ion battery [17]. In lithium-ion batteries, during lithiation/delitiation process, structural changes of HEO electrodes have been reported. Structural changes from rock salt to spinel and perovskite are the possible structural shifts reported in previous studies. The spinel structure has three-dimensional lithium-ion (Li^+) diffusion channels with enhanced conductivity, which supposes spinel-structured HEOs are more promising for the application of lithium-ion battery anode [18]. Schematical representation of change in the structure of spinel HEOs during the lithiation/delitiation process is shown below (Figure 16.4).

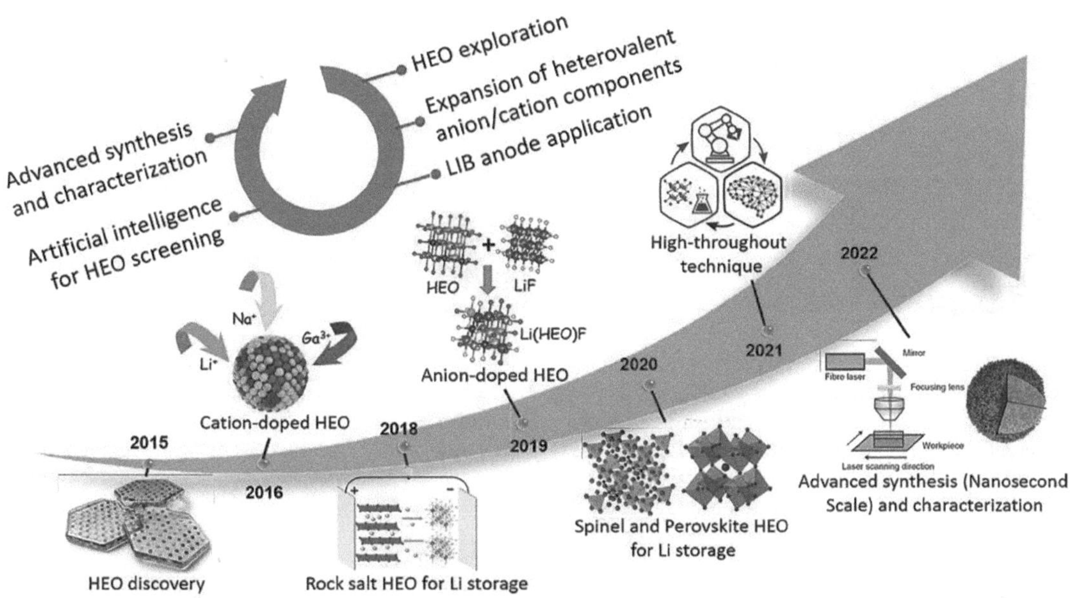

FIGURE 16.2 Development of HEO for LIB anode. Adapted from reference [13]. Copyright The Authors, some rights reserved; exclusive licensee [Wiley]. Distributed under a Creative Commons Attribution License 4.0 (CC BY) https://creativecommons.org/licenses/by/4.0/.

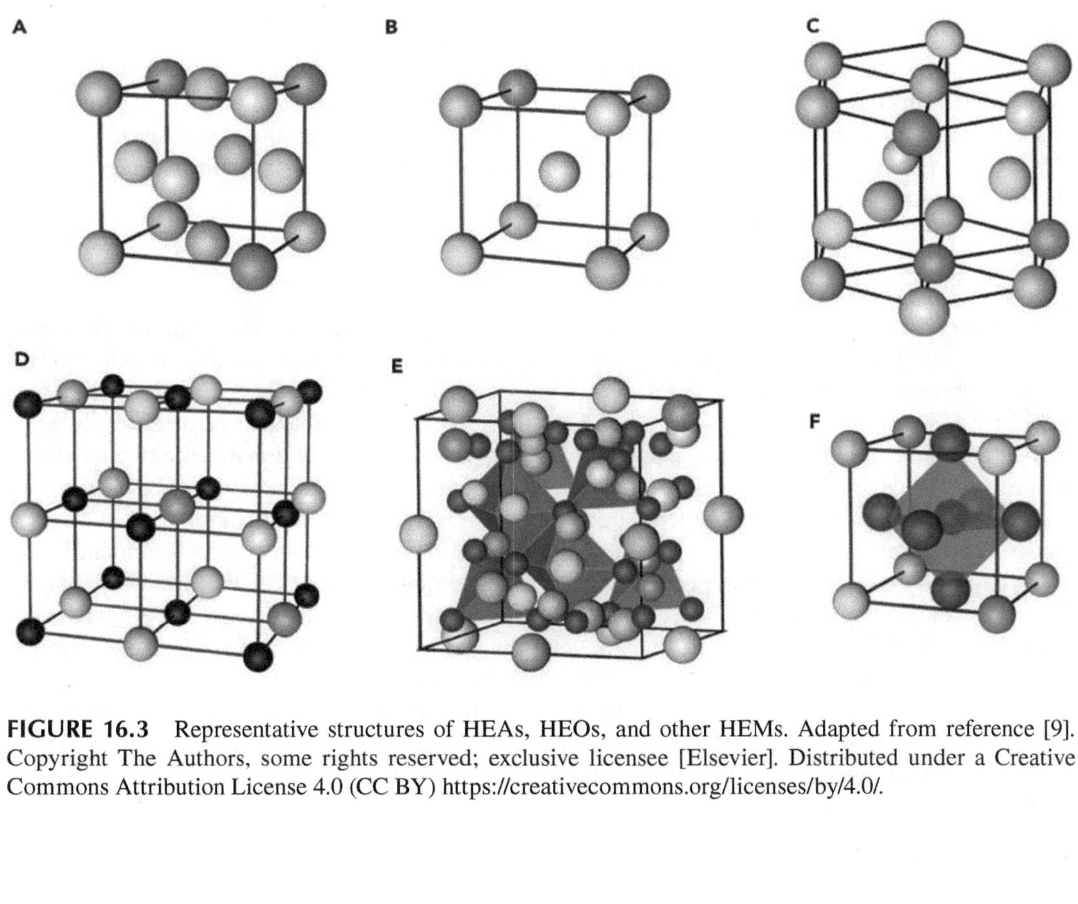

FIGURE 16.3 Representative structures of HEAs, HEOs, and other HEMs. Adapted from reference [9]. Copyright The Authors, some rights reserved; exclusive licensee [Elsevier]. Distributed under a Creative Commons Attribution License 4.0 (CC BY) https://creativecommons.org/licenses/by/4.0/.

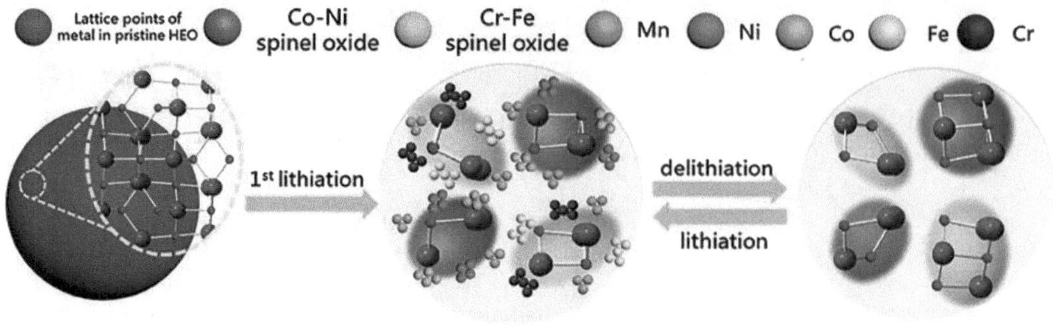

FIGURE 16.4 Schematics of the lithiation/delithiation mechanism of stabilizer-free Spinel-HEO. Adapted from reference [13]. Copyright The Authors, some rights reserved; exclusive licensee [Wiley]. Distributed under a Creative Commons Attribution License 4.0 (CC BY) https://creativecommons.org/licenses/by/4.0/.

16.3 CHARACTERIZATION METHODS OF HEOS

16.3.1 Physical Characterization Methods

Different physical characterization methods have been employed to confirm the material's composition, crystal structure, and morphology [19,20]. Characterization methods, such as Energy-Dispersive X-ray Spectroscopy (EDX) and Inductively Coupled Plasma-Optical Emission Spectrometry (ICP-OES), are used for the quantification of the elemental composition of the materials. The crystal structure and related phase change during synthesis and electrochemical cycling process can be traced by using X-ray Diffraction (XRD) and Transmission Electron Microscopy (TEM) [21,22]. Moreover, crucial information about the particle size and the d-spacing along with a particular lattice plane of the structure can be obtained [23,24]. The chemical environments of the materials such as the oxidation state of metals and oxygen vacancies can be sighted by using spectroscopic methods such as X-ray Photoelectron Spectroscopy (XPS) and X-ray Absorption Spectroscopy (XAS) [25,26]. Morphology of the synthesized materials can be obtained by using microscopic methods such as TEM and scanning electron microscopy (SEM)[27].

16.3.2 Electrochemical Characterization Methods

A secondary battery works through reversible electrochemical reactions (charging and discharging) processes. The reactions occur at the opposite electrodes (anode and cathode), which involves via electron exchange between the electrodes. To understand the battery chemistry, a reliable electrochemical setup is crucial [28]. For initial testing of new materials (electrodes or electrolytes), a half-cell (vs Li metal) coin-type cell can be used. The half-cell configuration may be two-electrode or three-electrode system. There are different electrochemical methods, which can be used to test the effectiveness of materials for battery applications. The first electrochemical testing can be done using cyclic voltammetry (CV) within a defined potential range (potential window) to identify the redox (charging and discharging) processes and the active materials that can be extracted from the electrodes (current) [29]. Electrochemical impedance spectroscopy (EIS) is also the other electrochemical method, which is crucial to study the resistance of the materials originating from different sources during electrochemical processes. Moreover, in rechargeable batteries, the battery capacity (energy density and power density) and capacity retention (lifetime), which are the most important parameters of battery performance can be measured using electrochemical methods. When promising results have been obtained using the half-cell, pouch cell (full cell) can be also employed to study the electrochemical feature of a battery.

16.4 MATERIALS DEVELOPMENT

Arc melting and mechanical alloying are two of the most often used techniques for producing high entropy alloys (HEA). HEAs made from nanocrystalline materials are frequently produced through mechanical alloying. To prepare, stoichiometric proportions of the metal powders are ball-milled, and then the mixture is sintered [30]. Quenching can be used to cool the material after heating since the rate of cooling is crucial to the creation of a single phase. De-alloying, which selectively leaches one element, can also be utilized to produce nano-porous HEAs with high catalytic activity. HEOs, on the other hand, must frequently be created using high-temperature techniques like solid-state sintering or spray pyrolysis [31]. For the creation of thin films, additional techniques include sol-gel combustion, co-precipitation, sonochemistry, carbothermal shock, or sputtering deposition. Metal oxides are ball-milled, formed into pellets, and then sintered using solid-state techniques. Spray pyrolysis, in contrast, is preferable since it has a shorter residence time at high temperatures. By passing an aerosol of metal precursors through a flame or reactor to cause fast breakdown and structural alteration, it is possible to create nanocrystalline powders [32]. Wang et al. [14] demonstrated that nebulized spray pyrolysis (NSP) may be used to create $Mg_{0.2}Co_{0.2}Ni_{0.2}Cu_{0.2}Zn_{0.2}O$ HEOs. Their study demonstrated how the HEO $Mg_{0.2}Co_{0.2}Ni_{0.2}Cu_{0.2}Zn_{0.2}O$ multi-cations synergy at the atomic and nanoscale during electrochemical reaction explains the "cocktail effect." Thus, it was determined through microstructure, elemental, and phase distribution analysis that the images of the cycled samples display a heterogeneous nanostructure while the intensity in high-angle annular dark field STEM (HAADF-STEM) has a homogeneous intensity distribution with only a few pores present (Figure 16.5). Researchers must consider the influence that specific elements and oxygen vacancies play in the characteristics of HEOs in addition to the synthesis process. By effectively oxidizing high-entropy FeCoNiCrMn alloy powders, $(FeCoNiCrMn)_3O_4$ HEO was created by Bin Xiao et al. and used as a new advanced anode material for LIBs. They claimed that a straightforward oxidation procedure was employed to create HEOs from HEA powder. In comparison to previous HEOs electrodes (i.e. $(FeCoNiCrMn)_3O_4$, (CoCuMgNiZn)O, $(MgCoNiZn)_{1-x}Li_xO$, and $(MgTiZnCuFe)_3O_4$), the $(FeCoNiCrMn)_3O_4$ HEO shown significantly improved rate performance and extremely extended cycle stability. The complete scope of potential material possibilities will become clear when researchers pinpoint the intricate relationship between material properties and HEM performance. For example, optimizing the calcination temperature of a high-entropy oxide, methodical research, and trial and error have typically been used to steer material discovery. These methods are appropriate when a field is still developing, and they will undoubtedly keep advancing our understanding through the method of incremental progress. One example of this is the expanding corpus of knowledge regarding high-entropy oxide spinels. However, a lot of work needs to be done before high-entropy materials can take the place of conventional Li-ion battery electrodes. It is currently unclear how much element choice, the "cocktail effect," entropy stabilization, particle size and morphology, etc., influence the electrochemical properties of HEMs. The Technology Readiness Level (TRL) of HEMs has been accelerated by researchers using computation-driven techniques to material discovery. As can be demonstrated in the work by Lun et al., density functional theory (DFT) and machine learning can assist in restricting the huge design space of candidates with desirable features. High-throughput techniques also deliver the experimental feedback required to confirm these computational techniques and finish the material design cycle.

Additionally, Rost et al. [33] created the non-metallic system (NiMgCuCoZn)O and reported that it possesses a single-phase rock-salt crystal structure. Equimolar concentrations of metal oxides were ground to create the ceramic, which was then heated, quenched, and quenched again. The discovery laid the groundwork for ESOs when it was discovered that low-temperature heat treatments may convert the single phase of entropy-driven oxides back to multiple phases. Additionally, high-temperature treatments could be used to achieve the ability to retain the single

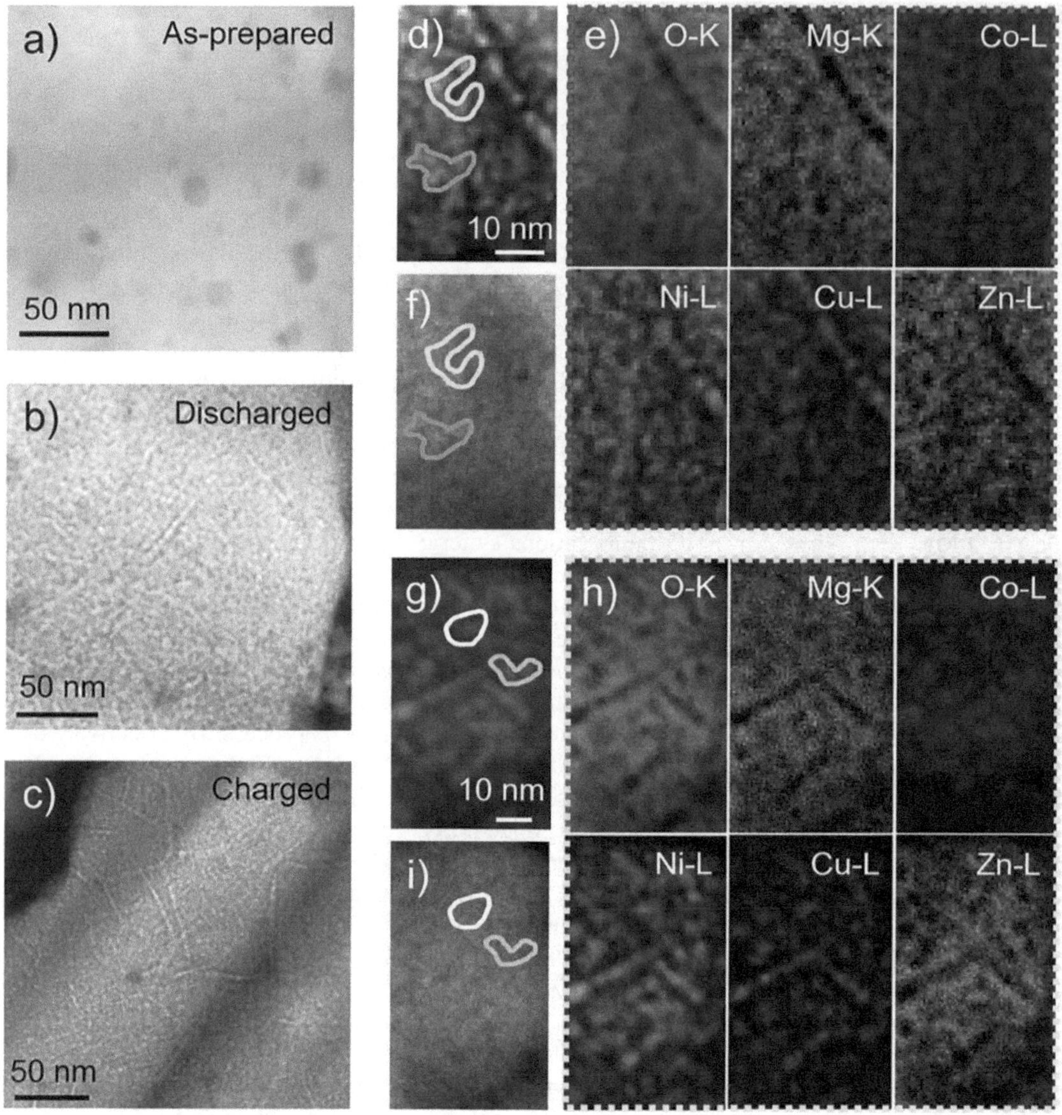

FIGURE 16.5 Analysis of the as-prepared and cycled samples' elemental distributions. Adapted from reference [14]. Copyright The Authors, some rights reserved; exclusive licensee [Nature]. Distributed under a Creative Commons Attribution License 4.0 (CC BY) https://creativecommons.org/licenses/by/4.0/.

phase once more. Mg, Co, Ni, Zn, and Cu are regarded as suitable candidates to be combined in a uniform high entropy oxide system among the elements in the periodic table. In contrast to the others, Cu and Zn have lower compatibility (different crystal structures), which restricts their stability. This could be seen as a barrier for possible applications, especially those at low to moderate temperature ranges because the stabilizing impact of entropy is dominating at high temperature. Systems in HEA are stable across a broad temperature range due to suitable chemical composition as well as entropy. To understand the novel class of oxides' thermal behavior, electrical conductivities, band gaps, and catalytic activity. The synthesis of (MgCoNiCuZn)O and its derivatives (Figure 16.6) has received a lot of interest. Methods used include co-precipitation, solid-state reaction, flame spray pyrolysis, modified solution combustion, and a NSP, synthesis technique.

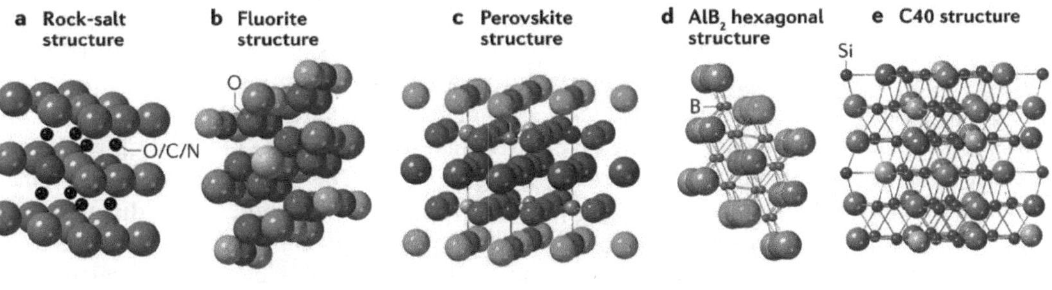

FIGURE 16.6 High-symmetry structures of high-entropy ceramics. Adapted from reference [34]. Copyright The Authors, some rights reserved; exclusive licensee [Nature]. Distributed under a Creative Commons Attribution License 4.0 (CC BY) https://creativecommons.org/licenses/by/4.0/.

16.5 HIGH-ENTROPY ANODES IN LI-ION BATTERY

16.5.1 Oxide Ceramics

HEOs are a novel category of oxide systems that have recently been developed and published, with initial demonstrations for transition metal-based HEO (TM-HEO), rare earth-based HEO (RE-HEO), and mixed HEO (TM-RE-HEO) [35]. HEOs are based on a brand-new, ground-breaking idea for entropy stabilization: to maintain a specific crystal structure that may be different from the conventional crystal structures of the component elements while increasing the configurational entropy of the resulting compounds. The initial report on this idea concerned metallic HEA [36]. Numerous publications show the study of HEA. According to several studies on TM-HEO, RE-HEO, and mixed TM-RE-HEO, high entropy stabilization in oxides with five or more cations at equiatomic quantities results in the production of single-phase rock-salt, fluorite, or perovskite structures [37]. These substances frequently exhibit intriguing and unexpected characteristics, such as extremely high Li-ion conductivities for solid-state electrolytes at room temperature in TM-HEO, extremely small and specialized band gaps in RE-HEO, and enormous dielectric constants in TM-HEO. The ability to produce unique qualities by utilizing the huge number of conceivable elemental combinations is the primary factor driving the increase in interest in HEO [38]. The availability of numerous synthesis and processing techniques, which have been proven to produce highly repeatable material systems, has helped the field of HEO expand quickly. In Li-ion batteries, graphite has historically been utilized as an anode material. Although its intercalation method makes it extremely robust, its 372 mAh/g intrinsic capacity is poor [39]. HEOs have been utilized as anode materials in batteries because of their capacity to host lithium ions through a conversion-type reaction. Although these materials frequently have capacities that are much higher than those of graphite, they have poor cycle stability. The first cycle Coulombic efficiency (CE) is frequently lower than 90% and, in extreme circumstances, as low as 50%. Numerous HEOs possess the ability to increase capacity with prolonged cycling. It is also known that older conversion-type materials like CoO and $ZnMn_2O_4$ exhibit the same characteristics [40]. The most common explanation for this phenomenon is reversible lithium storage within a polymeric layer that develops at the electrode surface. Other considerations include structural changes during cycling and the electrolyte's slow penetration of inner oxide pores. The rock salt ($(Co_{0.2}Cu_{0.2}Mg_{0.2}Ni_{0.2}Zn_{0.2})O$) is one of the HEOs for batteries that has received the greatest attention from researchers [41]. Different TM-HEO ($(Co_{0.2}Cu_{0.2}Mg_{0.2}Ni_{0.2}Zn_{0.2})O$) materials were developed by Abhishek et al. and used as anodes. A modular strategy for the systematic design of the electrode material is provided by the incorporation of several elements into HEO [42]. It is also demonstrated that, in contrast to conventional conversion materials, entropy-stabilized oxides have high-capacity retention and exhibit a de-/lithiation behavior. The novel effect is related to the stabilization of the lattice's configurational entropy, which preserves the structure of the original rock-salt mixture while acting as a long-term host matrix for the conversion cycles [43]. Further research into high entropy oxide electrode materials should be pursued to explore their full potential for energy storage applications considering these preliminary-yet encouraging-scarcely comprehensive results [44]. In the stable area (cycle numbers 60–400), the CE stabilizes between 99.4 and 99.95%. The capacity values for the 5-cation system were anticipated to be slightly different because of the heat treatment, as illustrated in Figure 16.7a. The standard terminology places 5-cation systems in the "high entropy" category and 4-cation systems in the "medium entropy" group. It should be emphasized that when contrasted to the "high entropy" oxides, all "medium entropy" oxides (MEO) had an entirely distinct and quite unstable electrochemical behavior (Figure 16.7b). The "medium entropy" oxides' electrochemical behavior, which initially showed high specific capacities before various degrees of deterioration with increasing numbers of cycles, corresponds favorably with that of proven conversion materials with large particle sizes [45]. The 5-cation high entropy oxide, in contrast, has a high-capacity value that does not decrease with an increase in the number of cycles. We draw the conclusion that the "high entropy" material exhibits a novel and fascinating electrochemical behavior that is likely connected to the stabilization of entropy from the capacity comparison between the 5-cation TM-HEO and the 4-cation oxides [33]. A report of such an observation has never been made before. The electrochemical performance of the 5-cation TM-HEO and three different 4-cation systems, each without Zn, Cu, or Co, is compared in Figure 16.7b.

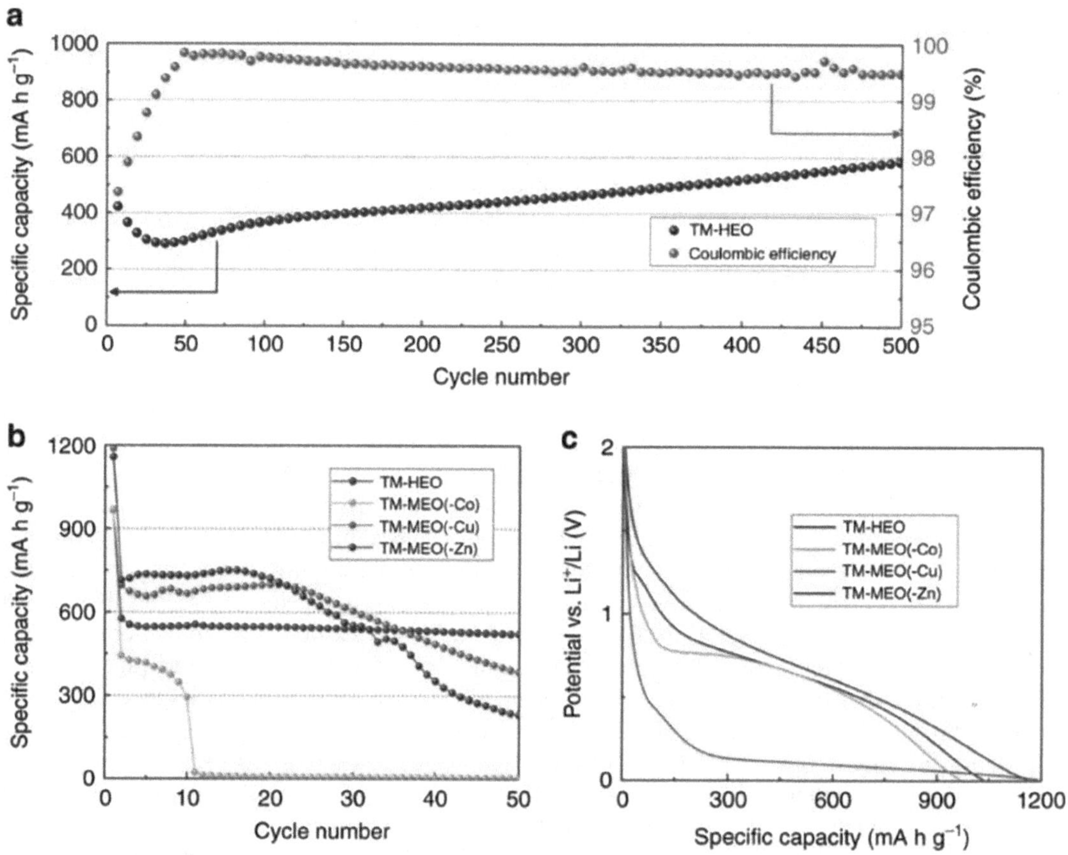

FIGURE 16.7 Specific capacities and lithiation profiles of TM-HEO and TM-MEO. Adapted from reference [46]. Copyright The Authors, some rights reserved; exclusive licensee [Nature]. Distributed under a Creative Commons Attribution License 4.0 (CC BY) https://creativecommons.org/licenses/by/4.0/.

16.5.2 Metallic Alloys

Despite being a relatively untapped area of research for the Li-ion battery anode, high-entropy alloys may be able to address issues with traditional silicon-based (Si) electrodes [47]. Its weak cycle stability is what makes silicon (and most alloy anodes) problematic. A volume increase of 300% occurs in conjunction with the alloying of lithium and silicon. This results in microscopic swelling, pulverization, and loss of electrical contact between the material and the current collector. Additionally, the electrolyte is consumed by the ongoing formation of solid-electrolyte-interface (SEI) [48]. The quick capacity decay of silicon and alloying materials in batteries is caused by these issues.

Numerous studies have been conducted on two or three multicomponent alloys that use Li as the negative electrode. Sn, Al, Ga, Bi, Ag, Zn, Sb, and Ge, are common elements that alloy with lithium in addition to Si [49]. As demonstrated by the eutectic Zn–Sn alloy, changing the compositional ratio of some systems can result in morphological modifications that enhance volumetric capacity. It is well known that a multicomponent alloy performs better in batteries than a straightforward binary alloy [50].

16.6 SUMMARY AND PERSPECTIVES

High-entropy materials have recently become more prevalent in the field of electrochemical energy storage. Due to their significantly higher capacity compared to the typical graphite anode, which is only capable of 372 mAh/g, HEOs have the potential to be employed as anode materials in Li-ion batteries. It must be emphasized, nonetheless, that these oxides typically have a lower electrochemical stability than conventional anodes and a higher first cycle irreversible capacity. Additionally, they typically operate at a higher potential than graphite (0.2 V–0.3 V) (1 V–2 V). This is excessive for most applications since it results in a significant reduction in battery output voltage when used with standard NMC layered cathodes (at about 4 V). Additionally, it is still unclear how non-faradaic and surface pseudocapacitive contributions to total capacity work. Numerous cations, many of which have good catalytic activity, could interact with the organic solvents and salts of the electrolytes and cause reactions. As an alternative, the creation of a highly stable SEI may be made possible by these unique electrode-electrolyte interactions. It is well recognized in the battery industry that disorder in cathode materials is a highly unfavorable characteristic. A well-known example of this is phase change and deterioration in $LiNiO_2$, which is brought on by considerable Li–Ni cation mixing. Other metal cations, such as Mn and Co, have been added successfully in the past, as seen in the instance of the commercial NMC111. These metal cations boosted stability in this case, and careful cation selection may further enhance structural stability. It is not yet known, though, whether entropy-stabilized cathodes are any less prone to deterioration. Additionally, some promising high-capacity cathodes, like DRXs, need to be processed under argon, which may restrict their viability on the market. However, the structure may be stabilized by HEO cathodes, allowing metal cations to have more stable sites with lithium-accepting energies. Higher capacities may result from the use of five metal cations, either alone or in combination with fluorination (as in the case of DRX), or from the use of multilayer structures and vacancies. There is not as much research on HE-cathodes as there is on HE-anodes. To fully investigate the possibilities of these new electrodes, more research is required. Despite the potential benefits, it is still important to emphasize the following points regarding the current issues and potential fixes for HEO.

- **Theoretical calculation:** While the high-entropy concept enables entirely new material compositions and specially tailored material properties, it is challenging to predict material behavior through straightforward chemical understanding due to the complex interactions of bound ions. Electronegativity, valence electron concentration, chemical compatibility, structural properties (ionic radii, structural displacement of ion pairs), and formation energy all interact to control lattice stabilization and distortion. To assess the synthesizability of various HEOs, researchers have created descriptors based on computational materials science and materials informatics, such as entropy forming ability (EFA value), Goldschmid tolerance factor, and size disorder factor. Computational design methods are based on established paradigms and may not consider truly unique materials. High-throughput investigations on HE-rock-salts, for example, may divert funds away from the identification of brand-new families of HE structures. High-quality data collecting, which is necessary for the successful use of machine learning, presents additional difficulties. In an ideal scenario, computational designs and more conventional experimental inquiry will be combined with high-throughput experimentation. For instance, the formation feasibility of new HEOs has been theoretically validated using DFT simulation based on the Special Quantum Structure method. However, due to the disorder of solid-solution structures and the complexity of multicomponent systems, DFT is not appropriate for big data virtual screening. Machine learning, which may be a preferable method for determining the synthesizability of HEOs, may quickly develop models by mapping the relationship between goal qualities and material property parameters. Additionally, by examining the conflicting relationship between entropy and enthalpy, in particular stable composition, and temperature range, as well as transition trends and critical conditions, thermodynamic studies and molecular dynamics simulations aid in the investigation of phase stability. Additionally, the microstructure, surface/interface states, and internal flaws of HEOs are all connected to their functional characteristics. Theoretical simulations provide a theoretical foundation for understanding the intricate redox process of HEOs and are useful in analyzing the electrochemical reaction kinetics of electrode materials at the atomic level.

- **Utilization of advanced characterization methods:** Currently EDS, XRD, and TEM are used primarily for the inaccurate and insufficient identification and characterization of HEOs. Through more dependable and exact techniques, it is necessary to further confirm the cation distribution and phase purity. Furthermore, the HEO's overall performance is not just a straightforward superposition of different oxides. Deconstructing the relationship between element combination, structure, and property/performance requires more sophisticated measurements. This largely encompasses equipment for operando characterization, neutron scattering/diffraction, atomic-resolution electron microscopy analysis, and technologies connected to synchrotron radiation. For instance, synchrotron X-ray technology can analyze the structural evolution mechanism, phase transition process, and charge compensation behavior of electrode materials by detecting the electronic and structural features of materials at various depths almost non-destructively through spectroscopy, scattering, and imaging functions. With the use of various characterization approaches, it is possible to determine the functional components of HEOs and expose the endogenous structure-activity correlation in the future.
- **Working mechanism challenges:** About HEO, there are a lot of unanswered questions. For instance, the lack of temperature-dependent reversible phase transition features in some HEO types (such as most S-HEOs) raises doubts about the applicability of the high-entropy effect in general. The solid-solution mechanism hasn't been verified, either. The multiphase mixture may be extremely miscible at high temperatures, or it may be based on a specific unary oxide as a matrix for successive inclusion. Understanding the structural stability of HEO depends on these results. Most papers also simply credit the electrochemical performance to the "cocktail effect" or multi-element synergy of high-entropy materials, but the precise synergistic process is yet unknown. The transformation of the active substances from spinel to rock salt to amorphous and the partially reversible redox reactions suggest that the lithium storage mechanism has changed. Lattice distortion is another intriguing subject. Because after undergoing volume expansion and transition

metal redox during electrochemical cycling, the already severely deformed HEO structure miraculously maintains its integrity. What is the tolerance upper limit for this stabilization of the lattice? These puzzles might be solved by unraveling the interconnections of atoms, lattices, metastable structures, stacking defects, and strain in HEOs.

- **Improving the Overall performance of HEOs:** The investigation of HEO anodes is still in its early stages overall. There is still room for improvement in a number of electrochemical parameters, including electrical conductivity, initial CE, and rate performance. Therefore, it is necessary to continually develop new classes of HEOs, particularly those with multiple Wyckoff sites. Additionally, polymeric anionic, polymeric cationic, or mixed anionic systems are interesting directions. For instance, HEO materials such as NASICON, oxychloride, fluorophosphate, dittmarite, and oxyfluoride, analog have been swiftly developed. These unusual HEO categories not only add to the existing high-entropy library but also show how anionic sites contribute positively to structural entropy, which can lead to fresh and creative approaches to the engineering design of materials. Beyond that, the development of HEO synthesis techniques, alterations to the morphology and structure of materials, and even coupling with new functional materials all hold great promise for accelerating the development of the next wave of energy storage technology innovation.

ACKNOWLEDGMENTS

This work reported in this paper was supported by Bahir Dar Energy Center, Bahir Dar Institute of Technology, Bahir Dar University, Ethiopia.

REFERENCES

[1] J. Yan, D. Wang, X. Zhang, J. Li, Q. Du, X. Liu, J. Zhang, X. Qi, A high-entropy perovskite titanate lithium-ion battery anode, *J. Mater. Sci.* 55 (2020) 6942–6951.

[2] M. Li, A. Su, Q. Qin, Y. Qin, A. Dou, Y. Zhou, M. Su, Y. Liu, High-rate capability of columbite CuNb2O6 anode materials for lithium-ion batteries, *Mater. Lett.* 284 (2021) 128915.

[3] Q. Wang, A. Sarkar, Z. Li, Y. Lu, L. Velasco, S.S. Bhattacharya, T. Brezesinski, H. Hahn, B. Breitung, High entropy oxides as anode material for Li-ion battery applications: A practical approach, *Electrochem. Commun.* 100 (2019) 121–125.

[4] P. Ghigna, L. Airoldi, M. Fracchia, D. Callegari, U. Anselmi-Tamburini, P. D'angelo, N. Pianta, R. Ruffo, G. Cibin, D.O. De Souza, E. Quartarone, Lithiation mechanism in high-entropy oxides as anode materials for li-ion batteries: An operando XAS study, *ACS Appl. Mater. Interfaces.* 12 (2020) 50344–50354.

[5] A.K. Worku, D.W. Ayele, Advances and Challenges in the Fabrication of Porous Metal Phosphate and Phosphonate for Emerging Applications, in: R. K. Gupta (Ed.), *Met. Phosphates Phosphonates Fundam. to Adv. Emerg.* Appl., Springer International Publishing, Cham, 2023: pp. 393–407.

[6] G.M. Tomboc, X. Zhang, S. Choi, D. Kim, L.Y.S. Lee, K. Lee, Stabilization, characterization, and electrochemical applications of high-entropy oxides: critical assessment of crystal phase-properties relationship, *Adv. Funct. Mater.* 32 (2022) 2205142.

[7] K.M. Abraham, How comparable are sodium-ion batteries to lithium-ion counterparts? *ACS Energy Lett.* 5 (2020) 3544–3547.

[8] R.Z. Zhang, F. Gucci, H. Zhu, K. Chen, M.J. Reece, Data-driven design of ecofriendly thermoelectric high-entropy sulfides, *Inorg. Chem.* 57 (2018) 13027–13033.

[9] M. Fu, X. Ma, K. Zhao, X. Li, D. Su, High-entropy materials for energy-related applications, *IScience.* 24 (2021) 102177.

[10] X. Qu, H. Huang, T. Wan, L. Hu, Z. Yu, Y. Liu, A. Dou, Y. Zhou, M. Su, X. Peng, H.H. Wu, T. Wu, D. Chu, An integrated surface coating strategy to enhance the electrochemical performance of nickel-rich layered cathodes, *Nano Energy.* 91 (2022) 106665.

[11] Y. Ma, Y. Ma, Q. Wang, S. Schweidler, M. Botros, T. Fu, H. Hahn, T. Brezesinski, B. Breitung, High-entropy energy materials: Challenges and new opportunities, *Energy Environ. Sci.* 14 (2021) 2883–2905.

[12] D. Bérardan, S. Franger, A.K. Meena, N. Dragoe, Room temperature lithium superionic conductivity in high entropy oxides, *J. Mater. Chem. A*. 4 (2016) 9536–9541.

[13] X. Liu, X. Li, Y. Li, H. Zhang, Q. Jia, S. Zhang, W. Lei, High-entropy oxide: A future anode contender for lithium-ion battery, *EcoMat*. 4 (2022) 1–14.

[14] K. Wang, W. Hua, X. Huang, D. Stenzel, J. Wang, Z. Ding, Y. Cui, Q. Wang, H. Ehrenberg, B. Breitung, C. Kübel, X. Mu, Synergy of cations in high entropy oxide lithium ion battery anode, *Nat. Commun*. 14 (2023) 1–9.

[15] Q. Wang, A. Sarkar, D. Wang, L. Velasco, R. Azmi, S.S. Bhattacharya, T. Bergfeldt, A. Düvel, P. Heitjans, T. Brezesinski, H. Hahn, B. Breitung, Multi-anionic and -cationic compounds: New high entropy materials for advanced Li-ion batteries, *Energy Environ. Sci*. 12 (2019) 2433–2442.

[16] Y.S. Lee, K.S. Ryu, Study of the lithium diffusion properties and high rate performance of TiNb6O17 as an anode in lithium secondary battery, *Sci. Rep*. 7 (2017) 0–4.

[17] D.B. Miracle, O.N. Senkov, A critical review of high entropy alloys and related concepts, *Acta Mater*. 122 (2017) 448–511.

[18] H. Guo, J. Shen, T. Wang, C. Cheng, H. Yao, X. Han, Q. Zheng, Design and fabrication of high-entropy oxide anchored on graphene for boosting kinetic performance and energy storage, *Ceram. Int*. 48 (2022) 3344–3350.

[19] R.Z. Zhang, M.J. Reece, Review of high entropy ceramics: design, synthesis, structure and properties, *J. Mater. Chem. A*. 7 (2019) 22148–22162.

[20] A.F. Alem, A.K. Worku, D.W. Ayele, N.G. Habtu, M.D. Ambaw, T.A. Yemata, Enhancing pseudocapacitive properties of cobalt oxide hierarchical nanostructures via iron doping, *Heliyon*. 9 (2023) e13817.

[21] E. Lökçü, Ç. Toparli, M. Anik, Electrochemical performance of (MgCoNiZn)1-xLixO high-entropy oxides in lithium-ion batteries, *ACS Appl. Mater. Interfaces*. 12 (2020) 23860–23866.

[22] A.K. Worku, D.W. Ayele, N.G. Habtu, M.D. Ambaw, Engineering nanostructured Ag doped α-MnO2 electrocatalyst for highly efficient rechargeable zinc-air batteries, *Heliyon*. 8 (2022) e10960.

[23] A.K. Worku, Engineering techniques to dendrite free Zinc-based rechargeable batteries, *Front. Chem*. 10 (2022) 1–15.

[24] G. Ambissa Begaw, D. Worku Ayele, A. Ketema Worku, T. Alemneh Wubieneh, T. Atnafu Yemata, M. Dagnew Ambaw, Recent advances and challenges of cobalt-based materials as air cathodes in rechargeable zn-air batteries, *Results Chem*. 5 (2023) 100896.

[25] N.G. Habtu, A.K. Worku, D.W. Ayele, M.A. Teshager, Z.G. Workineh, Facile preparation and electrochemical investigations of copper-ion doped α-MnO2 nanoparticles, *Lect. Notes Inst. Comput. Sci. Soc. Telecommun. Eng. LNICST*. 412 (2022) 543–553.

[26] A.F. Alem, A.K. Worku, D.W. Ayele, T.A. Wubieneh, A. Abebaw Teshager, T. Mihret Kndie, B.T. Admasu, M.A. Teshager, A.A. Asege, M.D. Ambaw, M.A. Zeleke, A.K. Shibesh, T.A. Yemata, Ag doped Co3O4 nanoparticles for high-performance supercapacitor application, *Heliyon*. 9 (2023) e13286.

[27] R. cheng Wang, Q. lin Pan, Y. hong Luo, C. Yan, Z. jiang He, J. Mao, K. Dai, X. wen Wu, J. chao Zheng, SnS particles anchored on Ti3C2 nanosheets as high-performance anodes for lithium-ion batteries, *J. Alloys Compd*. 893 (2022) 162089.

[28] T. Jin, X. Sang, R.R. Unocic, R.T. Kinch, X. Liu, J. Hu, H. Liu, S. Dai, Mechanochemical-assisted synthesis of high-entropy metal nitride via a soft urea strategy, *Adv. Mater*. 30 (2018) 1707512.

[29] A.C. Wagner, N. Bohn, H. Geßwein, M. Neumann, M. Osenberg, A. Hilger, I. Manke, V. Schmidt, J.R. Binder, Hierarchical structuring of NMC111-cathode materials in lithium-ion batteries: *An in-depth study on the influence of primary and secondary particle sizes on electrochemical performance, ACS Appl. Energy Mater*. 3 (2020) 12565–12574.

[30] J. Zhou, J. Zhang, F. Zhang, B. Niu, L. Lei, W. Wang, High-entropy carbide: A novel class of multicomponent ceramics, *Ceram. Int*. 44 (2018) 22014–22018.

[31] W.Y. Tang, M.H. Chuang, H.Y. Chen, J.W. Yeh, Microstructure and mechanical performance of new Al0.5CrFe1.5MnNi0.5 high-entropy alloys improved by plasma nitriding, *Surf. Coatings Technol*. 204 (2010) 3118–3124.

[32] D. Bérardan, S. Franger, D. Dragoe, A.K. Meena, N. Dragoe, Colossal dielectric constant in high entropy oxides, *Phys. Status Solidi - Rapid Res. Lett*. 10 (2016) 328–333.

[33] C.M. Rost, E. Sachet, T. Borman, A. Moballegh, E.C. Dickey, D. Hou, J.L. Jones, S. Curtarolo, J.P. Maria, Entropy-stabilized oxides, *Nat. Commun*. 6 (2015) 8485.

[34] C. Oses, C. Toher, S. Curtarolo, High-entropy ceramics, *Nat. Rev. Mater*. 5 (2020) 295–309.

[35] K. Chen, X. Pei, L. Tang, H. Cheng, Z. Li, C. Li, X. Zhang, L. An, A five-component entropy-stabilized fluorite oxide, *J. Eur. Ceram. Soc*. 38 (2018) 4161–4164.

[36] L. bo Tang, B. Zhang, T. Peng, Z. jiang He, C. Yan, J. Mao, K. Dai, X. wen Wu, J. chao Zheng, MoS2/SnS@C hollow hierarchical nanotubes as superior performance anode for sodium-ion batteries, *Nano Energy*. 90 (2021) 106568.
[37] J.S. Alzahrani, Z.A. Alrowaili, C. Eke, A.S. Altowyan, I.O. Olarinoye, M.S. Al-Buriahi, Radiation shielding ability and optical features of La2O3+TiO2+Nb2O5+WO3+X2O3 (X=B, Ga, and In) glass system containing high-entropy oxides, *Heliyon*. 9 (2023).
[38] Y. Liu, L. bo Tang, H. xin Wei, X. hui Zhang, Z. jiang He, Y. jiao Li, J. chao Zheng, Enhancement on structural stability of Ni-rich cathode materials by in-situ fabricating dual-modified layer for lithium-ion batteries, *Nano Energy*. 65 (2019) 104043.
[39] J. Zheng, Z. Yang, A. Dai, L. Tang, H. Wei, Y. Li, Z. He, J. Lu, Boosting Cell Performance of LiNi0.8Co0.15Al0.05O2 via Surface Structure Design, *Small*. 15 (2019) 1904854.
[40] X. Wang, X. Li, H. Fan, M. Miao, Y. Zhang, W. Guo, Y. Fu, Advances of entropy-stabilized homologous compounds for electrochemical energy storage, *J. Energy Chem*. 67 (2022) 276–289.
[41] M.G. Poletti, G. Fiore, F. Gili, D. Mangherini, L. Battezzati, Development of a new high entropy alloy for wear resistance: FeCoCrNiW0.3 and FeCoCrNiW0.3+5 at.% of C, *Mater. Des*. 115 (2017) 247–254.
[42] S. qi Yang, P. bo Wang, H. xin Wei, L. bo Tang, X. hui Zhang, Z. jiang He, Y. jiao Li, H. Tong, J. chao Zheng, Li4V2Mn(PO4)4-stablized Li[Li0.2Mn0.54Ni0.13Co0.13]O2 cathode materials for lithium ion batteries, *Nano Energy*. 63 (2019).
[43] S.H. Albedwawi, A. AlJaberi, G.N. Haidemenopoulos, K. Polychronopoulou, High entropy oxides-exploring a paradigm of promising catalysts: A review, *Mater. Des*. 202 (2021) 109534.
[44] Z. Li, K.G. Pradeep, Y. Deng, D. Raabe, C.C. Tasan, Metastable high-entropy dual-phase alloys overcome the strength-ductility trade-off, *Nature*. 534 (2016) 227–230.
[45] S.S. Nene, M. Frank, K. Liu, S. Sinha, R.S. Mishra, B.A. McWilliams, K.C. Cho, Corrosion-resistant high entropy alloy with high strength and ductility, *Scr. Mater*. 166 (2019) 168–172.
[46] A. Sarkar, L. Velasco, D. Wang, Q. Wang, G. Talasila, L. de Biasi, C. Kübel, T. Brezesinski, S.S. Bhattacharya, H. Hahn, B. Breitung, High entropy oxides for reversible energy storage, *Nat. Commun*. 9 (2018) 3400.
[47] B.S. Murty, S. Ranganathan, J.W. Yeh, P.P. Bhattacharjee, *High-entropy alloys*. NY, Elsevier (2019) 1–363. ISBN 978-0-12-800251-3
[48] B. Cantor, I.T.H. Chang, P. Knight, A.J.B. Vincent, Microstructural development in equiatomic multicomponent alloys, *Mater. Sci. Eng. A*. 375-377 (2004) 213–218.
[49] J. Gild, J. Braun, K. Kaufmann, E. Marin, T. Harrington, P. Hopkins, K. Vecchio, J. Luo, A high-entropy silicide: (Mo0.2Nb0.2Ta0.2Ti0.2W0.2)Si2, *J. Mater*. 5 (2019) 337–343.
[50] P. Sarker, T. Harrington, C. Toher, C. Oses, M. Samiee, J.P. Maria, D.W. Brenner, K.S. Vecchio, S. Curtarolo, High-entropy high-hardness metal carbides discovered by entropy descriptors, *Nat. Commun*. 9 (2018) 4980.

17 High Entropy Materials for Hydrogen Storage

Rekha Gaba and Ramesh Kataria

17.1 INTRODUCTION

The restricted availability of non-renewable energy resources and their detrimental effects on the environment have challenged scientists around the world to find a sustainable and impactful solution. The future of the globe depends on the sustainable growth of society by replacing the available toxic resources with alternative economical, environment-friendly energy resources. Around 1970, the concept of the hydrogen economy initially acquired momentum. Hydrogen gas is an abundantly available renewable source of energy and acts as a non-polluting, non-toxic, and environment-friendly energy carrier [1–3]. Hydrogen gas is a fuel that releases water vapors into the atmosphere as compared to fossil fuels enhancing global warming by liberating carbon dioxide. Due to its high energy density as compared to other available fuels, unlimited availability, and environmentally friendly nature, hydrogen gas has emerged as a potential candidate to satisfy the increasing energy demand of the whole world [4–6]. Hydrogen has also proved its compatibility with fuel cells and has been deployed in commercially available vehicles [7–8]. However, the application of hydrogen gas as a fuel is completely based on its production, storage, and utilization. Some safe and efficient ways of production of hydrogen gas and its storage are mandatory to establish a hydrogen economy worldwide.

Scientists around the globe are working to address the most crucial issues related to the storage of hydrogen gas. The biggest challenge of the hydrogen economy is to find an economical and profitable method of storage of this lightest molecule. Around 11 m^3 of volume is occupied by 1 kg of gas at room temperature and atmospheric pressure [9]. It is desirable to increase the storage density of hydrogen to attain futuristic economy goals. The hydrogen can be stored for stationary storage in its pure, molecular form as a gas or liquid form [10].

The three ways used to store hydrogen gas are [11]:

i. Compressed gas in high-pressure tanks
ii. Liquefied Hydrogen
iii. In solid form by absorption or chemical reaction with chemical compounds. Chemical bonding-based technology involves metal hydrides and chemical hydrides.

Further, it can also be absorbed into or onto a material through weak van der Waals bonds.

17.1.1 Compressed Hydrogen Storage

Hydrogen storage in high-pressure cylinders is the most common practice. Although a small volume of gas can be stored under high pressure as per the detailed analysis of the density-pressure graph for hydrogen gas [12], this approach is used by many countries for transportation and stationary utilization of hydrogen [13]. Four standard types of storage are used based on their material and strength (Figure 17.1):

DOI: 10.1201/9781003391388-17

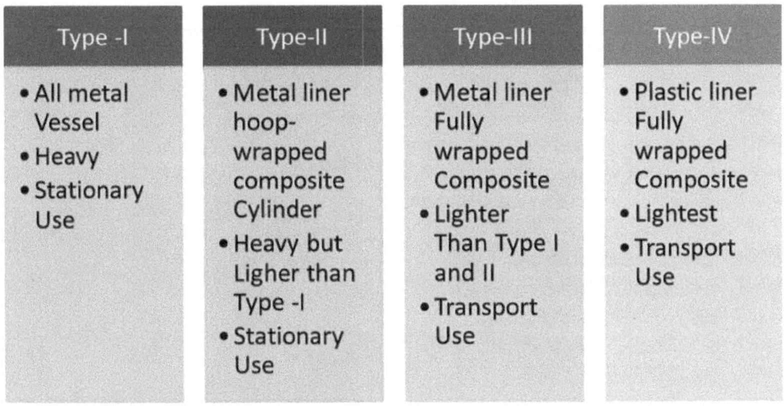

FIGURE 17.1 Standard types of storage based on the material and strength.

17.1.2 Hydrogen Storage in Liquid Form

The technology of storing hydrogen in liquid form refined the energy density of hydrogen. The complex structured double-walled with vacuum application vessels are used for the storage of liquid hydrogen. To minimize the heat transfer some additional materials are also added along with vacuum application.

17.1.3 Chemisorption Materials

The phenomenon of adsorption by lowering the temperature, enhancing pressure, and weakening the existing van der Waal forces between hydrogen could be implemented to achieve significant hydrogen storage density.

Many adsorbents like metal-organic framework [15], porous polymers, and carbon-based materials such as activated carbon, zeolites, etc. have been used as an adsorbent to store hydrogen. Heat management is the biggest challenge in the storage of hydrogen by using these adsorbents.

Other chemical hydrides like lithium borohydride, magnesium hydride, intermetallic hydrides, complex metal hydrides, and chemical hydrides [14] are developed as an efficient and low-risk alternative for hydrogen storage. Strong chemical bonds are formed in metal hydrides as compared to adsorption, which allows storing more hydrogen density at ambient conditions. Two processes can be used to extract hydrogen from these metal hydrides:

i. Thermolysis
ii. Hydrolysis

Most of the designed metal hydrides undergo thermolysis which is an endothermic process and brings significant results. On the other hand, intermetallic hydrides involve two elements for making an alloy, one element for strong hydrogen bonding and the other for weak hydrogen bonding. Different elements could be mixed in different ratios to prepare crystals for hydrogen storage. Despite their high cost, some intermetallic hydrides have also proved their utility for hydrogen storing capacity and long-term stability. Further, complex metal hydrides involve hydrogen as a part of an anion that binds with a cation. Alanates, borohydride and amides. Being light in weight of used elements, these hydrides provide high gravimetric storage capacities.

Furthermore, chemically bonded hydrides mostly liquid at room temperature like methanol, formic acid and ammonia formed by outsourcing hydrogen from water or natural gas have been used as an option for hydrogen storage. The schematic adsorption density is shown in Figure 17.2.

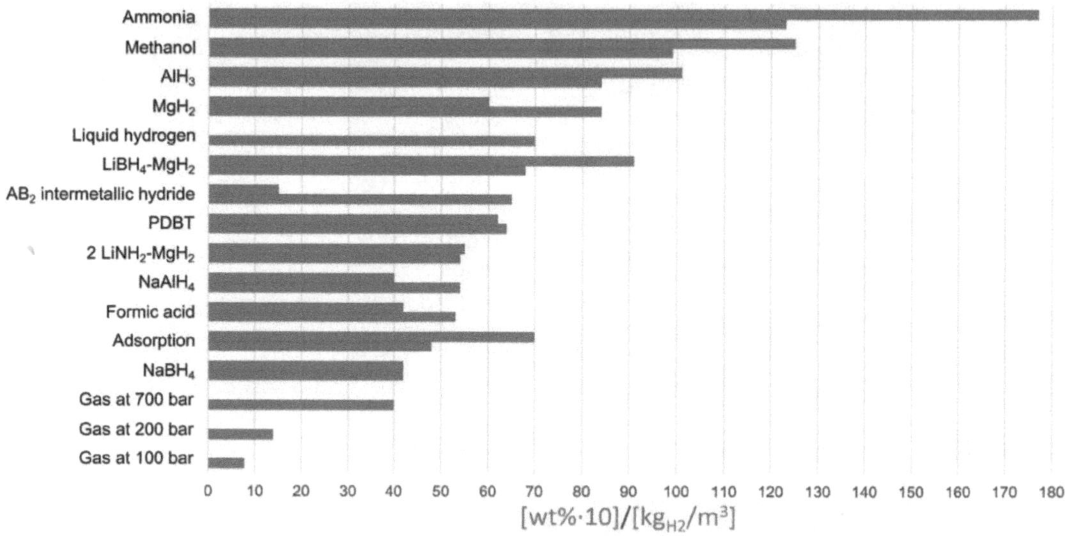

FIGURE 17.2 Volumetric and gravimetric hydrogen storage densities achieved by different methodologies. Adapted from reference [16]. Copyright Andersson *et al.,* some rights reserved; exclusive license Elsevier. Distributed under Creative Common License (CC-BY).

To overcome the difficulties in the implication of all the discussed possibilities of hydrogen storage, an old technique [17] of alloying has been used these days. Alloying has been done by dissociating hydrogen on metal by using Ni and Pd catalysts. $MgNi_2$ was the first discoverable alloy used for hydrogen storage in the mid-19th century. High-entropy alloys (HEAs), a new family of alloys that may create metal hydrides, have recently begun to be researched for hydrogen storage. Given that the kind of phase generated and chemical composition have a significant impact on the properties of metal hydrides, HEAs (with their wide range of compositions) have a high potential for creating effective materials for this application. Effective compositional design and synthesis are key factors in determining these alloys' potential. Further, it has been observed that alloys with body-centered cubic (BCC) structure show a remarkable capacity for hydrogen storage [18]. Although there are only a few research articles on the functioning of HEAs for storing hydrogen that have been published thus far, the number of HEA-related papers has been rising, indicating significant interest in the topic.

The available lattice strain favors the formation of AB_5-type to provide high hydrogen storage capacity and is named high entropy materials. This concept originated in 2004 [19], which represents the mixing of five or more elements in almost equal proportion. High entropy of mixing favors this multicomponent solid solution and BCC or Face-centered cubic (FCC) crystal structure. The lattice is typically deformed in HEA because of variations in the atomic radii of the used atoms. A stretched lattice produced by this distortion may be advantageous to produce hydrides [20]. Only a few limited studies are available for the storage of hydrogen by HEAs. In this chapter, the available literature for hydrogen storage by high entropy materials has been discussed.

17.2 METHODS FOR DESIGNING MULTI-ELEMENT ALLOY (THERMODYNAMIC PARAMETERS AND PHASE FORMATION)

The study of alloy design has been an old yet crucial area of study in metallurgy since from Bronze Age. The traditional approach to alloying is focused on the use of one or two elements as main alloying components, which has been shown to be effective but also limits the range of functional alloys that may be produced. The current development of HEAs for metallurgists substantially expands the concept of design and offers a lot of compositionally modified complex alloys for us to choose.

The basic idea behind this motivation is that by intermixing, the maximized configurational entropy of alloy could be attained, which aids in stabilizing the solid solution phase over other phases like intermetallic complex. Accordingly, the "high entropy" effect should be associated with alloys of a single-phase solid solution.

The atomic packing of genuine alloys, which can take the shape of the FCC lattice or the hexagonal close-packed lattice and in which the sizes of the component elements can vary greatly, proves that this is untrue. The idea of maximized configurational entropy of mixing out of equiatomic fraction is theoretically based on the implicit assumption that the alloy's constituent atoms are identical in size and tightly packed, as could be observed in an ideal gas, where configurational entropy of mixing is solely dependent on the arrangement of its constituent atoms. Total mixing entropy has four contributions: vibrational, configuration, magnetic dipole, and randomness. The major contributor configuration entropy overpowers other contributions. Boltzmann's could be the equation for calculating the configurational entropy in these multi-element alloys as:

$$\Delta S_{\text{conf}} = -k \ln w \tag{17.1}$$

Here, k is the Boltzmann constant and w represents the ways in which energy can be shared among various particles. For equiatomic alloys, it can be represented as (Figure 17.3):

$$\begin{aligned} \Delta S_{\text{conf}} &= -k \ln w \\ &= -R\left(\frac{1}{n}\ln\frac{1}{n}+\frac{1}{n}\ln\frac{1}{n}+ - - - - \frac{1}{n}\ln\frac{1}{n}\right) \\ &= R\ln\frac{1}{n} = R\ln n \end{aligned} \tag{17.2}$$

R is gas constant (8.314 J/K mol^{-1})

The microstructure and properties of HEAs depend on their lattice distortion, high entropy effect, cocktail effect, and sluggish diffusion. Lattice distortion affects the hardness, strength and reduces the thermal effect on the properties. High entropy can make a simpler structure by enhancing the formation of the solid phase. Formations of new phases require a cooperative diffusion among the different compositions of elements. It can affect growth, phase nucleation, and the morphology of newly diffusion-controlled phase transition. The cocktail effect enhances the properties of elements used to form HEA.

Phase prediction for the entropy alloys can be accomplished during the alloy design process using both thermodynamic factors and computer modeling of phase diagrams. The Gibbs free energy of every phase needs to be calculated for standardization. To create single-phase HEA compositions,

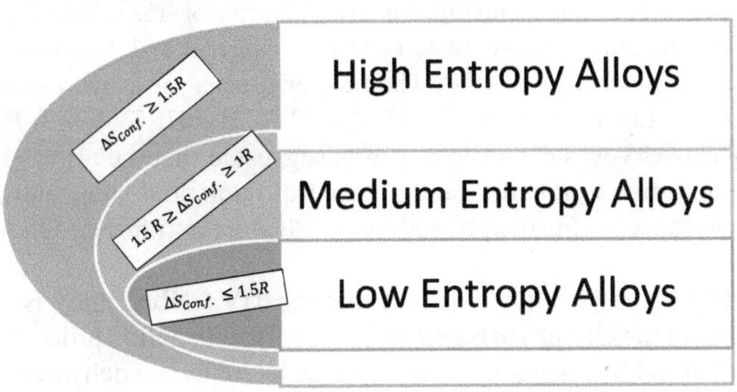

FIGURE 17.3 Thermodynamic parameters for designing multi-element alloys.

several techniques have been put forth, all of which are based on prior experience and research findings. Different parameters are checked as per Hume-Rothery criteria. These laws state that alloys having constituent elements with comparable atomic sizes, identical crystal structures, similar electronegativity, and as well as valency, are more likely to produce substitutional solid solutions. Based on the alloy composition, calculations are made for atomic size mismatch, Pauling electronegativity, the mixing enthalpy (H_{mix}), the parameter suggested by Yang and Zang [21], as well as the valence electron concentration (VEC), and other related thermodynamic quantities. Machine learning (ML) forecasts the hydrogenation characteristics of recently suggested HEAs using previously acquired data. Calphad is a software technology that uses thermodynamic model simulation to create a phase diagram. The link between structure and properties can be tracked theoretically using DFT (density functional theory), which enables the prediction of metal hydride properties and provides precise enthalpies of their formation. By using these computational techniques, HEAs are synthesized through ball milling, arc melting, and reactive milling [22].

Further, kinetic aspects have also been considered to predict phase behavior. The concept of viscosity as a function of temperature was applied to know the packing fraction, crystal structure, and atomic size of a unit cell using the solidification of liquid.

17.3 HYDROGEN ABSORPTION BY SOME SELECTIVE HEAs

The hydrogen storage capacity of HEAs is constrained since they are primarily made of mainly heavy transition metals and possess less reactivity toward hydrogen. The intricate interactions between the alloying elements have prevented attempts to greatly increase the hydrogen storage capacity of the corresponding alloys, despite the introduction of a few light elements that show the tendency to absorb hydrogen, such as Al, Mg, or Sc. The chemical composition of a HEA can be altered to dramatically increase its storage capacity, which gives rise to a large folk of materials blessed with a potential to absorb hydrogen reversibly at both low and high temperatures.

Twenty-one distinct HEAs for the TiVZrNbHf system framed with various elemental compositions, such as intermixing three elements for ternary, four different elements for quaternary, and five elements for quintary compositions, were examined for their hydrogen sorption capabilities [23]. The alloys' absorption kinetics was incredibly slowed down. Only one elemental composition featured a hexagonal structure with only a single-phase solution, and there were seventeen compositions with single-phase BCC structures. With maximal H/M ratios approaching 2, 15 of them showed the existence of either FCC structure or BCT metal hydrides. They also discovered that the synthesis of BCT metal hydrides was influenced by high concentrations of Zr and Hf as compared to FCC. They supported the hypothesis that it is crucial to keep an eye on the VEC and p parameters when designing HEAs for hydrogen storage applications in their conclusions.

The characteristics of ZrTiVCrFeNi have been explored [24], and it was found that 1.81 weight percentage of hydrogen could be stored there after being activated at 500°C and 100 pressures. The maximal hydrogen content at the same hydrogenation conditions got down to 1.56 wt% after applying supplementary heat treatments. During the investigation of $TiZrNbMoV_8$ at around at 50°C, it was discovered that the single-phase BCC-type material had less hydrogen storage (0.59 wt%) than that of multiphase material (2.3 wt%) when experimented at a hydrogen pressure of 85 bar at the same temperature. Later on, a quinary HEA of TiVZrNbHf [25] with H/M ratio of 2.5 and hydrogen-absorbing capacity was discovered. It was suggested that the alloy's lattice strain, allows it to absorb hydrogen at both tetrahedral and octahedral interstitial sites, and provides the alloy's high hydrogen storage capacity. Further, this alloy has shown the future of HEAs as storing materials for hydrogen.

Substitution of Hf with Fe [26] demonstrates that the TiVZrNbFe alloy is composed of a solid solution phase along with the C14 phase based on Nb. By hydrogenation process, FCC dihydride and C14 hydride are created and by reversing the process i.e. by following dehydrogenation, BCT monohydride with C14 phase is created. The variation in VEC between the two phases is what causes the differences in the behavior of hydrogen desorption among both created phases. According to testing on the alloy's hydrogen storage capabilities, it can absorb a 1.60 weight percentage of hydrogen under 1 MPa H_2 in about 100 seconds at 50°C. Based on two-step hydrogen absorption, the hydrogenation mechanism was determined to be the nucleation and development of hydrides, and its hydrogen absorption's apparent activation energy (EA) is reduced to 11.27 kJ/mol H_2. Four steps (Figure 17.4) may be distinguished in the corresponding hydride's decomposition process, which explores the ways of desorption of H_2.

The creation of brand-new HEA with BCC structures for hydrogen storage in the TiVNbCrMn system by using the CALPHAD method has been reported recently [27]. To create alloys, computational calculations were applied. By using the arc melting procedure, three different alloy compositions—$Ti_{27.5}V_{27.5}Nb_{20}Cr_{12.5}Mn_{12.5}$ $Ti_{32}V_{32}Nb_{18}Cr_9Mn_9$ and $Ti_{35}V_{35}Nb_{20}Cr_5Mn_5$ were created. These alloys are capable of absorbing hydrogen with high capacities of 3.38, 2.47, and 2.09 wt%, respectively, to produce FCC hydrides. For each alloy's composition, there were different activation kinetics and incubation times. Two alloys underwent hydrogen absorption/desorption cycles, with findings that are encouraging given that the alloys' maximal capacity has been preserved.

Pressure-Composition-Temperature, *scanning electron microscopy (SEM),* X-ray diffraction (XRD), energy dispersive X-ray spectroscopy (EDS), high-resolution transmission electron microscopy (HRTEM), X-ray photoelectron microscopy, etc. measurements were used to analyze the hydrogen retention properties of $Ti_{1.02}Cr_{1.0}Fe_{0.7-x}Mn_{0.3}Al_x$ $(0 \leq x \leq 0.1)$ HEAs with the primary C14 structure [28]. These alloys were made from plasma arc melting and subsequent heat treatment. Impacts of Ti replacement with Zr, Cr replacement with Fe, and their modified stoichiometry on the hydrogen holding capabilities of Ti–Zr–Cr–Fe alloys have shown [29] that the volume of cell and enthalpy steadily increased while the plateau pressure and hysteresis gradually diminishing with enhancing Zr concentration and 2 A/B ratio (1, 1.05, 1.1). While the hydrogen storage capacity somewhat decreased, the hydrogen plateau pressure in $Ti_{0.82+y}Zr_{0.20y}Cr_{0.9}Mn_{0.2}Fe_{0.8}V_{0.1}$$(y=0\text{–}0.06)$ alloys was raised. At 353 K, $Ti_{0.88}Zr_{0.14}Cr_{0.9}$ $Mn_{0.2}Fe_{0.8}V_{0.1}$ alloy was able to maintain dissociation pressures of 23.73 MPa, although the use of the valuable metal vanadium increased the price of the alloy's basic materials [30].

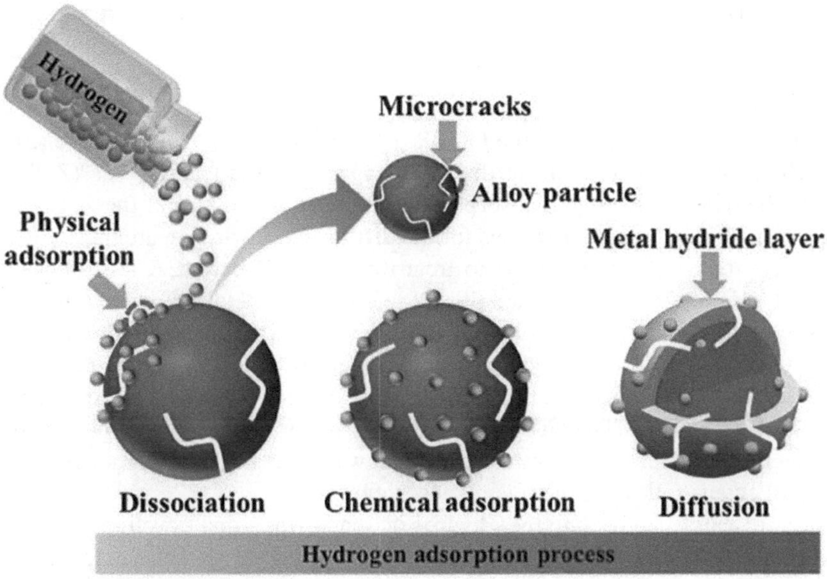

FIGURE 17.4 Hydrogen absorption process represented in literature. Adapted with permission [26]. Copyright 2023, Elsevier.

Further, the addition of Yttrium as $Ti_{1.08}Y_{0.02}Fe_{0.8}Mn_{0.2}$ mid entropy alloy resulted in a very high hydrogen storage capacity [31] (1.84 wt%) and outstanding kinetic characteristics at ambient temperature, it nevertheless requires an incubation period of roughly 1500s. But, by using electromagnetic induction melting, $Ti_{1.08}Y_{0.02}Fe_{0.8}Mn_{0.2}Zr_x$ (x=0, 0.02, 0.04, 0.06, 0.08) alloys were created. Energy dispersive spectrometer quantitative element analysis reveals that Ti element content is much greater than Fe element content in the second phase area containing Zr. The first-principle analysis of the Zr-doped TiFe system reveals that Zr is a more desirable Ti replacement than Fe. As a result, the Ti is replaced in part by the doping of Zr. When $x>0.04$, the alloy contains just pure TiFe phase, which limits the solubility of Zr in TiFe. The extra Zr precipitates occur when $x>0.4$, which lowers the TiFe alloy's capacity for reversible hydrogen absorption and desorption. The activation time and plateau pressure are both dramatically shortened by the addition of Zr [32].

Further, modifications in the microstructure of titanium-based alloys enhanced their hydrogen absorption capacity and quick kinetics. For example, $(ZrTiVFe)_xAl_y$ HEAs make them suitable hydrogen-absorbing materials. The influence of aluminum concentration on hydrogen storage capabilities and distribution of constituent elements at the micro level of $(ZrTiVFe)_xAl_y$alloys was further examined in another work [33] using equimolar as well as non-equimolar $(ZrTiVFe)_{80}Al_{20}$ and $(ZrTiVFe)_{90}Al_{10}$ alloys. The findings demonstrate that both alloys are made up primarily of the C14 Laves phase with minor amounts of tetragonal and hexagonal-close-packed phases. The vanadium element content in the C14 Laves phase likewise decreases with the increase in aluminum content, which is a result of the iron, vanadium, and zirconium element constituents of the C14 Laves phase having lower contents. The titanium element can mix along surplus aluminum and allow the formation of a tetragonal phase that surrounds the C14 phase, and as the tetragonal phase expands, the C14 phase becomes more refined. At normal temperature, $(ZrTiVFe)_{90}Al_{10}$ alloy captivated 1.3 wt % H_2, indicating improved hydrogenation kinetics and capacity. The enhanced C14 Laves phase, the addition of more vanadium elements to the C14 Laves phase, and the acute distortion in the lattice structure of the $(ZrTiVFe)_{90}Al_{10}$ alloy have improved the hydrogen storage property. It is discovered that the refined Laves phase makes it simple to activate the $(ZrTiVFe)_{80}Al_{20}$ alloy in just three cycles. Both alloys are hydrogenated, and the crystalline structure is left untouched.

But in another experiment [34], the VEC of the equimolar ZrTiVFe alloy was increased using the element Cu, and $(ZrTiVFe)_{1--x}Cu_x$ (x=0.05, 0.1, and 0.2) alloys were created. The results of examining the microstructural characteristics and hydrogen storage capabilities of these alloys ($(ZrTiVFe)_{0.90}Cu_{0.10}$ and $(ZrTiVFe)_{0.95}Cu_{0.05}$) show that they are primarily composed of the C14 Laves phase with minor amounts of the α-Titanium and α-Zirconium phases. The microstructure changes to the reticular $ZrTiCu_2$ phase around the C14 Laves phase when the Cu content reaches 20% at.%, and the Cu_8Zr_3 phase is created during the final stages of solidification. The $(ZrTiVFe)_{0.80}Cu_{0.20}$ alloy exhibits the quickest hydrogen absorption rate, which suggests that the $ZrTiCu_2$ and Cu_8Zr_3 phases may offer more substitute pathways for the diffusion of hydrogen atoms.

A high-energy ball milling was used to treat the MgAlTiFeNi HEA [35] in both Ar and H_2 atmospheres. It is demonstrated that when this alloy is ball milled (also known as mechanical alloying) under the environment of inert Argon gas, it develops a body-centered cubic structure; but, when milled under H_2 gas pressure (called reactive milling, or RM), it combines BCC, FCC, and Mg_2FeH_6 structures. Thermal studies and manometric measurements in Sievert's apparatus were used to assess the RM-MgAlTiFeNi samples' hydrogen storage behavior. The functional storage capacity of the RM–MgAlTiFeNi alloy for hydrogen was initially 0.87 weight percentage, rising to 0.94 weight percentage following the second absorption.

Systematic research [36] on the hydrogen capturing and cycling characteristics of $(VFe)_{60}(TiCrCo)_{40-x}Zr_x$ ($0 \leq x \leq 2$) alloys revealed that the BCC main phase, the C_{14} Laves phase, and the CeO_2 phase make up the alloys when x=0 and 0.5, while Zirconium-based FCC phase occurs in the alloys where x=1 and 2. The results also revealed that the amount of the Laves phase and the Zirconium-based FCC phase rise as more Zirconium is substituted for titanium. Zirconium substituted compositions were shown to have superior kinetics for hydrogen absorption, but as the substitution amounts rise, so do the hydrogen storage capacities.

The hydrogen storage capabilities along with the structural characterization of an A2B type alloy $MgTiNbCr_{0.5}Mn_{0.5}Ni_{0.5}$ and its other derivative, $Mg_{0.68}TiNbNi_{0.55}$, are presented [37] for the first time. When made by mechanical alloying (MA) or reactive milling (RM), these magnesium-containing multi-principal element alloys produce a major BCC phase (W-type, $Im\bar{3}m$) and major FCC hydride (Metal hydride with CaF_2-type structure). The hydrogen desorption from both RM samples from various hydrides is done in two steps. Although the $Mg_{0.68}TiNbNi_{0.55}$ composition has a uniform microstructure (fewer secondary phases), both alloys have a total gravimetric capacity of about 1.6 weight percentage H_2.

Another study [38] describes the production of $V_{0.3}Ti_{0.3}Cr_{0.25}Mn_{0.1}Nb_{0.05}$ HEA as well as its thermodynamic properties, initial hydrogenation kinetics, and cycling effect on the hydrogen storage properties. It was discovered that $V_{0.3}Ti_{0.3}Cr_{0.25}Mn_{0.1}Nb_{0.05}$ alloy crystallizes with a minor quantity of secondary phase in the BCC phase. Without an incubation period, the first hydrogenation is feasible at ambient temperature and has a maximum hydrogen storage capacity of 3.45 wt%. The reversible hydrogen desorption capacity is 1.78 weight percentage, and the desorption plateau pressure is 80.2 kPa, according to the pressure composition isotherm (P–C–I) at 298 K. The partial desorption at room temperature causes the hydrogen absorption capacity to decrease with cycling. Cycling causes an increase in the kinetics of hydrogen absorption.

Several quaternary and quintary HEAs that are connected to the ternary system of TiVNb are created to learn more about their structure and hydrogen storage capabilities by using manometric measurements in a Sievert's device, powder X-ray diffraction (PXD), scanning electron microscopy (SEM), thermogravimetric analysis (TGA), differential scanning calorimetry (DSC), and thermogravimetric analysis (TGA) [39]. Arc melting was used to make the alloy. All of the alloys that were produced had FCC-structured hydrides of metal showing hydrogen-to-metal ratios approaching 2, and they all created metal hydrides with BCC crystal structures. TiVZrNbTa shows no relationship between local lattice strain and hydrogen storage capacity. The hydrogen absorption capacity of TiVZrNbTa is not dependent on any sophisticated activation processes because TiVCrNb reveals quick hydrogen absorption kinetics and a hydrogen storage capacity of 1.96 wt%. The graph between the onset temperature T_{onset} and valence-electron concentration, VEC, shows a linear relationship (Figure 17.5) with the decrease in onset temperature. Their findings have led them to identify $TIVCrNbH_8$ as a material that has favorable thermodynamics for the solid-state storage of hydrogen. At ambient temperature and normal pressures, this HEA-based metal hydride shows a reversible hydrogen storage capacity of 1.96 weight percentage. Furthermore, it doesn't require any complicated activation processes to absorb hydrogen.

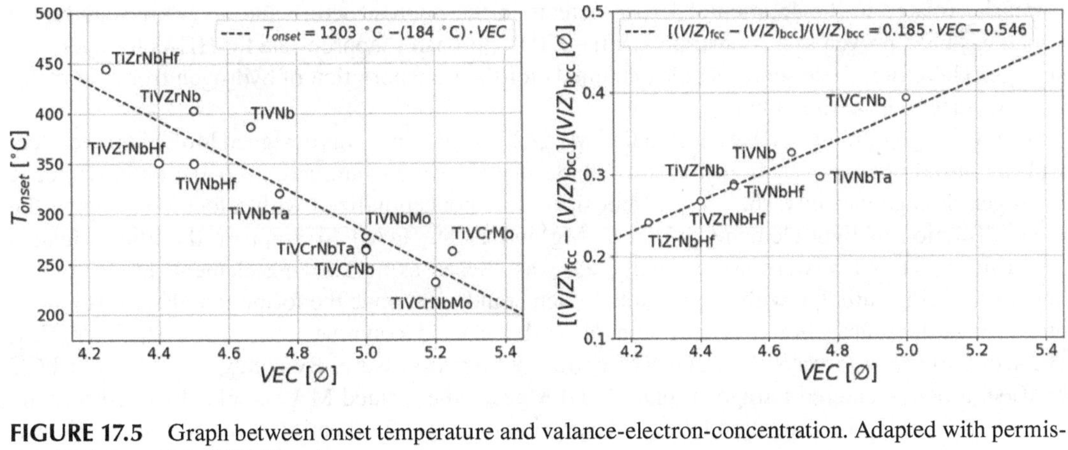

FIGURE 17.5 Graph between onset temperature and valance-electron-concentration. Adapted with permission [39]. Copyright 2019 Elsevier.

MgH_2 loaded with FeCoNiCrMn HEA [40] and its impact on Mg/MgH_2's hydrogen storage capabilities have also been reported. The HEA successfully reduced the dehydrogenation activation energy from 151.9 to 90.2 kJ mol^{-1} and exhibits significant catalytic activity toward hydrogen recombination and dissociation reactions. Currently, by using a vacuum electric arc furnace, the $Ti_{0.92}Zr,_{0.1}Cr_{1.6x}Mn_xFe_{0.4}$ ($x = 0.15$, 0.3, 0.45, and 0.6 at%) alloys were created [28]. For the primary hydrogen compressor, the impact of Manganese substitution on the plateau pressure regulation of the alloys was examined. In the case of $Ti_{0.92}Zr,_{0.1}Cr_{1.6x}Mn_xFe_{0.4}$ ($x = 0.15$, 0.3,0.45,0.6 at%) alloys' hydrogen absorption/desorption plateau pressures increased with increasing Mn content, but the slope factor of plateau reduced, and the enthalpy, entropy and Gibbs free energy change values also decreased. The $Ti_{0.92}Zr_{0.1}Cr_{1.0}Mn_{0.6}Fe_{0.4}$ alloy performance was remarkable.

Utilizing additive Laser Engineered Net Shaping (LENS) technology, the HEA TiZrNbMoV was synthesized [41]. They investigated how laser energy affected the derived microstructure and its potential for storing hydrogen as a result of their work. The TiZrNbMoV alloy showed a dual phase of a dendritic matrix with some non-melted Molybdenum particles and BCC solid-solution dendrites surrounded by an orthorhombic $NbTi_4$-type phase after being synthesized using low laser power. The alloy's maximal hydrogen capabilities were 0.59 weight percentage after synthesis and 0.61 weight percentage following successive heat treatment.

Most of the research to date on HEAs for storing hydrogen has been done on heavy transition metal-based alloys. A class of alloys known as lightweight HEAs is made up of one or more light elements (like aluminum, magnesium, lithium, titanium, and silicon) that have been partially integrated into a solution of solid to preserve a peak H/M ratio while raising the gravimetric capacity and lowering its molar mass. The Mg-based light metals are one of the most promising classes of materials.

Ball milling (BM), a MA technique, can be used in either a reactive or inert atmosphere. The high-energy ball milling (HEBM) technique is used to create a lightweight $MgZrTiFe_{0.5}Co_{0.5}Ni_{0.5}$ alloy. 24 hours were spent milling HEA at 600 rpm, 0.7 MPa in an inert Argon atmosphere, and 3.0 MPa in an H_2 environment. For the systems showing low alloying affinity (or positive mixing enthalpy), which have the propensity to separate undesirable phases, reactive ball milling procedures are utilized. Synthesis of $MgTiNbCr_{0.5}Mn_{0.5}Ni_{0.5}$ and $Mg_{0.68}TiNbNi_{0.55}$ alloys has also been used by using the ball milling technique [37]. During the synthesis of $Ti_{0.325}V_{0.275}Zr_{0.125}Nb_{0.275}$ alloy, no benefit of long milling was suggested [42].

To produce nano or ultra-fine grains in metallic materials, the HPT (High-Pressure Torsion) process is frequently used. Even the synthesis of immiscible systems with the potential to store hydrogen, like Mg–Ti and Mg–Zr, may be done well using this technique [43–45]. Additionally, attempts have been made to create non-metallic HEAs using this technique, including the MgTiVCrFe-H metal hydride for storing hydrogen, which is based on nitrides, oxides, and perhaps hydrides. In this case, an increase in catalytic activity and hydrogenation kinetics have been observed, which is primarily related to the creation of lattice defects in the microstructure and its grain boundaries, which has been proven to demonstrate that the HPT approach is appropriate for HEAs for hydrogen storage. Therefore, these serve as quick channels for the transportation of hydrogen from the material's surface to its entire volume.

A novel Mg-containing HEA, designated $Mg_{35}Al_{15}Ti_{25}V_{10}Zn_{15}$, was designed [46], characterized, and the initial phases of hydrogen absorption/desorption were examined. To boost its gravimetric hydrogen storage capacity, the alloy composition was conceptualized with a focus on raising the atomic fraction of light elements (35 at. % Mg and 15 at. % Al). To improve the alloy's hydrogen affinity, Ti and V were used. Finally, Zn was chosen as an alloying element due to its low negative mixing enthalpy with magnesium, which could encourage the formation of solid solutions and prevent the segregation of magnesium. HEBM was used to synthesize the $Mg_{35}Al_{15}Ti_{25}V_{10}Zn_{15}$ HEA in two separate ways: MA and RM under hydrogen pressure. By combining MgH_2, a FCC hydride, a body-centered tetragonal phase, and $MgZn_2$, the formed MA sample absorbed roughly 2.5 weight% of H_2. Investigation into the desorption process after heating revealed that MgH_2 was the first to desorb. The disintegration of the FCC hydride and BCT phase, which results in the creation of the BCC phase, completes the desorption sequence. The alloy created by RM, on the other hand, was made of a combination of an FCC hydride and MgH_2. The MgH_2 desorbs first when heated during the desorption process. The BCT phase is formed when the FCC hydride partially decomposes at lower temperatures (below 350°C). FCC hydride and BCT phases both occur at greater temperatures.

HEAs appear to be the most viable solution out of the available candidate materials appropriate for secure and efficient hydrogen capturing and storage because of their properties. The compositional and structural diversity of HEAs presents significant opportunities for further enhancing their hydrogen storage capabilities. An in-depth study in this area is just getting started, but the findings of numerous authors reveal promising advancements and point in the right areas for future research and development. For a successful hydrogen storage venture, various key factors play an important role which are temperature during the process, its energy, the capacity of hydrogen absorption, the kinetics of all these processes, and stability of the microstructure formed during the hydrogenation and dehydrogenation reversible cycles.

The development of processing techniques that activate the crystal lattice, is the next stage in enhancing the storage capacity of alloys. It includes designing the crystal lattice, which can raise the capacity of interstitial sites and decrease the energy needed for hydrogenation and dehydrogenation. Various techniques were employed in several studies, to improve the characteristics of elements used and resulted in the homogenization of an alloy's structure. The field of lightweight HEAs based on magnesium, aluminum, titanium, and several other elements is also intriguing, although it has received little research to date. A good contender appears to be the HEA $Mg_{0.10}Ti_{0.30}V_{0.25}Zr_{0.10}Nb_{0.25}$ [47], which absorbed 1.7 H/M (2.7 wt%) at ambient temperature.

Nowadays, TiVFeCuNb MPEA (multi-principal element alloy), a poorly crystalline alloy, has also been used to study hydrogenation/dehydrogenation properties. Although the alloy's poor hydrogen storage capacity (0.6 wt%, H/M equals 0.36) is one of its distinguishing characteristics, it is not necessarily a judgmental factor for the execution of the material. For instance, comparable alloys are intended to be used as catalysts in hydrogenation/dehydrogenation reactions where temperature and process kinetics are more important than storage capacity. Consequently, the information presented could be used as a springboard for creating new, imperfectly crystalline or amorphous MPEAs to use for hydrogen storage purposes [48].

17.4 CONCLUSION

The issue of hydrogen storage is one of the major difficulties in putting hydrogen technology into practice. For many hydrogen storage applications, metal hydrides are regarded as a long-term substitute that has been studied for many years. The never-ending quest to design and produce superior materials to fulfill the needs of next-generation engineering applications, notably for energy and transportation, will definitely present countless promising pursuits for materials scientists and engineers in terms of these alloys. High-entropy alloys, a novel class of alloys in this context, have opened up fresh and encouraging opportunities for studying metal hydrides HEAs have revolutionized the world of materials science, or more specifically, metallurgy; they are even expected to play a key role in the field of structural, and may be useful, materials for at least another ten years. We have outlined the fundamental ideas about HEAs as well as their potential use in storing hydrogen for upcoming applications. The most efficient and secure storage method for hydrogen is known to be solid-state storage. One of the best solid-state hydrogen storage materials will be the effective hydride-forming high-entropy materials or the strong hydride formers. Better hydrogen storage capacity might be achieved by strain-induced distorted lattices in high-entropy materials, which might encourage hydrogen atoms to occupy both the tetrahedral and octahedral sites. It may be suggested that the design of hydride-forming multicomponent HEAs through appropriate processing procedures may lead to better outcomes given the promise and prospects of HEAs as a result of their composition and structural properties. More studies are required to evaluate the potential of compositions with similar properties for storing hydrogen because there are so few publications on the subject.

REFERENCES

1. W. Qikun, Hydrogen storage by carbon nanotube and their films under ambient pressure, *Int. J. Hydrog. Energy* 27 (2002) 497–500.
2. J. O. Bockris and T. N. Veziroglu, A solar-hydrogen energy system for environmental compatibility. *Environ. Conserv., Cambridge Core*, 12 (1985), 105–118.
3. A. Züttel, A. Borgschulte and L. Schlapbach (Eds.), *Hydrogen as a future energy carrier* (2008).
4. Q. Sun, N. Wang, Q. Xu and J. Yu, Nanopore-supported metal nanocatalysts for efficient hydrogen generation from liquid-phase chemical hydrogen storage materials, *Adv. Mater.* 32 (2020) 2001818.
5. S. N. Talapaneni, G. Singh, I. Y. Kim, K. AlBahily, A. H. Al-Muhtaseb, A. S. Karakoti, E. Tavakkoli and A. Vinu, Carbon capture and conversion: Nanostructured carbon nitrides for CO_2 capture and conversion (Adv. Mater. 18/2020), *Adv. Mater.* 32 (2020) 2070142.
6. B. He, Q. Zhang, Z. Pan, L. Li, C. Li, Y. Ling, Z. Wang, M. Chen, Z. Wang, Y. Yao, Q. Li, L. Sun, J. Wang and L. Wei, Freestanding metal-organic frameworks and their derivatives: An emerging platform for electrochemical energy storage and conversion, *Chem. Rev.* 122 (2022) 10087–10125.
7. S. Liu, J. Liu, X. Liu, J. Shang, L. Xu, R. Yu and J. Shui, Hydrogen storage in incompletely etched multilayer Ti2CTx at room temperature, *Nat. Nanotechnol.* 16 (2021) 331–336.
8. J. Miyake, Y. Ogawa, T. Tanaka, J. Ahn, K. Oka, K. Oyaizu and K. Miyatake, Rechargeable proton exchange membrane fuel cell containing an intrinsic hydrogen storage polymer, *Commun. Chem.* 3 (2020).
9. L. Schlapbach and A. Züttel, Hydrogen-storage materials for mobile applications, *Nature* 414 (2001) 353–358.
10. V. Tietze, S. Luhr and D. Stolten, Bulk storage vessels for compressed and liquid hydrogen, hydrogen science and engineering : Materials, processes, *Sys. Technol.* (2016) 659–690.
11. W. I. F. David, Effective hydrogen storage: A strategic chemistry challenge, *Faraday Discuss.* 151 (2011) 399.
12. S. McWhorter, K. O'Malley, J. Adams, G. Ordaz, K. Randolph and N. T. Stetson, Moderate temperature dense phase hydrogen storage materials within the US Department of Energy (DOE) H_2 storage program: Trends toward future development, *Crystals* 2 (2012) 413–445.
13. S. Satyapal, J. Petrovic, C. Read, G. Thomas and G. Ordaz, The U.S. department of energy's national hydrogen storage project: Progress towards meeting hydrogen-powered vehicle requirements, *Catal. Today* 120 (2007) 246–256.
14. H. W. Langmi, J. Ren, B. North, M. Mathe and D. Bessarabov, Hydrogen storage in metal-organic frameworks: A review, *Electrochim. Acta* 128 (2014) 368–392.
15. H. Nazir, N. Muthuswamy, C. Louis, S. Jose, J. Prakash, M. E. Buan, C. Flox, S. Chavan, X. Shi, P. Kauranen, T. Kallio, G. Maia, K. Tammeveski, N. Lymperopoulos, E. Carcadea, E. Veziroglu, A. Iranzo and A. M. Kannan, Is the H2 economy realizable in the foreseeable future? Part II: H_2 storage, transportation, and distribution, *Int. J. Hydrog. Energy* 45 (2020) 20693–20708.
16. J. Andersson and S. Grönkvist, Large-scale storage of hydrogen, *Int. J. Hydrog. Energy* 44 (2019) 11901–11919.
17. J. J. Reilly and R. H. Wiswall, Reaction of hydrogen with alloys of magnesium and nickel and the formation of Mg2NiH4, *Inorg. Chem.* 7 (1968) 2254–2256.
18. E. Akiba and Y. Nakamura, Hydrogenation properties and crystal structures of Ti–Mn-V BCC solid solution alloys, *Met. Mater. Int.* 7 (2001) 165–168.
19. J.-W. Yeh, S.-K. Chen, S.-J. Lin, J.-Y. Gan, T.-S. Chin, T.-T. Shun, C.-H. Tsau and S.-Y. Chang, Nanostructured high-entropy alloys with multiple principal elements: Novel alloy design concepts and outcomes, *Adv. Eng. Mater.* 6 (2004) 299–303.
20. B. Cantor, I. T. H. Chang, P. Knight and A. J. B. Vincent, Microstructural development in equiatomic multicomponent alloys, *Mater. Sci. Eng.*: A 375-377 (2004) 213–218.
21. X. Yang and Y. Zhang, Prediction of high-entropy stabilized solid-solution in multi-component alloys, *Mater. Chem. Phys.* 132 (2012) 233–238.

22. K. Edalati, H. Shao, H. Emami, H. Iwaoka, E. Akiba and Z. Horita, Activation of titanium-vanadium alloy for hydrogen storage by introduction of nanograins and edge dislocations using high-pressure torsion, *Int. J. Hydrog. Energy* 41 (2016) 8917–8924.
23. G. Ek, M. M. Nygård, A. F. Pavan, J. Montero, P. F. Henry, M. H. Sørby, M. Witman, V. Stavila, C. Zlotea, B. C. Hauback and M. Sahlberg, Elucidating the effects of the composition on hydrogen sorption in TiVZrNbHf-based high-entropy alloys, *Inorg. Chem.* 60 (2020) 1124–1132.
24. I. Kunce, M. Polanski and J. Bystrzycki, Structure and hydrogen storage properties of a high entropy ZrTiVCrFeNi alloy synthesized using Laser Engineered Net Shaping (LENS), *Int. J. Hydrog. Energy* 38 (2013) 12180–12189.
25. M. Sahlberg, D. Karlsson, C. Zlotea and U. Jansson, Superior hydrogen storage in high entropy alloys, *Sci. Rep.* 6 (2016) 36770.
26. X. Ma, X. Ding, R. Chen, X. Chen, Q. Song and H. Cui, Study on microstructure and the hydrogen storage behavior of a TiVZrNbFe high-entropy alloy, *Intermetallics* 157 (2023) 107885.
27. L. Serrano, M. Moussa, J.-Y. Yao, G. Silva, J.-L. Bobet, S. F. Santos and K. R. Cardoso, Development of Ti-V-Nb-Cr-Mn high entropy alloys for hydrogen storage, *J. Alloys Compd.945* (2023) 169289.
28. Y. Yan, Z. Li, S. Zhou and Y. Wu, Effect of Mn substitution on plateau pressure regulation of the $Ti_{0.92}Zr_{0.1}Cr_{1.6-x}Mn_xFe_{0.4}$ alloys for primary hydrogen compressor, *J. Alloys Compd.* 952 (2023) 169935.
29. J. Li, Y. Guo, X. Jiang, S. Li and X. Li, Hydrogen storage performances, kinetics and microstructure of $Ti_{1.02}Cr_{1.0}Fe_{0.7-x}Mn_{0.3}$ Alx alloy by Al substituting for Fe, *Renew. Energ.* 153 (2020) 1140–1154.
30. Z. Cao, P. Zhou, X. Xiao, L. Zhan, Z. Jiang, M. Piao, S. Wang, L. Jiang and L. Chen, Studies on Ti-Zr-Cr-Mn-Fe-V based alloys for hydrogen compression under mild thermal conditions of water bath, *J. Alloys Compd.* 892 (2022) 162145.
31. Z. Han, Z. Yuan, T. Zhai, D. Feng, H. Sun and Y. Zhang, Effect of yttrium content on microstructure and hydrogen storage properties of TiFe-based alloy, *Int. J. Hydrog. Energy* 48 (2023) 676–695.
32. C. Li, X. Gao, B. Liu, X. Wei, W. Zhang, Y. Lan, H. Wang and Z. Yuan, Effects of Zr doping on activation capability and hydrogen storage performances of TiFe-based alloy, *Int. J. Hydrog. Energy* 48 (2023) 2256–2270.
33. X. Ma, X. Ding, R. Chen, W. Cao and Q. Song, Study on hydrogen storage property of (ZrTiVFe) Al high-entropy alloys by modifying Al content, *Int. J. Hydrog. Energy* 47 (2022) 8409–8418.
34. X. Ma, X. Ding, R. Chen, J. Zhang, Q. Song and H. Cui, Microstructural features and improved reversible hydrogen storage properties of ZrTiVFe high-entropy alloy via Cu alloying, *Int. J. Hydrog. Energy* 48 (2023) 2718–2730.
35. K. R. Cardoso, V. Roche, A. M. Jorge Jr, F. J. Antiqueira, G. Zepon and Y. Champion, Hydrogen storage in MgAlTiFeNi high entropy alloy, *J. Alloys Compd.* 858 (2021) 158357.
36. S. Yang, F. Yang, C. Wu, Y. Chen, Y. Mao and L. Luo, Hydrogen storage and cyclic properties of $(VFe)_{60}(TiCrCo)_{40-x}Zr_x$ $(0 \leq x \leq 2)$ alloys, *J. Alloys Compd.* 663 (2016) 460–465.
37. F. Marques, H. C. Pinto, S. J. A. Figueroa, F. Winkelmann, M. Felderhoff, W. J. Botta and G. Zepon, Mg-containing multi-principal element alloys for hydrogen storage: A study of the $MgTiNbCr_{0.5}Mn_{0.5}Ni_{0.5}$ and $Mg_{0.68}TiNbNi_{0.55}$ compositions. *Int. J. Hydrog. Energy* 45 (2020) 19539–19552.
38. J. Liu, J. Xu, S. Sleiman, F. Ravalison, W. Zhu, H. Liu, H. Cheng and J. Huot, Hydrogen storage properties of $V_{0.3}Ti_{0.3}Cr_{0.25}Mn_{0.1}Nb_{0.05}$ high entropy alloy, *Int. J. Hydrog. Energy* 47 (2022) 25724–25732.
39. M. M. Nygård, G. Ek, D. Karlsson, M. H. Sørby, M. Sahlberg and B. C. Hauback, Counting electrons - A new approach to tailor the hydrogen sorption properties of high-entropy alloys, *Acta Mater.* 175 (2019) 121–129.
40. H. Wan, X. Yang, S. Zhou, L. Ran, Y. Lu, Y. Chen, J. Wang and F. Pan, Enhancing hydrogen storage properties of MgH2 using FeCoNiCrMn high entropy alloy catalysts, *J. Mater. Sci. Technol.* 149 (2023) 88–98.
41. I. Kunce, M. Polanski and J. Bystrzycki, Microstructure and hydrogen storage properties of a TiZrNbMoV high entropy alloy synthesized using Laser Engineered Net Shaping (LENS), *Int. J. Hydrog. Energy* 39 (2014) 9904–9910.
42. J. Montero, C. Zlotea, G. Ek, J.-C. Crivello, L. Laversenne and M. Sahlberg, TiVZrNb multi-principal-element alloy: synthesis optimization, structural, and hydrogen sorption properties, *Molecules* 24 (2019) 2799.
43. K. Edalati, H.-W. Li, A. Kilmametov, R. Floriano and C. Borchers, High-pressure torsion for synthesis of high-entropy alloys, *Metals* 11 (2021) 1263.
44. F. Marques, M. Balcerzak, F. Winkelmann, G. Zepon and M. Felderhoff, Review and outlook on high-entropy alloys for hydrogen storage, *Energy Environ. Sci.* 14 (2021) 5191–5227.

45. M. O. de Marco, Y. Li, H.-W. Li, K. Edalati and R. Floriano, Mechanical synthesis and hydrogen storage characterization of MgVCr and MgVTiCrFe high-entropy alloy, *Adv. Eng. Mater.* 22 (2019) 1901079.
46. M. de B. Ferraz, W. J. Botta and G. Zepon, Synthesis, characterization and first hydrogen absorption/desorption of the $Mg_{35}Al_{15}Ti_{25}V_{10}Zn_{15}$ high entropy alloy, *Int. J. Hydrog. Energy* 47 (2022) 22881–22892.
47. J. Montero, G. Ek, M. Sahlberg and C. Zlotea, Improving the hydrogen cycling properties by Mg addition in Ti-V-Zr-Nb refractory high entropy alloy, *Scr. Mater.* 194 (2021) 113699.
48. M. Balcerzak, J. Ternieden and M. Felderhoff, Synthesis, thermal stability, and hydrogen storage properties of poorly crystalline TiVFeCuNb multi-principal element alloy, *J. Alloys Compd.* 943 (2023) 169142.

18 High Entropy Materials for Flexible Devices

Chenyang Shao and Yan Xing

18.1 INTRODUCTION

The rapid development of electronic technology, new energy, and modern medicine drives the demand for novel flexible devices, which not only provide convenience for people's daily lives but also greatly promote the progress of human society [1]. Typically, flexible devices should possess the characteristics of lightness, flexibility, stretchability and durability, human compatibility, electrochemical activity, etc., to have the opportunity to be applied in the field of biosensors [2], medical devices [3], solar cells [4], electronic skin [5], intelligent electronic fabrics [6] and electrodes [7]. Traditional electronic devices based on semiconductor wafers are not suitable for soft devices such as wearable sensors and electronic skins, due to their inherent rigidity [8,9]. To overcome these problems, flexible materials with excellent mechanical properties and reproducibility become the key to developing the devices. Currently, flexible devices are predominantly made of fabric and films based on metals and polymers, and great progress has been achieved. Meanwhile, such materials expose their drawbacks. For example, inorganic materials have poor tensile and flexural properties, while most organic polymers exhibit low conductivity. Additionally, high costs and energy-consumption manufacturing processes will also hinder practical applications [10]. Therefore, there emerges an urgent need to find novel materials for designing all sorts of flexible devices.

Recently, high entropy alloys (HEAs) and high entropy ceramics (HECs) have attracted continuous attention. The unique and tunable properties of such high entropy materials provide new opportunities for overcoming bottlenecks in various fields. In short, the concept of HEAs refers to a new class of materials containing five or more elements with an equal or near-equal atomic fraction [11]. Inherited from HEAs, HECs are defined as solid solutions of inorganic compounds with one or more Wyckoff sites shared by equal or near-equal atomic ratios of multi-principal elements [12]. Importantly, the popularity of high entropy materials is attributed to the four core effects, that is, the high entropy effect, sluggish diffusion effect, severe lattice distortion, and cocktail effect. The high entropy effect provides a high degree of freedom in composition and crystal structure design and improves thermodynamic stability. The sluggish diffusion effect hinders the phase transition even at high temperatures. The lattice distortion arising from the difference in atomic size and elastic properties results in enhancing strength and hardness. The cocktail effect contributes to unpredictable properties [13,14].

The combination of unique characteristics promotes the discovery of a significant number of high entropy materials with competitive properties, like high-temperature stability, low thermal conductivity, high ionic conductivity, high strength and fracture toughness, anti-corrosion, etc. [15–17]. As a result, high entropy materials demonstrate a bright prospect in a wide range of structural and functional applications [18–20]. Among them, the appealing mechanical and electrical properties of high entropy materials match well with the demands of materials for flexible devices used as electronics, energy cells, and biomedical applications [21]. Despite the rapid development of HEAs and HECs, the field of flexible devices made of high entropy materials is still in the infant stage for both academic research and practical applications. Although some advances have been achieved, the material families are quite limited, and the processing method is single. Moreover, the gap between the properties and practical application is still huge. Overall, the unique properties

DOI: 10.1201/9781003391388-18

of HEAs and HECs make them promising materials for flexible devices, and great efforts should be made to propel further development.

In this chapter, we summarize all the recent advances of high entropy materials in the field of flexible devices and this chapter is organized as follows. Section 18.2 discusses the choice of high entropy materials and existing material systems. In Section 18.3, processing methods employed for manufacturing flexible high entropy materials and assembling devices will be introduced. In Section 18.4, present and prospective applications of flexible devices are overviewed. In Section 18.5, future directions are suggested, and finally, conclusions are given.

18.2 CHOICE OF MATERIALS FOR FLEXIBLE DEVICES

Mechanical properties, electrical properties, and stability are key elements in flexible devices, guiding the development of materials. Nowadays, polymer-based, carbon-based, metal-based, and various composite systems have been designed with great progress achieved. Polymers are one of the ideal materials for flexible devices because of their flexibility, which is suitable for both substrate and active layers. When it comes to metals, better stability, thermal resistance, and electrical and magnetic conductivity attract continuous attention. Meanwhile, some metals, like Cu, Ag, etc., can provide an additional antibacterial effect, which is also appealing. To combine the advantages of different materials, a multilayer structure is commonly used. Since the discovery of high entropy materials, the construction of novel flexible devices by HEAs and HECs got off the road. However, in the dawning period, a lot of new high entropy systems have been discovered and proved to have great potential, as listed in Table 18.1.

18.2.1 HEAs

Taking advantage of the superior mechanical and electrical properties, the HEAs were first used in flexible device manufacturing compared with other high entropy materials. Wang et al. used magnetron sputtering to prepare CoCrFeMnNi nanocrystalline HEA films [22]. The resulting film has

TABLE 18.1
Summary of Recent Advances on High Entropy Materials for Flexible Devices

Composition	Phase	Structure	Preparation Method	Application	Ref
CoCrFeMnNi	FCC	Single layered film	Magnetron sputtering	Microelectromechanical systems	[22]
CrCoNi–Fe–Si–B	FCC	Nanocolumnar	Magnetron co-sputtering	Microelectromechanical systems	[26]
AlPdNiCuMo	FCC	Nanoporous film	Melt-spinning	Ethanol oxidation reaction	[27]
CoCrFeNi/Graphene	FCC	Multilayered film	Magnetron co-sputtering and chemical vapor deposition	Mechanical device	[28]
$Zr_{55}Cu_{30}Ni_5Al_{10}$	Amorphous	Transparent film	Ion beam deposition	Electronic skin	[29]
$Zr_{60}Al_{15}Ni_{10}Co_{15}$	Amorphous	Thin film	DC sputter deposition	Strain sensor /Electrode	[30]
$Al_{85}Ni_5Y_8Co_2$	Amorphous	Thin film	DC sputter deposition	Strain sensor /Electrode	[30]
$Ti_{1.1}V_{0.7}Cr_xNb_{1.0}Ta_{0.6}C_3T_z$	MAX phase	Multilayered	Vacuum filtration	Zn-ion hybrid supercapacitors /Li-ion battery	[31]
ZrNbMo-Cr-N	Amorphous	Multilayered film	Magnetron co-sputtering	Spectrally selective absorber	[32]
$(Ca_{0.35}Sr_{0.2}Ba_{0.15}Na_{0.2}Bi_{0.1})_3Co_4O_9$	Monoclinic	lamellar	Solid-state reaction	Thermoelectric devices	[33]

FIGURE 18.1 (a) TEM bright-field image of the as-deposited film showing the film thickness and growth direction, (b) the high-density nanotwins with different thicknesses (several atomic layers) in a columnar grain, (c) XRD pattern of the as-deposited film. Adapted with permission [22]. Copyright 2021 Elsevier.

a single FCC structure with a hardness of 8.5 GPa, fatigue resistance of 10^6 cycles, and electrical resistivity of 130 μΩ cm. As shown in Figure 18.1, the high-density nanotwins in CoCrFeMnNi HEA films suppressed the pore initiation and crack propagation, which contributed to a higher toughness. Combined with numerous grain boundaries and severe lattice deformation, the movement of electrons was delayed and achieved high electrical resistivity. The well-tuned mechanical and electrical properties in HEA films by microstructure design will contribute to their future application for flexible electronic devices in microelectromechanical systems. In addition, except for the pure HEAs systems, composite materials based on HEAs were also designed. For example, HEAs typically exhibit high ductility, but their shear strength does not exceed G/100, thus remaining far from the theoretical limit of G/10 [23]. Metallic glasses (MGs) demonstrate excellent yield shear strength of G/37 [24,25]. Wu et al. proposed a novel material design concept that combines amorphous metal glass and crystalline HEAs phases at the nanoscale [26]. This idea was realized by doping the glass-forming elements B and Si into a Cr–Fe–Co–Ni high entropy base alloy system during the magnetron co-sputtering process. The nano-sized amorphous phase regions wrapping the nanocolumnar grains containing high-density nanotwins lead to a material with a high yield strength of 4.1 GPa. These results demonstrated that nanoscale structural design not only enabled ductile materials to approach theoretical strength but also allowed for the adjustment of interesting properties such as soft magnetism and good thermal stability. Li et al. utilized a generic dealloying procedure to prepare the nanoporous AlPdNiCuMo films for flexible direct ethanol fuel cells (DEFCs) [27]. The HEA AlPdNiCuMo films used as anode in fuel cells showed good flexibility and stability in cyclic bending tests. Feng et al. prepared a nanolayered (NL) CoCrFeNi/graphene composite material using a magnetron co-sputtering method [28]. The study found that the NL CoCrFeNi/graphene composite exhibited both ultra-high strength of 4.73 GPa and over 20% compressive plasticity. Detailed electron microscopy observations and simulation results indicated that the single-layer graphene interface can effectively prevent crack propagation and stimulate dislocations to adapt to further deformation.

18.2.2 Amorphous Alloys

Due to the disordered atomic structure, amorphous alloys possess special properties, such as ultra-high elastic limitation, low-temperature coefficient of resistance, and good piezoresistive characteristics. Especially, the superior elastic property enables amorphous alloys to recover after deformation, while the low-temperature coefficient of resistance can effectively eliminate the thermal drift phenomenon caused by environmental temperature changes to achieve a wider working temperature range. These characteristics match well with the requirements of flexible materials, making amorphous HEAs potential for application in flexible electronic devices. Xian et al. prepared a $Zr_{55}Cu_{30}Ni_5Al_{10}$ amorphous HEA film on a polycarbonate substrate. The resulting multilayer films were flexible and could serve as electronic skin to measure finger bending degrees [29]. Cho et al. deposited amorphous Al–Co–Ni–Zr films on polyimide (PI) substrates by using DC magnetron sputtering to develop a strain sensor [30]. Under external bending strain, the resistance of the amorphous HEA films exhibited a linear trend with high sensitivity. After 100 cycles of bending, the resistance change rate remained constant, demonstrating good long-term stability and reliability.

18.2.3 HECs

Recent developments in high entropy ceramic open up new avenues for the construction of high-performance flexible devices. Distinct from HEAs, the diversity in crystal structure and electronic structure of HECs provides huge space for properties tuning through band structure engineering and phonon engineering. Aside from strengthening and hardening that has already been found in HEAs, new properties such as colossal dielectric constant, super ion conductivity, strong electromagnetic wave absorption, etc., have been discovered in HECs [12]. However, to design flexible devices based on HECs, flexibility is of vital importance. To address that problem, natural two-dimensional ceramics and fiber braids will be good choices. Etman et al. constructed a freestanding high entropy MXene film $Ti_{1.1}V_{0.7}Cr_xNb_{1.0}Ta_{0.6}C_3T_z$ (T_z=–F, –O, –OH) by vacuum filtration. The resulting films demonstrated a high degree of flexibility and excellent electrochemical performance as electrodes for Zn-ion hybrid supercapacitors and lithium-ion batteries [31]. He et al. prepared a high entropy ZrNbMo-Cr-N ceramic nanofilm by using a magnetron co-sputtering method [32]. By assembling HEC films with different metal content and selecting an appropriate stainless steel (SS) substrate, a flexible spectrally selective absorber device was constructed and demonstrated a high solar absorptance. In addition, Yang et al. synthesized a novel layered $(Ca_{0.35}Sr_{0.2}Ba_{0.15}Na_{0.2}Bi_{0.1})_3Co_4O_9$ high entropy ceramic using traditional solid-state reactions [33]. Compared with the original $Ca_3Co_4O_9$ sample, the higher carrier mobility resulting from *c*-axis orientation significantly improved the conductivity of the textured high entropy ceramic, resulting in a power factor of 0.27 mW/m·K^2. Then, multidimensional lattice defects induced by the high entropy effect can act as phonon scattering centers, effectively suppressing lattice thermal conductivity and achieving an ultralow thermal conductivity of 0.87 W/m·K at 973 K and a peak ZT value of 0.3, which is approximately 2.5 times that of $Ca_3Co_4O_9$.

Although HEAs, amorphous HEAs, and HECs have been developed for serving as the key components in flexible devices, up to now, the research studies on high entropy materials are still limited, and more efforts should be made to find new materials.

18.3 SYNTHESIS METHODS

18.3.1 Magnetron Sputtering

Sputter deposition has been a flexible, reliable, and effective coating method for several decades [34]. Initially, a direct current glow discharge or a direct current diode was used as the sputtering source, while selecting the targets based on product components. The working principle of magnetron sputtering involves electrons colliding with argon atoms under the influence of an electric field during their flight toward the substrate, resulting in ionization and the production of Ar^+ ions and new electrons. The new electrons fly toward the substrate while the Ar^+ ions are accelerated toward the cathode target by the electric field and bombard its surface with high energy, causing sputtering of the target material. Target atoms or molecules are then deposited on the substrate to form a film. Figure 18.2a shows a commonly used direct current planar sputtering structure where the cathode target is a circular or rectangular plate. Based on magnetron sputtering, various technologies, like radio frequency magnetron sputtering, co-sputtering, etc., have been developed to prepare films. For example, Wang et al. employed a radio frequency magnetron sputtering system and selected a CoCrFeMnNi target to synthesize CoCrFeMnNi nanocrystalline HEA films [22]. Differing from direct current magnetron sputtering (DC), the radio frequency magnetron sputtering

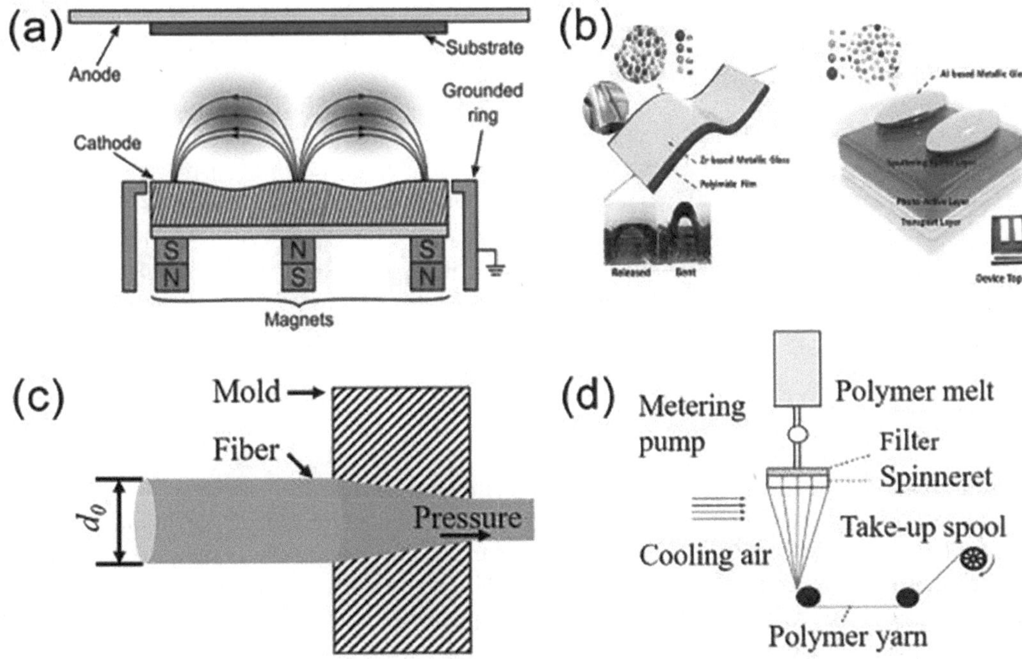

FIGURE 18.2 (a) A schematic of the DC planar magnetron discharge used for sputtering; (b) Metallic glass applications on thin film devices; (c) A schematic of fiber preparation by drawing method; (d) Schematic of a melt-spinning setup. Adapted with permission [34] for (a), Copyright 2020 Elsevier; [30] for (b), Copyright 2020 Elsevier; [36] for (d), Copyright 2015 Elsevier.

system used AC power. A fixed power of 150 W was applied and conducted at ambient temperature without substrate heating or cooling. The HEA films showed a single solid solution structure and fine morphology with a thickness of 1400 nm. Feng et al. used a magnetron co-sputtering technique to prepare a CoCrFeNi HEA film on the (100)-oriented Si wafers with a thickness of 625 nm. Then, multilayer devices were assembled followed by a chemical vapor deposition process. Cho et al. prepared a metallic glass sputtering target with designed atomic compositions ($Zr_{60}Al1_5Ni_{10}Co_{15}$ and $Al_{85}Ni_5Y_8Co_2$, respectively) by powder metallurgy processing using metallic glass powder at first [30]. Then, these two targets were used in the following DC sputter deposition process to produce the amorphous HEA thin films. Finally, the films with a thickness of 100 nm were deposited onto polyimide (PI) film with a thickness of 150 μm to build the strain sensor, as shown in Figure 18.2b. Wu et al. directly prepared a CrCoNi–Fe–Si–B crystal–glass high entropy nanocomposite by magnetron co-sputtering of CrCoNi and $Fe_{78}Si_9B_{13}$ alloy targets. By tuning the distance between the substrate and the target, the HEAs with varied compositions were produced.

18.3.2 Hot Drawing

The commonly used method for preparing HEA fibers is the drawing method, which converts casting HEAs into one-dimensional (1D) shape through hot forging and hot rotary forging. Subsequently, the diameter and length of such 1D HEAs can be further machined by repeatedly passing them through hard molds of different apertures using a drawing machine. Figure 18.2c shows a schematic diagram of the drawing process, where d_0 represents the initial diameter of the rod before drawing. HEA fibers prepared by the drawing method usually own good surface quality and dimensional accuracy. Moreover, due to the multiple deformation and annealing treatments during drawing, HEA fibers have a high degree of grain refinement, large dislocation density, and nanoscale precipitates, resulting in better mechanical strength. Zhang et al. [35] used a hot rotary forging method to prepare $Al_{0.3}$CoCrFeNi HEA fibers with diameters ranging from 1 to 3.15 mm. Structural analysis revealed that the main phase of the $Al_{0.3}$CoCrFeNi HEA fibers was FCC structure, and the second phase was the nanoscale Al–Ni-riched B2 phase that precipitated at grain boundaries. The $Al_{0.3}$CoCrFeNi HEA fibers with a diameter of 1.00 mm displayed remarkable tensile strength and ductility (1207 MPa and 7.8%) at 298 K, increasing to 1600 MPa and 17.5% at 77 K. Compared with cast and single-crystal $Al_{0.3}$CoCrFeNi, fibrous $Al_{0.3}$CoCrFeNi high entropy alloy exhibited a higher tensile strength, exceeding most bulk FCC and HCP structure HEAs, and even some BCC structure HEAs. In addition, the drastic reduction in lateral dimensions also made the HEA fibers highly flexible, showing a great application perspective in engineering and electrical devices.

18.3.3 Melt Spinning

Melt spinning is one of the simplest methods for fiber manufacturing [36], avoiding solvent-related issues. Thus, if the polymer precursor can provide a stable and desirable melt condition, melt spinning will be the preferred method. When polymer particles or chips serve as the starting material for melt spinning, they are first dried and then melted in an extruder. The homogenized and filtered melt was injected through a narrow channel into a quenching chamber where the fluid filament bundle solidified. Finally, the filament bundle is wound onto a tube roll (Figure 18.2d). Li et al. [27] conducted a melt spinning process with a tangent velocity of 45 m/s to prepare AlPdNiCuMo HEA ribbons with a thickness of 20–30 μm and a width of 3 mm. Then, followed by an etching procedure with a NaOH solution, the dense AlPdNiCuMo HEA ribbons became porous with high surface areas.

18.4 PROPERTIES AND APPLICATION

18.4.1 Electrodes

High entropy materials exhibit excellent stability, mechanical properties, and electrochemical activity and are promising candidates for electrodes in various batteries. Electrodes are crucial components of capacitors and batteries, as they play a significant role in determining the energy storage and stability of these devices. In recent years, a remarkable breakthrough in the field of flexible electrodes for batteries based on high entropy materials has been achieved. For example, as shown in Figure 18.3a, Li et al. prepared solid-state direct ethanol fuel cells by using a flexible nanoporous AlPdNiCuMo HEAs film as the anode [27]. The resulting anodes exhibited an electrochemical activity of 88.53 m^2/g_{Pd} and a mass activity of 2.67 A m/g_{Pd} in the ethanol oxidation reaction. Combined with a $(AlMnCo)_3O_4$ cathode, the fuel cells demonstrated an ultra-high energy density of 13.63 mWh/cm^2 with only 3 mL ethanol, which is outstanding compared with other similar solid-state energy devices. Moreover, during the continuous bending test, the discharge voltage of this flexible device was stable, as shown in Figure 18.3b. Etman et al. prepared a $Ti_{1.1}V_{0.7}Cr_xNb_{1.0}Ta_{0.6}C_3T_z$ (T_z=–F, –O, –OH) HE MXene thin films (Figure 18.3c), which was used as an electrode for Zn-ion hybrid supercapacitors [31]. The electrode exhibited an impressive capacity of up to 77 mAh/g (245 mAh/cm^3) and a current density of 0.5 A/g. Meanwhile, excellent stability was achieved, maintaining an efficiency of 99.9% and a capacity retention rate of 87% after 10,000 cycles (Figure 18.3d). Moreover, HE MXene films can also be used as negative electrodes for lithium-ion batteries with a capacity of 126 mAh/g (400 mAh/cm^3) at a current density of 0.01 A/g. This is a significant improvement over conventional lithium-ion battery electrodes and highlights the potential of high entropy alloys in revolutionizing battery technology.

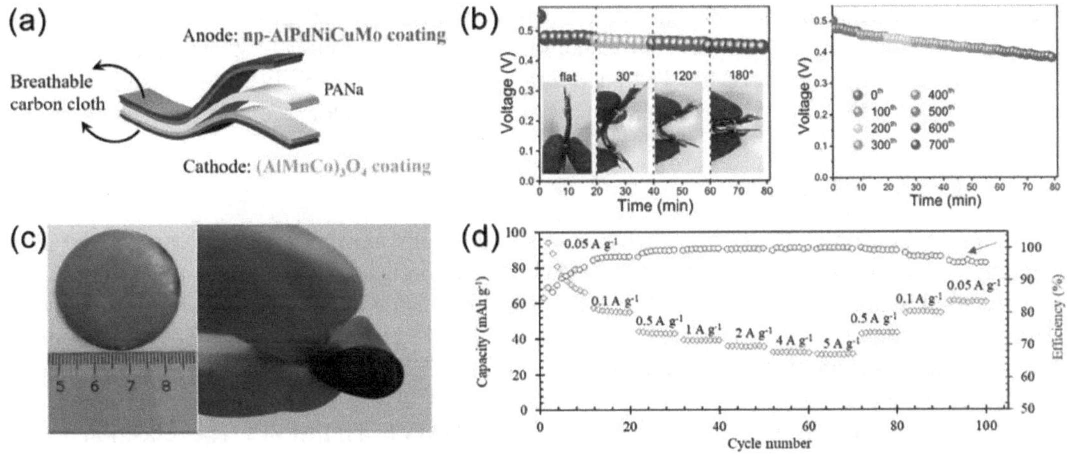

FIGURE 18.3 (a) Schematic explanation of the structure of flexible solid-state cell; (b) Discharge curves at 0.25 mA/cm^2 under consecutive bending from flat to 180° (left) and discharge curves at 0.25 mA/cm^2 undergoing various times of bending (right); (c) Photos of a HE MXene freestanding film; (d) Variation of discharge capacities and efficiency with the cycle number at different current densities. Adapted with permission [27] for (a,b), Copyright 2020 Wiley; [31] for (c,d), Copyright 2022 Elsevier.

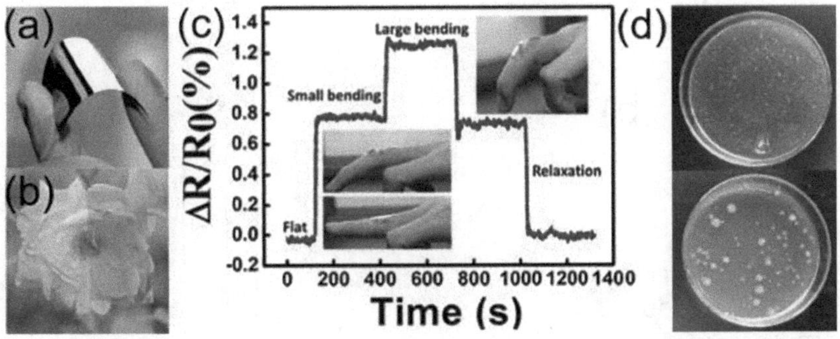

FIGURE 18.4 (a) The image of the MG skin; (b) Optical image illustrates that MG skin can be transparent; (c) Diagram illustrating MG skin used to monitor movements of hands at different bending degrees; (d) Optical images of *E. coli* on different materials after the antibacterial test. Up: PE control plate. Down: $Zr_{55}Cu_{30}Ni_5Al_{10}$ MG skin. Adapted with permission [29]. Copyright 2017 AIP.

18.4.2 Sensors

The optimized strain sensitivity, electrical conductivity, and reproducibility of HEAs make them suitable for stain sensors that have widespread applications in robots, health monitoring, and other fields. Xian et al. prepared an amorphous ZrCuNiAl HEA film by using an ion beam deposition method [29]. As shown in Figure 18.4a, the resulting HEA films were flexible and could be easily bent to 180°. The size of such film was 25 cm^2 and the thickness was 300 nm. When decreasing the thickness to 10 nm, the HEA films can be transparent (Figure 18.4b). Figure 18.4c exhibits the monitoring performance of ZrCuNiAl HEA films that stuck to human hands. Such HEA sensors can sensitively respond to bending and relaxing signals, showing a great application prospect in wearable medical devices and electronic skins for real-time monitoring of various physiological signals in humans. Moreover, benefiting from the metal Cu component, the ZrCuNiAl HEA films showed an additional antibacterial property, as shown in Figure 18.4d, the growth of Escherichia coli (*E. coli*) was effectively suppressed. Cho et al. reported ZrAlNiCo and AlNiYCo amorphous HEA films and characterized the performance used as strain sensors [30]. The flexible strain sensor was assembled by depositing the HEA films (100 nm) onto a polyimide (PI) film (150 µm). Under strain tests, the HEA film, especially ZrAlNiCo film, can make fast and stable responses to strain changes and can withstand long-term cyclic bending. The encouraging mechanical stability and reversible resistive response indicated that such HEA films were promising candidates for wearable strain sensors.

18.4.3 Energy Conversion

In addition to their application in electrodes and sensors, high entropy materials can also be used as energy conversion materials such as thermoelectric and optoelectric materials. For example, He et al. prepared high entropy ZrNbMo-Cr-N HEC nanofilms using magnetron sputtering for spectral selective absorption (SSA) devices [32]. As shown in Figure 18.5a, a novel three-layer structured SSA on a stainless steel (SS) substrate was designed. The Si_3N_4 layer was used as an antireflection layer, which decreases the surface reflection loss of sunlight. The intermediate layer was composed of two HEC films with different metal volume fractions (HMVF and LMVF). The reflection results (Figure 18.5b and c) confirmed that the resulting HEC films had a solar absorptance of 96.3% and a thermal emittance of 18.8% at 400°C, which could enable the harvest of solar energy. As shown in Figure 18.5d, the HEC films were angular insensitive and capable of absorbing solar energy over wide angular ranges, ensuring a promising application in the overall daytime. Additionally, the HEC films exhibit a solar thermal conversion efficiency of 85.4%. Detailed mechanism analysis suggested that the solar energy harvesting capability of the HEC film came from the synergistic effect of intrinsic absorption of high entropy nitride and substrate and the destructive interference effect (Figure 18.5e–h).

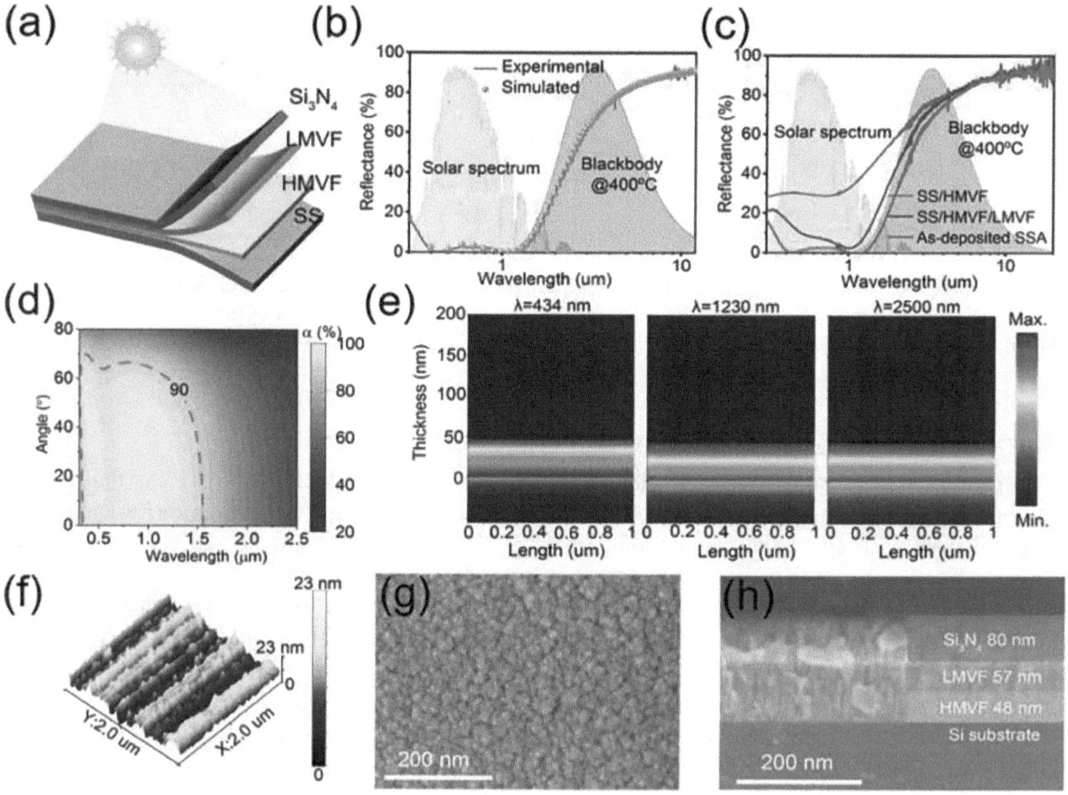

FIGURE 18.5 Fabrication of the ZrNbMo-Cr-N HECN SSA and its optical properties and microstructures. Adapted with permission [32]. Copyright 2022 Elsevier.

18.5 CONCLUSION AND FUTURE DIRECTIONS

In summary, this chapter has reviewed recent progress on high entropy materials for flexible devices. We have summarized the current material systems, and processing methods, followed by functional applications. Although in the infant stage, different families of high entropy materials, such as high entropy metals, amorphous alloys, and high entropy ceramics (HECs), have been developed and brought new opportunities for device construction. High entropy metals own excellent specific strength, exceptional ductility and fracture toughness, and superconductivity. By assembling film devices, they show great application prospects in the field of strain sensors, electronic skins, microelectromechanical systems, etc. Distinct from HEAs, the diversity in electronic and crystal structure of HECs endows them with newfangled properties such as super ion conductivity, high electrochemical activity, good adsorption, and catalytic properties. Flexible HEC devices have already displayed their abilities in fields such as batteries and capacitors. Due to the efforts of many research groups, studies on the design and construction of flexible devices based on high entropy materials are gradually increasing. Nowadays, advances in discovering new compositions, new microstructures, and processing technology have been obtained.

However, the reported high entropy materials for flexible devices are indeed very limited. Meanwhile, methods used to make flexible high entropy devices are limited, most of which involve magnetron sputtering, which will hinder practical applications and further development. More importantly, the small number of samples makes it difficult to delve into the understanding of the mechanism, as well as elucidate the relationship between the composition, microstructure, and performance. In the future, the comprehensive performance of flexible devices made by high entropy materials should be optimized and we propose the following directions, including (i) Demand-driven materials' design. More efforts should be paid to finding new high entropy materials based on the real needs for the development of flexible devices; (ii) Microstructure construction. In addition to two-dimensional thin films, fibers, and non-woven fabrics have controllable microstructures, large surface area, and excellent mechanical properties, which are also promising candidates for building flexible devices. Moreover, introducing heterojunction will bring new insight for tuning the performance; (iii) Novel and green manufacturing. New methods that are convenient, efficient, low-cost, and low energy consumption should be developed. Maybe, spin-coating, electrospinning, roll-to-roll, etc. are useful; (iv) Expanding application fields. The cocktail effect endows high entropy materials with optimized mechanical, electrical, and catalytic properties, providing devices with broader application prospects.

REFERENCES

[1] K. Nomura, H. Ohta, A. Takagi, T. Kamiya, M. Hirano, H. Hosono, Room-temperature fabrication of transparent flexible thin-film transistors using amorphous oxide semiconductors, *Nature.* 432 (2004) 488–492.

[2] W. Wu, L. Wang, Y. Yang, W. Du, W. Ji, Z. Fang, X. Hou, Q. Wu, C. Zhang, L. Li, Optical flexible biosensors: From detection principles to biomedical applications, *Biosens Bioelectron.* 210 (2022) 114328.

[3] X. Du and K. Zhang, Recent progress in fibrous high-entropy energy harvesting devices for wearable applications, *Nano Energy.* 101 (2022) 107600.

[4] S. Kim, V. Hoang, C. Bark, Silicon-based technologies for flexible photovoltaic (PV) devices: From basic mechanism to manufacturing technologies, *Nanomaterials.* 11 (2021) 2944.

[5] L. Mallory, Hammock, A. Chortos, T. Benjamin, T. Jeffrey, Z. Bao, 25th anniversary article: The evolution of electronic skin (E-Skin): A brief history, design considerations, and recent progress, *Adv Mater.* 25 (2013) 5997–6038.

[6] W. Weng, P. Chen, S. He, X. Sun, H. Peng, Smart electronic textiles, *Angew Chem Int Edit.* 55 (2016) 6140–6169.

[7] H. Zhang, F. Sun, G. Cao, D. Zhou, G. Zhang, J. Feng, S. Wang, F. Su, Y. Tian, Y. Liu, Y. Tian, Bifunctional flexible electrochromic energy storage devices based on silver nanowire flexible transparent electrodes, *Int J Extreme Manuf.* 5(2023), 015503.

[8] J. Rogers, T. Someya, Y. Huang, Materials and mechanics for stretchable electronics, *Science.* 327 (2010) 1603–1607.
[9] J. Song, S. Wang, T. Someya (ed), *Stretchable Electronics, Theory for Stretchable Interconnects* (2012) Weinheim Germany: Wiley-VCH.
[10] M. Zou, Y. Ma, X. Yuan, Y. Hu, L. Jie, J. Zhong, Flexible devices: From materials, architectures to applications, *J Semicond.* 39 (2018) 011010.
[11] E. George, D. Raabe, R. Ritchie, High-entropy alloys, *Nat Rev Mater.* 4 (2019) 515–534.
[12] H. Xiang, Y. Xing, F. Dai, H. Wang, L. Su, L. Miao, G. Zhang, Y. Wang, X. Qi, L. Yao, H. Wang, B. Zhao, J. Li, Y. Zhou, High-entropy ceramics: Present status, challenges, and a look forward, *J Adv Ceram.* 10 (2021) 385–441.
[13] A. Rogachev, Structure, stability, and properties of high-entropy alloys, *Phys Met Metallogr.* 121 (2020) 733–764.
[14] Y. Ye, Q. Wang, J. Lu, C. Liu, Y. Yang, High-entropy alloy: Challenges and prospects, *Mater Today.* 19 (2016) 349–362.
[15] A. Wright, J. Luo, A step forward from high-entropy ceramics to compositionally complex ceramics: A new perspective, *J Mater Sci.* 55 (2020) 9812–9827.
[16] J. Braun, C. Rost, M. Lim, A. Giri, D. Olson, G. Kotsonis, G. Stan, D. Brenner, J Maria, P. Hopkins, Charge-induced disorder controls the thermal conductivity of entropy-stabilized oxides, *Adv Mate.* 30 (2018) 1805004.
[17] T. Tu, J. Liu, Y. Wu, L. Zhou, G. Zhang, Synergistic effects of high-entropy engineering and particulate toughening on the properties of rare-earth aluminate-based ceramic composites, *J Adv Ceram.* 12 (2023) 861–872.
[18] Y. Xing, W. Dan, Y. Fan, X. Li, Low temperature synthesis of high-entropy $(Y_{0.2}Yb_{0.2}Sm_{0.2}Eu_{0.2}Er_{0.2})_2O_3$ nanofibers by a novel electrospinning method, *J Mater Sci Technol.* 103 (2022) 215–220.
[19] K. Wang, J. Zhu, H. Wang, K. Yang, Y. Zhu, Y. Qing, Z. Ma, L. Gao, Y. Liu, S. Wei, Y. Shu, Y. Zhou, J. He, Air plasma-sprayed high-entropy $(Y_{0.2}Yb_{0.2}Lu_{0.2}Eu_{0.2}Er_{0.2})_3Al_5O_{12}$ coating with high thermal protection performance, *J Adv Ceram.* 11 (2022) 1571–1582.
[20] D. Bérardan, S. Franger, D. Dragoe, A. Meena, N. Dragoe, Colossal dielectric constant in high entropy oxides, *Phys Status Solidi-R.* 10 (2016) 328–333.
[21] C. Oses, C. Toher, S. Curtarolo, High-entropy ceramics, *Nat Rev Mater.* 5 (2020) 295–309.
[22] Z. Wang, C. Wang, Y. Zhao, J. Kai, C. Liu, C. Hsueh, Nanotwinned CoCrFeMnNi high entropy alloy films for flexible electronic device applications, *Vacuum.* 189 (2021) 110249.
[23] A. Kelly, N. Macmillan, *Strong Solids* (1986) Walton Street: Oxford University Press.
[24] A. Greer, Metallic glasses, *Science.* 267 (1995) 1947–1953.
[25] W. Johnson, K. Samwer, A universal criterion for plastic yielding of metallic glasses with a $(T/Tg)^{2/3}$ temperature dependence, *Phys Rev Lett.* 95 (2005) 195501.
[26] G. Wu, S. Balachandran, B. Gault, W. Xia, C. Liu, Z. Rao, Y. Wei, S. Liu, J. Lu, M. Herbig, W. Lu, G. Demh, Z. Li, D. Raabe, Crystal-glass high-entropy nanocomposites with near theoretical compressive strength and large deformability, *Adv Mater.* 32 (2020) 2002619.
[27] S. Li, J. Wang, X. Lin, G. Xie, Y. Huang, X. Liu, H. Qiu, Flexible solid-state direct ethanol fuel cell catalyzed by nanoporous high-entropy Al–Pd–Ni–Cu–Mo anode and spinel $(AlMnCo)_3O_4$ cathode, *Adv Funct Mater.* 31 (2021) 2007129.
[28] X. Feng, K. Cao, X. Huang, G. Li, Y. Lu, Nanolayered CoCrFeNi/graphene composites with high strength and crack resistance, *Nanomaterials.* 12 (2022) 2113.
[29] H. Xian, C. Cao, J. Shi, X. Zhu, Y. Hu, Y. Huang, S. Meng, L. Gu, Y. Liu, H. Bai, W. Wang, Flexible strain sensors with high performance based on metallic glass thin film, *Appl Phys Lett.* 111 (2017) 121906.
[30] J. Cho, W. Jang, K. Park, D. Wang, Metallic amorphous alloy for long-term stable electrodes in organic sensors and photovoltaics, *Org Electron.* 84 (2020) 105811.
[31] A. Etman, J. Zhou, J. Rosen, $Ti_{1.1}V_{0.7}Cr_xNb_{1.0}Ta_{0.6}C_3T_z$ high-entropy MXene freestanding films for charge storage applications, *Electrochem Commun.* 137 (2022) 107264.
[32] C. He, P. Zhao, X. Gao, G. Liu, P. La, Efficient solar energy harvesting enabled by high-entropy ceramic nanofilms through a co-sputtering method, *J Alloy Compd.* 934 (2023) 167899.
[33] C. Yang, H. Wu, H. Song, X. Wang, S. Chen, X. Xu, L. Chen, Z. Zhao, L. Yu, L. Liu, Ultralow thermal conductivity and enhanced thermoelectric properties in a textured $(Ca_{0.35}Sr_{0.2}Ba_{0.15}Na_{0.2}Bi_{0.1})_3Co_4O_9$ high-entropy ceramic, *J Alloy Compd.* 940(2023) 168802.

[34] J. Gudmundsson, D. Lundin, D. Lundin, T. Minea, J. Gudmundsson (eds), *High Power Impulse Magnetron Sputtering, Introduction to Magnetron Sputtering* (2020) NY: Elsevier.
[35] D. Li, C. Li, T. Feng, Y. Zhang, G. Sha, J. Lewandowski, P. Liaw, Y. Zhang, High-entropy $Al_{0.3}CoCrFeNi$ alloy fibers with high tensile strength and ductility at ambient and cryogenic temperatures, *Acta Mater.* 123 (2017) 285–294.
[36] H. Qu, M. Skorobogatiy, T. Dias (ed), *Electronic Textiles, Conductive Polymer Yarns for Electronic Textiles* (2015) NY: Woodhead Publishing.

19 High Entropy Materials for Electrochemical Sensors

Rijith S, Sarika S, Athira S, and Sumi V S

19.1 OVERVIEW OF ELECTROCHEMICAL SENSORS

Electrochemical sensors are a class of analytical tools that measure the electrical signals generated by chemical reactions occurring at an electrode surface. These sensors have gained increasing importance in a wide range of fields, including environmental monitoring, biomedical diagnostics, and industrial process control. Electrochemical sensors offer several advantages over traditional sensing platforms, including high sensitivity, selectivity, and simplicity of use. The development of new materials and fabrication techniques has also enabled the creation of advanced electrochemical sensors with enhanced performance and capabilities. There are several types of electrochemical sensors available, each with its own unique features and advantages. These include potentiometric, amperometric, conductometric, and impedimetric sensors, as well as biosensors, gas sensors, and electrochemical imaging sensors.

Electrochemical sensors appeared in the second half of the 20th century and were widely used in commercial applications because of several factors. The use of sensors for signal acquisition, which is considered a clean model for analytical applications with no degeneration of wastes, for miniaturization in portable devices, and for fast analysis with low production cost, has popularized commercial applications. In association with nanotechnology, electrochemical sensors are becoming increasingly precise, selective, specific, and highly sensitive. Electrochemical method analysis can be divided into two main groups: interfacial methods and non-intrafacial methods. In the non-intrafacial method, electrical conductance is measured by using a cell of known size, with two equidistant electrodes. But in the case of the interfacial method, a disturbance of an electrical signal is measured based on the presence of an analyte on the electrode surface (sensor unit). Interfacial methods are divided into two main groups: static methods and dynamic methods. The static methods consist of electron transfer which occurs between the electrode and analyte. Dynamic methods involve current flow and can be used in analytical determination (sensors) and in the manufacturing of nanostructured materials.

Recent advances in materials science and micro-fabrication techniques have led to the development of novel electrochemical sensors with enhanced performance, such as nanomaterial-based sensors, 3D-printed sensors, and wearable sensors. In this chapter, we provide a comprehensive overview of the HEM-based electrochemical sensors, their working principles, and their applications. We also discuss the recent advances in the field and highlight the challenges and future directions for HEM-based electrochemical sensing.

19.2 INTRODUCTION TO HIGH ENTROPY MATERIALS

High entropy materials (HEM) are composed of multiple metallic elements which are characterized by a random, disordered atomic arrangement, which is different from the ordered atomic arrangement in traditional alloys. Thus, unlike traditional alloys that typically consist of one or two main elements with minor additions of other elements, HEMs can have five or more principal elements in their composition which are mixed in roughly equal proportions, resulting in a complex solid

DOI: 10.1201/9781003391388-19

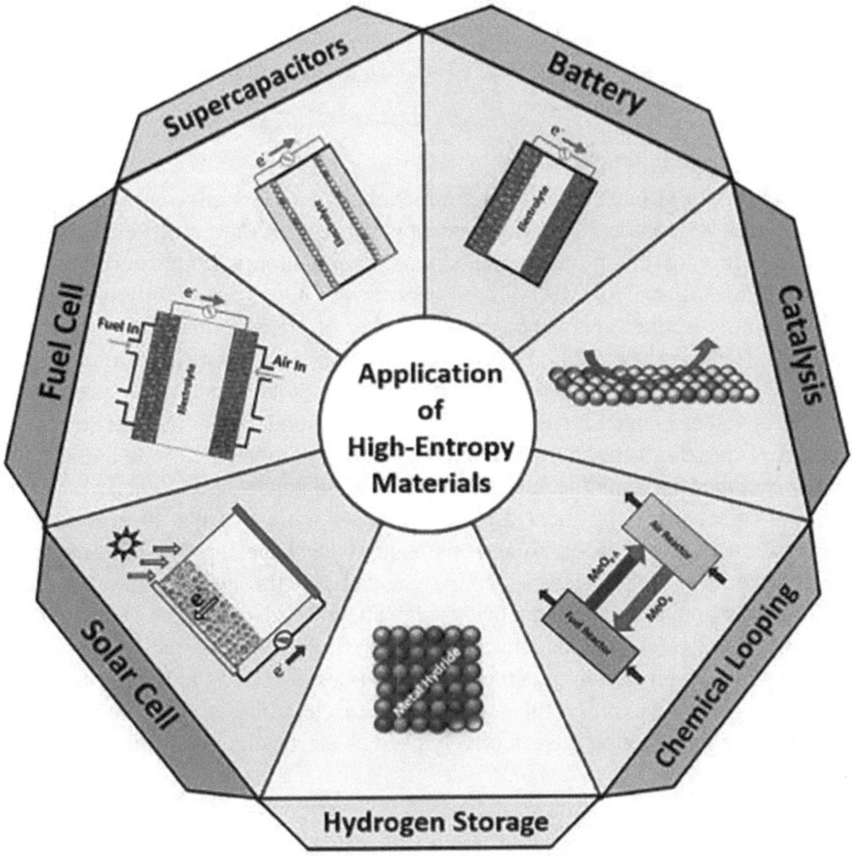

FIGURE 19.1 Various applications of high entropy materials. Adapted from reference [1]. Copyright The Authors, some rights reserved; exclusive licensee Royal Society of Chemistry. Distributed under a Creative Commons Attribution License 3.0 (CC BY).

solution with a high degree of entropy. This creates a complex, disordered atomic structure with high entropy, which can result in unique mechanical, physical, and chemical properties. HEAs can exhibit high strength, good ductility, high corrosion resistance, high degree of structural stability, good wear resistance, and more resistant to deformation and fracture, making them attractive for use in a wide range of applications, including catalysis, sensing, energy storage and conversion so on as depicted in Figure 19.1.

HEM has an essential role in electrochemical sensors due to their excellent mechanical properties, high thermal and chemical stability. Modern sensing systems focus on the manufacture of small sensors with more sensitivity and more selectivity and with low production and maintenance costs. HEMs that exhibit excellent current sensitivity can be manufactured by simple strategies. A group of HEMs have been used in electrochemical sensors and they can be classified based on their chemical compositions.

19.2.1 Classification of HEMs

There are two main types of HEMs: single-phase and multiphase. Single-phase HEMs have a homogeneous atomic structure, while multiphase HEMs have a mixture of two or more phases with different compositions and structures. Multiphase HEMs can exhibit improved mechanical properties due to the synergistic effects of different phases. The exact atomic structure of HEMs is still an area of active research, and various techniques such as X-ray diffraction, transmission electron microscopy, and neutron scattering are used to study their structure and properties. HEMs can be broadly categorized into metallic high entropy alloys (HEAs), ceramic high entropy compounds (HECs), and HEMs

19.2.1.1 Metallic High Entropy Alloys

An alloy that consists of at least five metal components and has a concentration that falls within the range of 5% to 35% is known as a high entropy alloy. HEAs are multi-component solid solution materials containing five or more elements in near equal atomic proportions with each element contributing to the overall structure as well as properties of alloys and stabilizing the structures by maximizing the configurational entropy. As the number of elements increases, its configurational entropy increases. HEAs can be applied to ceramics, polymers, and liquids of functional and structural materials. HEAs are further classified as high entropy solder and high entropy alloy brazing filler metal. HEAs solder is used for welding pure titanium and chromium–nickel–titanium stainless steel and HEAs brazing filler metal is used for welding cemented carbide and steel.

HEAs show excellent ferromagnetic, soft magnetic, and super-paramagnetic properties. HEAs can exhibit unique mechanical, physical, and chemical properties, such as high strength, good ductility, high corrosion rate, and good wear resistance. They have high irradiation resistance and corrosion resistance. Due to this reason, HEAs are used in the nuclear industry, especially for high-pressure vessels and vessels for nuclear fuels. They are also used as heat-resistant or wear-resistant coatings. Lightweight HEAs can be used as casing for mobile facilities, as battery anode materials, and in the transportation industry. Recent studies of HEMs, for example, $Al_{2.0}CoCrFeNi$ with near constant resistivity, make them useful for an electronic application. Recently, several HEAs have been developed which are effectively used as electrochemical sensors.

19.2.1.2 Ceramic HECs

HECs are similar to HEAs, but they are composed of multiple ceramic elements instead of metallic elements. HECs can have a complex, disordered atomic structure, which can lead to improved mechanical, thermal, and chemical properties. Some examples of ceramic HECs are ZrTiHfNbTa, MoNbTaTiV, and AlCrSiTiN. Both metallic HEAs and ceramic HECs have potential applications in a wide range of fields, including aerospace, energy, and biomedical industries.

19.2.1.3 High Entropy Materials

The incorporation of metal oxides, nitrides, and sulfides can increase the performance of HEM.

- High entropy oxides (HEOs): HEOs are a class of HEMs that have recently been investigated for gas-sensing applications. The HEO-based gas sensor exhibited high sensitivity and selectivity for hydrogen gas, attributed to the unique structure of the HEM. HEOs are complex oxides that offer a new paradigm in material science. They are formed by mixing five or more different metal cations in the same proportion and they have a single-phase crystal structure. $(MgNiCuCoZn)_{0.2}$ 0 is the first HEO, which has a rock salt structure. This structure is based on the configurational entropy value, which is exceptionally large in the case of a multi-component mixture, and it compensates for the unfavorable correspondent enthalpic contribution. The $(MgNiCuCoZn)_{0.2}O$ has wide applications in energy production and storage as an anode material in lithium-ion batteries and as large K dielectric materials in catalysis. Recent studies about the thermal properties of HEAs show an interesting correlation between low thermal conductivity and the disorder of the system. HEOs also have thin film deposition, sintering, mechanical, magnetic, optical, and electrochemical properties. The rock salt-based HEAs have many applications. This is due to its high lithium conductivities along with promising lithium storage capabilities. In HEOs, entropy stabilization not only affects the phase stability but also the functional properties.

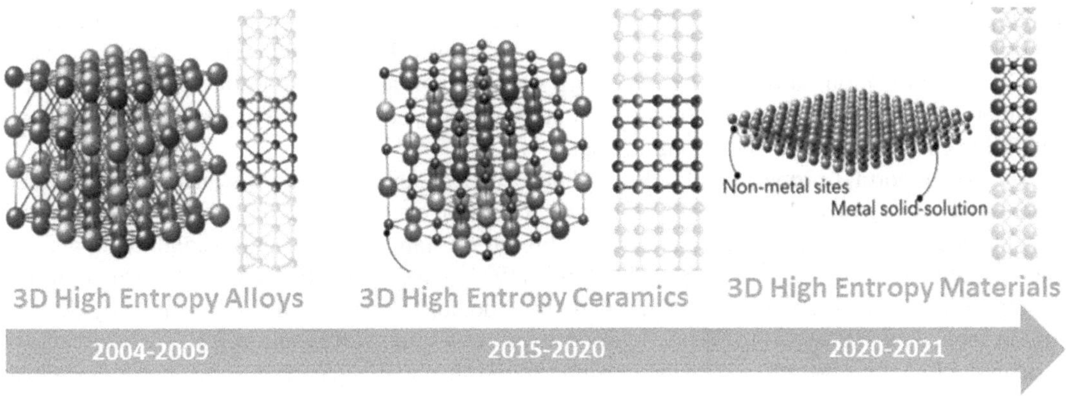

FIGURE 19.2 Schematic representation illustrating the progress in HEMs. Adapted from reference [2]. Copyright The Authors, some rights reserved; exclusive licensee Nature Portfolio. Distributed under a Creative Commons Attribution License 4.0 (CC BY).

- High entropy nitrides (HENs): HENs are another class of HEMs that have shown potential for gas-sensing applications. The HEN-based gas sensor for the detection of ammonia gas exhibits potential alternatives with other sensing materials. The HEN-based gas sensor exhibited high sensitivity and selectivity for ammonia gas, attributed to the unique surface properties of the HEM.
- High entropy sulfides (HESs): HESs are a relatively new class of HEMs that have shown promise for gas-sensing applications. In the last few decades, researchers have developed an HES-based gas sensor for the detection of hydrogen sulfide, which is used as high-temperature filter material. The HES-based gas sensor exhibited high sensitivity and selectivity for hydrogen sulfide gas, attributed to the unique surface chemistry of the HEM.

However, further research is needed to fully understand their properties and optimize their performance for specific applications. Research progress in various HEMs is depicted in Figure 19.2.

19.3 CHARACTERISTICS AND ADVANTAGES OF HEMS FOR ELECTROCHEMICAL SENSING

The need for sensing technology increases day by day due to its applications in our daily life, and different technologies have been used to develop better electrochemical sensors. Electrochemical sensing has emerged as an essential tool for the detection and quantification of various analytes in fields ranging from healthcare to environmental monitoring. Conventional electrochemical sensors have a shelf life of six months to one year, and the performance of electrochemical sensors largely depends on the material used as the sensing element. HEMs have recently gained significant attention as promising candidates for electrochemical sensing applications owing to their unique properties such as high surface area, high conductivity, long life, and excellent stability. These are structural materials having a unique structure with the co-existence of anti-site disordering and crystal periodicity.

The unique feature of HEMs is due to the presence of a large number of structural defects, which leads to the formation of high-density defect states and increased electronic conductivity. HEMs offer several advantages over traditional sensing materials such as metal oxides and carbon-based materials. One of the key advantages is their high surface area, which provides more active sites for analyte interaction, leading to enhanced sensing performance. Additionally, the high electronic conductivity of HEMs facilitates electron transfer, resulting in faster response times and improved sensitivity. HEM is used in hydrogen storage materials, electrochemical sensors, battery electrodes, and capacitors. HEMs have high-performance durability due to their entropy-driven stabilization.

The high ΔS_{mix} associated with the mixing of near equimolar solutions of five or more metal elements or ions favors the formation of HEMs. HEMs have well-known mechanical properties, entropic phase stabilization, lattice distortion, sluggish diffusion, and cocktail effect. Electrochemical sensors formed from HEM have shown excellent sensitivity and stability. This type of sensor shows maximum current response and a long-life span.

Another important property of HEMs is their excellent stability and resistance to degradation. Traditional sensing materials often suffer from degradation over time, leading to decreased sensitivity and accuracy. HEMs, on the other hand, exhibit high stability under harsh conditions, making them suitable for long-term sensing applications. HEMs have been explored for the detection of a variety of analytes such as glucose, hydrogen peroxide, and heavy metals. A recent study demonstrated the use of HEMs for the detection of glucose in human serum and the HEM-based sensor exhibited a linear response over a wide concentration range and a detection limit of 2.2 μM, which is comparable to the state-of-the-art glucose sensors.

One of the key advantages of HEMs is their ability to avoid the formation of brittle intermetallic compounds, which can limit the mechanical properties of traditional alloys. Instead, the random atomic arrangement in HEMs creates a high degree of structural stability, making these materials more resistant to deformation and fracture. Top of FormBottom of FormHigh entropy materials represents a promising class of materials with unique properties and potential for a wide range of applications including structural materials, catalysts, sensors, and coatings. They have the potential to be lightweight, corrosion-resistant, and high strength, making them attractive for use in aerospace, automotive, and other industries. Research on HEMs is still in its early stages, and further research is needed to fully understand their properties and optimize their performance for specific applications. There is ongoing work to understand the fundamental mechanisms that govern their properties and behavior.

19.4 SYNTHESIS AND FABRICATION TECHNIQUES OF HEMS FOR ELECTROCHEMICAL SENSORS

HEMs have gained significant attention in recent years due to their unique properties and potential applications in various fields. One of the promising areas of application for HEMs is electrochemical sensors. HEMs can be used as electrode materials in electrochemical sensors due to their high surface area, excellent conductivity, and unique surface properties. However, the synthesis and fabrication of HEMs for electrochemical sensors require careful consideration of various factors such as composition, morphology, and surface properties. Figure 19.3. depicts several possibilities of phase-segregated heterostructures of HEMs.

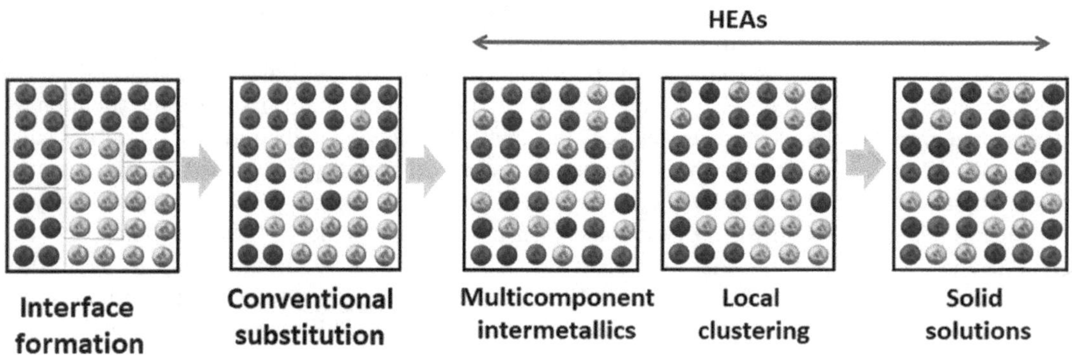

FIGURE 19.3 Schematic representation of various possibilities of formation of different degrees of ordered and disordered High Entropy Alloys. Adapted from reference [3]. Copyright The Authors, some rights reserved; exclusive licensee Nature Portfolio. Distributed under a Creative Commons Attribution License 4.0 (CC BY).

There are several techniques available for the synthesis and fabrication of HEMs for electrochemical sensors. One of the most commonly used techniques is mechanical alloying, which involves the milling of elemental powders in a ball mill to produce a homogeneous mixture. The high-energy milling process results in the formation of nanocrystalline grains and the development of a high degree of disorder, leading to the formation of HEMs. Another popular technique is the spark plasma sintering (SPS) method, which involves the application of high pressure and temperature to consolidate HEMs into dense materials. The SPS method has been shown to produce HEMs with enhanced mechanical properties and improved electrochemical performance.

Other techniques for the synthesis and fabrication of HEMs for electrochemical sensors include laser deposition, electrochemical deposition, and chemical vapor deposition. Laser deposition involves the use of a laser to melt and vaporize the HEMs, which then condense on a substrate to form a thin film. An electrochemical deposition involves the use of an electrolyte solution and an electrical current to deposit the HEMs onto a substrate. Chemical vapor deposition involves the use of a precursor gas that decomposes on a substrate to form the HEMs. The choice of synthesis and fabrication technique depends on several factors such as the desired composition, morphology, and properties of the HEMs, as well as the intended application. For example, the SPS method may be more suitable for producing bulk HEMs with enhanced mechanical properties, while laser deposition may be more suitable for producing thin films with tailored surface properties.

Non-equilibrium processes, such as the fast-moving bed pyrolysis (FMBP) approach and the carbothermal shock technique [4,5], can also be used for the fabrication of HEMs. The rapid heating-cooling process kinetically traps the thermodynamically advantageous high entropy phases. The formation of nanosized particles with increased surface areas is also ensured by the short growth period, which is measured in seconds or less. The second is to realize atomic mixing of various species by utilizing methods that produce local extreme conditions, such as ball milling and solvothermal synthesis [6,7]. It is important to note that these methods still operate under generally benign settings, which helps scale up implementations. The development of nanoscale HEMs at relatively low temperatures (200–300°C) can be facilitated by external variables in the liquid phase, such as solvent and ligand molecules in colloidal and polyol solutions [8,9]. The energy barrier for homogeneous nucleation and growth can be changed by the surface capping ligands, producing high entropy processes.

The synthesis and fabrication of HEMs for electrochemical sensors require careful consideration of various factors such as composition, morphology, and surface properties. Several techniques are available for the synthesis and fabrication of HEMs, and the choice of technique depends on the desired composition, morphology, and properties of the HEMs, as well as the intended application. The development of new synthesis and fabrication techniques is expected to further enhance the performance and expand the applications of HEMs in electrochemical sensors.

19.5 WORKING PRINCIPLE OF HEM-BASED ELECTROCHEMICAL SENSORS

The principle behind the electrochemical sensing behavior of HEM is based on its unique properties such as high thermal and chemical stability, high mechanical strength, and high resistance to corrosion. The HEMs are typically deposited on an electrode surface, which serves as the working electrode as shown in Figure 19.4. When the analyte of interest comes into contact with the HEMs present in the electrode surface, it induces a change in the electrical properties of the HEMs due to its interaction with the analyte which can be a result of various mechanisms such as adsorption, redox reactions or catalysis. These changes in the electrical properties can be detected by measuring the electrical current or potential at the working electrode using a reference electrode and counter electrode. The unique properties of HEMs such as their high surface area, porosity, and reactivity can enhance the interaction between the analyte and the HEMs, resulting in high sensitivity and selectivity of the sensor.

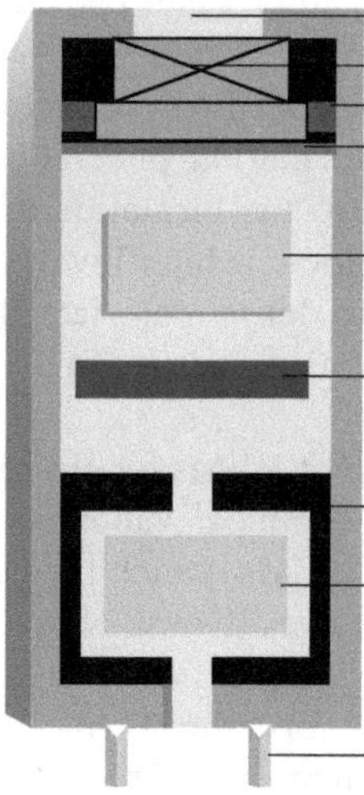

FIGURE 19.4 Schematic diagram of the instrumentation of HEM-based electrochemical sensors.

19.6 APPLICATIONS OF HEMS IN ELECTROCHEMICAL SENSING

The various applications of HEMs in electrochemical sensing include gas, chemical, and biological detection. HEMs have shown great potential as gas-sensing materials due to their high surface area, unique surface properties, and excellent conductivity. For instance, HEOs synthesized from SnO_2, ZnO, and TiO_2 have been shown to exhibit superior gas-sensing performance for gases like NO_2, H_2, CO, and CH_4, while high entropy metal chalcogenides formed from MoS_2 and WS_2 have also demonstrated promising gas-sensing properties for gases like NH_3 and NO_2. Researchers have also reported that HEMs can offer enhanced sensitivity, selectivity, and stability over traditional gas sensors.

HEMs can be employed as a chemical sensing material due to their unique surface properties and high surface area. For instance, high entropy metal-organic frameworks (MOFs) have shown excellent chemical sensing properties for chemicals such as ammonia, acetone, and formaldehyde. Similarly, HEMs based on graphene and its derivatives have demonstrated promising chemical sensing properties for various chemicals, including glucose, dopamine, and hydrogen peroxide. The use of HEMs in chemical sensors has the potential to improve the selectivity, sensitivity, and stability of these sensors.

HEMs can also be used as biosensing materials due to their unique beneficial properties. HEMs based on carbon materials like graphene oxide and carbon nanotubes have demonstrated excellent biosensing properties for biomolecules like DNA, proteins, and enzymes. In addition, HEMs based on metal nanoparticles like gold and silver have also shown promising biosensing properties for various biomolecules. The use of HEMs in biosensors has the potential to improve the sensitivity, selectivity, and stability of these sensors.

HEMs have significant potential as electrode materials in electrochemical sensors for gas, chemical, and biological detection. The unique properties of HEMs such as high surface area, excellent conductivity, and unique surface properties make them promising candidates for sensing applications. The use of HEMs in electrochemical sensors has the potential to improve the sensitivity, selectivity, and stability of these sensors, making them highly attractive for future sensing applications.

19.7 PERFORMANCE OF HEM-BASED ELECTROCHEMICAL SENSORS: ELECTROCHEMICAL PARAMETERS AND SENSITIVITY

Electrochemical sensors are widely used in a variety of applications, including environmental monitoring, medical diagnosis, food safety, and industrial process control, due to their high sensitivity, selectivity, and real-time monitoring capability. One of the key factors that determine the performance of HEM-based electrochemical sensors is the electrochemical parameters, which include the electrode potential, current density, and reaction kinetics. In addition, the sensitivity of the sensor is a crucial parameter that determines the ability of the sensor to detect low concentrations of analytes. The performance of HEMs can be affected by the following factors.

19.7.1 Factors Affecting the Performance of HEMs

Composition of HEMs: To achieve high sensitivity, selectivity, and stability of the sensor, HEMs should be carefully designed and chosen based on their qualities. Different elemental compositions of HEMs might result in varied electrochemical activities and responses to certain analytes. Moreover, the structural stability and durability of HEMs can be affected by their composition, which is crucial for long-term sensor performance.

Morphology and structure of HEMs: The sensitivity and selectivity of the sensor can be impacted by the surface area, porosity, and crystal structure of HEMs.

- Surface modification: A proper surface modification can enhance the number of active sites and the affinity for the analytes. Thus, functionalizing the surface of HEMs can improve the sensitivity and selectivity of the sensor.
- Electrode configuration: The sensitivity and selectivity of the sensor can be impacted by the choice of electrode material, geometry, and size.
- Operating conditions: Operating conditions such as temperature, pH, and ionic strength can also affect the performance of electrochemical sensors. These interactions between the analyte and the sensor surface can be impacted by these parameters as well as the stability and reactivity of HEMs.
- Analyte characteristics: The analyte parameters like concentration, redox potential, and size can influence the sensor performance. Several variables may affect the interaction of the analyte with the HEMs, which may impact the sensor's sensitivity and selectivity.

19.7.2 Strategies for Designing Better HEM-Based Electrochemical Sensors

There are several strategies that can be employed for designing better high entropy-based electrochemical sensors. The composition of HEMs can be optimized to achieve desirable properties for electrochemical sensing. This can involve adjusting the relative proportions of constituent elements or adding additional elements to the material.

Another strategy is microstructure engineering to achieve desirable properties for electrochemical sensing. This can involve controlling the grain size, porosity, or surface area of the material. Also, the surface of HEMs can be modified to improve their electrochemical performance. This can involve coating the material with a thin layer of another material or to modifying the surface chemistry to enhance selectivity. Moreover, the HEMs can be integrated with other materials to improve their electrochemical performance. This can involve combining HEMs with other conductive materials, such as carbon nanotubes or graphene, or with other sensing materials, such as metal oxides or polymers. The architecture of the electrochemical sensor can be designed to optimize the performance of HEM. This can involve optimizing the geometry and dimension of the sensing electrode, as well as the design of the electrolyte and reference electrodes. Finally, modeling and simulation can be used to predict the performance of HEMs-based electrochemical sensors and optimize their design. This can involve using computational methods to stimulate the electrochemical behavior of the material and predict its performance under different conditions.

19.8 ADVANCES IN HEM-BASED ELECTROCHEMICAL SENSORS: NOVEL MATERIALS

In recent years, HEMs have gained significant attention in the field of electrochemical sensors due to their unique properties and advantages over conventional materials. Researchers have been exploring novel materials and techniques to enhance the performance of HEM-based electrochemical sensors. This article focuses on the latest advances in HEM-based electrochemical sensors, including the development of new materials and fabrication techniques, as well as the improvement of electrochemical parameters and sensitivity.

The HEMs can also be used as an electrolyte for the development of electrochemical sensors. For example, NASICON PHOSPHATES ($Na_3M_2(PO_4)_3$ and $NaMPO_4O_X$, M=Ti, V, Mn, Cr, and Zr) used as the sodium conducting solid electrolytes, high entropy perovskite-based electrolytes are used as a proton conductor in electrochemical sensors and high entropy liquid electrolyte for lithium batteries are also used for making sensors.

Recently, a HEM-rGO composite consisting of CoCrFeNiTiAl alloy nanoparticles and reduced graphene oxide (rGO) was successfully synthesized for electrochemical sensing applications, which showed superior electrocatalytic activity toward the detection of hydrogen peroxide, with a detection limit as low as 20 nM. A highly sensitive electrochemical sensor based on a HEM of CuFeNiCoCrAl alloy exhibited a high sensitivity toward the detection of glucose, with a linear range of 0.1–10 mM and a detection limit of 0.03 mM. Some recently reported novel HEMs and their synthetic routes are tabulated in Table 19.1.

Furthermore, researchers have also been exploring new fabrication techniques such as 3D printing to create HEM-based sensors with customized designs and high accuracy. For instance, the successful fabrication of a 3D-printed HEM-based electrochemical sensor for the detection of heavy

TABLE 19.1
Reported Novel High Entropy Materials, Their Structure, And Synthetic Routes

Materials	Structure	Synthetic route	Ref.
HEM			
$KNa(MgMnFeCoNi)F_3$	Perovskite	Ball milling & solvothermal	[10]
$[La(CrMnFeCoNi)O_3]_{3/4}$	Perovskite	Ball milling	[11]
$CrMnFeCoNiS_x$	Cubic	Carbothermal shock	[12]
$CoCrFeMnNiP_x$	Hexagonal	Eutectic synthesis	[13]
ZnCoCdNiCu/MIM	ZIF	Ball milling	[14]
MnFeCoNiCu/BDC	MOF	Ambient solution phase method	[15]
HEO			
$NiMgCuZnCoO_x$	Rock salt	Template	[16]
$MoNiCuZnCoO_x$	Rock salt	Solid-state synthesis	[17]
$NiMgCuFeCoO_x$-Al_2O_3	Rock salt and amorphous	Nonhydrolytic sol-gel and ball milling	[18]
$CuCeO_x$-$NiMgCuZnCoO_x$	Fluorite and rock salt	Ball milling	[19]
$(CoCuFeMnNi)_3O_4$	Spinel	Solvothermal	[20]
$ZrCeMgLaYHfTiMnPdO_x$	Fluorite	Carbothermal shock	[21]
HEA			
CoMoFeNiCu	fcc	Carbothermal shock	[22]
IrPdPtRhRu	fcc	DFT and machine learning	[23]
PtPdRhNi	fcc	Carbothermal shock	[24]
AgIrPdPtRu	fcc	Sputtering	[25]
AlCuNiPtMn	fcc	Dealloying	[26]
CrMnFeCoNi	fcc	Laser ablation	[27]

metals in water. The sensor demonstrated a high sensitivity and selectivity toward the detection of lead ions, with detection limit as low as 0.22 μM. Overall, these recent advancements in novel materials and fabrication techniques for HEM-based electrochemical sensors have shown great promise in enhancing their performance and expanding their application potential.

19.9 COMPARISON OF HEM-BASED ELECTROCHEMICAL SENSORS WITH OTHER SENSING PLATFORMS

HEM-based electrochemical sensors have several advantages over other sensing platforms which include,

i. High sensitivity and selectivity: HEM-based electrochemical sensors have high sensitivity and selectivity toward specific analytes due to their unique composition and structure.
ii. Cost-effective: They are relatively simple and cost-effective compared to other sensing platforms, such as mass spectrometry and optical sensors.
iii. Real-time and on-site monitoring: They are suitable for real time and on-site monitoring of analytes, which is crucial in many applications, such as environmental monitoring and medical diagnostics.
iv. High stability: They have high stability, which allows for long-term sensing applications.
v. Easy to modify: They can be easily modified to enhance their performance, such as through the addition of functional groups or nanoparticles.

In contrast, other sensing platforms may require complex instrumentation, expensive equipment, and may not be suitable for on-site or real-time monitoring. For instance, mass spectrometry may require the use of vacuum conditions and specialized equipment, while optical sensors may require precise alignment and calibration. Overall, HEM-based electrochemical sensors offer several advantages over other sensing platforms, making them a promising option for a variety of applications. However, it is important to continue evaluating and comparing their performance to ensure their optimal use in various applications.

19.10 CHALLENGES AND FUTURE DIRECTIONS FOR HEM-BASED ELECTROCHEMICAL SENSORS

Despite the significant progress made in the development of HEM-based electrochemical sensors, there are still several challenges that need to be addressed. One of the main challenges includes synthesis reproducibility. The high entropy nature of these materials makes it challenging to control their composition and structure, which can lead to inconsistent sensor performance. Even though they are stable at a wide range of temperatures, they can be susceptible to degradation and instability over time, which can affect their sensitivity and selectivity. Thus, the long-term stability of the system is challenging and needs to be reconsidered. Also, the interference from other analytes present in the sample can affect their accuracy and precision. Furthermore, there is a need for standardization and validation protocols for HEM-based electrochemical sensors to ensure their reliable performance and comparability across different studies. Thus, HEM-based electrochemical sensors also have some drawbacks, which include limited availability, complex synthesis process, limited selectivity, surface passivation, and limited research. HEMs are still relatively new materials, and the process of synthesizing them is expensive, which makes their use in electrochemical sensors costly. While HEM-based sensors offer high sensitivity, their selectivity toward specific analytes is limited; this can lead to false positives and inaccurate results. The surface of HEMs tends to passivate quickly, which limits their performance in long-term applications. The research on HEMs is still in its early stages, and there is limited information available on their long-term stability and reliability.

Future directions for HEM-based electrochemical sensors include developing new synthesis strategies to improve the reproducibility and control over the composition and structure of HEM-based electrochemical sensors. Also, the research should focus on improving the stability and durability of HEM-based electrochemical sensors to extend their lifespan and improve their performance over time. Moreover, improving the selectivity and sensitivity of HEM-based electrochemical sensors toward specific analytes, as well as reducing interference from other analytes in the sample. More research should explore new applications for these sensors to address unmet needs and challenges in different fields. Thus, addressing the challenges and pursuing future directions for HEM-based electrochemical sensors will contribute to the development of more reliable, sensitive, and selective sensing platforms with broad applications in various fields.

19.11 CONCLUSION

The HEM-based electrochemical sensors have shown great potential for a wide range of applications due to their unique properties such as high entropy, compositional diversity, high thermal stability, excellent mechanical properties, and high corrosion resistance and tunability. These sensors have demonstrated high sensitivity, selectivity, and stability for detecting various analytes, including gases, chemicals, and biomolecules.

Despite the challenges in synthesis reproducibility, long-term stability, lack of standardization in HEM synthesis and characterization, limited understanding of the fundamental electrochemical properties of HEMs, high cost of production, and interference from other analytes, significant progress has been made in developing HEM-based electrochemical sensors. Future research directions should focus on developing new synthesis strategies, enhancing stability and durability, improving selectivity and sensitivity, and exploring new applications for these sensors. HEM-based electrochemical sensors hold great promise for addressing unmet needs and challenges in various fields, such as environmental monitoring, medical diagnostics, and food safety. They have the potential to revolutionize the field of electrochemical sensing and pave the way for the development of more reliable and sensitive sensing platforms. Overall, the future of HEM-based electrochemical sensors is bright, and continued research and development in this field will contribute to the advancement of science and technology and improve our ability to detect and monitor various analytes in real time.

REFERENCES

[1] A. Amiri, R. S. Yassar, Recent Progress of High entropy materials for energy storage and conversion. *Journal of Materials Chemistry A*, **9**, 782–823 (2021).

[2] S. K. Nemani, M. Torkamanzadeh, B. C. Wyatt, V. Presser, B. Anasori. Functional two-dimensional high-entropy materials. *Communication Materials* **4**:**16**, 1–7 (2023).

[3] Sun, Y. & Dai, S. High-entropy materials for catalysis: a new frontier. *Sci. Adv.* **7**, 1600 (2021).

[4] Y. Yao, Z. Huang, P. Xie, S. D. Lacey, R. J. Jacob, H. Xie, F. Chen, A. Nie, T. Pu, M. Rehwoldt, D. Yu, M. R. Zachariah, C. Wang, R. Shahbazian-Yassar, J. Li, L. Hu, Carbothermal shock synthesis of high-entropy-alloy nanoparticles. *Science* **359**, 1489–1494 (2018).

[5] S. Gao, S. Hao, Z. Huang, Y. Yuan, S. Han, L. Lei, X. Zhang, R. Shahbazian-Yassar, J. Lu, Synthesis of high-entropy alloy nanoparticles on supports by the fast-moving bed pyrolysis. *Nat. Commun.* 11, 2016 (2020).

[6] H. Xu, Z. Zhang, J. Liu, C.-L. Do-Thanh, H. Chen, S. Xu, Q. Lin, Y. Jiao, J. Wang, Y. Wang, Y. Chen, S. Dai, Entropy-stabilized single-atom Pd catalysts via high-entropy fluorite oxide supports. *Nat. Commun.* **11**, 3908 (2020).

[7] N. L. N. Broge, M. Bondesgaard, F. Søndergaard-Pedersen, M. Roelsgaard, B. B. Iversen, Autocatalytic formation of high-entropy alloy nanoparticles. *Angew. Chem. Int. Ed.* **59**, 21920–21924 (2020).

[8] C. R. McCormick, R. E. Schaak, Simultaneous multi cation exchange pathway to high entropy metal sulfide nanoparticles. *J. Am. Chem. Soc.* **143**, 1017–1023 (2021).

[9] D. Wu, K. Kusada, T. Yamamoto, T. Toriyama, S. Matsumura, S. Kawaguchi, Y. Kubota, H. Kitagawa, Platinum-group-metal high-entropy-alloy nanoparticles. *J. Am. Chem. Soc.* **142**, 13833–13838 (2020).

[10] T. Wang, H. Chen, Z. Yang, J. Liang, S. Dai, High-entropy perovskite fluorides: A new platform for oxygen evolution catalysis. *J. Am. Chem. Soc.* **142**, 4550–4554 (2020).
[11] T. Wang, J. Fan, C.L. Do-Thanh, X. Suo, Z. Yang, H. Chen, Y. Yuan, H. Lyu, S. Yang, S. Dai, Perovskite oxide-halide solid solutions: A platform for electrocatalysts. *Angew. Chem. Int. Ed.* **60**, 9953–9958 (2021).
[12] M. Cui, C. Yang, B. Li, Q. Dong, M. Wu, S. Hwang, H. Xie, X. Wang, G. Wang, L. Hu, High-entropy metal sulfide nanoparticles promise high-performance oxygen evolution reaction. *Adv. Energy Mater.* **11**, 2002887 (2021).
[13] X. Zhao, Z. Xue, W. Chen, Y. Wang, T. Mu, Eutectic synthesis of high-entropy metal phosphides for electrocatalytic water splitting. *ChemSusChem* **13**, 2038–2042 (2020).
[14] W. Xu, H. Chen, K. Jie, Z. Yang, T. Li, S. Dai, Entropy-driven mechanochemical synthesis of polymetallic zeolitic imidazolate frameworks for CO2 fixation. *Angew. Chem. Int. Ed.* **58**, 5018–5022 (2019).
[15] X. Zhao, Z. Xue, W. Chen, X. Bai, R. Shi, T. Mu, Ambient fast, large-scale synthesis of entropy-stabilized metal-organic framework nanosheets for electrocatalytic oxygen evolution. *J. Mater. Chem. A* **7**, 26238–26242 (2019).
[16] D. Feng, Y. Dong, L. Zhang, X. Ge, W. Zhang, S. Dai, Z. Qiao, Holey lamellar high-entropy oxide as an ultra-high-activity heterogeneous catalyst for solvent-free aerobic oxidation of benzyl alcohol. *Angew. Chem. Int.* Ed. **59**, 19503–19509 (2020).
[17] C. Deng, P. Wu, L. Zhu, J. He, D. J. Tao, L. Lu, M. He, M. Hua, H. Li, W. Zhu, High-entropy oxide stabilized molybdenum oxide via high temperature for deep oxidative desulfurization. *Appl. Mater. Today* **20**, 100680 (2020).
[18] Z. Zhang, S. Yang, X. Hu, H. Xu, H. Peng, M. Liu, B. P. Thapaliya, K. Jie, J. Zhao, J. Liu, H. Chen, Y. Leng, X. Lu, J. Fu, P. Zhang, S. Dai, Mechanochemical nonhydrolytic sol-gelstrategy for the production of mesoporous multimetallic oxides. *Chem. Mater.* **31**, 5529–5536 (2019).
[19] H. Chen, K. Jie, C. J. Jafta, Z. Yang, S. Yao, M. Liu, Z. Zhang, J. Liu, M. Chi, J. Fu, S. Dai, An ultrastable heterostructured oxide catalyst based on high-entropy materials: A new strategy toward catalyst stabilization via synergistic interfacial interaction. *Appl. Catal. B Environ.* **276**, 119155 (2020).
[20] D. Wang, Z. Liu, S. Du, Y. Zhang, H. Li, Z. Xiao, W. Chen, R. Chen, Y. Wang, Y. Zou, S. Wang, Low-temperature synthesis of small-sized high-entropy oxides for water oxidation. *J. Mater. Chem. A* **7**, 24211–24216 (2019).
[21] T. Li, Y. Yao, Z. Huang, P. Xie, Z. Liu, M. Yang, J. Gao, K. Zeng, A. H. Brozena, G. Pastel, M. Jiao, Q. Dong, J. Dai, S. Li, H. Zong, M. Chi, J. Luo, Y. Mo, G. Wang, C. Wang, R. Shahbazian-Yassar, L. Hu, Denary oxide nanoparticles as highly stable catalysts for methane combustion. *Nat. Catal.* **4**, 62–70 (2021).
[22] P. Xie, Y. Yao, Z. Huang, Z. Liu, J. Zhang, T. Li, G. Wang, R. Shahbazian-Yassar, L. Hu, C. Wang, Highly efficient decomposition of ammonia using high-entropy alloy catalysts. *Nat. Commun.* **10**, 4011 (2019).
[23] T. A. A. Batchelor, J. K. Pedersen, S. H. Winther, I. E. Castelli, K. W. Jacobsen, J. Rossmeisl, High-entropy alloys as a discovery platform for electrocatalysis. *Joule* **3**, 834–845 (2019).
[24] Y. Yao, Z. Huang, T. Li, H. Wang, Y. Liu, H. S. Stein, Y. Mao, J. Gao, M. Jiao, Q. Dong, J. Dai, P. Xie, H. Xie, S. D. Lacey, I. Takeuchi, J. M. Gregoire, R. Jiang, C. Wang, A. D. Taylor, R. Shahbazian-Yassar, L. Hu, High-throughput, combinatorial synthesis of multimetallic nanoclusters. *Proc. Natl. Acad. Sci.* **117**, 6316–6322 (2020).
[25] T. A. A. Batchelor, T. Löffler, B. Xiao, O. A. Krysiak, V. Strotkötter, J. K. Pedersen, C. M. Clausen, A. Savan, Y. Li, W. Schuhmann, J. Rossmeisl, A. Ludwig, Complex-solidsolution electrocatalyst discovery by computational prediction and high-throughputexperimentation. *Angew. Chem. Int. Ed.* **60**, 6932–6937 (2021).
[26] S. Li, X. Tang, H. Jia, H. Li, G. Xie, X. Liu, X. Lin, H.-J. Qiu, Nanoporous high-entropy alloys with low Pt loadings for high-performance electrochemical oxygen reduction. *J. Catal.* **383**, 164–171 (2020).
[27] T. Löffler, F. Waag, B. Gökce, A. Ludwig, S. Barcikowski, W. Schuhmann, Comparing the activity of complex solid solution electrocatalysts using inflection points of voltammetric activity curves as activity descriptors. *ACS Catal.* **11**, 1014–1023 (2021).

20 Thermohydraulic Performance and Entropy Generation Analysis of Nanofluids in Heat Exchanger

Anitha S, Priyadharshini P, and Vanitha M Archana

Nomenclature		Subscripts	
B0	Magnetic field	bf	Base fluid
C_p	Specific heat, J kg^{-1} K^{-1}	c	Cold
E	Exergy	c,i	Cold inlet
Ha	Hartmann number	c,o	Cold outlet
L	Length, mm	Thnf	Ternary hybrid nanofluid
$\dot{m}$	Mass flow rate, kg s^{-1}	f	Fluid
n	Shape factor,-	hnf	Hybrid nanofluid
Nu	Nusselt number, -	h	Hot
K	Thermal conductivity, W m^{-1} K^{-1}	h,i	Hot inlet
$\dot{S}_{gen}$	Entropy generation	h,o	Hot outlet
Δp	Pressure drop, pa	tot	Total
T	Temperature, °C (or) K	a	Ambient
u, v, w	Velocity components, m s^{-1}	r	Radial
$\dot{Q}$	Heat convection rate	nf	Nanofluid
Greek Symbols		s	Nanoparticle (solid particle)
ρ	density, kg m^{-3}	ave	Average
φ	volume fraction, %		
μ	viscosity, kg m^{-1}s^{-1}		

20.1 INTRODUCTION

Heat exchangers (HEs) play a vital role in industries from smaller to bigger scale. It accesses the thermal flow between two mediums which are at different temperatures. The present-day industrial domains focus strongly on heat exchange mechanisms, which are essential to various sectors like machinery and aerospace [1]. High-efficiency and low-resistance heat exchange systems contain more potential applications as science and technology advancement. HEs have many applications such as production, process, marine industries, food processing units, and so on. In every application, energy consumption and thermal performance play an important role. A lot of energy is required to process the HE in terms of pumping, heat recovery, separation of fluids, and so on.

 DOI: 10.1201/9781003391388-20

Therefore, it is important to evaluate the energy consumption as like their thermal performance [2]. There are many ideas are proposed to enhance the thermal performance of systems such as:

i. Addition of turbulators such as fins, bend, plates, conical rings, and so on.
ii. Applying external forces like magnetic and electrical forces.
iii. Altering the geometry of the system in order to increase the contact area between hot fluid and coolant.
iv. Adding nanoparticles to the coolant which is termed nanofluids.

Numerous studies have been carried out to examine the aforementioned variables, and they have produced intriguing findings that are helpful for the heat transfer industries as well as becoming a focal point for future studies. The above-said points are very helpful to increase the overall performance of the system. But, in the need to increase its performance, the nonentity point is that the augmentation in energy and entropy of HEs. By adding turbulators, the systems (HE) under consideration can efficiently enhance the heat transfer rate, and the variations in energy consumption are very high. Twisted tube is inserted into HE to vary thermal boundary layer of the fluid flow [3]. The detailed investigation of the turbulators and the generation of entropy using these turbulators is discussed in the following section.

Another way to improve the thermal performance (TP) is to add external forces like magnetic or electrical force to the flow field. Especially, the research community on thermal engineering and mathematics is more focused on investigating the TP of HE systems with the addition of an external magnetic field. It is highlighted that magnetic fields cause an enormous increase in the velocity of the fluid flow and so these fluids act like a solid. Consequently, the TP of the fluid enhances and so is the performance of HEs [4]. In the following section, the role of magnetic field on the minimization, maximization of entropy generation is discussed.

Recent research analyzed the magnetic field on TP of the system. Magnetic field is imposed on the wall periodically [5]. They used various nanofluids (which will be discussed in the forthcoming section) as coolants. They observed that with lower Hartmann number (Ha), Nusselt number (Nu) is higher. The Nu and the Ha are formulated as follows:

$$\mathrm{Nu}_{\mathrm{ave}} = -\left(\frac{k_{\mathrm{nf}}}{k_{\mathrm{bf}}}\right)\int_0^1 \frac{\partial \theta}{\partial Y}\partial X \tag{20.1}$$

$$\mathrm{Ha} = \mathrm{LB}_0\sqrt{\frac{\sigma_{\mathrm{bf}}}{\mu_{\mathrm{bf}}}} \tag{20.2}$$

The above is an example of how the TP of systems is governed by the magnetic field. The next innovative way to increase the TP of HEs is to add nanoparticles to the coolants (traditional coolants such as water, biofluids, oil, and so on) and it is termed nanofluids. The detailed discussion on nanoparticles, and mathematical modeling of nanofluids in HEs are discussed in the following section. Many researchers focused on analyzing the TP, exergy, and entropy generation of the system using nanofluids as a coolant. For example, Anitha et al. [6] analyzed the TP of a double tube HE numerically analyzed with Al_2O_3–Cu/water hybrid nanofluid. Heat is transferred from hot fluid to the coolant. By using nanofluid, the initial temperature (75°C), reduces to 45°C, whereas it is 54°C, with the usage of water.

20.2 HEs AND THEIR TYPES

HEs are used in a variety of applications such as marine engineering, food processing unit, production process, environmental engineering, and space applications. These HEs are categorized based on the criteria namely, re-generators, process of transferring heat (direct/indirect process), design (tubes, fins, and surfaces), mechanism (single/two-phase), and fluid flow (parallel/counter/cross flows).

The often used HEs are listed here. U-bend tube HE, helical HE, shell and tube HE, pillow plate HE, tubular HE, double-pipe HE, HE with louvered winglet tape, helically coiled tube-in tube HE, plate HE, finned tube HE, mini channel HE, co-current HE, shell and helically coiled finned tube HE, HE with helical twisted tapes, plate fin HE, fin and plate tube HE, double-pipe pin fin HE, rib grooved HE, cross HE, curved rectangular HE, and double pipe with perforated turbulators. The working principle, methodology, applications, and specifications of important and most used HEs in industries are discussed in the following section.

20.2.1 Helical HE

The helical HE contains two lids and a helical tube with rectangular cross-sections as shown in Figure 20.1 [7].

The upper and lower surfaces are covered with these lids. Two different fluids (hot fluid and coolant) flow in this HE which are at different temperatures. These fluids, flow in two different tubes. The coolant transfers heat from the hot fluid. One fluid uses the inner tube and another uses the outer tube. The flow of one fluid originates from the inlet and ends in the outlet. Another fluid enters the HE through a hole which is at the bottom cover of the system and flows in the radial direction. The rectangular cross-section allows the fluid to move freely so the heat transfer performance will be higher. To avoid corrosion or any kind of similar damage this HE is made of stainless steel. This HE is compact, easy maintenance and lower installation cost and it is used in industries in which high operating pressure and temperature gradient are involved.

20.2.2 Double-Pipe Heat Exchanger

The double-pipe HE contains more than one tube, as shown in Figure 20.2 [8]. The length of the outer tube is smaller than the inner tube. There will be two fluids: one is hot and another one is coolant. Heat is transferred from hot fluid to coolant. In this HE, gas or low viscous fluid can only be used, very small in size (5 times smaller than shell and tube HE), easy to fabricate, low maintenance and so it is popular among the small scale industries. The advantage of this HE is that it provides good efficiency with lower initial costs.

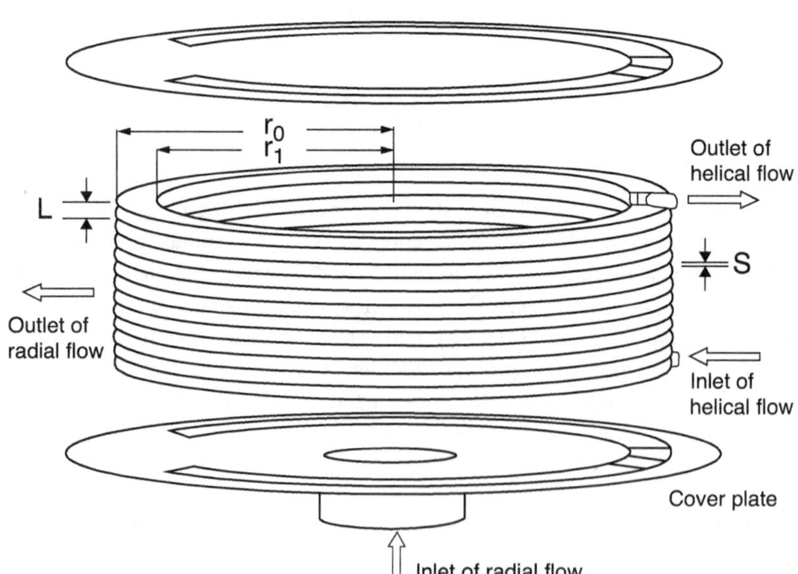

FIGURE 20.1 Schematic of helical heat exchanger. Reprinted from Jung et al. 2012. Adapted with permission [7]. Copyright (2012), Elsevier.

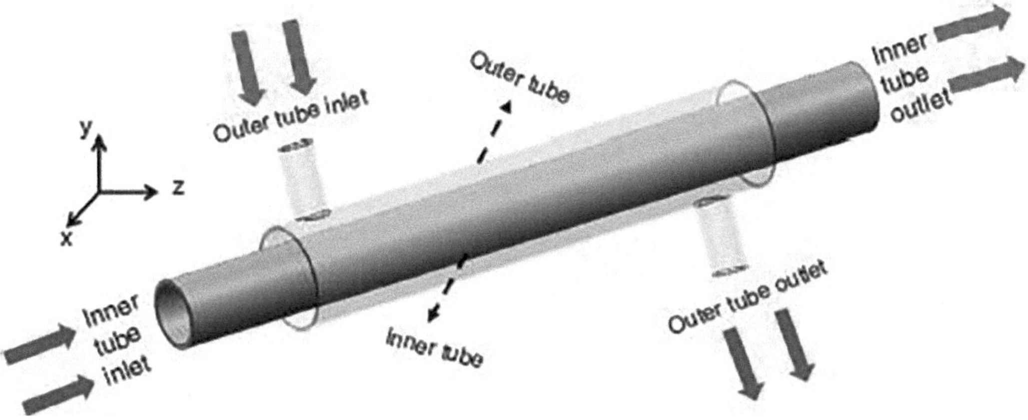

FIGURE 20.2 Double-pipe heat exchanger. Adapted with permission [8]. Copyright (2022), Springer.

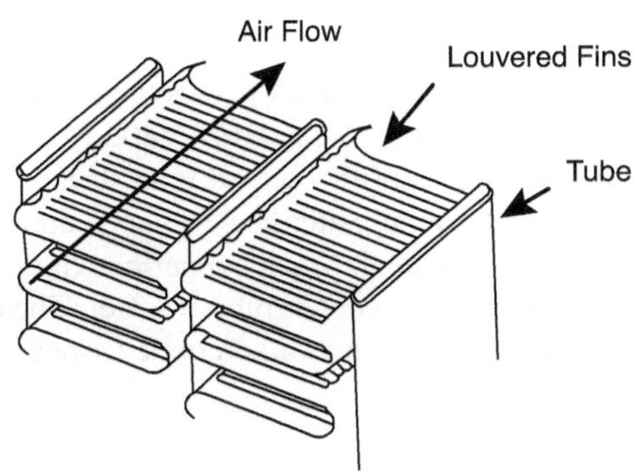

FIGURE 20.3 Schematic of louvered winglet tape heat exchanger. Adapted with permission [10]. Copyright (2006), Elsevier.

It can be also used in high pressure with high temperature. This exchanger is made of concentric pipes that are connected by mechanical closures. It is further divided into two types: (a) counter flow and (b) parallel flow. Generally, this HE is made of Galvanized iron.

20.2.3 Heat Exchanger with Louvered Winglet Tape

The louvered winglet fin HE is a type of compact HE, and it is shown in Figure 20.3. It is commonly used in automotive industry, air conditional industry, and so on. This type of HE has been used for a decade [9]. The various components of this HE are optimized by several researchers across the globe. Each louver in the HE increases the contact area between the hot oil and the coolant and therefore the TP of this HE is an effective one. There are a variety of models that are used in this louvered fin HE. The winglet that is fitted in the HE increases the boundary layer of the fluid which increases the overall heat transfer coefficient of the HE [10]. Variations in tube length drastically alter the TP of HE. It is shown that 31% increase in TP when compared to other HEs.

20.2.4 Shell and Helically Coiled Tube-in Tube Heat Exchanger

Shell and helically coiled HE is the one, which take small space and provides higher TP compared to other traditional HEs. It contains two concentric tubes which are bent in helical shape and it is shown in Figure 20.4 [11].

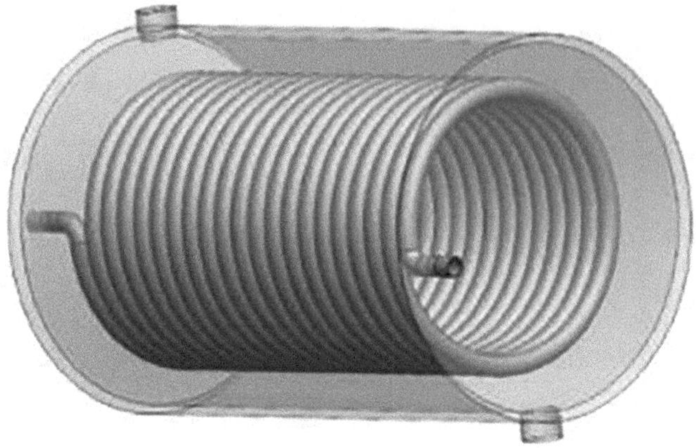

FIGURE 20.4 Schematic of shell and helically coiled heat exchanger. Adapted with permission [11]. Copyright (2023), Elsevier.

One fluid passes through the pipe whereas another passes the annulus space between the spaces. It also has a compact structure compared to other HEs. Various researchers evaluated the heat transfer performance with different shapes of the helical tube structure. It is broadly used in heat recovery systems, nuclear power plants, solar water heater plants, and so on. Generally, these HEs are made up of copper metal or stainless steel. Since copper has higher heat transfer rate most of this will be fabricated by using copper. Also, copper tube is softer. The coil is used to carry, pull or push loads. This spring has numerous uses such as automobile suspension systems, cooling valves, and so on.

20.3 MATHEMATICAL MODELING OF NANOFLUIDS AND ITS GENERATIONS

As discussed in the previous section, the innovative way to improve the performance of HE is to varying the coolant. That is adding nanosized particles (NP) in the conventional coolants namely water, oil, and so on. This section clearly explains the formation, mathematical models, and generation of nanofluids. This idea of adding metal particles to the coolant (to increase the TP of the considered system) was first originated by Maxwell. Initially, Maxwell added micro-sized particles to the coolants, but it caused problems such as higher pressure drop, erosion, and corrosion. Therefore, the significant TP cannot be attained. In the following years, the research community suggested to work on NP to be added in the coolants and its work well too. There are numerous studies that evaluated the TP, energy consumption and the entropy generation, and exergy loss both numerically and experimentally. Coming to the point of generation, as technology develops there are lots of improvement needed to be done for the practical deployment. Nanofluids are not an exemption for this new technology development. Yes, these nanofluids are developed into hybrid nanofluids as well as ternary hybrid nanofluid. Nanofluid is a term used to describe the dispersion of a single type of nanoparticle in a fluid. Hybrid nanofluids are suspensions of two different types of nanoparticles in a fluid, while ternary hybrid nanofluids are suspensions of three different types. In comparison with unitary nanofluids, hybrid nanofluids establish interesting rheological behavioral patterns and thermophysical properties. Also, this class of two-phase NP suspended in a clear fluid has the potential to attain the stability of a single-phase nanofluid. The desirable characteristics of these acceptable and suitable nanofluids include higher thermal conductivity at low nanoparticle concentrations [12]. Initially, this innovation was conducted by Jana et al. [13] through an experimental setup. Heat conduction processes practically become more effective in a hybrid nanofluid with the upsurge volume percentage of nanoparticles. Suresh et al. [14] proved this novel concept earlier. In thermal systems with high temperatures, such as solar power, air conditioning applications, biomedical, generator cooling, space, electronic cooling, automotive industry, and nuclear system cooling, hybrid nanomaterials are effective in providing cooling performance. A HE with an exterior flow was investigated by Karimi and Afrand [15]. In this study, water and MgO–CNT/Engine oil hybrid nanofluid acted as radiator fluids while the external fluid was chosen to be air. The findings indicated that vertical pipe radiators have superior efficacy of up to 10% greater when compared to horizontal pipe radiators. To examine the impact of nanoparticle concentration on pressure loss and heat transfer properties, Bhattad et al. [16] conducted numerical and experimental assessments on a plate HE employing hybrid nanoparticles (Al_2O_3+CNT) suspended in water as the coolant. With only a small increase in pumping power, it was found that using this nanofluid improves heat transfer coefficient by 39.16%. In current years, researchers focused on a brand-new class of Tri/Ternary/Tripartite hybrid nanofluid, which provides better TP at lower volume concentrations when compared to mono nanofluid and hybrid nanofluid [17–20]. This fluid is created by dispersing three distinct nanoparticles of the same or various sizes and shapes in base fluids as traditional fluids. Three-phase nanofluids are a slightly different category of nanofluids with numerous potential applications in energy transfer, such as defense, medicine, acoustics, and naval structures. Many academicians and experimenters carried out their research in improving the heat transfer rate by assuming different physical geometries with technical configurations subject to the ternary hybrid nanofluid characteristics in diverse circumstances [21,22]. The mathematical modeling of these nanofluids can be written as follows. The analysis of entropy production, heat transfer, and energy consumption in HEs heavily relies on the thermophysical physical parameters. The properties are listed below: density, thermal conductivity, viscosity, and specific heat capacity.

Mathematical modeling of nanofluids [23]:

Properties	Thermophysical properties
ρ (density)	$\rho_{nf} = \varphi_S \rho_S + (1-\varphi_S)\rho_f$
μ (viscosity)	$\mu_{nf} = \mu_f (1+2.5\,\varphi_S)$
K (thermal conductivity)	$\frac{K_{nf}}{K_f} = \frac{K_S+(n-1)K_f-(n-1)\varphi_S(K_f-K_S)}{K_S+(n-1)K_f+\varphi_S(K_f-K_S)}$
C_p (specific heat)	$(C_p)_{nf} = (1-\varphi_S)(\rho C_p)_f + \varphi_S(\rho C_p)_S$

Mathematical modeling of hybrid nanofluids [23]:

Properties	Thermophysical properties
ρ (density)	$\rho_{hnf} = \varphi_{S_1}\rho_{S_1} + \varphi_{S_2}\rho_{S_2} + (1-\varphi_S)\rho_{nf}$
μ (viscosity)	$\mu_{hnf} = \mu_{nf}(1+2.5(\varphi_{S_1}+\varphi_{S_2}))$
K (thermal conductivity)	$\frac{K_{hnf}}{K_{bf}} = \frac{K_{S_2}+(n-1)K_f-(n-1)\varphi_{S_2}(K_{bf}-K_{S_2})}{K_{S_2}+(n-1)K_{bf}+\varphi_{S_2}(K_{bf}-K_{S_2})};$ $\frac{K_{bf}}{K_f} = \frac{K_{S_1}+(n-1)K_f-(n-1)\varphi_{S_1}(K_f-K_{S_1})}{K_{S_1}+(n-1)K_f+\varphi_{S_1}(K_f-K_{S_1})}$
ρC_p (specific heat)	$(C_p)_{hnf} = (1-\varphi_S)(\rho C_p)_{nf} + \varphi_{S_1}(\rho C_p)_{S_1} + \varphi_{S_2}(\rho C_p)_{S_2}$

Mathematical modeling of ternary hybrid nanofluids [24]:
Density and specific heat:

$$\rho_{\text{Thnf}} = \left(1 - \varphi_{S_1} - \varphi_{S_2} - \varphi_{S_3}\right)\rho_{bf} + \varphi_{S_1}\rho_{S_1} + \varphi_{S_2}\rho_{S_2} + \varphi_{S_3}\rho_{S_3}$$

$$\rho C_{p\text{Thnf}} = \left(1 - \varphi_{S_1} - \varphi_{S_2} - \varphi_{S_3}\right)\left(\rho C_p\right)_{bf} + \varphi_{S_1}\left(\rho C_p\right)_{S_1} + \varphi_{S_2}\left(\rho C_p\right)_{S_2} + \varphi_{S_3}\left(\rho C_p\right)_{S_3}$$

Viscosity and thermal conductivity (for spherical shaped np)

$$\frac{\mu_{\text{Thnf }1,2}}{\mu_{\text{bf}}} = 1 + 2.5\varphi + 6.2\ \varphi^2$$

$$\frac{k_{\text{Thnf }1,2}}{k_{\text{bf}}} = \frac{k_1 + 2k_{\text{bf}} + 2\varphi\left(k_1 - k_{\text{bf}}\right)}{k_1 + 2k_{\text{bf}} - \varphi\left(k_1 - k_{\text{bf}}\right)}$$

Viscosity and thermal conductivity (for cylindrical shaped np)

$$\frac{\mu_{\text{Thnf }3}}{\mu_{\text{bf}}} = 1 + 13.5\varphi + 904.4\ \varphi^2$$

$$\frac{k_{\text{Thnf }3}}{k_{\text{bf}}} = \frac{k_2 + 3.9k_{\text{bf}} + 3.9\varphi\left(k_2 - k_{\text{bf}}\right)}{k_2 + 3.9k_{\text{bf}} - \varphi\left(k_2 - k_{\text{bf}}\right)}$$

Viscosity and thermal conductivity:

$$\mu_{\text{Thnf}} = \frac{\mu_{\text{Thnf},1}\varphi_{S_1} + \mu_{\text{Thnf},2}\varphi_{S_2} + \mu_{\text{Thnf},3}\varphi_{S_3}}{\varphi_{S_1} + \varphi_{S_2} + \varphi_{S_3}}$$

$$k_{\text{Thnf}} = \frac{k_{\text{Thnf},1}\varphi_{S_1} + k_{\text{Thnf},2}\varphi_{S_2} + k_{\text{Thnf},3}\varphi_{S_3}}{\varphi_{S_1} + \varphi_{S_2} + \varphi_{S_3}}$$

20.4 THERMO - HYDRAULIC PERFORMANCE AND ENTROPY GENERATION ANALYSIS OF NANOFLUIDS IN HE

20.4.1 Mathematical Modeling of Entropy Generation in HEs

Exergy loss (E) is the maximum amount of work that may be done, and the operation should end when the ambient temperature is attained. The equation that follows is used to compute E [25].

$$E = T_{\text{ambient}}\left[\dot{m}C_{pc}\ln\left(\frac{T_{c,o}}{T_{c,i}}\right) + \dot{m}C_{ph}\ln\left(\frac{T_{h,o}}{T_{h,i}}\right)\right] \tag{20.3}$$

The dimensionless exergy loss (S'), which is found from the exergy loss, is written as follows:

$$S' = \frac{\text{Exergy loss } (E)}{(\dot{m}C_p)_{\min}(T_{h,i} - T_{c,i})} \tag{20.4}$$

Exergetic efficiency (η) is evaluated by using the following equation:

$$\eta = \frac{E_c}{E_h} \times 100. \tag{20.5}$$

where E_c and E_h are found using the following equations:

$$E_c = T_{\text{ambient}}\ C_{pc}\ \ln\left(T_{c,o}/T_{c,i}\right) \tag{20.6}$$

$$E_h = T_{\text{ambient}}\ C_{ph}\ \ln\left(T_{h,o}/T_{h,i}\right) \tag{20.7}$$

The total entropy generation $(\dot{S}_{\text{gen}})_T$ is the total of heat transfer $((\dot{S}_{\text{gen}})_{\text{HT}})$ and the friction factor $((\dot{S}_{\text{gen}})_{\text{FF}})$ of the HE and it is given as

$$(\dot{S}_{\text{gen}})_{\text{T}} = (\dot{S}_{\text{gen}})_{\text{HT}} + (\dot{S}_{\text{gen}})_{\text{FF}} \tag{20.8}$$

where T is the total, HT is heat transfer and FF denotes the friction factor.

$$(\dot{S}_{\text{gen}})_{\text{HT}} = \dot{m}C_{pc}\ln\left(T_{c,o}/T_{c,i}\right) + \dot{m}C_{ph}\ln\left(T_{h,o}/T_{h,i}\right) \tag{20.9}$$

and

$$(\dot{S}_{\text{gen}})_{\text{FF}} = \left[\left(\frac{\Delta P_c}{P_c}\right)\dot{m}_c \frac{\ln\left(T_{c,o}/T_{c,i}\right)}{T_{c,o} - T_{c,i}}\right] + \left[\left(\frac{\Delta P_h}{P_h}\right)\dot{m}_h \frac{\ln\left(T_{h,o}/T_{h,i}\right)}{T_{h,o} - T_{h,i}}\right] \tag{20.10}$$

The entropy generation in each regenerative HE component is calculated over the course of the fluid flow. We can regulate the net entropy generation for a given period by adding together all of the entropy generation that has been calculated so far. The irreversibility (entropy generation) idea based on the second law of thermodynamics may be used in HE analysis. The amount of entropy produced by these irreversible processes can be used to gauge the HE's efficiency. The efficiency of the HE improves when entropy is reduced. One may determine the extent to which a system's HE, for example, contributes to the destruction of the system's useful energy (exergy) by measuring the entropy generation in that component. There is little doubt that energy is lost as heat moves through a space. Heat transfer is one of the many factors that cause entropy to develop in thermal systems. The main mechanisms that create entropy in closed systems include heat conduction, chemical reaction, mass transfer, and fluid friction. Entropy rises as frictional irreversibility rises, which is a well-known fact.

Entropy production rate is a measurement of the irreversibility that occurs during a specific process. To calculate the overall generation of entropy in a thermofluidic system at the system level, the general equation for a control volume medium is presented as follows. [26]:

$$\frac{ds}{dt} = \sum \dot{m}s_{\text{in}} - \sum \dot{m}s_{\text{out}} + \sum \frac{\dot{Q}}{T} + \dot{S}_{\text{gen,tot}} \tag{20.11}$$

This equation 20.11 provides the real-time entropy generation rate of a process, as may be seen on the left. The first and second terms on the RHS show the total entropy flow into and out of the system, respectively, whereas the third term displays the rate at which entropy is generated as a result of internal heat creation. Under simplifying assumptions like steady state and no heat generating rate in the system, the difference between entropy flow out of and into the system is the overall entropy generation rate. Equation (20.11) reduces to equation (20.12) as follows if the process under examination is a microchannel using air and R134a as the refrigerant:

$$\dot{S}_{\text{gen,tot}} = \sum_{\text{segment}} \left[\dot{m}_a \left(s_{a,\text{out}} - s_{a,\text{in}} \right) + \dot{m}_r \left(s_{r,\text{out}} - s_{r,\text{in}} \right) \right] \tag{20.12}$$

Using partial differential equations, the entropy generation rate of the HE is specified, and from this, the entropy at the given place in the system is calculated. Consider, for instance, a straightforward 2-d microchannel in which a fluid is moving while absorbing heat from the walls. The fluid's partial differential equation for transferring energy is expressed as follows in equation 20.13.

$$\rho C_p u \frac{\partial T}{\partial x} = k \frac{\partial^2 T}{\partial y^2} + \mu \left(\frac{du}{dy} \right)^2 \tag{20.13}$$

The following equation is the entropy balancing equation for a control volume and defines the entropy generation rate $\left(\dot{S}_{\text{gen}}\right)$ which is related to process losses and it is depicted in equation 20.14 [27].

$$\dot{S}_{\text{gen}} = \sum \dot{m}_e s_e - \sum \dot{m}_i s_i + \sum \left(\frac{\dot{Q}}{T} \right)_e - \sum \left(\frac{\dot{Q}}{T} \right)_i + \frac{dS_{cv}}{dt} \tag{20.14}$$

where T is the temperature at which heat fluxes cross the process boundary, S is the entropy, $\dot{S}_{\text{gen}}$ is the rate at which entropy is generated, and S is the total amount of entropy. The entropy balance equation can be expressed as follows (equation 20.15) under steady-state conditions:

$$\sum \dot{m}_i s_i + \sum \left(\frac{\dot{Q}}{T} \right)_i + \dot{S}_{\text{gen}} = \sum \dot{m}_e s_e + \sum \left(\frac{\dot{Q}}{T} \right)_e \tag{20.15}$$

Remember that in real life, the rate of entropy generation that results from internal irreversibility is always larger than entropy generation.

Maximization of power production resembles the minimization of the entropy production rate [28].

20.4.2 Entropy Generation in HE

Numerous researchers have worked in evaluating the thermal optimization, of various HE. Recently this same interest has been given to analyzing the entropy generation of HEs. Studies show that the main reason and the source for the entropy generation in a system that is closed is due to chemical reaction, friction factor, heat and mass transfer, and pressure loss. Also, the generation of entropy is the function of frictional irreversibility. In particular, Mohsen et al. analyzed [29] the thermal optimization and entropy generation of shell and tube HE equipped with various models of cross-sectional baffles. They have considered the volume fraction varies from 0 to 2%. Since the HE contains baffles, the flow field turned turbulent, and therefore the Reynolds number (it is formulated in equation 20.16) is varied from 25,000 to 45,000.

Reynolds number:

$$Re = \frac{\rho UL}{\mu} \tag{20.16}$$

where in the above equation, ρ – density, U –velocity, L - characteristic length of the system. μ - dynamic viscosity.

They have used copper/aluminum oxide hybrid nanofluids as a coolant. The tube wall is maintained constant temperature 400 K. With the mathematical model that is explained in the previous section, the entropy generation is analyzed. This work mainly compared the performance of shell and tube HE.

With these two different types of baffles, the entropy generation is evaluated.

Total entropy generation is defined as follows mathematically:

$$\dot{S}_{\text{gen}} = \dot{S}_{\text{gen},T} + \dot{S}_{\text{gen},f} = \frac{\dot{Q}^2}{\text{Nu}\pi k T_{\text{in}} T_{\text{out}} L} + \frac{8 f \dot{m}^3 L}{\rho^2 \pi^2 D_h^5 (T_{\text{out}} - T_{\text{in}})} \ln \frac{T_{\text{out}}}{T_{\text{in}}} \tag{20.17}$$

It is said that the variance in the entropy is caused because of the variations in the flow speed of the fluid flow. In addition, the variations in the volume fraction of the nanofluids highly influence the entropy generation of the fluid.

Another interesting result is to evaluate the entropy generation in helical HE with various rib profiles using Al_2O_3-water nanofluid. The impact of ribs and coil revolutions is analyzed on TP and entropy generation of helical HE [30]. Different types of ribs and coils revolutions are used in this study and the different streamlines of the fluid flow for various ribs are found. It is shown how the fluid behavior changes with the changes in the rib configurations.

The variations in the entropy generations for these different rib coil revolutions are explained. As like the other parameters such as friction factor, entropy generation also increases with the variations in the coil revolutions and it is shown in Figure 20.5. This is due to flow variations in the flow field. The flow variations induce chaotic motion (the turbulent motion) and subsequently energy of the HE varies and these changes are shown in the entropy generations as well. In conclusion, the lower rib size reduces the entropy generation and the higher rib size increases the entropy generation.

Jatau et al [31] have discussed how to minimize the generation of entropy in a U-bend tube heat-exchanging system. The uniform structured mesh is considered.

With the usage of this U-bend tube HE obviously the TP of HE will be higher but the problem is with the higher need in pressure drop. They have aimed to have an optimal design for the U-bend tube HE by minimizing the entropy production. Therefore for this purpose, the U-bend tube hear exchanger's length and diameter are varied. By varying these lengths and diameters, entropy generation is calculated, and it is shown in Figure 20.6(i) and (ii), respectively.

The HE's entropy generation varies depending on the mass flow rate. From 100 kg/m^2/s to 500 kg/m^2/s, the flow rate can be changed. The HE's entropy generation rises as the mass flow rate does as well. It should be observed that when length and diameter increased, the formation of entropy began to diminish. This is brought about by a decrease in pressure drop and an improvement in heat transfer efficiency. In a chamber with a wavy surface filled with nanofluid, In the presence of viscous

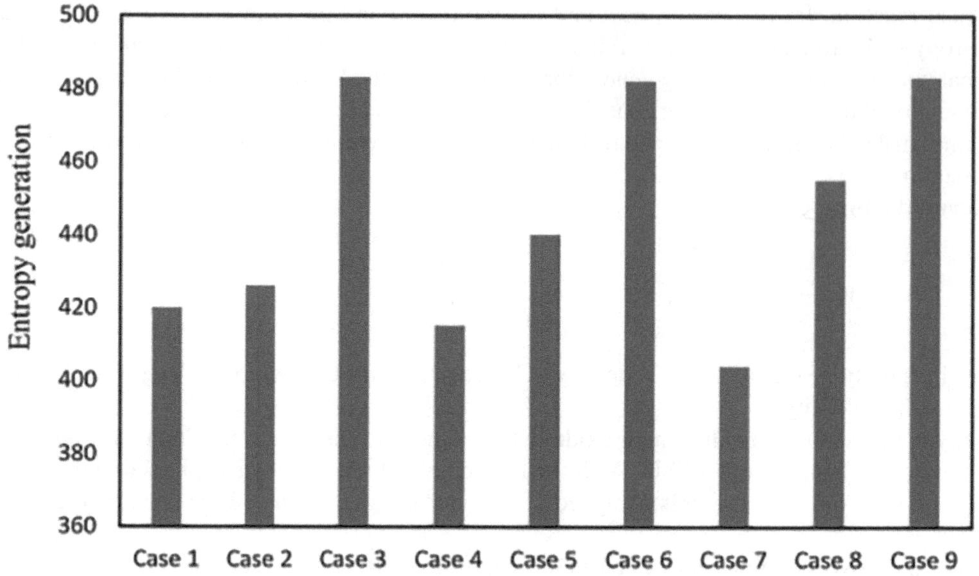

FIGURE 20.5 Entropy generation of helical heat exchanger with the various ribs and revolutions. Adapted with permission [30]. Copyright (2022), Elsevier.

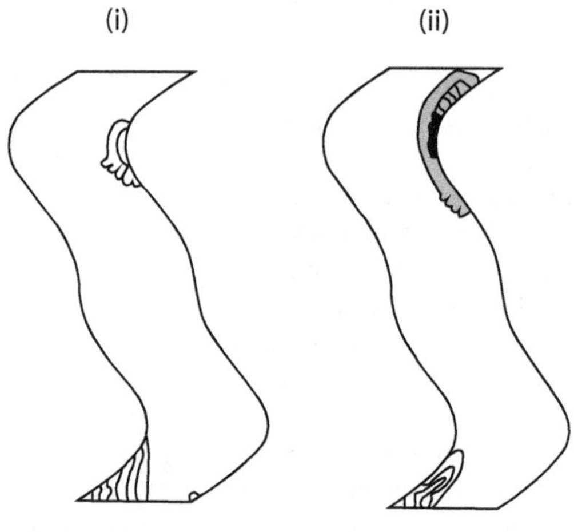

FIGURE 20.6 Entropy generation in terms of Rayleigh number (a) Ra = 1* 10^2 (b) Ra = 1* 10^6. Adapted with permission [32]. Copyright (2016), Elsevier.

dissipation, Cho [32] studied the entropy generation and natural convection heat transmission. The local entropy generation is caused by the Rayleigh numbers on the wavy-surface cavity at the values 1* 10^2 and 1* 10^6, and it is shown in Figure 20.6. The relationship between buoyancy and viscosity forces multiplied by the proportion of momentum to thermal diffusivities is described by the Rayleigh number (Ra), a dimensionless number. Conduction regulates heat flow within the cavity in this study because the thermal-induced buoyant impact is insufficient at low Rayleigh numbers. (Ra =1* 10^2). On the other hand, the high Rayleigh number (Ra = 1* 10^6) has a significant impact on the wavy surface cavity's thermally generated buoyancy. Consequently, the convection effect predominates in heat transfer. For both the lower and higher values of the Rayleigh number, the isotherms are more closely concentrated in the cavity's left-lower crest area and right-upper trough area. As a result, these particular regions of the cavity experience a higher temperature gradient.

As a result, the development of local entropy is increased by the irreversibility of heat transport. This research also shows that higher fluid friction irreversibility and higher magnetic field irreversibility are implied by higher Eckert numbers.

With the influences of wavy wall geometry and porous media, Cho [33] observed the interpretation of heat transfer by convection and the amount of entropy production of nanofluid. Figure 20.7 shows how the distribution ratio (χ) and overall entropy generation vary for various Rayleigh numbers (Ra). Here, a higher value of χ distributes the greater entropy generation, which is denoted by S_t. Hence, the entropy generation enhances an increasing irreversibility distribution ration. Furthermore, this study demonstrates that the Bejan number declined due to the augmented fluid friction irreversibility driven by a stronger convective effect. The thermal optimization of a counter-flow HE for the turbulent flow of the nanofluid was examined by Bahmani et al. [34]. The summary of the results shows that when NP concentration or Reynolds number improves, the Nu and thermal optimization is increased. Rabbani and Languri showed an exergy analysis on nanofluids in a shell and tube HE [35]. The study's findings demonstrated that, for each velocity of the fluid flow situations, utilizing graphene oxide nanofluids as the hot fluid minimizes exergy loss in the HE. Pourhoseini et al. [36] reported the effects of utilizing a nanofluid on the heat transfer coefficient of a plate HE. The results demonstrated that the overall function of HE is improved with increasing nanofluid fraction or mass flow rate. Similar to this, many researchers have created NP suspended in a base fluid for use in HEs [37–41].

Afridi et al. [42] explored the entropy generation rate within the boundary layer flow through a curved stretched surface in the existence of frictional heating on entropy generation and viscous dissipation, and curvature parameters. To find the higher entropy generation, two cases are considered, namely Al_2O_3/water as nanofluid and Cu–Al_2O_3/water as hybrid nanofluid. This approach shows that the single-phase nanoparticle (Cu) suspended in the base fluid generates less entropy than the two-phase (Cu–Al_2O_3) nanoparticles. Finally, in comparison with the case of flow across a curved surface, the flow through a surface case implies low entropy generation. This is one of the markable points that is noticed here.

Sohail et al. [43] attempted the entropy-optimized flow of three-phase nanoparticles dispersed in a base fluid (TiO_2+Fe2O3+SiO_2/WEG) over an exponentially stretching wall. In this study, the findings state that the increasing values of the radiation parameter boost the entropy generation.

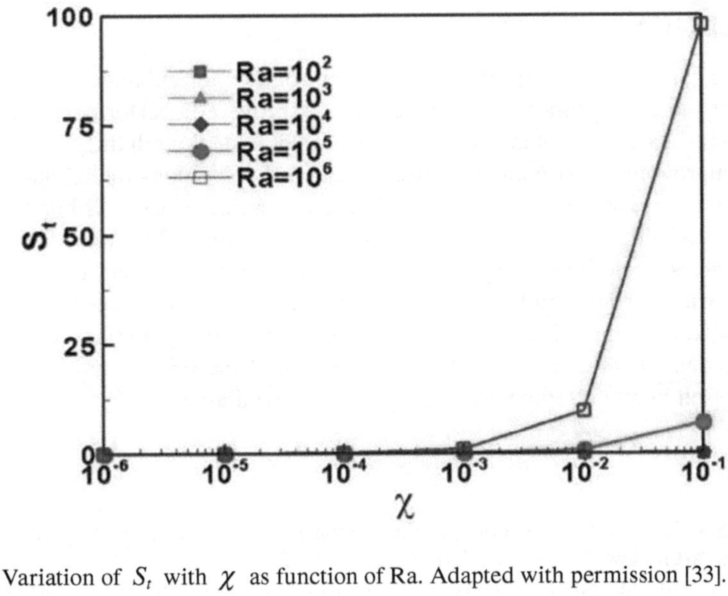

FIGURE 20.7 Variation of S_t with χ as function of Ra. Adapted with permission [33]. Copyright (2020), Elsevier.

Physically, the higher amount of radiation enriches the transport rate of thermal in the flow area, which improves disorders. Also, the markable results found that the entropy rate and thermal profile through radiation have an identical trend impact. Sahoo [44] highlighted the thermal and dynamic analysis of an automobile radiator by means of an emerging water-based coolant made of NP of various shapes, including cylinder-shaped, blade-shaped, and platelet-shaped ternary hybrid nanofluid (CNT+Al_2O_3+Gr). An irreversibility and entropy generation of the system steadily become higher which is closer to 12.24% and 19.50% for the variations of Reynolds number and volume fraction nanoparticles of ternary hybrid nanofluid from 1 to 3%. The thermohydraulic performance of these new CuO/MgO/TiO_2 ternary hybrid nanofluids with entropy generation in micro/mini channel heat sinks was described by Souby et al [45]. At Re=200, the CuO/MgO/TiO2-water fluid showed the largest reduction in thermal resistance. It is primarily caused by enhanced unique thermophysical characteristics of ternary hybrid nanofluid. The adoption of ternary/binary hybrid nanofluids as coolants for cooling electronics applications may be encouraged by this practical drop in thermal resistance.

Entropy scrutiny is an elaborate technique designed to explicitly demonstrate the 2nd law of thermodynamics and evaluate the irreversibility of thermal optimization. Research on the entropy evaluation of a component or a system of industry productions has been documented, and the findings are wonderful. However, the process from one area to another is a typical and inevitable transitory phenomenon for heaters. Bejan [46] outlined entropy generation for HEs and implemented it to manage the volume of an open system. According to Bejan, entropy generation less than zero indicates more acute quality, while entropy generation equal to zero signifies an excellent quality. When investigating the entropy generation of a HE in an experimental setup, Cheng [47] looked researched the efficiency of common heat transfer methods and HEs– from a thermodynamic angle. In this investigation, several kinds of HEs were looked at in terms of minimizing entropy generation, and it was shown that entropy resistance minimization is a different approach for evaluating HEs. The dynamical responses of the heater throughout a period of transient operations were examined by Wang et al. [48]. Additionally, they discussed the rate of real-time entropy generation brought on by heat conduction in metals as well as heat transfer between the work medium and metal surfaces. The findings show that the entire quantity of heat transfer in the heater is disabled by the increase in heat resistance between the working fluid and metal surface.

20.5 CONCLUSION

This book chapter examines several HE forms and approaches for enhancing TP. This chapter has made it clear that there are numerous approaches to improve HE performance, and that there is active research being done on the subject all over the world. Even though there are many approaches to recommend improving performance, current research has demonstrated that one of the most crucial ways is to minimize entropy formation. In the upcoming years, it is hoped that the perfect fabrication of HEs will depend on minimizing entropy formation. The HE's TP is influenced by a number of variables, including the system's shape, the coolants utilized, outside influences, and more. Similar to what was learned from this book chapter, it is known that the pressure drop, friction factor, extended fins, and other factors affect how much entropy a HE produces. Additionally, numerous studies demonstrate that the entropy generation of the HE is significantly influenced by fluid velocity, Nusselt number, temperature, length, and tube diameter.

REFERENCES

[1] W. Ajeeb, S.M.S. Murshed, Nanofluids in compact heat exchangers for thermal applications: A State-of-the-art review, *Therm. Sci. Eng. Prog*, 30, 101276, 2022.

[2] S. Gungor, Experimental comparison on energy consumption and heat transfer performance of corrugated H-type and L-type brazed plate heat exchangers. *Int. Commun. Heat Mass Transf.*, 144, 106763, 2023.

[3] H. Maddah, R. Aghayari, M. Farokhi, S. Jahanizadeh, K. Ashtary, Effect of twisted-tape turbulators and nanofluid on heat transfer in a double pipe heat exchanger, *J. Eng.*, 2014 1–9, 2014. Article ID 920970. https://doi.org/10.1155/2014/920970.

[4] B. M'hamed, N. A. C Sidik, M.N.A.W.M. Yazid, R. Mamat, G. Najafi, G.H.R. Kefayati, A review on why researchers apply external magnetic field on nanofluids, *Int. Commun. Heat Mass Transf.*, 78, 60–67, 2016.

[5] Md. Shariful Alam, S. Sultana Keya, U. Salma, S. M. Chapal Hossain, Md. Masum Billah, Convective heat transfer enhancement in a quarter-circular enclosure utilizing nanofluids under the influence of periodic magnetic field, *Int. J. Thermofluids*, 16, 100250, 2022.

[6] S. Anitha, T. Thomas, V. Parthiban, M. Pichumani, What dominates heat transfer performance of hybrid nanofluid in single pass shell and tube heat exchanger? *Adv. Powder. Technol.*, 30(12), 3107–3117, 2019.

[7] J. Y. San, C. H. Hsu, S. H. Chen, Heat transfer characteristics of a helical heat exchanger, *Appl. Therm. Eng.*, 39, 114–120, 2012.

[8] S. Anitha, M. Pichumani, Numerical analysis on heat transfer performance of industrial double-tube heat exchanger using CNT: Newtonian/non-Newtonian hybrid nanofluids, *J. Therm. Anal. Calorim.*, 147(17), 9603–9624, 2022.

[9] N. Jayranaiwachira, P. Promvonge, C. Thianpong, Pitak Promthaisong, Sompol Skullong, Effect of louvered curved-baffles on thermohydraulic performance in heat exchanger tube, *Case Stud. Therm. Eng.*, 42, 102717, 2023.

[10] P. A. Sanders, K.A. Thole, Effects of winglets to augment tube wall heat transfer in louvered fin heat exchangers, *Int. J. Heat Mass Transf.*, 49(21-22), 4058–4069, 2006.

[11] A. B.Çolak, D. Akgul, H. Mercan, A. S. Dalkılıç, S. Wongwises, Estimation of heat transfer parameters of shell and helically coiled tube heat exchangers by machine learning, *Case Stud. Therm. Eng.*, 42, 102713, 2023.

[12] F. Mabood, T.A. Yusuf, W.A. Khan, Cu-Al2O3-H2O hybrid nanofluid flow with melting heat transfer, irreversibility analysis and nonlinear thermal radiation, *J. Therm. Anal. Calorim.*, 143(2), 973–984, 2020.

[13] S. Jana, A. Salehi-Khojin, W.-H. Zhong, Enhancement of fluid thermal conductivity by the addition of single and hybrid nano-additives, *Thermochim. Acta*, 462(1-2), 45–55, 2007.

[14] S. Suresh, K.P. Venkitaraj, P. Selvakumar, Synthesis, Characterisation of Al2O3-Cu Nano Composite Powder and Water Based Nanofluids. *Adv Mat Res.*, 328-330, 1560–1567, 2011.

[15] A. Karimi, M. Afrand, Numerical study on thermal performance of an air-cooled heat exchanger: Effects of hybrid nanofluid, pipe arrangement and cross section, *Energy Convers. Manag.*, 164, 615–628, 2018.

[16] A. Bhattad, J. Sarkar, P. Ghosh, Discrete phase numerical model and experimental study of hybrid nanofluid heat transfer and pressure drop in plate heat exchanger, *Int. Commun. Heat Mass Transf.*, 91, 262–273, 2018.

[17] R.R. Sahoo, V. Kumar, Development of a new correlation to determine the viscosity of ternary hybrid nanofluid, *Int. Commun. Heat Mass Transf.*, 111, 2020.

[18] S. Kashyap, J. Sarkar, and A. Kumar, Performance enhancement of regenerative evaporative cooler by surface alterations and using ternary hybrid nanofluids. *Energy*, 225, 2021.

[19] A.S. Adnan, G. Kamel, Zehba A Raizah, E. Tageldin, Thermal efficiency in hybrid (Al2O3-CuO/H2O) and tri-hybrid (Al2O3-CuO-Cu/H2O) nanofluids between converging/diverging channel with viscous dissipation function: Numerical analysis, *Front. Chem.*, 10, 960369, 2022.

[20] R. R. Sahoo, Thermo-hydraulic characteristics of radiator with various shape nanoparticle-based ternary hybrid nanofluid, *Powder Technol.*, 370, 19–28, 2020.

[21] A.I. Ramadhan, W.H. Azmi, R. Mamat, Experimental investigation of thermo-physical properties of tri-hybrid nanoparticles in water-ethylene glycol mixture. *Walailak J. Sci. & Tech.*, 18(8), 9335, 2021.

[22] T. Elnaqeeb, I.L. Animasaun, N.A. Shah, Ternary-hybrid nanofluids: significance of suction and dual-stretching on three-dimensional flow of water conveying nanoparticles with various shapes and densities. *Zeitschrift für Naturforschung A*, 76(3), 231–243, 2021.

[23] S. Anitha, K. Loganathan, M. Pichumani, Approaches for modelling of industrial energy systems: correlation of heat transfer characteristics between magnetohydrodynamics hybrid nanofluids and performance analysis of industrial length-scale heat exchanger, *J. Therm. Anal. Calorim.*, 144(5), 1783–1798, 2020.

[24] M. Arif, P. Kumam, W. Kumam, Z. Mostafa, Heat transfer analysis of radiator using different shaped nanoparticles water-based ternary hybrid nanofluid with applications: A fractional model, *Case Stud. Therm. Eng.*, 31, 101837, 2022.

[25] S. Anitha, M. Shasthick, B. Senthilkumar, Sheikholeslami, P. Chandramohan, M. Pichumani, How the estimation of entropy generation and exergy loss of hybrid nanofluids governs the thermal performance of heat exchanger? *Int. J. Fluid Mechanics Res.*, 1, 2022. 1–24.

[26] M. Ozturk, Ibrahim Dincer, Geothermal Energy Conversion, 472–542.

[27] M. Ozturk, I. Dincer, 4.11 Geothermal Energy Conversion, In Comprehensive Energy Systems, Elsevier Inc., Cambridge, MA, 474–544, 2018. eBook ISBN: 978012820896.

[28] S Bucsa, D Dima, A Serban, M-F Stefanescu, V Popa, A Dobrovicescu, Heat exchanger design based on minimum entropy generation, *IOP Conf. Series: Mater. Sci. Eng.,* 595 (2019) 012020.

[29] M. Tavakoli, M.R. Soufivand, Performance evaluation criteria and entropy generation of hybrid nanofluid in a shell-and-tube heat exchanger with two different types of cross-sectional baffles, *Eng. Anal. Bound. Elem.*, 150, 272–284, 2023.

[30] M.J. Hasan, A.A. Bhuiyan, Investigation of thermal performance and entropy generation in a helical heat exchanger with multiple rib profiles using Al2O3-water nanofluid, *Case Stud. Therm. Eng.*, 40, 102514, 2022.

[31] T. Jatau, T. Bello-Ochende, Minimization of entropy generation in U-bend tube heat exchanger during flow boiling of R134a, *Int J Therm Sci.*, 185, 108032, 2023.

[32] C.C. Cho, Influence of magnetic field on natural convection and entropy generation in cu-water nanofluid-filled cavity with wavy surfaces, *Int. J. heat mass transf.*, 101, 637–647, 2016.

[33] C.C. Cho, Effects of porous medium and wavy surface on heat transfer and entropy generation of Cu-water nanofluid natural convection in square cavity containing partially-heated surface, *Int. Commun. Heat Mass Transf.*, 119, 104925, 2020.

[34] M.H. Bahmani, G.H. Sheikhzadeh, M. Zarringhalam, O.A. Akbari, A.A.A.A. Alrashed, G.H. Ahmadi Sheikh Shabani, M. Goodarzi, Investigation of turbulent heat transfer and nanofluid flow in a double pipe heat exchanger, *Adv Powder Technol*, 29,273–82, 2018.

[35] M. Rabbani Esfahani, E. Mohseni Languri, Exergy analysis of a shell-andtube heat exchanger using graphene oxide nanofluids, *Exp Therm Fluid Sci.*, 83, 100–106, 2017.

[36] S.H. Pourhoseini, N. Naghizadeh, H. Hoseinzadeh, Effect of silver-water nanofluid on heat transfer performance of a plate heat exchanger: An experimental and theoretical study, *Powder Technol*, 332, 279–286, 2018.

[37] M.F. Al-Rjoub, A.K. Roy, S. Ganguli, R.K. Banerjee, Enhanced heat transfer in a micro-scale heat exchanger using nano-particle laden electro-osmotic flow, *Int. Commun. Heat Mass Transf.*, 68, 228–235, 2015.

[38] R. Ellahi, M. Hassan, A. Zeeshan, Shape effects of nanosize particles in Cu-H2O nanofuid on entropy generation, *Int. J. Heat Mass Transf.*, 8, 1449–1456, 2015.

[39] N. Kumar, S.S. Sonawane, Experimental study of Fe2O3/water and Fe2O3/ethylene glycol nanofluid heat transfer enhancement in a shell and tube heat exchanger, *Int. Commun. Heat Mass Transf.*, 78, 277–284, 2016.

[40] V. Kumar, A.K. Tiwari, S.K. Ghosh, Effect of variable spacing on performance of plate heat exchanger using nanofluids, *Energy*, 114, 1107–1119, 2016.

[41] A.E. Kabeel, T. Abou El Maaty, Y. El Samadony, The effect of using nano-particles on corrugated plate heat exchanger performance, *Appl. Therm. Eng.*, 52(1), 221–229, 2013.

[42] M.I. Afridi, T.A Alkanhal, M. Qasim, Iskander Tlili, Entropy generation in Cu-Al2O3-H2O hybrid nanofluid flow over a curved surface with thermal dissipation. *Entropy*, 21(10), 941, 2019.

[43] S. A. Khan, T. Hayat, A. Alsaedi, Thermal conductivity performance for ternary hybrid nanomaterial subject to entropy generation, *Energy Reports*, 8, 9997–10005, 2022.

[44] R. R. Sahoo, Heat transfer and second law characteristics of radiator with dissimilar shape nanoparticle-based ternary hybrid nanofluid, *J. Therm. Anal. Calorim.*, 22, 1–3, 2020.

[45] M.M. Souby, M. Bargal, Y.P. Wang, Thermohydraulic performance improvement and entropy generation characteristics of a microchannel heat sink cooled with new hybrid nanofluids containing ternary/binary hybrid nanocomposites, *Energy Sci. Eng.*, 9(12), 2493–2513, 2021.

[46] A. Bejan, *Entropy Generation Through Heat and Fluid Flow*, Wiley, New York, 103, 1982.

[47] X. Cheng, Entropy resistance minimization: An alternative method for heat exchanger analyses, *Energy*, 58, 672–678, 2013.

[48] C. Wang, M. Liu, Y. Zhao, Y. Qiao, J. Yan, Entropy generation analysis on a heat exchanger with different design and operation factors during transient processes, *Energy*, 330–342, 2018.

21 Tribological Properties of High Entropy Materials

Qing Zhou, Mingda Xie, Shuo Li, and Haifeng Wang

21.1 INTRODUCTION

The design and development of advanced alloy surfaces may be crucial to control and decrease friction and wear in drivetrains for safety-critical applications since engineering equipment is utilized in hostile environments with heavy loads, cyclic stresses, and wide temperature fluctuations. In various industries, an enormous amount of research effort has been devoted to developing anti-friction and wear-resistant materials for applications in various environmental conditions [1,2]. High-entropy materials (HEMs), first proposed by Yeh and Cantor et al. independently [3,4], are a relatively new member of the structural material family that contain multi-principal elements. Superior properties such as high hardness/strength, excellent ductility, good wear, and corrosion resistance [5,6], resulting from this innovative alloy design paradigm, make HEMs one of the research highlights. The combination of their superior properties over traditional alloys, originating from specific characteristics including sluggish diffusion, lattice torsion, and the cocktail effect [7], endows HEMs with vast potential to serve as wear and friction-reducing materials.

By appropriately controlling the constituent elements and thermomechanical treatments, the microstructure and associated properties of HEMs can be fundamentally altered, producing a variety of exceptional properties [8]. In general, strength and ductility are prioritized in practical engineering applications, so their related mechanical properties such as tribological properties are extensively researched. The wear process, involving shear deformation, adhesion, oxidation, fracture, and other mechanisms [9], reduces mechanical performance and degrades the surface integrity of component parts. Therefore, the development of alloys with good wear resistance has significant social and economic implications. In tribological systems, the coefficient of friction (COFs) and wear rate are frequently employed to assess the wear resistance of materials. At the same time, the microstructure traits of the worn surface and subsurface are taken into consideration while analyzing the wear mechanism. Numerous factors are reported to influence the tribological features of HEMs. In this chapter, by focusing on the microstructure evolution of the worn surface and methodically revealing the wear mechanism, the tribological properties of HEMs and their composites are introduced.

21.2 TRIBOLOGICAL PROPERTY OF BULK HEMS

21.2.1 Tribological Property at Room Temperature

In general engineering applications, friction contact between mechanical components at room temperature is ubiquitous. The tendency to form simple phases is an important feature of HEMs [5], and one of the classical FCC HEM systems is CoCrFeMnNi, which is referred to as Cantor alloy. Well known for their high tensile ductility and fracture toughness at room temperature [10,11], the tribological properties of Canter alloy and its derivatives have been extensively explored.

Dollmann et al. [12] reported the deformation behavior of CoCrFeMnNi in a dry, reciprocating tribological contact under a mild normal load. After a single stroke, the subsurface material can be divided into three separate layers, which are characterized by nanocrystalline grains, twins, and

DOI: 10.1201/9781003391388-21

bands of localized dislocation motion. In addition, oxide-rich layers were found after several sliding cycles [7], intermixing with the nanocrystalline layer. Fragmentation of twins and dislocation rearrangement leads to the nanocrystalline layer underneath the worn surface, and all these features form the gradient deformation microstructure in HEM.

Zhang et al. [11] systematically investigated the room-temperature dry sliding tribological behavior of CoCrFeMnNi HEM against silicon nitride (Si_3N_4). Figure 21.1a–c shows the typical worn surface morphology at 5 N, 15 N, and 20 N. The delamination caused by plastic deformation with minor furrow, surface scratches, and tribofilm was seen on the worn surface, and some microcracks appeared on the tribo-layer when the load increased to 20 N. During the friction process, the oxide glazing layer was created by friction heat and reactive oxidation, which has a particular anti-wear function. It is a distinctive feature of diffusion and oxidative wear for HEM.

Figure 21.1(d) shows the cross-section morphologies of the CoCrFeMnNi HEM along the sliding direction after sliding against a Si_3N_4 ceramic ball under 15 N. After wear tests, the subsurface deformation layer can be clearly identified from cross-section morphologies, indicating that the near-surface has undergone various degrees of severe plastic deformation. The EDS mapping in Figure 21.1(d) verifies the presence of a tribo-layer, which acts as a lubricant in the friction process. The lubricating oxides reduce the contact area during the friction process, which to a certain extent lowers the friction coefficient and wear rate [12,13]. The soft Cantor alloy base can be strengthened by the dynamic microstructural development that occurs during tribological processes, with the result that wear resistance will improve.

Soft single-phase FCC HEMs have the disadvantages of low hardness and relatively inferior wear resistance. To overcome these shortcomings and improve their tribological performance, strategies to modify the microstructures and constituent elements of Cantor alloy-based systems have been proposed. Zhang et al. [14] investigated the tribological behavior of CoCrFeNiV and parallel experiments were conducted with CoCrFeNiMn. The addition of the V element precipitated the hard sigma (σ) phase, enduing CoCrFeNiV outstanding hardness (975±11 HV) and wear resistance as compared to CoCrFeNiMn with a single FCC structure. Liu et al. [15] successfully

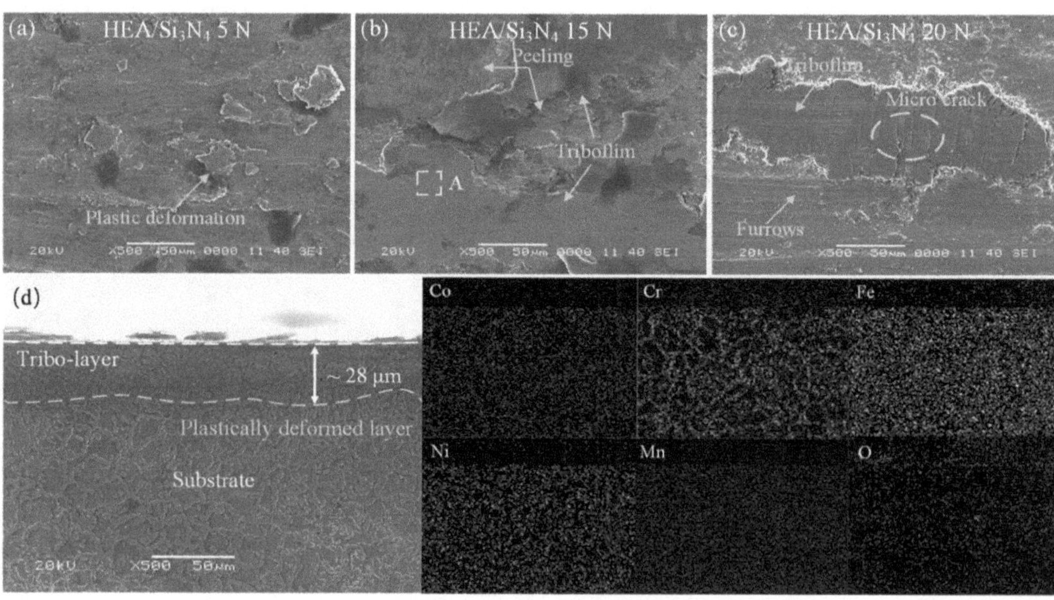

FIGURE 21.1 Worn surface morphologies of the CoCrFeNiMn HEM sliding against (a–c) Si3N4; Cross-section morphologies of the (d) CoCrFeNiMn HEM sliding against Si3N4 under 15 N. Adapted with permission [11]. Copyright (2022) Elsevier.

fabricated $CoCrNiNb_x$ alloys by using casting and injection technique. The CoCrNi FCC matrix develops a lamellar eutectic microstructure after Nb addition, which introduces the Laves phase. The lowest COF and wear rate is obtained for $CoCrNiNb_{0.385}$ with full eutectic structure. The co-deformation between FCC and Laves phases in eutectic microstructure reduces the local plastic mismatch between the two constituent phases, therefore postponing the crack/fracture initiation on worn surfaces.

In addition, it is often adopted to add Al [16], Ti [17] to the HEMs system to induce the precipitation of hard reinforcement phase. The load-bearing capacity of materials can be increased, and wear resistance improved as a result of hard precipitates in a soft matrix which can prevent plastic flow and dislocation motion. In general, BCC HEMs are stronger and harder than FCC HEMs. Wu et al. [16] investigated the adhesive wear behaviors of $Al_xCoCrCuFeNi$ alloys with different Al contents prepared by arc melting. By adjusting Al content, the microstructure transits from a simple ductile FCC phase to an FCC+BCC dual-phase structure. The volume fraction of the BCC phase and the hardness value both rise with an increase in Al content, which lowers the wear rate. The hard BCC phase helped to resist plastic deformation caused by abrasive wear and thus decreased wear loss.

21.2.2 Tribological Property at Elevated Temperatures

In addition to the aforementioned plastic deformation of the subsurface at room temperature, wear at elevated temperatures is a complicated process that also includes phase change, grain growth, thermal softening, and surface oxidation [12]. HEMs are widely considered good material candidates for high-temperature applications.

Joseph et al. [18] analyzed that even in the stable Cantor alloy, the interaction of the temperature environment and wear-induced deformation can promote the second phase's precipitation. Dynamically precipitated hard σ phase increased wear resistance and decreased plastic plow at moderate temperatures. In addition to subsurface microstructural evolution, which is dominant at room temperature, high-temperature oxidation behavior is a crucial aspect determining the overall wear performance of HEMs. Zhang et al. [14] reported that the wear rate and COF of CoCrFeNiV decreased substantially as temperature increased from 400°C to 800°C, benefiting from oxidation products and σ phase. Additionally, the formation of V_2O_5 can be an active lubricious component during high-temperature friction and wear process, leading to a lower wear rate value than CoCrFeNiMn at 800°C.

Between room temperature and 300°C, Pan et al. [19] investigated the wear response of fine-grained CoCrNi alloy compared with Inconel alloy 718 equivalents, paying particular attention to the change in the wear mechanism and the evolution of the subsurface microstructure caused by sliding. With the increase of temperature, the wear mechanism changes from abrasive wear to adhesive wear as well as oxidative wear, and wear rates monotonically decrease. The ND-SD cross-sectional microstructure (Figure 21.2a1–a4) of worn pins at 150°C shows no glaze layer formation, confirming direct metallic tribo-pair contacts. The improved wear resistance is mainly ascribed to the thermal softening of the mating material and the increased contribution of adhesive wear. However, at 200°C, a glaze layer with a thickness of 5–6 µm (Figure 21.2b1–b4) is formed. The deformation layer formed after the wear at 300°C was deeper than at 200°C. At 200°C and above, the reduced wear rates and COFs are associated with the formation of glaze layer. The CoCrNi alloy also developed a gradient microstructure with a high density of dislocation cells, nanoscale deformation twins, and numerous stacking faults. Therefore, the decreased wear rates and COFs at high temperatures are attributed to this gradient microstructure as well as the presence of a glaze layer. Chuang et al. [17] reported that $Co_{1.5}CrFeNi_{1.5}Ti$ and $Al_{0.2}Co_{1.5}CrFeNi_{1.5}Ti$ alloys have at least two times higher wear resistance than typical wear-resistant steels, even with comparable hardness. The softening rate with temperature is significantly slower than that of traditional alloys, especially at temperatures above 800°C. Since wear-induced frictional heating can cause a local high temperature of hundreds of degrees, the exceptional anti-softening properties of HEMs are often credited with their superior tribological performance.

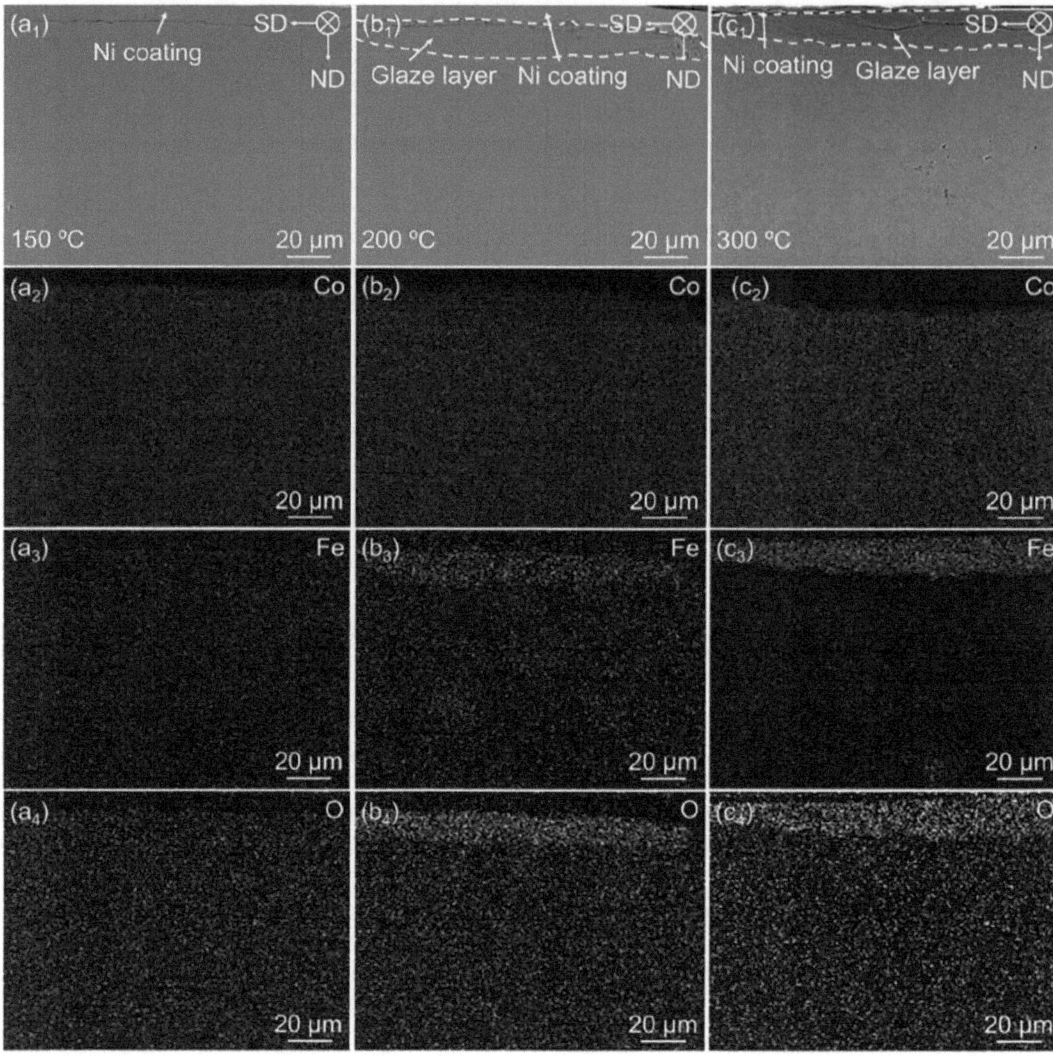

FIGURE 21.2 SEM images and corresponding EDX elemental maps of ND-SD cross-sections of the CoCrNi MEA pins after sliding wear at different temperatures. Adapted with permission [19]. Copyright (2019) Elsevier.

21.3 TRIBOLOGICAL PROPERTIES OF HIGH-ENTROPY COATINGS/FILMS

Several methods have been used in the last ten years to deposit HEM coatings and films, including magnetron sputtering, thermal spraying, laser cladding, and others. Due to their extensive coverage area, less material consumption, and variety of application scenarios, HEM coatings and films have received a lot of attention, which show excellent performance in high-temperature oxidation resistance, corrosion resistance, and superior wear resistance. Therefore, high-entropy coatings have been suggested as a prominent method to alleviate high-temperature tribological problems because of their exceptional combinations of mechanical and thermal capabilities [20,21]. This section provides a comprehensive overview of HEM coatings/films in tribological applications.

21.3.1 Thick Coatings

Generally, HEM coatings usually thicker than 10 μm can be prepared by laser cladding and spraying techniques. Each method has a unique impact on the coatings' characteristics, which might result in coatings with various morphologies and thicknesses. At present, HEM coatings are mainly classified into 3D transition metal coatings and refractory metal coatings, as well as doped with other elements to enhance specific functions. The following factors can be taken into consideration when designing the preliminary material system for the improvement of tribological properties.

Tuning the phase structure through proper composition modification is one effective method for enhancing the tribological properties of HEM coatings. This is expected to increase the coating's strength and yield better anti-plastic deformation ability as well as wear resistance. The most frequently used elements in this aspect are Ti, Al, V, etc., which can make the initial FCC structure in a supersaturated and metastable state, tending to decompose and transform into phases with better deformation resistance. Wang et al. [20] prepared $(CoCrFeMnNi)_{85}Ti_{15}$ HEA coating on Q235 steel substrate via plasma cladding, its microstructure, hardness, and high temperature tribological performance were investigated. Compared with the single FCC CoCrFeMnNi HEM coating, the $(CoCrFeMnNi)_{85}Ti_{15}$ HEM coating, made up of FCC, BCC solid solutions, and an intermetallic sigma phase, improves its hardness six times. For tribological performance, the CoCrFeMnNi coating surface showed a large number of furrows during wear, while $(CoCrFeMnNi)_{85}Ti_{15}$ is dominated by fatigue wear, and formed a layer of lubricating oxide, thus improving the wear resistance. Cui et al. [22] developed the $FeCoCrNiMnAl_x$ coatings on 4Cr5MoSiV steel by laser cladding and systematically investigated the microstructure and properties of the coatings. It was claimed that the addition of Al element promoted phase transition in $FeMnCrNiCoAl_x$ cladding layers from FCC to FCC+BCC. The grain size of $FeCoCrNiMnAl_x$ cladding layer gradually reduced as the Al concentration increased, thereby reinforcing the alloys in a way of *H–P* effect. In sum, Al element significantly improved the high-temperature wear performance of the $FeMnCrNiCoAl_x$ cladding layers.

By incorporating non-metallic elements, ceramic particles are created in the coating to serve as dispersion-strengthening. Yang et al. [23] successfully fabricated $FeCoCrNiMoSi_x$ (x=0.5, 1.0, 1.5) HEM coatings on the Q235 steel substrate by laser cladding. The microstructure of the coating materials mainly consists of Fe-rich FCC and FeMoSi phases, where the Si addition induced severe distortion in crystal lattice. In comparison with the as-cast FeCoCrNiMo HEM, the $FeCoCrNiMoSi_x$ coatings offer much higher microhardness, lower coefficient of frictions, and wear rates. It was shown in Figure 21.3 that Q235 steel exhibits many scratches and large grooves along the sliding direction and small adhesion parts and deformation exist in the interface featured, indicating adhesive and abrasive wear. As shown in Figure 21.3f and g, the HEM coatings with Si additions up to 1.0, a combination of abrasive and adhesive wear is the main wear mechanism. The coating tends to partially break during a friction test when the Si concentration reaches 1.5 (see Figure 21.3h), indicating a delamination wear mechanism, which worsens the wear resistance. Liu et al. [24] prepared $AlCoCrFeNiSi_x$ (x=0, 0.1, 0.2, 0.3, 0.4, 0.5) HEM coatings on the surface of AISI 304 stainless steel via laser cladding technology. The coating was mostly composed of an ordered BCC Al–Ni nanoparticle phase and a disordered BCC Fe–Cr matrix phase. As the Si content raised, the COF and wear rate of the coatings were dramatically lowered, and the main wear mechanism evolved from adhesive wear, abrasive wear and delamination wear to oxidation wear.

Like laser cladding, thermal spraying is also advantageous and practical for creating coatings with a lower degree of dilution and required qualities. The HEMs produced by thermal spray methods have excellent mechanical properties and good microstructural stability. Currently, high-velocity oxygen-fuel spraying (HVOF), cold spraying (CS), and plasma spraying (PS) are the three main thermal spray coating techniques utilized to produce the HEM coatings [21]. Lobel et al. [25] successfully developed $AlCoCrFeNiTi_{0.5}$ HEM coatings by HVOF. The creation of two significant BCC phases is also revealed by phase analysis; the wear behavior was investigated in a wide temperature

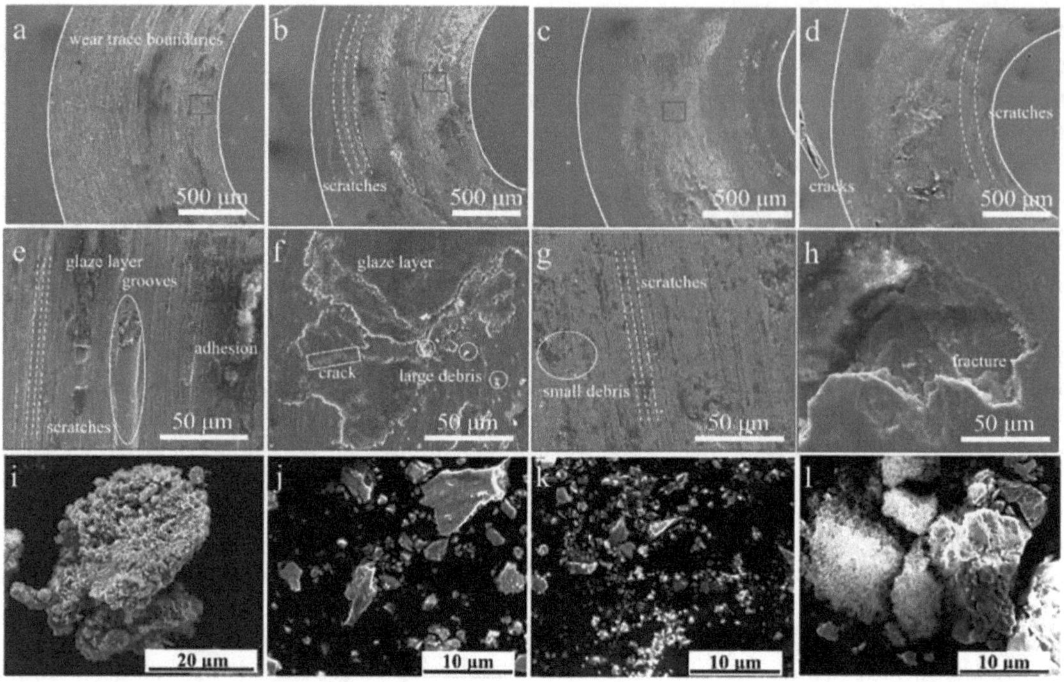

FIGURE 21.3 (a–d) The SEM images of worn surface of Q235 steel, $FeCoCrNiMoSi_{0.5}$, $FeCoCrNiMoSi_{1.0}$ and $FeCoCrNiMoSi_{1.5}$ coatings, respectively, e-h: the amplified characteristic morphologies from the marked red boxes in panels (a–d); (i–l): the morphologies of wear debris in panels (a–d). Adapted with permission from [23]. Copyright (2022) Elsevier.

range under reciprocating conditions. Below 500°C, abrasive wear is the main wear mechanism of the metallic coatings, and the formation of loose oxides was seen. The amount of oxide covering the surface increases as the temperature rises. An oxide layer that is compact forms at high temperatures (over 800°C), and the wear resistance increases significantly after initially declining with the rise in temperature.

In addition, the properties of HEM coatings prepared by thermal spraying can be improved from the aspects of material design and process improvement. $FeCoNiCrSiAl_x$ (x=0.5, 1.0, 1.5) HEM coatings were prepared by atmospheric plasma spray [26]. The BCC phase dominates the microstructure of HEM coatings, which also contains modest amounts of FCC phase, independently of the Al concentration. However, the main phases in heat-treated $FeCoNiCrSiAl_x$ HEMCs are FCC+$Cr_3Ni_5Si_2$, FCC+BCC+$Cr_3Ni_5Si_2$, and BCC, respectively. The microhardness and wear resistance of the as-sprayed coatings are greatly improved after heat treatment. The heat-treated $FeCoNiCrSiAl_{1.0}$ HEMC exhibits the lowest wear rate, which is attributed to the formation of BCC and $Cr_3Ni_5Si_2$ phases.

21.3.2 Thin Films

In comparison to thick coatings, thin films fabricated by vapor deposition techniques, which are frequently used to produce thin films by transforming target materials onto the surface of substrates, often have higher microhardness and favorable frictional characteristics. Recently, it has been reported that physical vapor deposition and chemical vapor deposition techniques, including vacuum arc deposition, reactive magnetron sputtering, and direct current magnetron sputtering, can be used to fabricate high entropy films with desirable surface attributes [27]. In this section, the authors summarize the results of various studies on the tribological properties of high entropy films prepared by magnetron sputtering.

Alvi et al. [28] synthesized the CuMoTaWV film by means of magnetron sputtering method with a single partially spark-plasma-sintered targe, confirming the ability of magnetron sputtering to fabricate nanocrystalline films with small grain sizes and exceptional mechanical properties. The CuMoTaWV refractory high entropy material (RHEM) film showed an average hardness and nanopillar compressive strength of 19±2.3 and 10±0.8 GPa, respectively. The remarkable mechanical properties of the CuMoTaWV RHEM film were ascribed to the nanocrystalline particle size and grain boundary-controlled plastic deformation, which also leads to excellent COF and wear resistance at RT. By annealing at 300°C, the wear behavior and adhesion on a steel substrate of CuMoTaWV film were further improved, suggesting that RHEM films may be useful for wearable and nanopillar applications.

Composition control to improve the tribological properties of thin films is also suitable for magnetron sputtering deposition films. Zhao et al. [29] prepared the $FeCoNiCrMnMo_x$ HEM films by direct current (DC) magnetron co-sputtering. The phase structure eventually shifts from a single FCC phase to a combination of FCC and BCC phases as the Mo content rises, making the films denser. From no Mo to x = 1, The grain size shrinks by 40%, the lattice constant rises, the hardness increases by 41% from 8.5 GPa to 12 GPa, and the COF decreases from 0.5 to 0.3. Fan et al. [30] prepared VAlTiCrMo-based high entropy coatings with BCC structure by magnetron sputtering, and the tribological properties were improved by regulating Mo content. In contrast to the relatively stable VAlTiCr coating, the addition of Mo encouraged the uphill migration and segregation of solid solution metals in the coatings during friction, producing a notably stratified oxide scale at 700°C, which can promote the formation of oxide like V_2O_5, $Al_2(MoO_4)_3$ that significantly reduces the friction coefficient.

21.4 HIGH ENTROPY COMPOSITES

21.4.1 Addition of Hard Phases

21.4.1.1 Bulk HEMs

To improve wear performance, plenty of research on HEM matrix composites, typically consisting of a strengthening phase, has been reported. Among all strengthening phases, ceramic phases, including TiC, SiC, Cr_3C_2 as well as oxides, have attracted tremendous attention in the field of tribological modification [31].

Gu et al. [32] introduced SiC (0.1, 0.5, 1, and 3 vol.%) particles into $Fe_{50}Mn_{30}Co_{10}Cr_{10}$ composites, and the effect of SiC content on microstructure characteristics, mechanical and wear properties of HEM were studied. The outcomes demonstrated that SiC refined the grain sizes and stabilized the γ phase, And the best wear friction ability was achieved by 1 vol.% SiC addition, where the optimized SiC content was crucial for carrying load during friction. To improve the wear resistance of CoCrNi alloy, Cai et al. [33] introduced TiC by arc melting. A large number of needled and blocky TiC particles precipitated in the composites as a result of the dissolution and precipitation process produced by TiC. The yield strength of $CoCrNi/(TiC)_x$ composites increased from 108 to 1371 MPa when compared to the CoCrNi alloy, leading to significantly improved wear resistance. Zhu et al. [34] also reported the excellent wear resistance of FeCoNiCuAl–TiC. With an increase in TiC content, the microhardness of FeCoNiCuAl–TiC composites rises from 467 to 768 HV, and the COF increases first and then decreases, while the wear loss continually decreases, which is mostly ascribed to the reinforcement of the hard TiC phases and the finer microstructures.

Zhou et al. [35] prepared $(FeCoCrNi)_{1-x}(WC)_x$ (x = 3–11 at.%) HEM composites by spark plasma sintering mechanically alloyed powder. The microstructures of HEM composites consist of four phases: FCC matrix phase, WC-type carbide, $M_{23}C_6$-type, and M_7C_3-type carbide. The COF increases first with the content of WC increasing to 7% and then drops. Guo et al. [36]

used Cr_3C_2 particles with varying contents (10 wt.%, 20 wt.%, and 40 wt.%) to strengthen the classical Cantor alloy by spark plasma sintering, and the composites consisting of the FCC and Cr_7C_3 phases were formed. The tiny Cr_7C_3 phase can reinforce the composite by preventing dislocation motion and emission. With the increase of Cr_3C_2 content, the volume proportion of tiny Cr_7C_3 particles, which can reinforce the composite by preventing dislocation motion and emission, rises. The hardness of HEM increases from 181.2 HV to 681.6 HV by increasing the content of Cr_3C_2, and the combination of optimized mechanical and tribological properties are found in CoCrFeNiMn-10% Cr_3C_2 composite (Figure 21.4). The grooves and furrows disperse on the worn surface of CoCrFeNiMn, which imply the abrasive wear during testing at 200°C (Figure 21.4a and b). There are also craters and patches, suggesting that adhesive wear also takes place. When the temperature reaches 400°C, fine wear debris begins to build up on the worn surface, indicating that abrasive wear is the primary wear process of CoCrFeNiMn (Figure 21.4c and d). By contrast, the worn surfaces of CoCrFeNiMn-10% Cr_3C_2 composite are smoother at 200°C and 400°C (Figure 21.4e–h). The wear mechanism of the composite is abrasive wear at 200°C and 400°C, as seen by the apparent furrows and strewn wear debris. Therefore, adding hard particles, which can prevent internal cracks in HEM and guarantee its mechanical stability under wear, as the second phase can significantly improve the hardness and wear resistance of HEM, and the tribological properties of HEM composites are greatly influenced by the type and quantity of hard particles.

21.4.1.2 Coatings and Nitride Films

Compared with only adding a single element for composition modification, the addition of hard ceramic particles, such as metal carbide or nitride, will not affect the stable structure of the HEA coatings during the preparation process [21]. Cai et al. [37] successfully fabricated the FeMnCrNiCo+x(TiC) composite HEM coatings by laser cladding. As a result, the FeMnCrNiCo+x(TiC) coatings' grains were effectively refined, and the combined effects of refined grain and higher dislocation density boosted the cladding layers' resistance to plastic deformation. Zhu et al. [38] in-situ synthesized micro-nano TiN–Al_2O_3 multiphase ceramics in CoCrFeNiMn high-entropy alloys coating by plasma cladding. In comparison to pure HEMs coating, the hardness of the TiN–Al_2O_3/CoCrFeNiMn HEMs composite coating was 17.6% greater, and the wear volume of the composite coating was 12.5% less. The ceramic TiN–Al_2O_3 particles significantly reduce adhesive wear and increase the steady-state friction of the HEMs during sliding. Similar phenomena have also been observed in the Al_2O_3 particles reinforced FeCoNiCrMn HEM composite coatings [39]. Because of work hardening, refined grain, and evenly dispersed Al_2O_3 particles, the composite coatings had a harder surface than the pure FeCoNiCrMn coating. The development of tribo-layers, which can successfully retard material loss, is responsible for the improved wear resistance.

Researchers desire to increase hardness to meet the requirements of wear resistance due to the existence of Archard's theory, thus HEM nitride and carbide films have attracted the attention of experts in the field of friction. Zhang et al. [40] prepared the (CrNbTiAlV)N_x high-entropy nitride films via magnetron sputtering. The mechanical performances of (CrNbTiAlV)N_x films are improved by nitrogen doping, the hardness is up to 49.95 GPa. The COF of CrNbTiAlV film is increased by furrow and debris on the surface of wear track, while nanocrystalline film exhibits smaller wear loss and lower COF. Feng et al. [41] prepared multielement CrTaNbMoV nitride film at different $N_2/(Ar+N_2)$ flow ratios (R_N). With increasing R_N from 0% to 40%, the alloy CrTaNbMoV film changes from BCC structure to FCC structure. At $R_N=20\%$, the hardness and modulus reached maximum values of 21.6 GPa and 241.5 GPa, respectively, before slightly declining. The COF of nitride films is decreased at different levels compared with alloy film. Jhong et al. [42] successfully fabricated (CrNbSiTiZr)C_x coatings by reactive radio-frequency magnetron sputtering using an equimolar CrNbSiTiZr alloy target under CH_4-containing Ar flow. The values of COFs and wear rates visibly dropped as the carbon content of coatings rose, and the minimum COF (0.07) of carbide coatings appeared at 87.8 at.% carbon concentration.

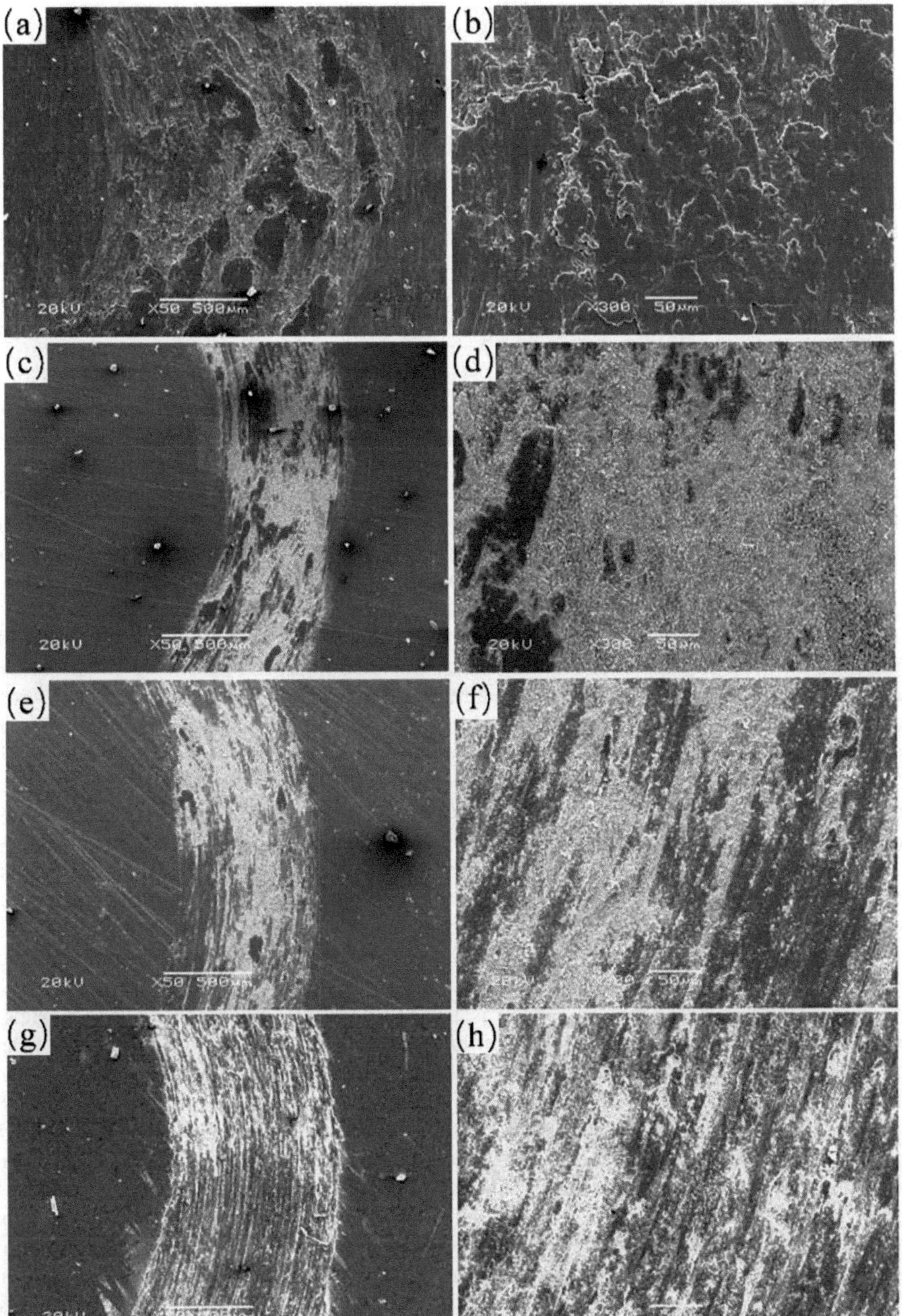

FIGURE 21.4 SEM images of worn surface of CoCrFeNiMn HEM at 200°C (a and b), 400°C (c and d) CoCrFeNiMn-10% Cr_3C_2 composite at 200°C (e and f) and 400°C (g and h). Adapted with permission from [36]. Copyright (2020) Elsevier.

21.4.2 Design of Self-Lubrication HEMs

21.4.2.1 Bulk High Entropy Composites

Reinforcing phases can be added to HEM to increase its hardness and wear resistance, but this often sacrifices the ductility of the material. To increase the adaptability of HEMs, solid lubricants are therefore introduced into various substrates to significantly reduce COF and wear rate. There are various kinds of solid lubricants, such as soft metals, TMDs, fluorides, and carbon-based materials.

Soft metals have multiple slip planes and have special properties like low COF over a wide working temperature range, and Ag and Cu are the soft metals that are most frequently utilized as solid lubricants. Geng et al. [43] successfully prepared an AlCoCrFeNi HEM-based composite reinforced with different contents of solid lubricant Ag by Spark Plasma Sintering technique. Compared with AlCoCrFeNi HEM, the AlCoCrFeNi HEM-Ag composites sliding against Inconel 718 alloys are about one-third lower than that of the AlCoCrFeNi HEM at RT –600°C, due to the formation of a lubricated Ag film, which avoids serious wear or adhesion transfer of the sliding pair. Verma et al. [44] studied the influence of Cu on microstructure evolutions in $CoCrFeNiCu_x$ (x=0, 0.2, 0.4, 0.6, 0.8, and 1.0) HEM systems, as well as wear resistance at room and elevated temperatures. Cu is discovered to be segregated at grain boundaries, which reduces grain size. At room temperature and at elevated temperatures, wear rate reduces with increasing Cu concentration. What's more, CoCrFeNiCu HEM systems have greater wear resistance at 600°C than that of room temperature, which was explained by the self-lubricating glazing Cu oxide layer that forms at high temperatures. To sum up, the addition of soft metals as solid lubricants can provide excellent lubricity.

In addition to the preparation of self-lubricating HEM through composition regulation by adding soft metals, the wear resistance of the material can also be improved by adding solid lubricants. Zhang et al. [45] synthesized the CoCrFeNi HEM matrix self-lubricating composite with Ni-coated graphite powder and Ni-coated MoS_2 by spark plasma sintering. In sliding wear from room temperature to 400°C, the COF and wear rates of the composites are lower than those of pure CoCrFeNi HEM, benefiting from the synergistic lubricating effect of graphite and MoS_2. However, the lubrication effect of graphite and MoS_2 on HEM is not obvious at temperature above 400°C, which may be attributed to the decomposition of MoS_2. Zhang et al. [13] added Ag and BaF_2/CaF_2 eutectic to the CoCrFeNi HEM matrix to obtain good lubrication in a wide temperature range from room temperature to 800°C. At temperatures from room temperature to about 400°C, Ag significantly contributed to the reduction of the COF of composite. The combination lubricating effects of Ag, BaF_2/CaF_2 eutectic, and different metal oxides that developed on the composite surface at high temperatures were primarily responsible for the good lubricity from 400°C to 800°C.

Carbon and carbon-based materials, such as graphite, graphene, and graphene nanoplatelets are used as popular solid lubricant additives in self-lubricating materials. Its layered structure, which is similar to that of MoS_2, and WS_2, makes graphite easier to shear and decreases friction between contact surfaces in relative motion. Xiao et al. [46] systematically investigated the effect of carbon content on the microstructure, hardness, scratch resistance, and wear resistance of the $CoCrFeMnNiC_x$ (x=0, 0.3, 0.6, 0.9, 1.2) HEMs. The carbon alloyed $CoCrFeMnNiC_x$ HEMs have a dual-phase structure comprised of FCC phase and M_7C_3 phase. The precipitation of M_7C_3 hard carbide and low porosity increases the hardness of $CoCrFeMnNiC_x$ HEMs, improving the wear and scratch resistances. The abrasive mechanism becomes more prevalent as the carbon content rises, which also increases the concentration of M_7C_3 carbides, and thus promotes the wear resistance. Wang et al. [47] prepared FeCoCrNiAl HEM matrix composite containing graphite nanoplates by spark plasma sintering. The addition of graphite nanoplates can drastically lower the COF from 0.5–0.8 of the pure HEM to 0.27–0.16, together with a reduction in wear. Excellent tribological properties are attributed to in-situ graphene, friction-induced oxide layers containing nanoscale wear fragments, and graphene on the friction interface (Figure 21.5a–c). Figure 21.5d depicts the formation of graphene and the self-lubrication mechanism of the HEM-graphite (HEM-G) composite. During the

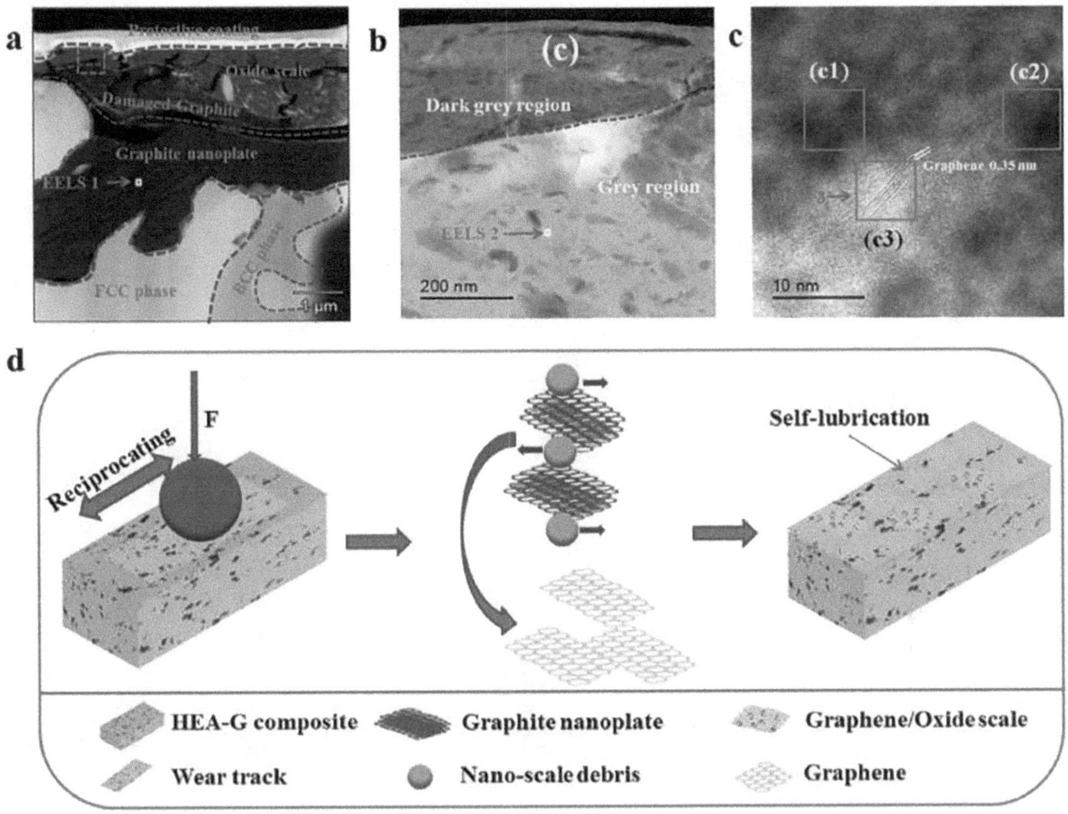

FIGURE 21.5 (a) HAADF image of cross section of wear trajectory of HEM-G composite material tested at 30 N, (b) BF image of the marked dotted red box in Fig. 5a, (c) HRTEM of the marked dotted red box in Fig. 5b, (d) schematic illustration of the self-lubrication mechanism of the HEM-G composite and the formation of GO from graphite nanoplate. Adapted with permission from [47]. Copyright (2021) Elsevier.

friction process, some tiny wear debris may immerse into the graphite nanoplates and function like the balls in the ball milling process to promote the exfoliation of graphite nanoplates to produce graphene, which lower the COF and wear rate by acting as a protective covering on worn surfaces.

21.4.2.2 Self-Lubrication Coatings and Films

Designing self-lubricating composites appears to be an efficient technique to improve the tribological properties of HEM coatings and films. Shi et al. [48] prepared AlCoCrFeNi high-entropy alloy matrix coatings with Ag and BaF_2/CaF_2 eutectic by atmospheric plasma spray. At room temperature, the lubricant effectively inhibits abrasion behavior and eminently decreases the wear rate by 10 times. At 800°C, the principal oxidation products change from Fe_3O_4, Cr_2O_3, and Al_2O_3 to CrO_2, CrO_3, FeO, NiO, and Al_2O_3, and the introduction of the lubricating phase reduces the COF by nearly one time. Ji et al. [49] studied the influence of MoS_2 and WC doping on the tribological properties of FeCoCrNiCu HEA coating respectively. The addition of MoS_2 and WC particles is equally dispersed within the FCC solid solution matrix. MoS_2 will form a lubricating layer that will significantly reduce the COF. Under high load, the stratification and fracture of the brittle layer will speed up lubrication failure. The addition of WC has no obvious effect on the reduction in COFs, but it can significantly improve the wear resistance.

Our group [50] prepared NbMoWTa/Ag composite film with different sublayer thicknesses by magnetron sputtering. When the sublayer thickness is reduced from 100 to 2.5 nm, the deformation mechanism transited from classical *Hall–Petch* strengthening to coherent strengthening, increasing

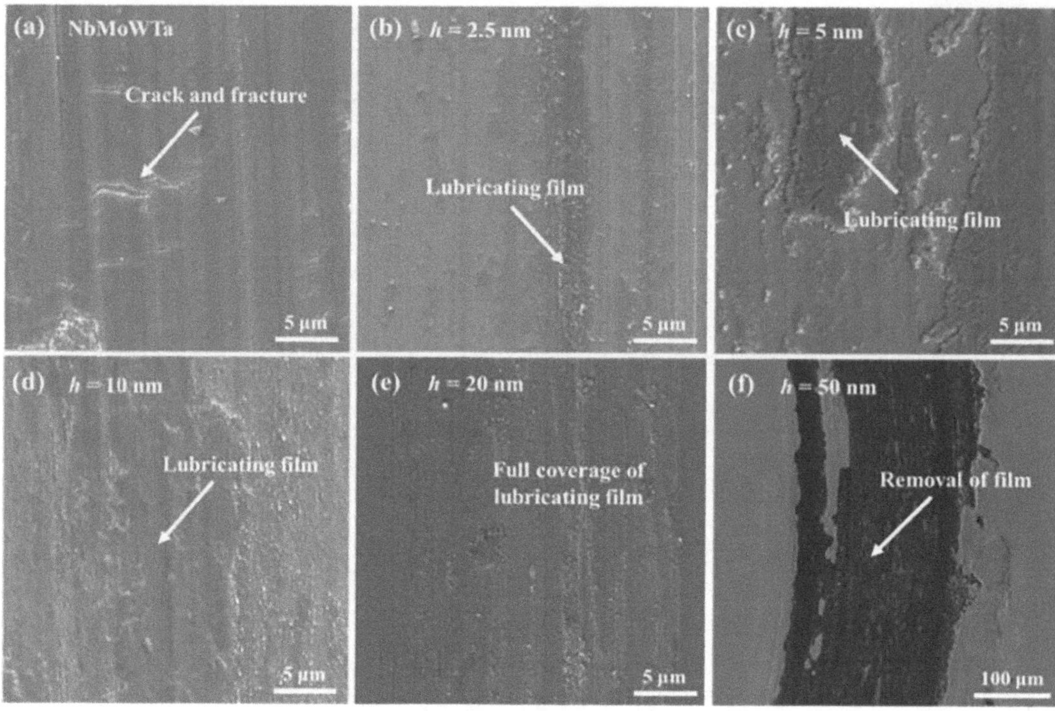

FIGURE 21.6 SEM micrographs of the worn surfaces of (a) monolithic NbMoWTa and NbMoWTa/Ag multilayers with h of (b) 2.5 nm, (c) 5 nm, (d) 10 nm, (e) 20 nm, and (f) 50 nm. Adapted with permission from [50]. Copyright (2021) American Chemical Society.

the hardness of NbMoWTa/Ag multilayer monotonously giving the film a high load-bearing capacity. As shown in Figure 21.6, distinct patches of Ag-rich films emerge on the worn surface during the gradual wear process, which can be helpful in enhancing the lubricating property. As h grew, a more complete lubricating film was seen on the friction surface, and at $h = 20$ nm, the multilayer film had almost total film coverage. The durable lubricating film formed by the addition of Ag reduces the shear force between the film and the antibody, thus preventing adhesive wear of the material and overcoming the brittleness of the NbMoWTa film, thus greatly improving its tribological properties.

21.5 SUMMARY AND RESEARCH OUTLOOK

After nearly 20 years of exploration and development, researchers have carried out in-depth research on the influences of elements composition, preparation process, and interface structure on the microstructure and tribological properties of HEM bulk and coatings/films. In this chapter, the microstructure and tribological properties of HEM and its composites are reviewed, and the possible strategies and design criteria for obtaining excellent tribological properties are discussed. The main contents and future research prospects can be concentrated in the following aspects:

i. HEMs exhibit excellent tribological properties at room temperature and high temperature, which can be attributed to the dynamic microstructural evolution of their subsurface, oxide layer formation, and excellent resistance to softening. Although FCC HEMs show inferior hardness and wear properties when compared with BCC HEMs, they still represent a promising group of materials for specific applications, such as those requiring a certain degree of malleability.

ii. By introducing hard particles, the hardness and wear resistance of HEM can be improved effectively. In addition to the inherent lubricity and anti-wear properties of HEM, self-lubricating HEM with solid lubricants exhibits low COF and wear rate, proving the potential application of HEMs in reducing friction and wear.
iii. The status and properties of HEM coatings/films for tribological applications are reviewed. Special attention is paid to the mechanical properties, friction and wear characteristics of HEM coatings and films. Similar to HEA bulk materials, adding hard ceramic particles and solid lubricants can effectively improve their tribological properties. In particular, ultrahigh hardness has been achieved in some HEM nitride films and coatings, effectively improving the wear resistance.
iv. However, the research of HEA self-lubricating system is far less than that of improving its wear resistance. Referring to the design strategy from existing self-lubricating material, the development of new composite material systems based on HEM to meet different extreme environmental conditions is required.
v. By utilizing first principles, molecular dynamics, thermodynamics, and other theoretical techniques in the composition design of HEM, novel advanced alloy systems should be explored for performance breakthrough. At the same time, technologies more suitable for large-scale industrial production should be developed to promote the application of HEM bulks and films in actual industry.

ACKNOWLEDGMENTS

The authors would like to thank the National Key Research and Development Program of China (2022YFB3809000), Natural Science Foundation of China (No. 52175188), Key Research and Development Program of Shaanxi Province (2023-YBGY-434), State Key Laboratory for Mechanical Behavior of Materials (20222412), and the Fundamental Research Funds for the Central Universities (3102019JC001)

REFERENCES

[1] D. Hua, Q. Xia, W. Wang, Q. Zhou, S. Li, D. Qian, J. Shi, and H. Wang, Atomistic insights into the deformation mechanism of a CoCrNi medium entropy alloy under nanoindentation, *International Journal of Plasticity*. 142 (2021) 102997.
[2] M. Khadem, O.V. Penkov, H.-K. Yang, and D.-E. Kim, Tribology of multilayer coatings for wear reduction: A review, *Friction*. 5 (3) (2017) 248–262.
[3] D.B. Miracle, High entropy alloys as a bold step forward in alloy development, *Nat Commun*. 10 (1) (2019) 1805.
[4] J.W. Yeh, S.K. Chen, S.J. Lin, J.Y. Gan, T.S. Chin, T.T. Shun, C.H. Tsau, and S.Y. Chang, Nanostructured high-entropy alloys with multiple principal elements: Novel alloy design concepts and outcomes, *Advanced Engineering Materials*. 6 (5) (2004) 299–303.
[5] M.-H. Tsai, and J.-W. Yeh, High-entropy alloys: A critical review, *Materials Research Letters*. 2 (3) (2014) 107–123.
[6] X.-Z. Wang, Y. Wang, Z. Huang, Q. Zhou, and H. Wang, Tribocorrosion behavior of CoCrNi medium entropy alloy in simulated seawater, *Metals*. 12 (3) (2022) 401.
[7] Z. Liu, C. Patzig, S. Selle, T. Höche, P. Gumbsch, and C. Greiner, Stages in the tribologically-induced oxidation of high-purity copper, *Scripta Materialia*. 153 (2018) 114–117.
[8] Z. Li, K.G. Pradeep, Y. Deng, D. Raabe, and C.C. Tasan, Metastable high-entropy dual-phase alloys overcome the strength-ductility trade-off, *Nature*. 534 (7606) (2016) 227–230.
[9] L. Bai, S. Wan, G. Yi, Y. Shan, S.T. Pham, A.K. Tieu, Y. Li, and R. Wang, Temperature-mediated tribological characteristics of 40CrNiMoA steel and Inconel 718 alloy during sliding against Si3N4 counterparts, *Friction*. 9 (5) (2020) 1175–1197.
[10] C. Nagarjuna, K. Yong Jeong, Y. Lee, S. Min Woo, S. Ig Hong, H. Seop Kim, and S.-J. Hong, Strengthening the mechanical properties and wear resistance of CoCrFeMnNi high entropy alloy fabricated by powder metallurgy, *Advanced Powder Technology*. 33 (4) (2022) 103519.

[11] M. Zhang, Z. Jiang, M. Niu, Y. Sun, and X. Zhang, Tribological behavior of CoCrFeNiMn high-entropy alloy against 304, Al2O3 and Si3N4 counterparts, *Wear.* 508-509 (2022) 204471.

[12] A. Dollmann, A. Kauffmann, M. Heilmaier, C. Haug, and C. Greiner, Microstructural changes in CoCrFeMnNi under mild tribological load, *J Mater Sci.* 55 (26) (2020) 12353–12372.

[13] A. Zhang, J. Han, B. Su, and J. Meng, A novel CoCrFeNi high entropy alloy matrix self-lubricating composite, *Journal of Alloys and Compounds.* 725 (2017) 700–710.

[14] M. Zhang, X. Zhang, M. Niu, Z. Jiang, H. Chen, and Y. Sun, High-temperature tribological behavior of CoCrFeNiV high-entropy alloys: A parallel comparison with CoCrFeNiMn high-entropy alloys, *Tribology International.* 174 (2022) 107736.

[15] Y. Liu, F. Zhang, Z. Huang, Q. Zhou, Y. Ren, Y. Du, and H. Wang, Mechanical and dry sliding tribological properties of CoCrNiNb medium-entropy alloys at room temperature, *Tribology International.* 163 (2021) 107160.

[16] J.-M. Wu, S.-J. Lin, J.-W. Yeh, S.-K. Chen, Y.-S. Huang, and H.-C. Chen, Adhesive wear behavior of AlxCoCrCuFeNi high-entropy alloys as a function of aluminum content, *Wear.* 261 (5-6) (2006) 513–519.

[17] M.-H. Chuang, M.-H. Tsai, W.-R. Wang, S.-J. Lin, and J.-W. Yeh, Microstructure and wear behavior of AlxCo1.5CrFeNi1.5Tiy high-entropy alloys, *Acta Materialia.* 59 (16) (2011) 6308–6317.

[18] J. Joseph, N. Haghdadi, M. Annasamy, S. Kada, P.D. Hodgson, M.R. Barnett, and D.M. Fabijanic, On the enhanced wear resistance of CoCrFeMnNi high entropy alloy at intermediate temperature, *Scripta Materialia.* 186 (2020) 230–235.

[19] S. Pan, C. Zhao, P. Wei, and F. Ren, Sliding wear of CoCrNi medium-entropy alloy at elevated temperatures: Wear mechanism transition and subsurface microstructure evolution, *Wear.* 440-441 (2019) 32–44.

[20] J. Wang, B. Zhang, Y. Yu, Z. Zhang, S. Zhu, X. Lou, and Z. Wang, Study of high temperature friction and wear performance of (CoCrFeMnNi)85Ti15 high-entropy alloy coating prepared by plasma cladding, *Surface and Coatings Technology.* 384 (2020) 125337.

[21] D. Luo, Q. Zhou, Z. Huang, Y. Li, Y. Liu, Q. Li, Y. He, and H. Wang, Tribological behavior of high entropy alloy coatings: *A Review, Coatings.* 12 (10) (2022) 1428.

[22] Y. Cui, J. Shen, S.M. Manladan, K. Geng, and S. Hu, Wear resistance of FeCoCrNiMnAlx high-entropy alloy coatings at high temperature, *Applied Surface Science.* 512 (2020) 145736.

[23] Y. Yang, Y. Ren, Y. Tian, K. Li, L. Bai, Q. Huang, Q. Shan, Y. Tian, and H. Wu, Microstructure and tribological behaviors of FeCoCrNiMoSix high-entropy alloy coatings prepared by laser cladding, *Surface and Coatings Technology.* 432 (2022) 128009.

[24] H. Liu, S. Sun, T. Zhang, G. Zhang, H. Yang, and J. Hao, Effect of Si addition on microstructure and wear behavior of AlCoCrFeNi high-entropy alloy coatings prepared by laser cladding, *Surface and Coatings Technology.* 405 (2021) 126522.

[25] M. Löbel, T. Lindner, and T. Lampke, High-temperature wear behaviour of AlCoCrFeNiTi0.5 coatings produced by HVOF, *Surface and Coatings Technology.* 403 (2020) 126379.

[26] J.-K. Xiao, Y.-Q. Wu, J. Chen, and C. Zhang, Microstructure and tribological properties of plasma sprayed FeCoNiCrSiAlx high entropy alloy coatings, *Wear.* 448-449 (2020) 203209.

[27] P. Patel, A. Roy, N. Sharifi, P. Stoyanov, R.R. Chromik, and C. Moreau, Tribological Performance of High-Entropy Coatings (HECs): A review, *Materials (Basel).* 15 (10) (2022) 3699.

[28] S. Alvi, D.M. Jarzabek, M.G. Kohan, D. Hedman, P. Jenczyk, M.M. Natile, A. Vomiero, and F. Akhtar, Synthesis and mechanical characterization of a CuMoTaWV high-entropy film by magnetron sputtering, *ACS Appl Mater Interfaces.* 12 (18) (2020) 21070–21079.

[29] Y. Zhao, X. Zhang, H. Quan, Y. Chen, S. Wang, and S. Zhang, Effect of Mo addition on structures and properties of FeCoNiCrMn high entropy alloy film by direct current magnetron sputtering, *Journal of Alloys and Compounds.* 895 (2022).

[30] J. Fan, X. Liu, J. Pu, and Y. Shi, Anti-friction mechanism of VAlTiCrMo high-entropy alloy coatings through tribo-oxidation inducing layered oxidic surface, *Tribology International.* 171 (2022) 107523.

[31] M. John, and P.L. Menezes, Self-lubricating materials for extreme condition applications, *Materials (Basel).* 14 (19) (2021) 5588.

[32] Y.-l. Gu, M.-l. Yi, Y. Chen, J. Tu, Z.-m. Zhou, and J.-r. Luo, Effect of the amount of SiC particles on the microstructure, mechanical and wear properties of FeMnCoCr high entropy alloy composites, *Materials Characterization.* 193 (2022) 112300.

[33] Y. Cai, Y. Tong, Y. Hu, H. Huang, X. Zhang, M. Hua, S. Xu, Y. Mei, C. Ma, and Z. Li, Wear-resistant TiC strengthening CoCrNi-Based high-entropy alloy composite, *Materials (Basel).* 14 (16) (2021) 4665.

[34] T. Zhu, H. Wu, R. Zhou, N. Zhang, Y. Yin, L. Liang, Y. Liu, J. Li, Q. Shan, Q. Li, and W. Huang, Microstructures and tribological properties of TiC reinforced FeCoNiCuAl high-entropy alloy at normal and elevated temperature, *Metals*. 10 (3) (2020) 387.

[35] R. Zhou, G. Chen, B. Liu, J. Wang, L. Han, and Y. Liu, Microstructures and wear behaviour of (FeCoCrNi)1-x(WC)x high entropy alloy composites, *International Journal of Refractory Metals and Hard Materials*. 75 (2018) 56–62.

[36] Z. Guo, A. Zhang, J. Han, and J. Meng, Microstructure, mechanical and tribological properties of CoCrFeNiMn high entropy alloy matrix composites with addition of Cr3C2, *Tribology International*. 151 (2020) 106436.

[37] Y. Cai, L. Zhu, Y. Cui, M. Shan, H. Li, Y. Xin, and J. Han, Fracture and wear mechanisms of FeMnCrNiCo+x(TiC) composite high-entropy alloy cladding layers, *Applied Surface Science*. 543 (2021) 148794.

[38] S. Zhu, Y. Yu, B. Zhang, Z. Zhang, X. Yan, and Z. Wang, Microstructure and wear behaviour of in-situ TiN-Al2O3 reinforced CoCrFeNiMn high-entropy alloys composite coatings fabricated by plasma cladding, *Materials Letters*. 272 (2020) 127870.

[39] Y. Zou, Z. Qiu, C. Huang, D. Zeng, R. Lupoi, N. Zhang, and S. Yin, Microstructure and tribological properties of Al2O3 reinforced FeCoNiCrMn high entropy alloy composite coatings by cold spray, *Surface and Coatings Technology*. 434 (2022) 128205.

[40] C. Zhang, X. Lu, C. Wang, X. Sui, Y. Wang, H. Zhou, and J. Hao, Tailoring the microstructure, mechanical and tribocorrosion performance of (CrNbTiAlV)N high-entropy nitride films by controlling nitrogen flow, *Journal of Materials Science & Technology*. 107 (2022) 172–182.

[41] X. Feng, K. Zhang, Y. Zheng, H. Zhou, and Z. Wan, Chemical state, structure and mechanical properties of multi-element (CrTaNbMoV)Nx films by reactive magnetron sputtering, *Materials Chemistry and Physics*. 239 (2020) 121991.

[42] Y.-S. Jhong, C.-W. Huang, and S.-J. Lin, Effects of CH4 flow ratio on the structure and properties of reactively sputtered (CrNbSiTiZr)Cx coatings, *Materials Chemistry and Physics*. 210 (2018) 348–352.

[43] Y. Geng, H. Tan, J. Cheng, J. Chen, Q. Sun, S. Zhu, and J. Yang, Microstructure, mechanical and vacuum high temperature tribological properties of AlCoCrFeNi high entropy alloy based solid-lubricating composites, *Tribology International*. 151 (2020) 106912.

[44] A. Verma, P. Tarate, A.C. Abhyankar, M.R. Mohape, D.S. Gowtam, V.P. Deshmukh, and T. Shanmugasundaram, High temperature wear in CoCrFeNiCux high entropy alloys: The role of Cu, *Scripta Materialia*. 161 (2019) 28–31.

[45] A. Zhang, J. Han, B. Su, P. Li, and J. Meng, Microstructure, mechanical properties and tribological performance of CoCrFeNi high entropy alloy matrix self-lubricating composite, *Materials & Design*. 114 (2017) 253–263.

[46] J.-K. Xiao, H. Tan, J. Chen, A. Martini, and C. Zhang, Effect of carbon content on microstructure, hardness and wear resistance of CoCrFeMnNiCx high-entropy alloys, *Journal of Alloys and Compounds*. 847 (2020) 156533.

[47] L. Wang, Y. Geng, A.K. Tieu, G. Hai, H. Tan, J. Chen, J. Cheng, and J. Yang, In-situ formed graphene providing lubricity for the FeCoCrNiAl based composite containing graphite nanoplate, *Composites Part B: Engineering*. 221 (2021) 109032.

[48] P. Shi, Y. Yu, N. Xiong, M. Liu, Z. Qiao, G. Yi, Q. Yao, G. Zhao, E. Xie, and Q. Wang, Microstructure and tribological behavior of a novel atmospheric plasma sprayed AlCoCrFeNi high entropy alloy matrix self-lubricating composite coatings, *Tribology International*. 151 (2020) 106470.

[49] X. Ji, J. Zhao, H. Wang, and C. Luo, Sliding wear of spark plasma sintered CrFeCoNiCu high entropy alloy coatings with MoS2 and WC additions, *The International Journal of Advanced Manufacturing Technology*. 96 (5-8) (2017) 1685–1691.

[50] D. Luo, Q. Zhou, W. Ye, Y. Ren, C. Greiner, Y. He, and H. Wang, Design and characterization of self-lubricating refractory high entropy alloy-based multilayered films, *ACS Appl Mater Interfaces*. 13 (46) (2021) 55712–55725.

22 High Entropy Materials for Thermoelectric Applications

Wenjie Li, Hangtian Zhu, Lavanya Raman, Soumya Sridar, and Bed Poudel

22.1 INTRODUCTION

Thermoelectric (TE) materials and devices, based on the direct energy conversion between heat and electricity, have been considered promising alternatives to meet the challenges of the global energy dilemma. There are three primary TE effects defined as the Seebeck effect for power generation, the Peltier effect for refrigeration, and the Thomson effect for reversible cooling or heating (Figure 22.1a–c) [1]. The energy conversion efficiency of a TE material is determined by the dimensionless figure of merit (zT), $zT = \frac{S^2\sigma T}{\kappa}$ where S is the Seebeck coefficient, σ is electrical conductivity, T is absolute temperature, and κ is the total thermal conductivity, composed of the electronic contribution (κ_{el}) and lattice contribution (κ_L). In general, a high power factor ($S^2\sigma$) and a low κ are required to maximize the zT, which typically occurs at carrier concentration (n_H) between 10^{19} and 10^{21} carriers/cm^3, a regime between common metals and semiconductors, on the other word, heavily doped semiconductors (Figure 22.1d). However, enhancing zT value is a formidable task, primarily due to the intricate interplay among these key TE parameters (S, σ, κ). Consequently, it becomes crucial to find ways to decouple electron and phonon transport behaviors. Various effective

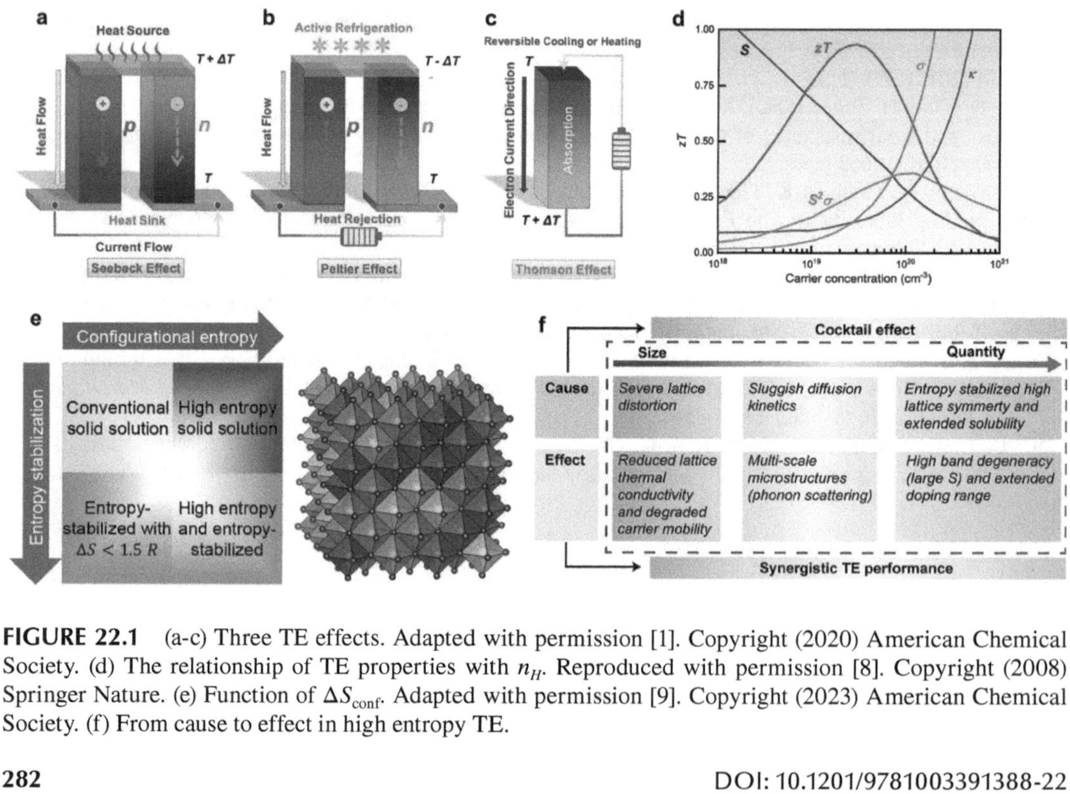

FIGURE 22.1 (a-c) Three TE effects. Adapted with permission [1]. Copyright (2020) American Chemical Society. (d) The relationship of TE properties with n_H. Reproduced with permission [8]. Copyright (2008) Springer Nature. (e) Function of ΔS_{conf}. Adapted with permission [9]. Copyright (2023) American Chemical Society. (f) From cause to effect in high entropy TE.

 DOI: 10.1201/9781003391388-22

electronic perturbation. However, by doping elements w
match, it is possible to diminish the overall scattering ef
the type of constituent elements, their amounts, the order
are alloyed [12–14].

Isoelectronic alloying on different sublattices introduc
mechanisms that act orthogonally. Moreover, phase sep
occurs due to the strain resulting from atomic size misı
cies, dislocations, and stacking faults, can further contrib
of grain boundaries by refining the grain size and develo
with the formation of secondary phases, also aids in re
performance. These strategies collectively work toward e
materials.

22.2.1 High Entropy Thermoelectric Alloys

22.2.2.1 Chalcogenide-Based TE Alloys

Traditional high-performance chalcogenide-based comp
Cu_2Se, have smaller band gaps (E_g) and m^* compared t
their bonds, resulting in higher μ_H. The heavy atomic w
reducing κ. Among these compounds, PbTe is particular
power generation. Recently, a high peak zT of 2.0 at 9
[15] and 1.8 at 900 K for $Pb_{0.89}Sb_{0.012}Sn_{0.1}Se_{0.5}Te_{0.25}S_{0.25}$ [1
introducing alloying elements such as Ge, Pb, and Mn o
900 K was achieved for $(Sn_{0.7}Ge_{0.2}Pb_{0.1})_{1-x}Mn_{1.1x}Te$ (x=0
is known for its high concentration of Ge vacancies a
Through high-entropy-induced band convergence and p
800 K were achieved in $Ge_{0.84}In_{0.01}Pb_{0.1}Sb_{0.05}Te_{0.997}I_{0.003}$ I
$b_{0.12}Bi_{0.01}Te$ [18].

One highlighting feature of HE TE alloys is the hig
$Na_{0.025}Cd_{0.04}Se_{0.5}S_{0.25}Te_{0.25}$, the hierarchical structure wi
propagation of phonons across all frequencies (Figure
approaches the theoretical minimum value of 0.31 W/m-
$Te_{0.25}S_{0.25}$, a significantly reduced κ_L of 0.3 W/m-K was
strains, which increased from 0.16% to 0.45%, and the
presence of intertwined line defects and strain clusters in
a low κ_L of 0.32 W/m-K [12]. Furthermore, in $Ge_{0.61}Ag_{0.}$
disorder leads to localized phonons and a phonon mode s
agation of transverse phonons and increases anharmoni
to ~0.3 W/m-K [18].

Materials with a high entropy of mixing tend to cry
alloying of Sn in PbTe-based alloys helps stabilize the c
convergence of multiple phase peaks into a single (200) p
alloys, the stability of the high-temperature cubic phase c
by alloying In/Pb/Sb/I due to the reduction of phase tran
introduction of $AgBiSe_2$ in GeTe-based alloys can stabili
the addition of $AgSbSe_2$ stabilizes a new rhombohedral p
of elements in GeTe- and SnTe-based alloys is greatly e
with the elimination of impurity phases [12,18].

Though high entropy alloying can often lead to a degr
electrons, it also facilitates band convergence, which can

approaches have been explored thus far to increasing zT, including but not limited to the solid solution, nano-inclusion, band degeneracy and hierarchical microstructure [2,3].

In high entropy alloys (HEAs), the factors contributing to the overall entropy change (ΔS) are configurational, electronic randomness, magnetic dipole, and vibrational. Among them, the configurational entropy (ΔS_{conf}) dominates the other three factors [4]. The ΔS_{conf} per mole for the formation of a solid solution from n elements with a mole fraction of x_i is written as follows:

$$\Delta S_{conf} = -R\sum_{i=1}^{n} x_i \ln x_i = -R\ln\frac{1}{n} = R\ln n \tag{22.1}$$

where R is the gas constant. The empirical threshold of ΔS_{conf} for HEAs exceeds 1.5R. As depicted in Figure 22.1e, the conventional solid solutions exhibit low levels of configurational entropy and entropy stabilization. When the number of constituent elements increases, the increase in ΔS_{conf} leads to a decrease in Gibbs energy and stabilizes simple crystal structure with high symmetry, provided the increase in entropy outweighs that of enthalpy. Moreover, employing entropy as the driving force allows for an expanded phase space to optimize TE performance.

For TE materials, the fundamental effects of high entropy (HE) can be harnessed from the following perspectives (Figure 22.1f):

i. Short-range disorder and multiscale microstructures: Severe lattice distortion and sluggish diffusion lead to effective phonon scattering, resulting in a significant reduction in κ_L.
ii. Extended solid solubility and more stable, high symmetry crystal structures: This leads to the formation of overlapping or multiple bands near the Fermi level, resulting in a higher density of states and effective mass (m^*), and thus enhancing S [5–7].
iii. ΔS_{conf}: An increase in ΔS_{conf} may lead to a reduction in the phase transition temperature. Consequently, higher symmetric crystal structures are maintained over a wide range of temperature, extending the stability region [6].

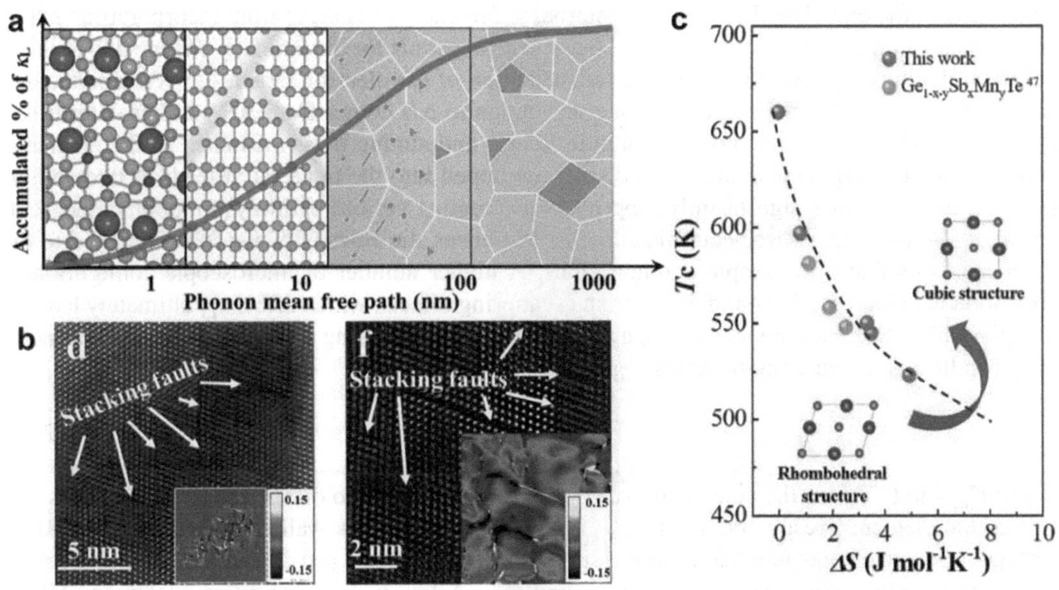

FIGURE 22.2 (a) Phonon mean free path, accumulated κ_L and microstructure and (b) stacking faults leads to tense lattice strains. Adapted from [15]. Copyright B Jiang et al., some rights reserved; exclusive license [Springer Nature]. Distributed under a Creative Commons Attribution License 4.0 (CC BY). (c) ΔS_{conf} determined Phase transition [17]. Adapted with permission [17]. Copyright (2019) Royal Society of Chemistry.

Therefore, entropy optimization serves as a viable ap
electrical transport and κ_L reduction. This approach sh
of future materials in a complex compositional space.
TE materials has the potential to offer valuable insigh

22.2 ENTROPY ENGINEERING OF THERM

The state-of-the-art entropy engineering strategies fo
ply increasing ΔS_{conf}. The focus is on compensating
reducing κ_L in the following ways:

- **Increasing entropy.** Co-doping with elements
as well as varying vibration states in multiple
ΔS_{conf}. Additionally, developing hierarchical m
liquid-like ionic migration tendencies can also
- **Compensating for reduced μ_H.** In cases wher
the Regel limit, the utilization of multiple elem
taining μ_H. This is relevant for materials with in
tion consideration, choose elements with a sma
to minimize scattering potential and the influe
tuned by alloying the sites with low conduction
edges in p-type materials [10].
- **Reduce κ_L.** The energy minimization describe
by two terms: the change in enthalpy ΔH and ch
can be written as [5]

$$\Delta H = \Delta H_s +$$

where ΔH_s is the enthalpy change due to internal stra
ent elements with varying atomic size and mass. ΔH_c
nal ionic field energy resulting from the redistributi
ΔH_c is typically negligible, and the major contributio
by factors such as the shear modulus, atomic radius,
When elements with varying masses and sizes are d
in the strain field which significantly suppress κ_L by c
tering of phonons during isoelectronic alloying. Mor
number of possible microscopic configurations. A h
corresponds to increased phonon disorder and scatter
to a glass-like state and, in some cases, a minimum κ
(Γ_{cal}) can be calculated with the following equation [1

$$\Gamma_{cal} = \Gamma_{mass} +$$

where Γ_{mass} and Γ_{size} are the contributions to scatterin
size of the elements, respectively. The Γ_{size} depends
among various elements and nature of the sublattice
ent phase interfaces can increase both Γ_{mass} and Γ_{siz}
with a large atomic mass mismatch and a small size
. For example, in half-Heusler alloys (hH), there is
match between group V and VI elements, resulting i

The increase of Cd-doping causes the valence band maximum to shift downward, reducing the energy offset between the light and heavy valence bands and achieving band convergence. This results in an enhanced S and improved electrical transport properties [16]. Similarly, despite the significantly higher n_H, a high S in SnTe-based HEAs was maintained, because of band convergence and the large m^* induced by 20 mol% higher solubility of Mn [12]. In $Ge_{0.61}Ag_{0.11}Sb_{0.13}Pb_{0.12}Bi_{0.01}Te$ HEAs, the attenuation of Anderson localization in electrons and the reduction of phase boundary scattering contribute to the well-maintained high $S^2\sigma$. The increased electron delocalization resulting from increased crystal symmetry further contributes to improved band convergence and electrical properties. Consequently, a high zT of 2.7 at 750 K was achieved [18].

22.2.2.2 Half-Heusler (hH) TE Alloys

The hH alloys are a class of compounds characterized by their XYZ crystallization with cubic *F-43m* structure, where X and Y mainly present transition elements (X=Ti, Zr, Hf, Nb, Ta or La; Y=Fe, Co or Ni) and Z is a p-block element (Z=Sn, Pb, Sb or Bi) (Figure 22.3a). The electrical properties of hH alloys are strongly influenced by the donation of electrons from X-site to the $(YZ)^{4-}$ sublattice, which results in a close-shell configuration [21]. These materials exhibit large $S^2\sigma$, excellent mechanical properties, and thermal stability. However, the widespread adoption of hH alloys has been hindered by their intrinsically high κ_L, ~10 W/m-K for pristine compounds. To optimize TE properties of hH compounds, substitution can be performed on any of the X, Y, or Z positions, allowing control over n_H and the suppression of κ_L. Among these positions, isovalent substitution on the X-site by manipulating the entropy has been found to have minimal detrimental effects on electrical properties leading to high TE performance. For instance, in $Nb_{1-x}(Hf_{0.25}Zr_{0.25}Mo_{0.25}V_{0.25}Ti_{0.25})_xFeSb$ alloys, a zT of 0.88 was achieved at 873 K [22]. Fine-tuning the entropy on Zr-site in

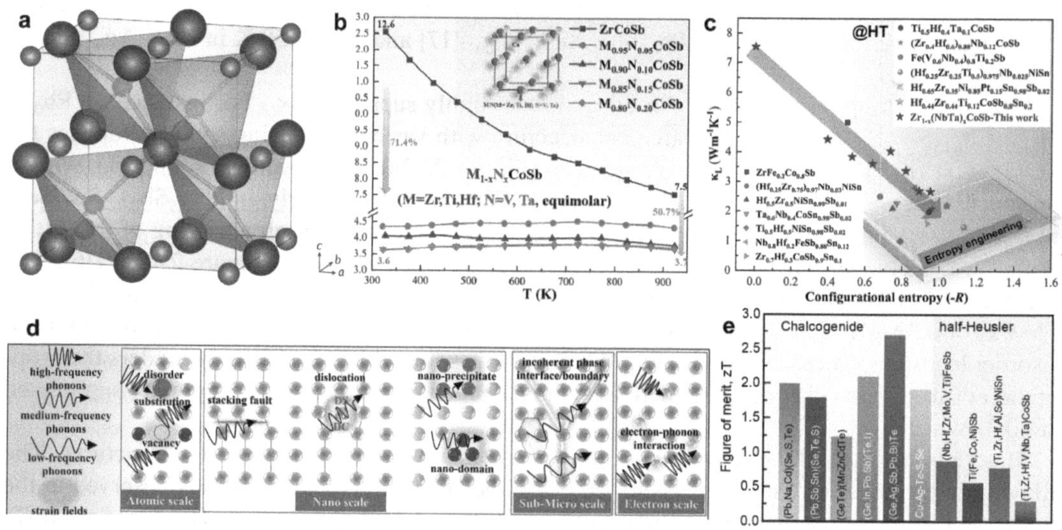

FIGURE 22.3 (a) hH structure. Adapted with permission [21]. Copyright (2016) Springer Nature. (b) Reduction of κ in ZrCoSb-based hH alloys. Adapted with permission [25]. Copyright (2022) Elsevier. (c) ΔS_{conf} dependent κ_L in high entropy hH-based alloy and (d) phonon scattering mechanism for the medium/high entropy hH alloys. Adapted with permission [13]. Copyright (2023) Elsevier. (e) The zT of state-of-the-art entropy engineered alloys [5,10,12,16–18,22–24].

$(Ti_{0.33}Zr_{0.33}Hf_{0.33}Al_{0.005}Sc_{0.005})NiSn$ alloy allowed for the optimization of n_H and κ_L, resulting in a significant enhancement of zT [23]. By increasing the entropy in triple $(Zr, Ti, Hf)_{0.9}(V, Ta)_{0.1}CoSb$, κ was reduced by ~71.4% compared to that of pristine compound (Figure 22.3b).

The semiconductor behavior of hH alloys relies on achieving a close-shell configuration with either 8 or 18 valence electron count (VEC), which makes it crucial to adhere to the VEC rule when considering aliovalent substitution. Aliovalent substitution often introduces a high concentration of ionization impurities, leading to strong scattering of carriers. Consequently, it is generally restricted due to the detrimental effects caused by such impurities. Luo et al. [10] realized a notable reduction in κ_L by increasing entropy and inducing significant lattice distortion in the $Ti(Fe_{1/3+x}Co_{1/3}Ni_{1/3-x})Sb$ alloy through random mixing of Fe/Co/Ni. Entropy engineering in $(Ti_{0.166}Zr_{0.166}Hf_{0.166}V_{0.166}Nb_{0.166}Ta_{0.166})FeSb$ allowed for the introduction of aliovalent substitution, enhancing entropy and regulating the electrical properties. The σ of $MFe_{1-x}Co_xSb$ demonstrated a transition from p-type behavior ($x<0.4$) to n-type behavior ($0.4<x<1$). These samples exhibited effectively suppressed κ_L of ~2 W/m-K at 300 K [24]. As a gene-like performance indicator, the increase of ΔS_{conf} in a TE material promotes the generation of numerous lattice defects as phonon scattering center and thereby suppress κ_L to approach the theoretical limit (Figure 22.3c and d). Figure 22.3e summarizes the peak zT of representative HE TE alloys [5,10,12,16–18,22–24].

22.2.2 High Entropy Thermoelectric Ceramics

Ceramics, including metal oxides, sulfides and carbides, could be promising TE materials due to their diverse crystal structures, local site symmetries, and excellent thermal and chemical stability at high temperatures. In 2015, Rost et al. [26] developed high-entropy oxides by incorporating five cations in equimolar proportions in a rock salt MgO–CoO–NiO–CuO–ZnO oxide to convert the multiphase into a single-phase structure. It was demonstrated that κ can be significantly suppressed by increasing ΔS_{conf}, with an additional factor of two reduction achieved by introducing a sixth constituent element (Figure 22.4a). Additionally, high-entropy perovskites (HEPs), high-entropy sulfides (HESs), and high-entropy carbides (HECs) have been developed, particularly for low κ (Figure 22.4b).

Over the years, researchers relied on Goldschmidt's tolerance factor (t_D) to design novel ABO_3 perovskite structures as: [27,28]

$$t_D = \frac{r_A + r_O}{\sqrt{2}(r_B + r_O)} = \frac{\left[\prod_{i=1}^{12}(d_{A-O})\right]^{1/12} + r_O}{\sqrt{2}\left\{\left[\prod_{i=1}^{6}(d_{B-O})\right]^{1/6} + r_O\right\}} \quad (22.4)$$

where r_A, r_B, r_O, $d_{A/B-O}$ are the radii of cations A, B and anion O, and geometric average of the cation-O distances, respectively. A cubic phase is likely to be stable when $0.9 \le t_D \le 1$. An orthorbombic/rhombohedral and tetrahedral/hexagonal phase may form when $t_D<0.9$ and $t_D>1$, respectively (Figure 22.4c). It could be very interesting to further validate the structural field map for HEPs because of their high symmetry structure leading to high degeneracy and m^* for high S. However, the high κ_L caused by strong metal-oxygen ion bonds limits the enhancement of their TE performance. Banerjee et al. [27] reported a single-phase solid solution $Sr(Fe_{0.2}Ti_{0.2}Mo_{0.2}Nb_{0.2}Cr_{0.2})O_3$ HEPs with synergistically increased S and σ. The presence of multiple oxidation states in the five transition metals led to Anderson localization of charge carriers, resulting in a non-degenerate Fermi gas model in the temperature-dependent electrical conductivity behavior. Furthermore, the HEPs strategy induces more anharmonicity by lowering the mean free path of phonons via enhanced multiphonon scattering to successfully reduce κ_L and an ultralow κ of 0.7 W/m-K at 1100

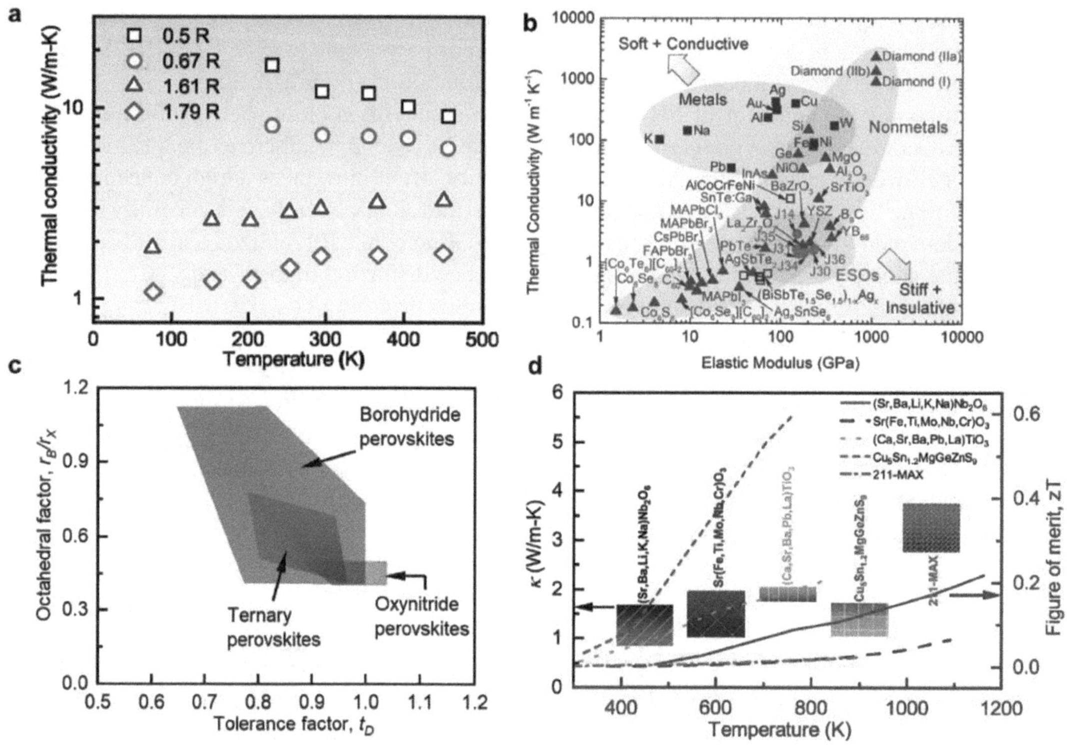

FIGURE 22.4 (a) Reduction of κ with increase of ΔS_{conf} and (b) κ and elastic modulus of various ceramics. Reproduced and adapted with permission [33], Copyright (2018) Wiley-VCH. (c) Structure field map for perovskites. (d) κ and zT of selected high entropy ceramics [27,29–32].

K was achieved (Figure 22.4d). Zheng et al. [29] investigated $(Ca_{0.2}Sr_{0.2}Ba_{0.2}Pb_{0.2}La_{0.2})TiO_3$ HEPs with pure cubic phase. A disorder parameter (Γ) accounting for the different mas and atomic volume of the different elements, can be used to estimate the relaxation rate $1/\tau_M$, is given by:

$$\Gamma = \chi(1-\chi)\left[\left(\frac{\Delta M}{M}\right)^2 + \varepsilon\left(\frac{\Delta\delta}{\delta}\right)^2\right] \tag{22.5}$$

where χ, ΔM, $\Delta\delta$ and ε represent the atomic fraction of impurity atoms, mass fluctuation between the two atoms, difference in average interatomic spacing, and an adjustable parameter used to estimate the contribution from bond strength, respectively. Benefiting from the large lattice distortion and the huge mass fluctuation among the elements on A-site, the κ was almost ten times lower than that of $SrTiO_3$-based counterparts. As a results, a minimum κ_l of 1.17 W/m-K at 923 K were achieved (Figure 22.4d).

The simple tungsten bronze structure, such as AM_2O_6 (A is alkaline metals and B is alkaline earth metals), is a complex crystal structure, composed of a regular or distorted corner-sharing MO_6 octahedron framework, resulting in the formation of pentagonal (A1), tetragonal (A2) and trigonal (*C*) interstitial sites. Jana et al. [30], for the first time, reported a rare-earth-free single-phase solid solution of novel high entropy $(Sr_{0.2}Ba_{0.2}Li_{0.2}K_{0.2}Na_{0.2})Nb_2O_6$. An ultralow κ had been reached resulted from multiphonon scattering due to the presence of five cations in A-site, and an enhanced Seebeck coefficient had been realized by the strong electron scattering in multivalent metal ions.

CuS based compounds have ample compositional variation and are low-cost and eco-friendly minerals with moderate power factor and low κ_L. Zhang et al. [31] reported a metallic $Cu_5Sn_{1.2}MgGeZnS_9$, where an additional Sn was utilized to tune n_H to maintain reasonable σ. The low κ of ~1 W/m-K was observed to result from weak Cu–S bonding and fine grain size produced by mechanical alloying, and a zT of 0.58 at 723 K was obtained (Figure 22.4d and e), similar to the values for ordered ternary and quaternary diamond-like sulfides. The MAX phases, generally with the chemical formula of $M_{n+1}AX_n$ (M, A, and X are an early transition metal, IIIA and IVA groups element, and carbon and/or nitrogen elements, respectively, and n_H is in the range of 1–3), are composed of near-close-packed layers of covalent M_6X octahedrons and intercalated with metal-like A atom layers, which allows it to compile both metal and ceramic properties such as good machinability, high electrical conductivity, and corrosion resistance. Liu et al. [32] developed HE $(Ti_{0.2}V_{0.2}Cr_{0.2}Nb_{0.2}Ta_{0.2})_2AlC$–$(Ti_{0.2}V_{0.2}Cr_{0.2}Nb_{0.2}Ta_{0.2})C$ (M_2AlC–MC) ceramics. The solid solution of Ti, V, Cr, Nb, and Ta atoms in the M-site of the 211-MAX configuration with a classic layered structure has been achieved, where the weight percent of M_2AlC was more than 90%. The κ has been reduced significantly as compared to other single MAX phase ceramics. More than 85% of κ is contributed from electron side, which indicates that there are still great opportunities to further optimize the materials design with lower n_H to reduce κ_{el} and improve zT performance.

22.3 COMPUTATIONAL-ASSISTED DESIGN OF HIGH ENTROPY THERMOELECTRIC MATERIALS

With the advancement of high-performance computing, computational design is a reliable approach for discovering new materials with advanced properties. As opposed to the trial-and-error experimental investigations involving high material costs, computational methods can guide an efficient material design by reducing the experimental evaluations. With a myriad of computational tools available for the design of TE materials, the extensively used techniques for the design and validation of the HE TE materials can be classified into three categories as elaborated below.

22.3.1 Ab-Initio Methods

The *ab-initio* methods based on density functional theory (DFT) have been extensively used for the design and validation of HE TE materials mainly for computing the enthalpy of formation and Gibbs energy, and calculation of density of states. DFT is a quantum-mechanical tool used for computing the electronic structure of atoms, molecules and periodic/non-periodic solid phases. The commonly used DFT packages for modeling the TE materials are Vienna Ab-Initio Simulation Package (VASP) [34] and Wien2K [35]. It is to be noted that all the DFT calculations are performed at 0 K. The enthalpy of formation (ΔH) of a compound can be calculated as $\Delta H\left(A_xB_y\right) = E_{\text{tot}}\left(A_xB_y\right) - xE_{\text{tot}}(A) - yE_{\text{tot}}(B)$, where E_{tot} is the total energy calculated using DFT. High-entropy TE materials require mixing of several elements in different lattice sites leading to a disordered alloy which can be generated using tools such as special quasirandom structures (SQS) [36] and Korringa-Kohn-Rostoker coherent potential approximation [37]. A drawback in generating the random/mixed structures is the large number of atoms in the supercell to mimic alloying, making them computationally intensive and necessitating higher computing power. An example of a 64-atom supercell SQS generated to perform DFT calculations for PbSnSeTe HE TE alloy with a face-centered cubic structure [38] is shown in Figure 22.5a.

To compute the thermodynamic properties above 0 K, quasi-harmonic approximation (QHA) that evaluates the volume dependence of phonon frequencies can be employed. The Gibbs energy can be estimated as $\Delta G = \Delta H - T\Delta S$, where the total entropy (ΔS) is a sum of ΔS_{conf} and vibrational (ΔS_{vib}) entropies. ΔS_{conf} can be estimated using the Boltzmann equation (equation 22.1) whereas ΔS_{vib} is computed using QHA as implemented in Phonopy code using the supercell approach [39].

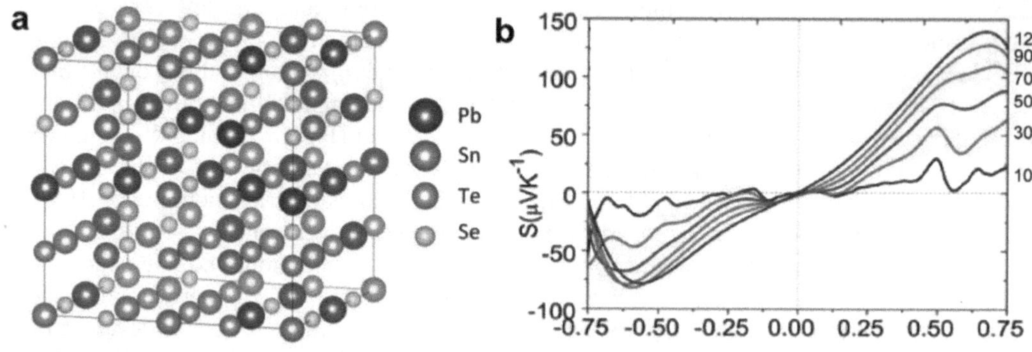

FIGURE 22.5 (a) 64-atom SQS supercell for PbSnSeTe high-entropy alloy. Adapted with permission [38]. Copyright M. Xia et al, some rights reserved; exclusive license [MDPI]. Distributed under a Creative Commons Attribution License 4.0 (CC BY). (b) Seebeck coefficient as a function of spin for $Ti_{0.75}HfMo_{0.25}CrGe$ TE material. Adapted with permission [42], Copyright (2018) AIP Publishing.

The calculated Gibbs energy is used to determine the phase stability which was implemented for PbSe-based HE TE material [16] (Figure 22.5b). It is to be noted that random alloys have a small fraction of negative phonon frequencies at 0 K which can be neglected [16], however, if significant number of negative frequencies are found, it implies that the calculation failed due to dynamic instability of the input structure. To evaluate zT, separate codes can be used to evaluate the required quantities that are conveniently interfaced with VASP and Wien2K. phonop3py [40] can be used for calculating the κ_L by estimating the phonon-phonon interactions. Green–Kubo theory can be employed for estimating κ_L combined with on-the-fly machine-learned force-fields module of VASP [38]. BoltzTraP (Boltzmann Transport Properties) [41] code can be employed for calculating quantities such as S (Figure 22.5c), κ_{el} and σ as implemented for $Ti_{0.75}HfMo_{0.25}CrGe$ thermoelectric material [42]. These essential computational tools for evaluating zT have made ab-initio based techniques as an effective tool to design high-performance HE TE materials.

22.3.2 CALculation of PHAse Diagrams (CALPHAD) Method

Phase diagrams serve as a roadmap for materials design. CALPHAD method [43] is defined as a computer-assisted modeling procedure with experimental and theoretical data as input. Though it was initially devised for phase diagram determination, it is being extended into estimating thermophysical properties. Its major advantage is the ease of predicting the properties of multicomponent systems by extrapolation of properties from the bounding lower-order systems. For the design of HE TE alloys, the CALPHAD method can be utilized for designing the microstructure and evaluating thermophysical properties. For microstructure design, it provides inputs such as phase stability to achieve compositional uniformity (Figure 22.6a), solubility limits for maximum doping concentration and a solidification path for understanding the final microstructure [44]. The phase diagram of pseudo-ternary (Hf,Ti,Zr)NiSn system was deduced using the CALPHAD method to understand the phase separation for improving the thermoelectric properties in the range of 500–1300 K [45] (Figure 22.6a). Moreover, phase boundary mapping involving the study of different thermodynamic states arising due to differing chemical potential within different phase regions of a phase diagram can be applied to understand the defect formation energies and the related properties and has been implemented for Zintl phase ($Ca_9Zn_{4+x}Sb_9$) [46].

CALPHAD approach can be used to predict the thermophysical properties required to evaluate zT. Currently, thermophysical databases are available for computing κ and electrical

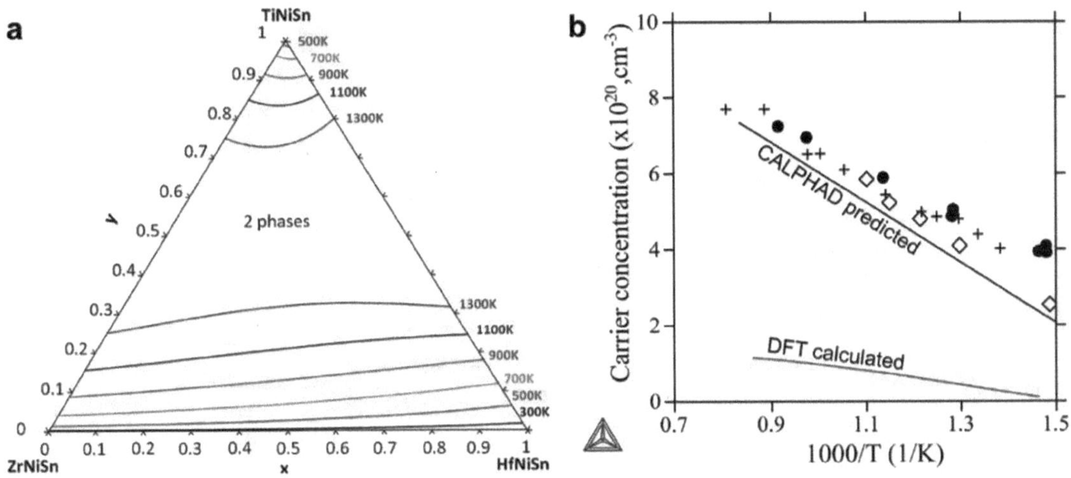

FIGURE 22.6 (a) Isotherms from 500–1300 K for HfNiSn–TiNiSn–ZrNiSn system. Adapted with permission [45]. Copyright (2017) Elsevier. (b) Carrier concentration of Pb-rich PbSe calculated using CALPHAD and DFT. Replotted with permission [47]. Copyright (2019) Springer Nature.

resistivity. Though these properties can be evaluated using ab-initio methods, the CALPHAD method outshines in simulating the composition- and temperature-dependent properties coupled with phase behavior. Besides, other properties such as n_H [47] can be evaluated using this method which will complement the multicomponent TE material design (Figure 22.6b). At present, CALPHAD-type assessments are available for estimating carrier concentration in Pb–Se systems [47] which can be extended to multicomponent systems (Figure 22.6b), however, more efforts are required to implement it for other TE systems. Currently, the lack of adequate thermodynamic and thermophysical databases limits the prospect of exploring the effectiveness of the CALPHAD method for the design of HE TE materials. For designing these multicomponent alloys, efforts are required to develop databases for lower-order (binary and ternary) systems with high fidelity that can be effectively used for high-throughput screening for the discovery of new HE TE materials.

22.3.3 Data-Driven Methods

With artificial intelligence, the current trend for new material discovery is using approaches such as machine learning and data-driven screening in the multi-element chemical space. A large volume of data needs to be screened for the effective implementation of these techniques. There are a few reports pertaining to the design of TE compounds using data-driven models [48], however, efforts toward the design of HE TE materials are scarce. Cation-sulfur bond length for diamond-like compounds containing Cu and S identified by screening ICSD (Inorganic Crystal Structure Database) was used as a chemical descriptor for the design of a HES with good thermoelectric performance as shown in Figure 22.7a [31]. Similarly, optimal composition ranges with high thermoelectric performance were discovered for Cu(S, Se, Te) solid solution using a data-driven searching approach by predicting the contouring diagrams from known experimental data (Figure 22.7b) [49]. More efforts in the direction of data-driven methods are crucial for the accelerated discovery of new HE TE materials with enhanced properties.

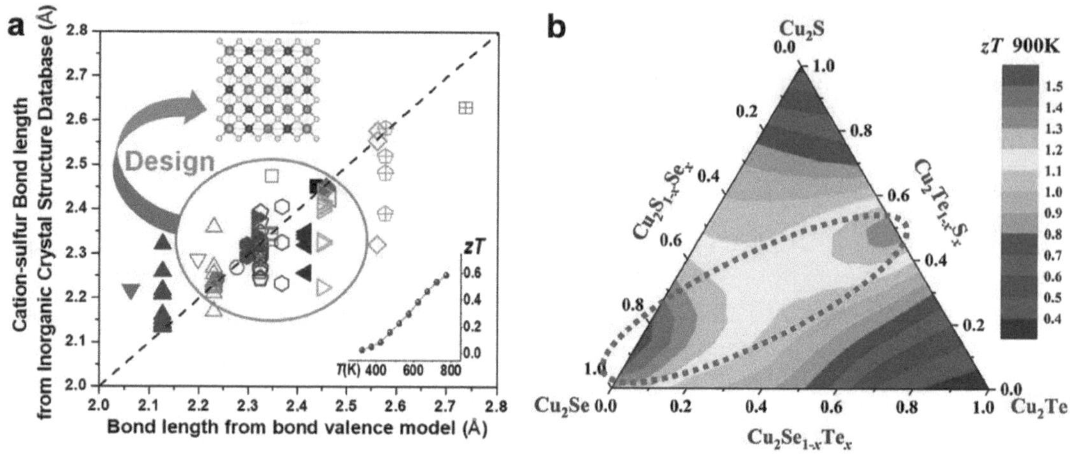

FIGURE 22.7 (a) Data-driven design of diamond-like TE compounds with Cu and S. Adapted with permission [31]. Copyright R. Z. Zhang et al, some rights reserved; exclusive license [American Chemical Society]. Distributed under a Creative Commons Attribution License 4.0 (CC BY). (b) Contour diagram for zT of Cu(S,Se,Te) solid solutions. Adapted with permission [49]. Copyright (2022) Elsevier.

22.4 CONCLUSIONS AND PERSPECTIVES

This chapter provides a comprehensive review of the application of entropy engineering in improving TE performance. It highlights the importance of decoupling electron and phonon transport to simultaneously reduce thermal conductivity and enhance power factor for achieving high zT. Entropy engineering offers a promising approach to address this challenge by leveraging lattice distortion, sluggish diffusion kinetics, high lattice symmetry, and extended solubility effects. The review shows the success of entropy engineering in alloys and ceramics, particularly in chalcogenide-based materials, where the enhancement of zT has been achieved through careful manipulation of entropy-related effects. Furthermore, the chapter recognizes the limited utilization of data science in exploring new high entropy TE materials. The involvement of data science holds promises in developing new theories, innovative structural design, and manipulation of band structure. To summarize, this chapter provides a critical and concise appraisal of the advancements and potential of entropy engineered thermoelectric strategies. It sheds light on the importance of entropy engineering in achieving high-performance thermoelectric materials.

REFERENCES

[1] X.-L. Shi, J. Zou, Z.-G. Chen, Advanced thermoelectric design: From materials and structures to devices, *Chem. Rev.* 120 (2020) 7399–7515.

[2] W. Li, B. Poudel, A. Nozariasbmarz, R. Sriramdas, H. Zhu, H.B. Kang, S. Priya, Bismuth telluride/half-heusler segmented thermoelectric unicouple modules provide 12% conversion efficiency, *Adv. Energy Mater.* 10 (2020) 2001924.

[3] W. Li, B. Poudel, R.A. Kishore, A. Nozariasbmarz, N. Liu, Y. Zhang, S. Priya, Toward high conversion efficiency of thermoelectric modules through synergistical optimization of layered materials, *Adv. Mater.* 35 (2023) 2210407.

[4] B.S. Murty, J.-W. Yeh, S. Ranganathan, P.P. Bhattacharjee, *High-Entropy Alloys*, 2nd ed., Elsevier, 2019.

[5] R. Liu, H. Chen, K. Zhao, Y. Qin, B. Jiang, T. Zhang, G. Sha, X. Shi, C. Uher, W. Zhang, L. Chen, Entropy as a gene-like performance indicator promoting thermoelectric materials, *Adv. Mater.* 29 (2017) 1702712.

[6] Y. Huang, S. Zhi, S. Zhang, W. Yao, W. Ao, C. Zhang, F. Liu, J. Li, L. Hu, Regulating the configurational entropy to improve the thermoelectric properties of (GeTe)1−x(MnZnCdTe3)x alloys, *Materials (Basel).* 15 (2022) 6798.

[7] Z. Zhang, K. Zhao, H. Chen, Q. Ren, Z. Yue, T.-R. Wei, P. Qiu, L. Chen, X. Shi, Entropy engineering induced exceptional thermoelectric and mechanical performances in Cu2-Ag Te1-2S Se, *Acta Mater.* 224 (2022) 117512.

[8] G.J. Snyder, E.S. Toberer, Complex thermoelectric materials, *Nat. Mater.* 7 (2008) 105–114.

[9] S.S. Aamlid, M. Oudah, J. Rottler, A.M. Hallas, Understanding the role of entropy in high entropy oxides, *J. Am. Chem. Soc.* 145 (2023) 5991–6006.

[10] P. Luo, Y. Mao, Z. Li, J. Zhang, J. Luo, Entropy engineering: A simple route to both p- and n-type thermoelectrics from the same parent material, *Mater. Today Phys.* 26 (2022) 100745.

[11] J. Yang, G.P. Meisner, L. Chen, Strain field fluctuation effects on lattice thermal conductivity of ZrNiSn-based thermoelectric compounds, *Appl. Phys. Lett.* 85 (2004) 1140–1142.

[12] L. Hu, Y. Zhang, H. Wu, J. Li, Y. Li, M. Mckenna, J. He, F. Liu, S.J. Pennycook, X. Zeng, Entropy engineering of SnTe: Multi-principal-element alloying leading to ultralow lattice thermal conductivity and state-of-the-art thermoelectric performance, *Adv. Energy Mater.* 8 (2018) 1802116.

[13] R. Chen, Q. Jiang, L. Jiang, R. Min, H. Kang, Z. Chen, E. Guo, X. Yang, T. Wang, Entropy-driven multiscale defects enhance the thermoelectric properties of ZrCoSb-based half-Heusler alloys, *Chem. Eng. J.* 455 (2023) 140676.

[14] X. Zhang, M. Huang, H. Li, J. Chen, P. Xu, B. Xu, Y. Wang, G. Tang, S. Yang, Ultralow lattice thermal conductivity and improved thermoelectric performance in a Hf-free half-Heusler compound modulated by entropy engineering, *J. Mater. Chem. A.* 11 (2023) 8150–8161.

[15] B. Jiang, Y. Yu, H. Chen, J. Cui, X. Liu, L. Xie, J. He, Entropy engineering promotes thermoelectric performance in p-type chalcogenides, *Nat. Commun.* 12 (2021) 3234.

[16] B. Jiang, Y. Yu, J. Cui, X. Liu, L. Xie, J. Liao, Q. Zhang, Y. Huang, S. Ning, B. Jia, B. Zhu, S. Bai, L. Chen, S.J. Pennycook, J. He, High-entropy-stabilized chalcogenides with high thermoelectric performance, *Science* 371 (2021) 830–834.

[17] Y. Qiu, Y. Jin, D. Wang, M. Guan, W. He, S. Peng, R. Liu, X. Gao, L.-D. Zhao, Realizing high thermoelectric performance in GeTe through decreasing the phase transition temperature via entropy engineering, *J. Mater. Chem. A.* 7 (2019) 26393–26401.

[18] B. Jiang, W. Wang, S. Liu, Y. Wang, C. Wang, Y. Chen, L. Xie, M. Huang, J. He, High figure-of-merit and power generation in high-entropy GeTe-based thermoelectrics, *Science* 377 (2022) 208–213.

[19] S. Roychowdhury, T. Ghosh, R. Arora, U. V. Waghmare, K. Biswas, Stabilizing n-type cubic GeSe by entropy-driven alloying of AgBiSe 2 : Ultralow thermal conductivity and promising thermoelectric performance, *Angew. Chemie Int. Ed.* 57 (2018) 15167–15171.

[20] Z. Huang, S.A. Miller, B. Ge, M. Yan, S. Anand, T. Wu, P. Nan, Y. Zhu, W. Zhuang, G.J. Snyder, P. Jiang, X. Bao, High thermoelectric performance of new rhombohedral phase of GeSe stabilized through alloying with AgSbSe 2, *Angew. Chemie Int. Ed.* 56 (2017) 14113–14118.

[21] W.G. Zeier, J. Schmitt, G. Hautier, U. Aydemir, Z.M. Gibbs, C. Felser, G.J. Snyder, Engineering half-Heusler thermoelectric materials using Zintl chemistry, *Nat. Rev. Mater.* 1 (2016) 16032.

[22] J. Yan, F. Liu, G. Ma, B. Gong, J. Zhu, X. Wang, W. Ao, C. Zhang, Y. Li, J. Li, Suppression of the lattice thermal conductivity in NbFeSb-based half-Heusler thermoelectric materials through high entropy effects, Scr. *Mater.* 157 (2018) 129–134.

[23] D. Rabin, A. Meshulam, D. Fuks, Y. Gelbstein, High entropy alloy on single sub-lattice in MNiSn compound: Stability and thermoelectric properties, *J. Alloys Compd.* 874 (2021) 159940.

[24] K. Chen, R. Zhang, J.-W.G. Bos, M.J. Reece, Synthesis and thermoelectric properties of high-entropy half-Heusler MFe1–xCoxSb (M=equimolar Ti, Zr, Hf, V, Nb, Ta), *J. Alloys Compd.* 892 (2022) 162045.

[25] R. Chen, H. Kang, R. Min, Z. Chen, E. Guo, X. Yang, Z. Tian, T. Wang, Entropy engineering induced low thermal conductivity in medium-entropy (Zr, Ti, Hf)CoSb triple half-Heusler compounds, *Materialia.* 23 (2022) 101453.

[26] C.M. Rost, E. Sachet, T. Borman, A. Moballegh, E.C. Dickey, D. Hou, J.L. Jones, S. Curtarolo, J.-P. Maria, Entropy-stabilized oxides, *Nat. Commun.* 6 (2015) 8485.

[27] R. Banerjee, S. Chatterjee, M. Ranjan, T. Bhattacharya, S. Mukherjee, S.S. Jana, A. Dwivedi, T. Maiti, High-entropy perovskites: an emergent class of oxide thermoelectrics with ultralow thermal conductivity, ACS Sustain. *Chem. Eng.* 8 (2020) 17022–17032.

[28] W. Li, E. Ionescu, R. Riedel, A. Gurlo, Can we predict the formability of perovskite oxynitrides from tolerance and octahedral factors? *J. Mater. Chem. A.* 1 (2013) 12239.

[29] Y. Zheng, M. Zou, W. Zhang, D. Yi, J. Lan, C.-W. Nan, Y.-H. Lin, Electrical and thermal transport behaviours of high-entropy perovskite thermoelectric oxides, *J. Adv. Ceram.* 10 (2021) 377–384.

[30] S.S. Jana, T. Maiti, Designing rare earth-free high entropy oxides with a tungsten bronze structure for thermoelectric applications, *Mater. Horizons.* 10 (2023) 1848–1855.

[31] R.-Z. Zhang, F. Gucci, H. Zhu, K. Chen, M.J. Reece, Data-driven design of ecofriendly thermoelectric high-entropy sulfides, *Inorg. Chem.* 57 (2018) 13027–13033.

[32] C. Liu, Y. Yang, Z. Zhou, C. Nan, Y. Lin, (Ti 0.2 V 0.2 Cr 0.2 Nb 0.2 Ta 0.2) 2 AlC-(Ti 0.2 V 0.2 Cr 0.2 Nb 0.2 Ta 0.2)C high-entropy ceramics with low thermal conductivity, *J. Am. Ceram. Soc.* 105 (2022) 2764–2771.

[33] J.L. Braun, C.M. Rost, M. Lim, A. Giri, D.H. Olson, G.N. Kotsonis, G. Stan, D.W. Brenner, J. Maria, P.E. Hopkins, Charge-induced disorder controls the thermal conductivity of entropy-stabilized oxides, *Adv. Mater.* 30 (2018) 1805004.

[34] G. Kreese, Furthmuller J, Efficiency of ab-initio total energy calculations for metals and semiconductors using a plane-wave basis set, *Comput. Mater. Sci.* 6 (1996) 15–50.

[35] P. Blaha, K. Schwarz, F. Tran, R. Laskowski, G.K.H. Madsen, L.D. Marks, WIEN2k: An APW+lo program for calculating the properties of solids, *J. Chem. Phys.* 152 (2020) 074101.

[36] A. van de Walle, P. Tiwary, M. de Jong, D.L. Olmsted, M. Asta, A. Dick, D. Shin, Y. Wang, L.-Q. Chen, Z.-K. Liu, Efficient stochastic generation of special quasirandom structures, *Calphad.* 42 (2013) 13–18.

[37] H. Akai, Fast Korringa-Kohn-Rostoker coherent potential approximation and its application to FCC Ni-Fe systems, *J. Phys. Condens. Matter.* 1 (1989) 8045–8064.

[38] M. Xia, M. Record, P. Boulet, Investigation of PbSnTeSe high-entropy thermoelectric alloy: A dft approach, *Materials (Basel).* 16 (2023) 235.

[39] A. Togo, I. Tanaka, First principles phonon calculations in materials science, *Scr. Mater.* 108 (2015) 1–5.

[40] K. Mizokami, A. Togo, I. Tanaka, Lattice thermal conductivities of two SiO2 polymorphs by first-principles calculations and the phonon Boltzmann transport equation, *Phys. Rev. B.* 97 (2018) 224306.

[41] G.K.H. Madsen, D.J. Singh, BoltzTra P. A code for calculating band-structure dependent quantities, *Comput. Phys. Commun.* 175 (2006) 67–71.

[42] H. Ma, J. Li, G. Yang, Y. Yang, X. Mao, C. Li, F. Yin, Improved thermoelectric property of Ti0.75HfMo0.25CrGe by doping Ti2CrGe Heusler alloy with Hf and Mo: Confirmation of entropy "gene" in thermoelectric materials design, *J. Appl. Phys.* 124 (2018) 235104.

[43] H. Lukas, S.G. Fries, B. Sundman, *Computational thermodynamics: the Calphad method*, Cambridge university press, 2007.

[44] X. Li, Z. Li, C. Chen, Z. Ren, C. Wang, X. Liu, Q. Zhang, S. Chen, CALPHAD as a powerful technique for design and fabrication of thermoelectric materials, *J. Mater. Chem. A.* 9 (2021) 6634–6649.

[45] A. Berche, J.C. Tédenac, P. Jund, Scripta Materialia Phase separation in the half-Heusler thermoelectric materials (Hf, Ti, Zr) NiSn, *Scr. Mater.* 139 (2017) 122–125.

[46] S. Ohno, U. Aydemir, M. Amsler, J.-H. Pöhls, S. Chanakian, A. Zevalkink, M.A. White, S.K. Bux, C. Wolverton, G.J. Snyder, Achieving zT < 1 in Inexpensive Zintl Phase Ca9Zn4+xSb9 by phase boundary mapping, *Adv. Funct. Mater.* 27 (2017) 1606361.

[47] M.C. Peters, J.W. Doak, J.E. Saal, G.B. Olson, P.W. Voorhees, Using first-principles calculations in CALPHAD models to determine carrier concentration of the binary PbSe semiconductor, *J. Electron. Mater.* 48 (2019) 1031–1043.

[48] X. Wang, Y. Sheng, J. Ning, J. Xi, L. Xi, D. Qiu, J. Yang, X. Ke, A critical review of machine learning techniques on thermoelectric materials, *J. Phys. Chem. Lett.* 14 (2023) 1808–1822.

[49] Z. Zhang, H. Chen, T.R. Wei, K. Zhao, X. Shi, Data-driven discovery of high-performance multicomponent solid solution thermoelectric materials, *Mater. Today Energy.* 28 (2022) 101070.

23 High Entropy Materials for Thermal and Electromagnetic Protection

Huimin Xiang and Yanchun Zhou

23.1 THERMAL CONDUCTIVITY OF HIGH-ENTROPY MATERIALS

Generally, the heat is dissipated through the movement of electrons and phonons (vibration of the lattice) in a material, and the origin of a thermal obstacle is the scattering of these energy-bearing media. The scattering sources of these media in a material include the particles themselves and imperfections, such as impurities, grain boundaries, and defects. For high-entropy materials (HEMs), the main structural features are the severely distorted lattice and complex chemical composition, leading to mass disorder and force constant fluctuations [1,2]. These structure features modify the movement of electrons and phonons in HEMs and reshape their thermal conductivity. Thus, to give a clear review of the thermal conductivity of HEMs, we will discuss the electronic structure and lattice dynamics features of HEMs first.

23.1.1 Electronic Structure and Lattice Dynamics

To capture the influence of lattice distortion and chemical complexity on the electronic structure of high-entropy alloys (HEAs), Troparevsky *et al.* utilized Korringa–Kohn–Rostoker coherent-potential approximation (KKR–CPA) to calculate the configurationally averaged density of states (DOS) and Bloch spectral function (BSF) of two five-component HEAs, CoCrFeMnNi and MoNbTaVW [3]. They observed severe smearing for occupied states just below the Fermi energy. Similar smearing near the Fermi level was also observed in $Al_{0.5}TiZrPdCuNi$ HEA by Odbadrakh *et al.* in the calculations of BSF [4], as demonstrated in Figure 23.1. Given the significance of conductive "free" electrons near the Fermi level in energy transportation, high electrical resistance resulting from smearing-induced lifetime shortening of electrons would be a natural speculation. Experimentally, a drastic decrease in electrical conductivity, indeed, has been observed in $Al_xCoCrFeNi$ HEAs and Ni-based HEAs, where the resistance of these HEAs was at least one magnitude higher than that

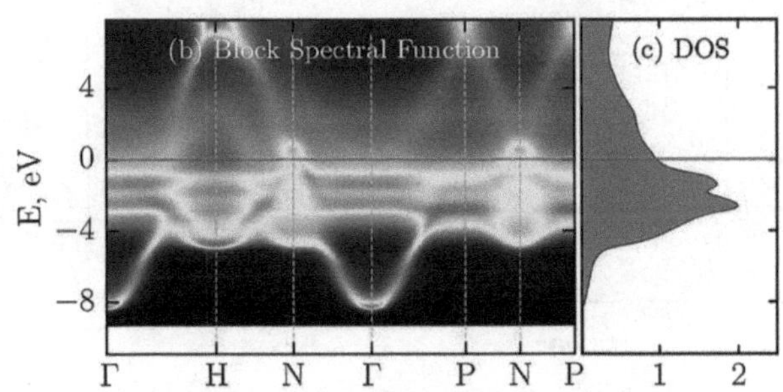

FIGURE 23.1 Calculated electronic structure of the $Al_{0.5}TiZrPdCuNi$ crystalline random alloys. (a) Block Spectral Function, and (b) DOS. The smearing in the DOS is quite obvious. Adapted with permission [4]. Copyright (2019), AIP Publishing

DOI: 10.1201/9781003391388-23

of single-component metals [5,6]. Yet, no theoretical proofs on the electronic structure smearing of conductive high-entropy ceramics (HECs) have been reported, and reduction in electrical conductivity of high-entropy borides, nitrides, and carbides has been discovered [7,8], implying augment in the scattering of conductive electrons.

As regards the lattice dynamics of HEMs, Körmann *et al.* constructed the special quasi-random structures (SQS) for the equimolar 2-, 3-, 4-, and 5-component alloys containing 128, 54, 128, and 125 atoms with the elements of Hf, Mo, Nb, Ta, Ti, V, and W [9]. They observed very strong broadening, reaching widths of up to a few THz, in the mid and high-frequency region (above ~ 4 THz), as demonstrated in Figure 23.2. They also proved that the number of constituents was less important in determining the phonon spectra than a general impression. Similarly, Esters *et al.* found broadening in the total and projected phonon DOS of high-entropy carbides using a partial occupation (POCC) algorithm where statistically weighted structures were used to produce ensembles determining physical properties [10]. They concluded that the phonon broadening in the carbon was the consequence of force constant disorder, whereas, for metals, it was governed by both force constant and mass fluctuations. These results demonstrate that the broadening of phonon dispersion might be a common structure feature in HEMs.

In general, smearing in the electronic structure and phonon dispersion is observed in HMEs due to their severe mass disorder and force constant fluctuations. Smearing-induced shortening of the lifetime of "free" electrons and phonons indicates a significant reduction of thermal conductivity in these distorted materials.

23.1.2 Thermal Conductivity

Compared with that of pure metals and solid solutions, dramatic deterioration of thermal conductivity in HEMs is a common observation in experiments [2,11], and the lattice distortion and chemical complexity are the most used explanations to comprehend this phenomenon. However, these qualitative explanations fail to describe the detailed transportation process of the energy-bearing media and blur the thermal behavior of materials with different conductivity mechanisms. As we know, the main heat carriers in conductors are conductive electrons and phonons, while for insulators,

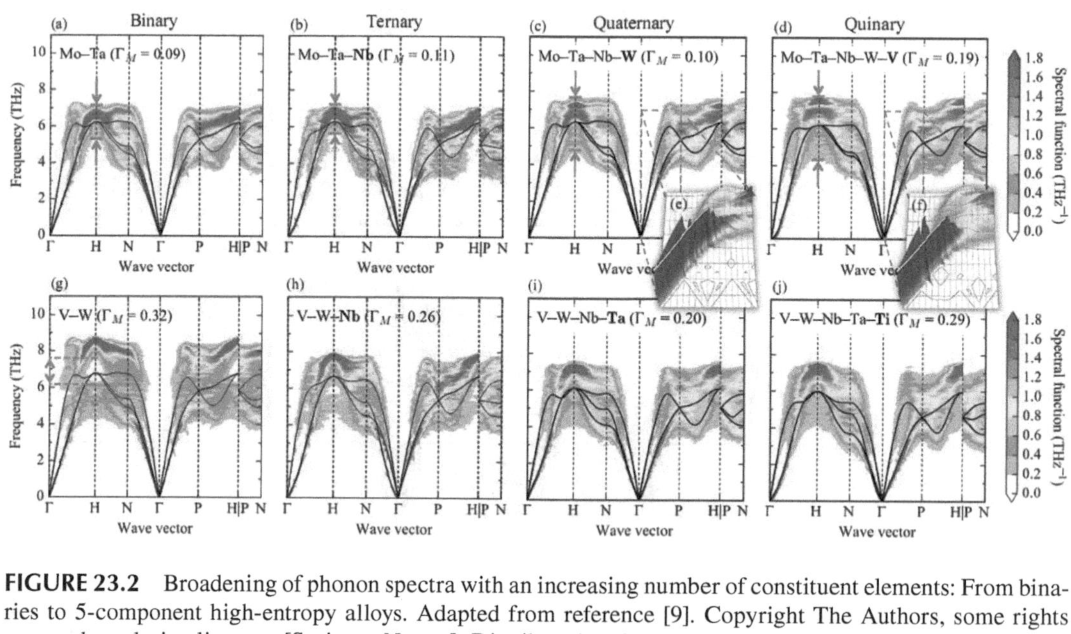

FIGURE 23.2 Broadening of phonon spectra with an increasing number of constituent elements: From binaries to 5-component high-entropy alloys. Adapted from reference [9]. Copyright The Authors, some rights reserved; exclusive licensee [Springer Nature]. Distributed under a Creative Commons Attribution License 4.0 (CC BY)

phonons are the main carriers. Although multicomponent alloying-induced smearing exists in the electronic structure and phonon dispersion, concurrently, the effect of smearing on the thermal transportation of electrons and phonons is different, as we will discuss below. Thus, it would be reasonable to discuss the thermal conductivity behavior of HEMs by their conductive nature.

For conductive HEMs, including HEAs and part of HECs, the thermal conductivity consists of contributions from electrons and phonons, which can be expressed as:

$$\kappa_{\text{total}} = \kappa_e + \kappa_L \tag{23.1}$$

where κ_L and κ_e is the lattice thermal conductivity and electronic thermal conductivity, respectively. For the estimation of κ_L and κ_e, the most common and convenient method is the Wiedemann-Franz law, where the electronic thermal conductivity is related to electric conductivity:

$$\kappa_e = L_0 \sigma T \tag{23.2}$$

where L_0 is the Lorentz constant with the ideal value of 2.44×10^{-8} W·Ω·K^{-2}, σ the electric conductivity, and T the temperature. After separating κ_L and κ_e, an interesting phenomenon is observed, that is, κ_L is comparable to κ_e or even higher at cryogenic temperatures [5,11,12]. While for metals, the majority of the heat is carried by electrons. Moreover, the lattice thermal conductivity (κ_L) seems more insensitive to the composites and structure disorder than κ_e, as demonstrated in Ni-based alloys by Lee *et al.* [12]. Chemical disorder appears to scatter electrons more effectively than phonons in HEAs. Till now, no rigorous explanation has been proposed. Some logical interpretations could be achieved by examining the electronic structures and phonon dispersions of HEAs. Zhang *et al.* [13] calculated the electronic structure of Ni, NiCo, NiFe, and NiCoFeCr by the KKR-CPA method. Their results indicated that the scattering of conductive electrons was intensified with the increase of chemical complexity. The Fermi energy wave vector broadening of the BSF is related to the inverse of the electron mean free path (MFP), thus, intensified smearing indicates confined electron MFP, and, consequently, high residual resistivity. According to the Wiedemann-Franz law, high residual resistivity leads to low electronic thermal conductivity. As regards phonon dispersion, heavy smearing is also observed when a chemical disorder is introduced [9]. However, the phonons being smeared are distributed in the mid and high-frequency region (above ~4 THz), while the long wavelength limit (near the center of Brillouin zone (BZ)) is nearly untouched. In metals, κ_L is mainly from the long wavelength phonons due to their large group velocity and small scattering rate [14]. Consequently, the dependence of lattice thermal conductivity of HEAs on chemical disorder is weak, adding more elements does not necessarily further decrease the lattice thermal conductivity [13].

Due to their large interatomic bonding and force constant, the majority of the heat is not necessarily carried by electrons in conductive ceramics [15]. Nevertheless, a significant reduction in the thermal conductivity of conductive HECs has also been observed in multiple material systems [8,16–18]. To the authors' knowledge, no research regarding the effect of chemical complexity on κ_L and κ_e, or electronic structure and phonon dispersion of these ceramics has been reported, making the exploration of the chemical-dependent thermal behavior of conductive HECs difficult. Presuming that the deduction used to discuss the thermal response of thermal conductivity of HEAs is also suitable to conductive HECs, that is, electrons near the Fermi surface and acoustic phonons at the BZ boundary are heavily scattered, the deterioration of thermal conductivity in conductive HECs is also from the broadening of conductive electrons. Verification of this conjecture requires solid experimental and theoretical proof.

Another interesting phenomenon regarding the thermal conductivity of conductive HEMs is their temperature dependence, which is positively correlated to temperature. Normally, for pure metal or conductive ceramics, their thermal conductivities decrease fast with temperature and approach a lower limit at high temperatures. This phenomenon has been observed in Ni-based

HEAs, $Al_xCoCrFeNi$, AlCrFeMnNi, and high-entropy carbide [5,8,11,19]. Several interpretations have been proposed, such as the augmentation of electronic MFP induced by low sensitivity of phonon concentration with respect to temperatures and large linear thermal expansion in HEAs [5], and increased phonon MFP at higher temperature due to the expansion of lattice [19]. Jin *et al.* separated the contribution of electrons and phonons to the total thermal conductivity of Ni-based HEAs [11]. They found that the lattice thermal conductivity of HEAs increased rapidly with temperature up to 50 K, and showed weak temperature dependence afterward, while the electronic contribution increased almost linearly with temperature. They attributed the temperature dependence of the thermal conductivity of these alloys to their large and nearly constant resistivity. Their explanation is consistent with the results of electronic structure and phonon dispersion. As discussed above, the electron states being heavily smeared distribute in the energy range of 2 eV in the vicinity of the Fermi level, indicating weak additional contribution from temperature-related electron-electron scattering and weak temperature-dependent resistivity in HEAs [12,13]. Thus, the contribution from electrons to total thermal conductivity is nearly linearly related to temperature according to Equation (23.2). As for lattice thermal conductivity, since the modes near the center of BZ are untouched, κ_L increases with temperature due to the consecutive thermal activation of these modes. On the other hand, with the temperature increasing further, the modes at the boundary of BZ are activated; however, the contribution from these modes is negligible since they are heavily smeared, leading to the weak temperature dependence of κ_L at higher temperatures. More direct proofs are needed to establish the relationship between thermal conductivity and the movements of electrons and phonons.

The thermal barrier in non-conductive HECs is also intensified, and deterioration of thermal conductivity is observed [2]. The temperature dependence of their thermal conductivity is rather weak above room temperature [20,21]. Several interpretations have been proposed. Braun *et al.* [20] investigated the thermal conductivity of (MgCoNiCuZn)O. They found that with the addition of a sixth cation (Sc, Sb, Sn, Cr or Ge), 50% reduction in thermal conductivity could be achieved. The results from a virtual crystal approximation (VCA) model suggested that the amorphous or glass-like thermal conductivity was attributed to disorder in interatomic force constants (IFC) rather than mass and size disorder. Wright *et al.* [22] investigated a series of 22 single-phase pyrochlores, including 18 compositionally complex (i.e., medium- and high-entropy) pyrochlore oxides and four additional "low-entropy" benchmark compositions. They found that the size disorder parameter was a more effective descriptor to forecast a reduction in thermal conductivity than the ideal mixing configurational entropy, and severe lattice distortion was significant in reducing the thermal conductivity. Till now, no general-accepted conclusion or explanation on the thermal conductivity reduction of these non-conductive HECs has been achieved. Delicate examinations of the phonon dispersion and thermal conductivity at cryogenic temperatures are needed to obtain a clearer picture of the thermal behavior of non-conductive HECs.

23.1.3 Predictions of Thermal Conductivity of HEMs

Theoretical predicting is appealing due to its ability in revealing the micro-mechanism of transportation, screening material systems, and tuning the thermal property. As discussed above, the structure and chemical composition in HEMs are quite complex, making it difficult to choose or define a suitable period cell in the simulations. The state-of-the-art methods within density functional theory (DFT) to deal with disordered solid solutions can be categorized into two classes: one is based on the coherent-potential approximation (CPA) [23]; the other is based on supercell models. Due to its ability to calculate of the interatomic force constant of HEMs, and consequently, phono dispersion and thermal conductivity, the supercell models are usually employed. Considering the computational demanding nature of DFT calculations, the SQS are introduced to provide a good representation of chemical randomness in small supercells [24], as Körmann *et al.* did in the exploration of phonon dispersion of high entropy alloys [9]. Nevertheless, using DFT-based methods to calculate the thermal conductivity of these disordered materials is still difficult and time-consuming in the evaluation of multiple scattering processes of phonons. Thus, molecular dynamics (MD) is always chosen in the theoretical estimation of the thermal conductivity of HEMs.

Caro *et al.* performed non-equilibrium MD calculations of phonon thermal conductivity in multiple component solid solutions represented by Lennard–Jones (LJ) potentials [25]. They found that the lattice thermal conductivity of multicomponent alloy was coupled with the mass and atomic size of the components. Heavier element was more efficient in decreasing the thermal conductivity if, simultaneously, it had smaller size. In the calculation of thermal conductivity of $(Mg_{0.2}Co_{0.2}Ni_{0.2}Cu_{0.2}Zn_{0.2})O$ as well as ones plus Sc, Sn, Cr, or Ge in equal-molar cation proportions, Lim *et al.* [26] constructed supercells containing a total of 30720 atoms. The thermal conductivity was calculated using the Green-Kubo method by MD simulations, and the interatomic potential was modeled using a pair sum of an exponential-6 function and long-range Coulomb interactions. They found that phonon scattering from charge disorder was responsible for the lowering of thermal conductivity with the addition of a sixth cation.

Although MD is efficient in thermal conductivity simulation, the accuracy of the simulation depends on the description precision of interatomic potentials on the atomic interaction. For HMEs, due to the chemical and bonding complexity, the interatomic interaction cannot be described precisely by classic potentials, and the accuracy of the simulations is questionable. To tackle this problem, machine learning (ML) potential is proposed. The description of ML potential on the locality of atomic interactions and atomic energies is more accurate than traditional MD potentials, and the accuracy of MD simulation based on these potentials improves substantially. Dai *et al.* [27,28] employed a deep leaning potential to investigate lattice thermal conductivity of high-entropy $(Zr_{0.2}Hf_{0.2}Ti_{0.2}Nb_{0.2}Ta_{0.2})C$ and $(Zr_{0.2}Hf_{0.2}Nb_{0.2}Ta_{0.2}Ti_{0.2})B_2$. They found that the predicted lattice thermal conductivity of these two HECs was rather low, which was consistence with experiments, as shown in Figure 23.3.

23.1.4 Applications as Thermal Protection Materials

Introducing multicomponent leads to the dramatic deterioration of thermal conductivity in HEMs. This appealing phenomenon makes these materials quite promising in the field of thermal protection from high-temperature damage, where low thermal conductivity is mandatory. Besides that, other superior properties are also found in these materials, such as good high-temperature stability

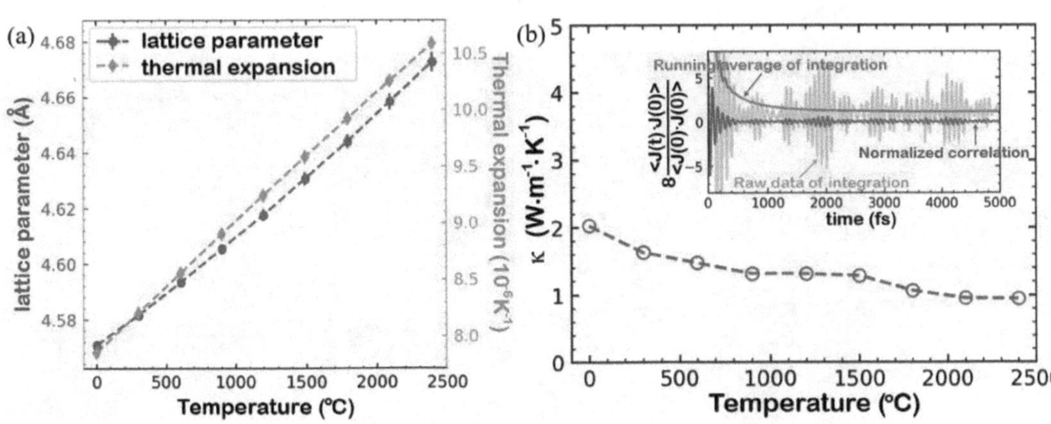

FIGURE 23.3 Temperature-dependent (a) lattice parameters and thermal expansion coefficients and (b) phonon thermal conductivity κ of HEMC $(Ti_{0.2}Zr_{0.2}Hf_{0.2}Nb_{0.2}Ta_{0.2})C$. The inset in (b) shows a typical heat current autocorrelation function, integration of autocorrelation function, and the running average of the integration. Adapted with permission [27], Copyright (2020), Elsevier

due to the high-entropy effect, and low grain growth rate from sluggish diffusion. The combination of these properties makes them perfect candidates for thermal/environmental barrier coatings (T/EBCs) used in turbine engines and thermal protection and insulation barriers used in hypersonic vehicles. In this section, the state-of-the-art progress regarding the application of HEMs as thermal protection materials will be reviewed.

23.1.4.1 T/EBCs Applications

The main requirements of T/EBCs include low thermal conductivity, good thermal stability, low grain growth rate, good water-vapor corrosion resistance, and good molten CMAS (calcium-magnesia-aluminosilicate) corrosion resistance. The state-of-the-art T/EBC material used in the Ni-based superalloy blade of aero-engines is Y_2O_3 stabilized ZrO_2 (YSZ). One of the main drawbacks of current T/EBCs is their stability at high temperatures. Long-term service at high temperatures leads to a phase transition, sintering and grain coarsening, resulting in failure of the coating. High-entropy strategy is proven to be an efficient way to overcome this problem. Zhao *et al.* [29] reported that high-entropy $(La_{0.2}Ce_{0.2}Nd_{0.2}Sm_{0.2}Eu_{0.2})_2Zr_2O_7$ exhibited lower thermal conductivity due to the enhanced scattering of phonons, and much slower grain growth rate that resulted from sluggish diffusion effect. Figure 23.4 compares the microstructure evolution of $(La_{0.2}Ce_{0.2}Nd_{0.2}Sm_{0.2}Eu_{0.2})_2Zr_2O_7$ and $La_2Zr_2O_7$ after annealing at 1500°C, which reveals that grain growth rate of high entropy $(La_{0.2}Ce_{0.2}Nd_{0.2}Sm_{0.2}Eu_{0.2})_2Zr_2O_7$ is much lower. Moreover, abnormal growth of grains is suppressed in high entropy $(La_{0.2}Ce_{0.2}Nd_{0.2}Sm_{0.2}Eu_{0.2})_2Zr_2O_7$. This phenomenon has also been discovered in high-entropy $(Y_{0.25}Yb_{0.25}Er_{0.25}Lu_{0.25})_2(Zr_{0.5}Hf_{0.5})_2O_7$ and $(Y_{0.2}Yb_{0.2}Lu_{0.2}Eu_{0.2}Er_{0.2})_3Al_5O_{12}$ systems [30,31]. A slower grain growth rate is beneficial to the long-term service of the coating by minimizing the thermal stress from sintering.

Another advantage of HEMs is the enhanced phase stability since cracks resulting from the volume change during the phase transition are the main failure mechanism of T/EBCs. Zhao *et al.* [32] employed multicomponent alloying to solve the decomposition of ZrP_2O_7 at high temperatures. They synthesized a novel solid solution $(TiZrHf)P_2O_7$, which might not be categorized as high-entropy material strictly, nevertheless, improved thermal stability of this solid solution was found. Unlike ZrP_2O_7, after annealing at 1550 °C for 3 h, no decomposition was detected in this novel multicomponent material. Sun *et al.* [33] designed and synthesized a multicomponent $(Gd_{1/6}Tb_{1/6}Dy_{1/6}Tm_{1/6}Yb_{1/6}Lu_{1/6})_2Si_2O_7$ disilicate. Compared with single-component rare-earth (RE)-disilicates, the melting point of this high-entropy disilicate was much higher and no phase transition was found before melting. These results demonstrate that high-entropy design is efficient in improving the thermal

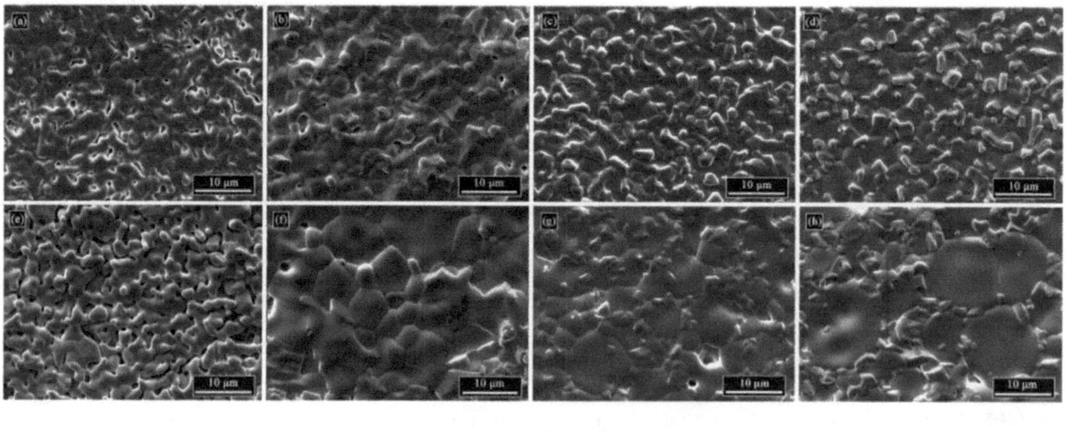

FIGURE 23.4 Microstructures of $(La_{0.2}Ce_{0.2}Nd_{0.2}Sm_{0.2}Eu_{0.2})_2Zr_2O_7$ and $La_2Zr_2O_7$ compacts after annealing at 1500 °C in air for (a): 1 h, (b): 6 h, (c): 12 h, (d): 18 h for $(La_{0.2}Ce_{0.2}Nd_{0.2}Sm_{0.2}Eu_{0.2})_2Zr_2O_7$ and (e): 1 h, (f): 6 h, (g): 12 h, (h): 18 h for $La_2Zr_2O_7$. Adapted with permission [29], Copyright (2019), Elsevier

stability of T/EBC and suppressing the possible phase transition, which is beneficial to high-temperature thermal protection applications.

For a suitable T/EBC, a certain resistance to CMAS corrosion is needed. Sun *et al.* investigated the corrosion mechanism of high entropy disilicate ($(Er_{0.25}Tm_{0.25}Yb_{0.25}Lu_{0.25})_2Si_2O_7$) in CMAS at 1500°C [34]. When in contact with molten CMAS, high-entropy disilicate dissolved and Ca–Si–RE containing apatite precipitated when the elements reached saturation. Nevertheless, the resistance of high entropy disilicate to CMAS was significantly higher than that of single constituents, and the improvements could be attributed to a slower dissolution rate and enhanced grain boundary stability, which resulted from the sluggish diffusion effect and lattice distortion. Sun *et al.* examined CMAS corrosion resistance of two RE bixbyite structured high-entropy oxides, $(Eu_{0.2}Er_{0.2}Lu_{0.2}Y_{0.2}Yb_{0.2})_2O_3$ and $(Sm_{0.2}Er_{0.2}Lu_{0.2}Y_{0.2}Yb_{0.2})_2O_3$ at 1250 and 1350°C for 4h [35]. Compared with other T/EBC materials and single constituents, the thickness of reaction layers for two high-entropy oxides was much thinner. The reaction mechanism was characterized by the diffusion of CMAS to HE RE_2O_3. The improvement in CMAS corrosion resistance is also observed in high-entropy fluorite and some medium-entropy oxides [36,37].

23.1.4.2 Ultrahigh Temperature Thermal Protection

Besides the T/EBCs, high-entropy materials with low thermal conductivity are also promising as thermal protection and insulation component in hypersonic vehicles. The service environment of hypersonic vehicles is harsh, most of the oxides cannot withstand such an extreme environment. Therefore, materials with higher melting points are preferred, i.e., ultrahigh temperature ceramics (UHTCs) including transition metal borides, nitrides, and carbides. It has been proven that high-entropy UHTCs have much lower thermal conductivity [8,16], however, it is still too high for thermal protection and insulation applications at ultrahigh temperatures. A new strategy is introduced to decrease the thermal conductivity of these HECs further. Zhou and his coworkers developed a novel *in-situ* reaction/partial sintering method to fabricate highly porous bulk materials [2]. They carefully design the reactions to ensure enormous gas release in the process of reduction. The released gas can not only increase the porosity but also block large shrinkage during the reaction. Thus, no pore-forming agent is needed and the reaction/partial sintering of the porous HECs can be achieved in a single step. Using this technique, Chen *et al.* prepared highly porous high-entropy carbide and diboride with ultralow thermal conductivity [38,39]. For example, they prepared a porous (ZrHfTiNbTa)C with a porosity of 80.99%, compressive strength of 3.45 MPa, and low room temperature thermal conductivity of 0.39 $W \cdot m^{-1} \cdot K^{-1}$. In addition, they found that this porous high-entropy carbide possessed incredible structure stability at ultrahigh temperatures. After reheating the porous sample to 1850°C, no phase transition, pore size, or volume shrinkage was found. Using the same technique, they designed and prepared porous high-entropy metal hexaborides with a high population of through-hole [40]. These hexaborides possessed higher permeability than other porous UHTCs, which guaranteed their application in active thermal protection.

In general, high-entropy materials possess lower thermal conductivity, higher phase and structure stability at high temperatures, and better corrosion resistance from molten salt than their single-component counterparts. The combination of these superior properties promises their applications as high-temperature thermal protection materials.

23.2 ELECTROMAGNETIC ATTENUATION PROPERTY

High-performance electromagnetic (EM) wave absorption materials are in high demand due to the worsening of electromagnetic interference and pollution, and/or the need for electromagnetic stealth in defense areas. In essence, the absorption of EM energy is implemented by the attenuation of electric or/and magnetic fields of EM wave. Thus, based on their attenuation mechanism, the EM wave absorbing materials can be roughly categorized into two types: dielectric loss-type and magnetic loss-type. However, problems emerge when these materials serve at high temperatures. For dielectric loss-type materials, such as carbon-based materials and silicon carbide, the limited absorption capacity and narrow effective absorption bandwidth (EAB) due to the poor impedance match hinder their application. Meanwhile, poor oxidation and corrosion resistance hinder their high-temperature applications. As regards magnetic loss-type ones, they suffer from a sudden decrease of permeability at high temperatures. Moreover, the stability of delicately designed nano-structures and interfaces at high temperatures is doubtful. Therefore, it is of significance to develop materials with highly efficient EM wave absorption performance, good thermal stability, and good resistance to oxidation and corrosion for high-temperature EM absorption applications. As we will discuss below, high-entropy materials might be the solution. Before we discuss the EM absorption potential of high-entropy materials, the effect of multicomponent alloying on the dielectric and magnetic properties of HEMs will be reviewed shortly. Considering the intense EM reflective effect of HEAs, the discussion will focus on HECs.

23.2.1 Dielectric Property

In HECs, there is more than one sublattice in the structure. The sublattice exhibits not only long-range periodicity with the distortion that influences the behaviors of electrons, dipoles, and band structure but also a compositional disorder that reduces short-range order resulting from the randomly distributed multi-elements, where plenty of flexible methods to tailor the dielectric properties of HECs are provided [2].

Bérandan *et al.* reported colossal dielectric constants in rock-salt-structured HEOs [41]. The relative permittivity of $(Mg,Co,Ni,Cu,Zn)_{0.95}Li_{0.05}O$ measured at 300 K and 20 Hz with an LCR bridge method was close to 1×10^3. They found that, regardless of the substitution atoms, the dielectric constant of materials in this HEC family was large. They contributed the large relative permittivity to dipolar fluctuations in nano-size domains, interface effects, electronic phase separation, defect dipoles, dipole relaxations, or hopping transport, which needed further verification.

The chemical composition is proven to influence the dielectric property of high-entropy materials greatly. Pu *et al.* examined the effect of multi-element doping in the A-site on the dielectric properties of perovskite [42]. They found that the dielectric peak of $(Na_{0.2}Bi_{0.2}Ba_{0.2}Sr_{0.2}Ca_{0.2})TiO_3$ was broad and shifted toward higher temperature with increasing frequency, and the maximum discharge energy density for this HEO was determined as 1.02 J/cm^3 under the electric strength of 145 kV/cm, which exceeded those of many single-phase perovskite ceramics. While for $(Bi_{0.2}Na_{0.2}K_{0.2}Ba_{0.2}Ca_{0.2})TiO_3$, the discharge energy storage density decreased to a value of 0.684 J/cm^3 [43].

In spite of a few types of research concerning the dielectric property, it is still proven that controlling the chemical composition and lattice distortion is an efficient approach to tuning the dielectric property of HECs in a large range.

23.2.2 Magnetic Property

Investigations on magnetic properties of HECs are still in their nascent beginning. Mao *et al.* examined the magnetic properties of rock-salt $(Co_{0.2}Cu_{0.2}Mg_{0.2}Ni_{0.2}Zn_{0.2})O$ HEC [44]. These nanocrystalline powders exhibited long-range antiferromagnetic behavior below Néel temperature (T_N = 106 K), which could be well understood by the super-exchange interactions in the rock-salt HEO. Such long-range magnetic ordering in rock-salt HEC was confirmed by Jimenez-Segura *et al.*, where they found antiferromagnetic order below 120 K despite the structural disorder of randomly distributed magnetic ions [45]. Meisenheimer *et al.* examined the role of disorder and composition on the exchange anisotropy of permalloy/$(Mg_{0.25(1-x)}Co_xNi_{0.25(1-x)}Cu_{0.25(1-x)}Zn_{0.25(1-x)})O$ heterostructures [46]. They found that the disorder, exchange field, and magnetic anisotropy could be tuned by changing the composition of the oxide. And the exchange bias observed at low temperatures could reach 10 times of that observed in a CoO/Py control sample.

Besides the rock-salt oxides, the magnetic properties of perovskite-type high-entropy oxides have also been explored. Witte *et al.* comprehensively investigated the magnetic properties of RE and transition metal (TM) based perovskite-type high-entropy oxides [47,48]. They found that the impact of the chemical disorder on the magnetic properties of these compounds was largely different. In general, the disorder on the A-sites did not drastically change the nature of the magnetic, where canted antiferromagnetic super-exchange prevailed in the TM sublattice. While for those with configurational disorder on the B-sites, the nature of the magnetic exchange interaction was intrinsically changed. The magnetic moments of the different transition metals coupled either parallel or antiparallel depending on the Goodenough–Kanamori rules, leading to frustrated and non-collinear magnetism with antiferro- and ferromagnetic contributions and a comparably large magnetic anisotropy.

23.2.3 Electromagnetic Attenuation Applications

As stated above, high-performance electromagnetic wave absorption materials with good high-temperature stability are in high demand. Previous research has demonstrated the capability of high-entropy design in tailoring the dielectric and magnetic properties of HEMs, therefore, rational design of chemical composition and structure would be a feasible approach to realize good impedance match and attenuation ability in single-phase material. Efforts have been made in several promising material systems.

Ma *et al.* [49] designed and prepared three high-entropy spinel-type ferrite ceramics including $(Mg_{0.2}Mn_{0.2}Fe_{0.2}Co_{0.2}Ni_{0.2})Fe_2O_4$, $(Mg_{0.2}Fe_{0.2}Co_{0.2}Ni_{0.2}Cu_{0.2})Fe_2O_4$, and $(Mg_{0.2}Fe_{0.2}Co_{0.2}Ni_{0.2}Zn_{0.2})Fe_2O_4$ to cope the poor dielectric loss capacity of ferrite. Indeed, they found enhanced dielectric loss in these materials from the interfacial polarization of nano-domains and the resonance absorption between tetrahedral site and octahedral site ions, and better impedance matching in the frequency range of 7.8–18 GHz. The HECs they designed show much better electromagnetic wave absorption properties than other nano-ferrites and ferrite-based composites with EAB larger than 6.12 GHz. Starting from perovskite $BaTiO_3$, Zhao *et al.* [50] designed a high-entropy $(Ca_{0.2}Sr_{0.2}Ba_{0.2}La_{0.2}Pb_{0.2})TiO_3$ composition. Significant grain boundaries, discontinuous lattice fringes, and lattice deficiencies were ubiquitous, indicating the generation of ultra-dense entropy-driven structural defects in high-entropy titanate perovskite. Stronger strains were found in the lattice of the high-entropy perovskites along the ε_{xx} and ε_{yy} directions in comparison with those of their low-entropy and medium-entropy counterparts. Moreover, electron holography analysis demonstrated enhanced charge distribution around grain boundaries and strain-concentrated location in high-entropy perovskites, which was advantageous in enhancing dielectric polarization loss. Benefiting from these advantages, high-entropy $(Ca_{0.2}Sr_{0.2}Ba_{0.2}La_{0.2}Pb_{0.2})TiO_3$ exhibited a remarkably enhanced absorption capability, with the EAB improved almost two times compared to $BaTiO_3$.

Other than oxides, Zhou *et al.* have done impressive pioneer works on high-entropy UHTCs with good EM wave absorptions for their good high-temperature stability [51–56]. Normally, UHTCs are not ideal electromagnetic attenuation materials due to their relatively high electrical conductivity. To realize electromagnetic attenuation in UHTCs, a high-entropy design is employed by Zhou and his coworkers. By alloying Cr in high-entropy borides and carbides, a very impressive EM wave absorption ability was obtained [53,56], where EAB was larger than 10 GHz. As shown in Figure 23.5, high-entropy $(Cr_{0.2}Zr_{0.2}Hf_{0.2}Nb_{0.2}Ta_{0.2})B_2$ exhibits a good combination of EMW absorption properties with RL_{min} of –56.2 dB (at 8.48 GHz with a film thickness of 2.63 mm) and the EAB of 11.0 GHz (film thickness of 2.2 mm), which is superior to most other EM wave absorption materials. They attributed the electromagnetic attenuation ability to the enhanced magnetic loss and improved impedance match due to the addition of Cr. They also proved that the ground magnetic state of CrB_2 and CrB_2-based solid solutions were ferromagnetic by first principles [51]. In contrast, the addition of Ti, a nonmagnetic element, degenerates the EM absorption properties greatly, indicating the important role of Cr in high-entropy EM absorption UHTCs. Moreover, since

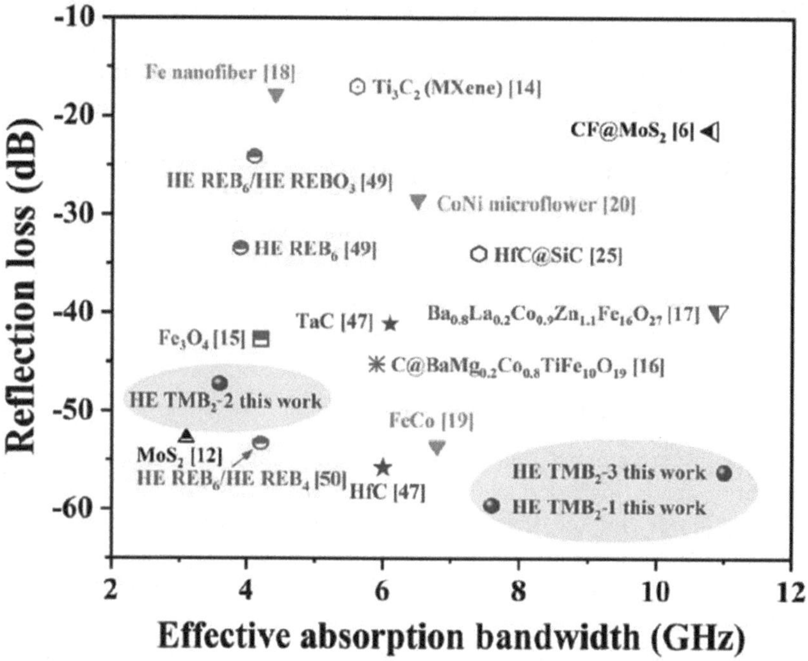

FIGURE 23.5 Comparison of optimal optimized EAB and RL_{min} values of high-entropy materials and other EMW absorbers. Adapted from reference [53]. Copyright The Authors, some rights reserved; exclusive licensee [Springer Nature]. Distributed under a Creative Commons Attribution License 4.0 (CC BY)

no need to design delicate nanostructures, it is much easier to fabricate the high-entropy UHTCs using solid-state reactions with better high-temperature stability and oxidation resistance than other composites made of nano-materials.

23.3 SUMMARY AND PERSPECTIVES

In this chapter, potential applications of HEMs in high-temperature thermal and electromagnetic protection are briefly introduced. Beneficial from chemical disorder and the "cock-tail" effect, high-entropy materials overwhelm their single-component counterparts, such as good thermal stability, low thermal conductivity, slow grain growth rate, and good corrosion resistance. Moreover, considering the expanded solubility of different elements in HEMs, the design space of HEMs is huge, which gives a great opportunity for demand-driven property design in HEMs.

The research on HEMs is still in its nascent beginning, especially for HECs. There may be still a long period before these materials are applied in real life since multi-factors should be considered, such as cost, and lifespan. Nevertheless, the fascinating properties originating from a high-entropy design make HEMs promising candidates in many industrial applications. The discussion in this chapter provides a valuable perspective on the possible applications in thermal and electromagnetic protection, and it can be expected that more and more works will emerge in the search for HEMs with better performance.

REFERENCES

1 J. Yeh, Alloy design strategies and future trends in high-entropy alloys, *JOM* 65 (2013) 1759–1771.
2 H. Xiang, F. Dai, Y. Zhou, *High-entropy materials: from basics to applications*, 2023, Weinheim, Germany: WILEY-VCH GmbH.

3 M.C. Troparevsky, J.R. Morris, M. Daene, Y. Wang, A.R. Lupini, G.M. Stocks, Beyond atomic sizes and Hume-Rothery rules: Understanding and predicting high-entropy alloys, *JOM* 67 (2015) 2350–2363.

4 K. Odbadrakh, L. Enkhtor, T. Amartaivan, D.M. Nicholson, G.M. Stocks, T. Egami, Electronic structure and atomic level complexity in $Al_{0.5}$TiZrPdCuNi high-entropy alloy in glass phase, *J. Appl. Phys.* 126: (2019) 095104.

5 H. Chou, Y. Chang, S. Chen, J. Yeh, Microstructure, thermophysical and electrical properties in Al_xCoCrFeNi ($0 \leq x \leq 2$) high-entropy alloys, *Mater. Sci. Eng. B* 163 (2009) 184–189.

6 Y. Zhang, G.M. Stocks, K. Jin, C. Lu, H. Bei, B.C. Sales, L. Wang, L.K. Béland, R.E. Stoller, G.D. Samolyuk, M. Caro, A. Caro, W.J. Weber, Influence of chemical disorder on energy dissipation and defect evolution in concentrated solid solution alloys, *Nat. Comm.* 6 (2015) 8736.

7 H. Zhang, B. Zhao, F. Dai, H. Xiang, Z. Zhang, Y. Zhou, $(Cr_{0.2}Mn_{0.2}Fe_{0.2}Co_{0.2}Mo_{0.2})B$: A novel high-entropy monoboride with good electromagnetic interference shielding performance in K-band, *J. Mater. Sci. Technol.* 77 (2021) 58–65.

8 T. Wen, B. Ye, M.C. Nguyen, M. Ma, Y. Chu, Thermophysical and mechanical properties of novel high-entropy metal nitride-carbides. *J. Am. Ceram. Soc.* 103 (2020) 6475–6489.

9 F. Körmann, Y. Ikeda, B. Grabowski, M.H.F. Sluiter, Phonon broadening in high entropy alloys, *npj Comput. Mater.* 3 (2017) 36.

10 M. Esters, C. Oses, D. Hicks, M.J. Mehl, M. Jahnátek, M.D. Hossain, J.-P. Maria, D.W. Brenner, C. Toher, S. Curtarolo, Settling the matter of the role of vibrations in the stability of high-entropy carbides, *Nat. Commun.* 12 (2020) 5747.

11 K. Jin, B.C. Sales, G.M. Stocks, G.D. Samolyuk, M. Daene, W.J. Weber, Y. Zhang, H. Bei, Tailoring the physical properties of Ni-based single-phase equiatomic alloys by modifying the chemical complexity, *Sci. Rep.* 6 (2016) 20159.

12 J.I. Lee, H.S. Oh, E.S. Park, Manipulation of σ_y/κ ratio in single phase FCC solid-solutions, *J. Appl. Phys.* 109 (2016) 061906.

13 Y. Zhang, G.M. Stocks, K. Jin, C. Lu, H. Bei, B.C. Sales, L. Wang, L.K. Béland, R.E. Stoller, G.D. Samolyuk, M. Caro, A. Caro, W.J. Weber, Influence of chemical disorder on energy dissipation and defect evolution in concentrated solid solution alloys. *Nat. Comm.* 6 (2015) 8736.

14 Z. Tong, S. Li, X. Ruan, H. Bao, Comprehensive first-principles analysis of phonon thermal conductivity and electron-phonon coupling in different metals, *Phys. Rev. B* 100 (2019) 144306.

15 H. Xiang, J. Wang, Y. Zhou, Theoretical predictions on intrinsic lattice thermal conductivity of ZrB2, *J. Eur. Ceram. Soc.* 39 (2019) 2982–2988.

16 X. Yan, L. Constantin, Y. Lu, J.-F. Silvain, M. Nastasi, B. Cui, $(Hf_{0.2}Zr_{0.2}Ta_{0.2}Nb_{0.2}Ti_{0.2})C$ high-entropy ceramics with low thermal conductivity, *J. Am. Ceram. Soc.* 101 (2018) 4486–4491.

17 B. Ye, T. Wen, M.C. Nguyen, L. Hao, C. Wang, Y. Chu, First-principles study, fabrication and characterization of (Zr0.25Nb0.25Ti0.25V0.25)C high-entropy ceramics, *Acta Mater.* 170 (2019) 15–23.

18 C.M. Rost, T. Borman, M.D. Hossain, M. Lim, K.F. Quiambao-Tomko, J.A. Tomko, D.W. Brenner, J.-P. Maria, P.E. Hopkins, Electron and phonon thermal conductivity in high entropy carbides with variable carbon content, *Acta Mater.* 196 (2020) 231–239.

19 C. Lu, S. Lu, J. Ye, W. Hsu, Thermal expansion and enhanced heat transfer in high-entropy alloys, *J. Appl. Cryst.* 46 (2013) 736–739.

20 J.L. Braun, C.M. Rost, M. Lim, A. Giri, D.H. Olson, G.N. Kotsonis, G. Stan, D.W. Brenner, J.-P. Maria, P.E. Hopkins, Charge-induced disorder controls the thermal conductivity of entropy-stabilized oxides, *Adv. Mater.* 30 (2018) 1805004.

21 M. Ridley, J. Gaskins, P. Hopkins, E. Opila, Tailoring thermal properties of multi-component rare earth monosilicates, *Acta Mater.* 195 (2020) 698–707.

22 A.J. Wright, Q. Wang, S.T. Ko, K.M. Chung, R. Chen, J. Luo, Size disorder as a descriptor for predicting reduced thermal conductivity in medium- and high-entropy pyrochlore oxides, *Scr. Mater.* 181 (2020) 76–81.

23 L. Vitos, Total-energy method based on the exact muffin-tin orbitals theory, *Phys. Rev. B* 64 (2001) 014107.

24 A. Zunger, S. Wei, L.G. Ferreira, J.E. Bernard, Special quasirandom structures, *Phys. Rev. Lett.* 65 (1990) 353–356.

25 M. Caro, L.K. Béland, G.D. Samolyuk, R.E. Stoller, A. Caro, Lattice thermal conductivity of multi-component alloys, *J. Alloy Compd.* 648 (2015) 408–413.

26 M. Lim, Z. Rak, J.L. Braun, G.N. Kotsonis, P.E. Hopkins, J.-P. Maria, D.W. Brenner, Influence of mass and charge disorder on the phonon thermal conductivity of entropy stabilized oxides determined by molecular dynamics simulations, *J. Appl. Phys.* 125 (2019) 055105.

27 F. Dai, B. Wen, Y. Sun, H. Xiang, Y. Zhou, Theoretical prediction on thermal and mechanical properties of high entropy (Zr0.2Hf0.2Ti0.2Nb0.2Ta0.2)C by deep learning potential, *J. Mater. Sci. Technol.* 43 (2020) 168–174.

28 F. Dai, Y. Sun, B. Wen, H. Xiang, Y. Zhou, Temperature dependent thermal and elastic properties of high entropy (Ti0.2Zr0.2Hf0.2Nb0.2Ta0.2)B2: Molecular dynamics simulation by deep learning potential, *J. Mater. Sci. Technol.* 72 (2021) 8–15.

29 Z. Zhao, H. Xiang, F. Dai, Z. Peng, Y. Zhou, (La0.2Ce0.2Nd0.2Sm0.2Eu0.2)2Zr2O7: A novel high-entropy ceramic with low thermal conductivity and sluggish grain growth rate, *J. Mater. Sci. Technol.* 35 (2019) 2647–2651.

30 Z. Zhao, H. Chen, H. Xiang, F. Dai, X. Wang, W. Xu, K. Sun, Z. Peng, Y. Zhou, (Y0.25Yb0.25Er0.25 Lu0.25)2(Zr0.5Hf0.5)2O7: A defective fluorite structured high entropy ceramic with low thermal conductivity and close thermal expansion coefficient to Al2O3, *J. Mater. Sci. Technol.* 39 (2020) 167–172.

31 H. Chen, Z. Zhao, H. Xiang, F. Dai, W. Xu, K. Sun, J. Liu, Y. Zhou, High entropy (Y0.2Yb0.2Lu0.2Eu 0.2Er0.2)3Al5O12: A novel high temperature stable thermal barrier material, *J. Mater. Sci. Technol.* 48 (2020) 57–62.

32 Z. Zhao, H. Xiang, F. Dai, Z. Peng, Y. Zhou, (TiZrHf)P2O7: An equimolar multicomponent or high entropy ceramic with good thermal stability and low thermal conductivity, *J. Mater. Sci. Technol.* 35 (2019) 2227–2231.

33 L. Sun, Y. Luo, X. Ren, Z. Gao, T. Du, Z. Wu, J. Wang, A multicomponent γ-type (Gd1/6Tb1/6Dy1 /6Tm1/6Yb1/6Lu1/6)2Si2O7 disilicate with outstanding thermal stability, *Mater. Res. Lett.* 8 (2020) 424–430.

34 L. Sun, Y. Luo, Z. Tian, T. Du, X. Ren, J. Li, W. Hu, J. Zhang, J. Wang, High temperature corrosion of (Er0.25Tm0.25Yb0.25Lu0.25)2Si2O7 environmental barrier coating material subjected to water vapor and molten calcium-magnesium-aluminosilicate (CMAS), *Corr. Sci.* 175 (2020) 108881.

35 Y. Sun, H. Xiang, F. Dai, X. Wang, Y. Xing, X. Zhao, Y. Zhou, Preparation and properties of CMAS resistant bixbyite structured high-entropy oxides RE2O3 (RE=Sm, Eu, Er, Lu, Y, and Yb): Promising environmental barrier coating materials for Al2O3f/Al2O3 *composites*, *J. Adv. Ceram.* 10 (2021) 596–613.

36 A.J. Wright, C.Y. Huang, M.J. Walock, A. Ghoshal, M. Murugan, J. Luo, Sand corrosion, thermal expansion, and ablation of medium- and high-entropy compositionally complex fluorite oxides, *J. Am. Ceram. Soc.* 104 (2021) 448–462.

37 S. Qiu, H. Xiang, F. Dai, H. Wang, M, Huang, C. Wan, Q. Meng, J. Li, X. Wang, Y. Zhou, Medium-entropy (Me,Ti)0.1(Zr,Hf,Ce)0.9O2 (Me=Y and Ta): Promising thermal barrier materials for high-temperature thermal radiation shielding and CMAS blocking, *J. Mater. Sci. Technol.* 123 (2022) 144–153.

38 H. Chen, H. Xiang, F. Dai, J. Liu, Y. Zhou, Porous high entropy (Zr0.2Hf0.2Ti0.2Nb0.2Ta0.2)B2: a novel strategy towards making ultrahigh temperature ceramics thermal insulating, *J. Mater. Sci. Technol.* 35 (2019) 2404–2408.

39 H. Chen, H. Xiang, F. Dai, J. Liu, Y. Lei, J. Zhang, Y. Zhou, High porosity and low thermal conductivity high entropy (Zr0.2Hf0.2Ti0.2Nb0.2Ta0.2)C, *J. Mater. Sci. Technol.* 35 (2019) 1700–1705.

40 H. Chen, Z. Zhao, H. Xiang, F. Dai, J. Zhang, S. Wang, J. Liu, Y. Zhou, Effect of reaction routes on the porosity and permeability of porous high entropy (Y0.2Yb0.2Sm0.2Nd0.2Eu0.2)B6 for transpiration cooling, *J. Mater. Sci. Technol.* 38 (2020) 80–85.

41 D. Bérandan, S. Franger, D. Dragoe, A.K. Meena, N. Dragoe, Colossal dielectric constant in high entropy oxides, *Phys. Status Solidi RRL* 10 (2016) 328–333.

42 Y. Pu, Q. Zhang, R. Li, M. Chen, X. Du, S. Zhou, Dielectric properties and electrocaloric effect of high-entropy (Na0.2Bi0.2Ba0.2Sr0.2Ca0.2)TiO3 ceramic, *Appl. Phys. Lett.* 115 (2019) 223901.

43 J. Liu, K. Ren, C. Ma, H. Du, Y. Wang, Dielectric and energy storage properties of flash-sintered high-entropy (Bi0.2Na0.2K0.2Ba0.2Ca0.2)TiO3 ceramic, *Ceram. Int.* 46 (2020) 20576–20581.

44 A. Mao, H. Xiang, Z. Zhang, K. Kuramoto, H. Yu, S. Ran, Solution combustion synthesis and magnetic property of rock-salt (Co0.2Cu0.2Mg0.2Ni0.2Zn0.2)O high-entropy oxide nanocrystalline powder, *J. Magn. Magn. Mater.* 484 (2019) 245–252.

45 M.P. Jimenez-Segura, T. Takayama, D. Bérardan, A. Hoser, M. Reehuis, H. Takagi, N. Dragoe, Long-range magnetic ordering in rocksalt-type high-entropy oxides, *Appl. Phys. Lett.* 114 (2019) 122401.

46 P.B. Meisenheimer, T.J. Kratofil, J.T. Heron, Giant enhancement of exchange coupling in entropy-stabilized oxide heterostructures, *Sci. Rep.* 7 (2017) 13344.

47 R. Witte, A. Sarkar, R. Kruk, B. Eggert, R.A. Brand, H. Wende, H. Hahn, High-entropy oxides: An emerging prospect for magnetic rare-earth transition metal perovskites, *Phys. Rev. Mater.* 3 (2019) 034406.
48 R. Witte, A. Sarkar, L. Velasco, R. Kruk, R.A. Brand, B. Eggert, K. Ollefs, E. Weschke, H. Wende, H. Hahn, Magnetic properties of rare-earth and transition metal based perovskite type high entropy oxides, *J. Appl. Phys.* 127 (2020) 185109.
49 J. Ma, B. Zhao, H. Xiang, F. Dai, Y. Liu, R. Zhang, Y. Zhou, High-entropy spinel ferrites MFe2O4 (M=Mg, Mn, Fe, Co, Ni, Cu, Zn) with tunable electromagnetic properties and strong microwave absorption, *J. Adv. Ceram.* 11 (2022) 754–768.
50 B. Zhao, Z. Yan, Y. Du, L. Rao, G. Chen, Y. Wu, L. Yang, J. Zhang, L. Wu, D.W. Zhang, R. Che, High-entropy enhanced microwave attenuation in titanate perovskites, *Adv. Mater.* 35 (2023) 2210243.
51 Y. Zhou, B. Zhao, H. Chen, H. Xiang, F. Dai, S. Wu, W. Xu, Electromagnetic wave absorbing properties of TMCs (TM=Ti, Zr, Hf, Nb and Ta) and high entropy (Ti0.2Zr0.2Hf0.2Nb0.2Ta0.2)C, *J. Mater. Sci. Technol.* 74 (2021) 105–118.
52 W. Zhang, B. Zhao, H. Xiang, F. Dai, S. Wu, Y. Zhou, One-step synthesis and electromagnetic absorption properties of high entropy rare earth hexaborides (HE REB6) and high entropy rare earth hexaborides/borates (HE REB6/HE REBO3) composite powders, *J. Adv. Ceram.* 10 (2021) 62–77.
53 W. Zhang, F.-Z. Dai, H. Xiang, B. Zhao, X. Wang, N. Ni, R. Karre, S. Wu, Y. Zhou, Enabling highly efficient and broadband electromagnetic wave absorption by tuning impedance match in high-entropy transition metal diborides (HE TMB2), *J. Adv. Ceram.* 10 (2021) 1299–1316.
54 W. Zhang, B. Zhao, N. Ni, H. Xiang, F.-Z. Dai, S. Wu, Y. Zhou, High entropy rare earth hexaborides/tetraborides (HE REB6/HE REB4) composite powders with enhanced electromagnetic wave absorption performance, *J. Mater. Sci. Technol.* 87 (2021) 155–166.
55 H. Chen, B. Zhao, Z. Zhao, H. Xiang, F.-Z. Dai, J. Liu, Y. Zhou, Achieving strong microwave absorption capability and wide absorption bandwidth through a combination of high entropy rare earth silicide carbides/rare earth oxides, *J. Mater. Sci. Technol.* 47 (2020) 216–222.
56 W. Zhang, H. Xiang, F.-Z. Dai, B. Zhao, S. Wu, Y. Zhou, Achieving ultra-broadband electromagnetic wave absorption in high-entropy transition metal carbides (HE TMCs), *J. Adv. Ceram.* 11 (2022) 545–555.

Index

FSC
www.fsc.org
MIX
Papier | Fördert
gute Waldnutzung
FSC® C083411

Kolibri 360 GmbH
Ferdinand-Jühlke-Straße 7
99095 Erfurt, Deutschland
produktsicherheit@kolibri360.de